Carbohydrate-Based Vaccines: From Concept to Clinic

ACS SYMPOSIUM SERIES **1290**

Carbohydrate-Based Vaccines: From Concept to Clinic

A. Krishna Prasad, Editor
Pfizer Vaccines Research and Development
Pearl River, New York

Sponsored by the
ACS Division of Carbohydrate Chemistry
ACS Division of Cellulose and Renewable Materials

American Chemical Society, Washington, DC

Distributed in print by Oxford University Press

Library of Congress Cataloging-in-Publication Data

Names: Prasad, A. Krishna, editor. | American Chemical Society, issuing body. | American Chemical Society. Division of Carbohydrate Chemistry, sponsoring body. | American Chemical Society. Cellulose and Renewable Materials Division, sponsoring body.
Title: Carbohydrate-based vaccines : from concept to clinic / A. Krishna Prasad, editor.
Other titles: Carbohydrate-based vaccines (Prasad)
Description: Washington DC : American Chemical Society, 2018. | Series: ACS symposium series ; 1290 | Includes bibliographical references and index.
Identifiers: LCCN 2018032978 (print) | LCCN 2018033236 (ebook) | ISBN 9780841233294 (ebook) | ISBN 9780841233379 (alk. paper)
Subjects: | MESH: Vaccines | Glycoconjugates | Carbohydrates--immunology
Classification: LCC QR189 (ebook) | LCC QR189 (print) | NLM QW 805 | DDC 615.3/72--dc23
LC record available at https://lccn.loc.gov/2018032978

The paper used in this publication meets the minimum requirements of American National Standard for Information Sciences—Permanence of Paper for Printed Library Materials, ANSI Z39.48n1984.

Distributed in print by Oxford University Press

Foreword

The ACS Symposium Series was first published in 1974 to provide a mechanism for publishing symposia quickly in book form. The purpose of the series is to publish timely, comprehensive books developed from the ACS sponsored symposia based on current scientific research. Occasionally, books are developed from symposia sponsored by other organizations when the topic is of keen interest to the chemistry audience.

Before agreeing to publish a book, the proposed table of contents is reviewed for appropriate and comprehensive coverage and for interest to the audience. Some papers may be excluded to better focus the book; others may be added to provide comprehensiveness. When appropriate, overview or introductory chapters are added. Drafts of chapters are peer-reviewed prior to final acceptance or rejection, and manuscripts are prepared in camera-ready format.

As a rule, only original research papers and original review papers are included in the volumes. Verbatim reproductions of previous published papers are not accepted.

ACS Books Department

Contents

Indexes

Dedication

This book is dedicated in honor of

Dr. David Hamilton Smith, M.D. (1932-1999)

Visionary, Leader, Scientist, Physician, Entrepreneur, Conservationist, Family Man, and Founder of Praxis Biologics (now part of Pfizer Vaccines)

Figure . Photograph of Dr. David Smith is reproduced from ref (2). Copyright (1999), Oxford University of Press (License number 4344770987480, Date 09 May, 2018).

"The changes for which he worked, in fields ranging from conservation to public health, are still present today. They are examples for all of us to follow. David Smith (*1*) willed something even more valuable than his bequest - the legacy and example of himself. Never losing hold of his vision, he made his dream a reality and brought the harsh chapter of spinal meningitis to a dramatic close. In just 10 years after brining the vaccine to the market, the number of cases of spinal meningitis plummeted from 20,000 in 1987 to 81 in 1997. A true success story. In 1996, Dr. Smith was awarded both the Lasker Prize (*2*), one of the most prestigious awards in the country, and the Pasteur Award from the World Health Organization. The recognition for years of work and sacrifice were well deserved (*3*, *4*)."

References

1. Robbins, J. B.; Schneerson, R.; Anderson, P.; Smith, D. H. *JAMA, J. Am. Med. Assoc.* **1996**, *276*, 1181–1185.
2. Insel, R. A.; Smith, A. L. *J. Infect. Dis.* **1999**, *180*, 577–578.
3. D. H. Smith *Conservation Research Fellowship Home* https://conbio.org/mini-sites/smith-fellows/about-the-program/david-h.-smith/ (accessed 9 May 2018).
4. D. H. Smith *Center for Vaccine Biology and Immunology* https://www.urmc.rochester.edu/cvbi/about/history.aspx (accessed 9 May 2018).

Foreword

This volume evolved from a gathering held as part of the 2017 American Chemical Society National Meeting. The conference included scientists, from academia and industry, involved in the design and development of carbohydrate-based vaccines. When members of the general public think of vaccines, they immediately focus in on "classic" viral vaccines such as measles, mumps and polio. Vaccines to prevent bacterial infections do not immediately come to mind, partly due to the lower incidence of severe bacterial disease. When the impact of bacterial infections is considered, it is usually in the context of diseases resulting from exposure to bacterial toxins. Of course, highly effective toxoid-based vaccines to prevent diseases such as diphtheria and tetanus have been available for decades.

However, there are a number of bacterial infections that can present a spectrum of disease manifestations, some of which can be quite severe. This is particularly the case for infants and young children who have not yet developed full immunological competency, as well as older adults who experience age-related immunological decline. These severe manifestations typically involve systemic infections that are life-threatening and often result in long-lasting morbidities.

Many of these pathogenic bacteria are characterized by a carbohydrate capsule against which an effective antibody-mediated response is directed. Unfortunately, immature immune systems cannot readily mount a response to a pure carbohydrate-based vaccine and, even in adults, the immune responses to pure carbohydrates are typically poor and relatively short-lived.

It was first demonstrated many decades ago that the conjugation of bacterial polysaccharides to immunogenic carrier proteins (such as bacterial toxoids) can render poorly immunogenic carbohydrates capable of eliciting effective and long-lasting immune responses. However, it was not until the development and licensure of the *Haemophilus influenzae* type B vaccine in the 1980s that the powerful potential of the technology became apparent. The development of the vaccine was based on the definition of novel chemical conjugation processes that could be reproducibly carried out at manufacturing scale, as well as on the development of novel analytical methods that permitted consistent at-scale production. The technology was advanced further by the development of a highly effective conjugated vaccine for prevention of disease caused by specific serotypes of *Streptococcus pneumoniae*. Further advances were incorporated in the design of follow-on generations of *S. pneumoniae* vaccines and in the development of vaccines to prevent severe meningococcal disease. The global impact of these vaccines on mortality and morbidity, particularly among young children, cannot be overstated.

The future holds considerable promise as continuing improvements in carbohydrate-protein conjugation chemistries make it possible to develop lower-cost vaccines, particularly for control of *S. pneumoniae* disease, that can be used more broadly in lower-income countries. In addition, as conjugation science and associated manufacturing processes advance, the potential exists for the development of highly effective and low-cost vaccines against bacterial diseases that place a heavy burden on populations in many lower- and developing middle-income countries. These include improved vaccines for prevention of typhoid as well as the potential development of vaccines to prevent group B streptococcal disease in infants. The promise of vaccines based on carbohydrate conjugation chemistry for prevention of severe bacterial disease worldwide continues to hold true.

Emilio A. Emini
Bill & Melinda Gates Foundation

Chapter 1

Introduction

A. Krishna Prasad*

Pfizer Vaccines Research and Development, 401 N. Middletown Rd., Pearl River, New York 10965, United States
***E-mail: krishna.prasad1@pfizer.com.**

A vast body of clinical experience has supported the introduction of several conjugate vaccines and related multi-antigen vaccines into routine use across the globe resulting in a dramatic impact on public health. Research and development efforts continue to translate this impact and find scientific and novel technological innovations to target other infectious diseases and address unmet medical needs. The primary motivation of this book is to provide a comprehensive forum for the presentation of various components that define the development and control strategies to produce safe and stable glycoconjugate vaccines that elicit consistent and robust immunogenic responses. The book aims to cover the complete landscape comprising immunological mechanisms; pre-clinical considerations; clinical history; design; development; characterization; conformational modeling; formulation; and chemistry, manufacturing and control (CMC) of glycoconjugate vaccines. These disciplines are closely tied to regulatory aspects leading to the commercial licensure of vaccines with a final chapter providing guidance on "lessons learned" to achieve this goal. All the chapters have been put together by leaders in industry, academia, government and non-profit organizations, with decades of experience in vaccine development and licensure.

All approaches at a higher level are suspect until confirmed at the molecular level."

Francis Crick in "*What Mad Pursuit*" (1988)

"The ultimate aim of the modern movement in biology is, in fact, to explain all biology in terms of physics and chemistry."

Francis Crick in "*Of Molecules and Men*" (1966)

The past three decades have witnessed the development and regulatory approval of glycoconjugate vaccines against several medically important bacterial pathogens, including *Haemophilus influenzae* type b, *Streptococcus pneumoniae, Neisseria meningitides* and *Salmonella typhi*. Immunologic protection against these and many other bacterial diseases is mediated through opsonophagocytic antibodies directed against the surface carbohydrates that define the bacterial serogroup or serotype and serve as virulence factors. These vaccines are composed of bacterial capsular polysaccharides chemically conjugated to immunogenic carrier proteins. Given that the diseases caused by these bacterial pathogens are most pronounced in infants and young children, the development of the glycoconjugate vaccine technology has had a considerable impact on public health.

Many of the chapters in this volume were assembled as a follow-up from a symposium entitled "*Carbohydrate-based vaccines and adjuvants*" which took place at the 254th American Chemical Society National Meeting held in Washington, DC (August 2017). The symposium was sponsored by Pfizer and the Carbohydrate (CARB) and Biotechnology (BIOT) divisions. This book, therefore, reflects the importance of this field toward design, development, manufacture and licensure of the complex carbohydrate-based (glycoconjugate) vaccines.

The book has been organized into thirteen chapters, which cover a comprehensive landscape including the clinical history, design, development, chemistry, manufacturing and control (CMC) aspects, pre-clinical assays, adjuvants and the various approaches used to develop carbohydrate-based vaccines.

Chapter 2 reviews the clinical experience of glycoconjugates from the ground-breaking monovalent *Haemophilus influenzae* type b vaccine to contemporary multi-valent (pneumococcal and meningococcal) and multi-component (glycoconjugate vaccines containing protein antigens) products and their positive public health impact. The chapter provides a chronological as well as comprehensive narrative of novel and ground breaking clinical trials, particularly those leading to marketing approvals for new categories of vaccines. Multi-valent pneumococcal polysaccharide conjugate vaccines have been demonstrated to be effective in preventing pneumococcal pneumonia, invasive pneumococcal disease and otitis media in children. A 7-valent conjugate vaccine, including pneumococcal polysaccharides from seven serotypes individually conjugated to CRM_{197} carrier protein was licensed for use in infants and children < 5 years of age in 2000. The introduction of this first generation pneumococcal conjugate vaccine resulted in a substantial decrease in invasive pneumococcal disease (IPD) in children, less than 5 years of age, in the USA. A 10-valent pneumococcal conjugate (*Synflorix®*), using non-typable *Haemophilus influenzae* protein D, tetanus toxoid and diphtheria toxoid as carrier proteins, was later approved for immunization in infants and young children, outside the United States in 2009. This was followed by the regulatory approval of a 13-valent pneumococcal conjugate vaccine in 2009 in European Union and in early 2010, in the USA. In addition, as a result of a vaccine "herd immunity" effect, fewer pneumococcal infections have been observed in older adults more than 65 years of age. The vaccine has also demonstrated direct protective effects in adults against community-acquired pneumonia, resulting in recommended routine use of the vaccine in the United States, for both infants and older adults.

A meningococcal glycoconjugate vaccine against meningococcal serogroup C bacterial meningitis was introduced first in the UK in 1999, followed subsequently in many countries around the world. The impact against serogroup C disease has been notable. Similarly, the introduction of a serogroup A meningococcal polysaccharide-tetanus toxoid conjugate vaccine (MenAfriVac®) in the "meningitis belt" of sub-Saharan Africa resulted in the near elimination of the group A epidemics among children and young adults in this part of the world. Several multi-valent glycoconjugate vaccines covering serogroups A, C, Y and W135 (Menactra®, Menveo® and Nimenrix®) have also been developed and approved to date.

Salmonella Typhi is the major cause of enteric fever in lower income countries. Recently, WHO announced the prequalification of a typhoid conjugate vaccine (Typbar-TCV) manufactured by the Indian firm Bharat Biotech, initially licenced for use in India and Nepal.

In a recent Phase 1 clinical study, a multicomponent 4-antigen *Staphylococcus aureus* vaccine, composed of capsular polysaccharide conjugates of serotypes 5 and 8 (CP5 and CP8), with two additional recombinant proteins clumping factor (rmClfA), and recombinant manganese transporter protein C (rMntC) were evaluated to confirm safety and immunogenicity of SA4Ag in a surgical population. This vaccine candidate is now in an efficacy study examining its potential to prevent ortheopedic nosocomial infections.

Chapter 3, covers the immunological aspects related to glycoconjugate vaccines, with emphasis on T-cell immunity elicited by glycoconjugate vaccines. Purified "free" polysaccharides are T cell-independent immunogens and, as such, are poorly immunogenic in infants. By contrast, the carrier proteins used for conjugation are T-cell dependent immunogens and the covalent linkage of the polysaccharides to these proteins causes the anti-polysaccharide immune response to also become T cell-dependent. This renders the conjugated polysaccharides immunogenic for infants.

Chapter 4 describes the strategies to produce a glyconjugate vaccine that can stimulate a potent and specific immune response to a saccharide antigen. This chapter discusses several key elements that need to be considered during the design and development of conjugate vaccines. This includes preservation of critical immunogenic saccharide epitopes and other features such as selection of an effective and well-tolerated carrier protein as well as the stability of the conjugate in the drug substance as well as drug product. The key design features necessary to develop the “optimal conjugate vaccine construct” and the "toolbox" to achieve these ends are also covered in this chapter. Due to the distinguishing, sometimes subtle, structural differences resulting in variations in physicochemical attributes that differ by each saccharide antigen (in multi-valent vaccines), the conjugation to the carrier protein needs to be carried out by the specific customized process chemistry. Keeping in mind this variability in polysaccharide reactivity, the parameters that affect the final conjugate structural attributes also need to be identified based on the specific polysaccharide antigen structure at each stage of the process (fermentation, purification, activation/conjugation). The final optimal conjugate construct for each polysaccharide serotype is determined, therefore, by a number of factors to generate the optimal conjugate construct for robust immunogenicity and stability.

Chapter 5 provides the commercial process development and manufacturing considerations of glyconjugate vaccines. It provides comprehensive coverage of the key elements required for process/product control strategy approaches during various clinical phases through to licensure including statistical process control and performance qualification necessary for a full process understanding. Central to this approach is the development of a control strategy to demonstrate consistency, robustness testing and transfer of commercially viable process technologies into manufacturing facilities, along with the production of clinical trial material.

Multi-valent carbohydrate-based vaccines to prevent bacterial infection caused by several serogroups of *Neisseria meningitidis* have been licensed or are currently in clinical development. Chapter 6 discusses the various conjugation routes employed. Other key attributes, such as the chemical structures and functional groups involved in the covalent coupling of polysaccharides to carrier proteins for the preparation of tetravalent meningococcal serogroup A, C, W, Y glycoconjugates vaccines, have been described in depth.

Coverage of a vaccine could potentially be expanded when specific antigens have been demonstrated to provide cross-protection against infection by structurally closely related, non-vaccine strains. However, structural similarity between carbohydrate antigens is not a reliable predictor of cross-protection. Conformational analysis can provide a mechanistic insight into clinical observations on cross-protection and may further indicate the importance of specific structural features, such as non-saccharide substituents. This may augment vaccine design and development strategies toward potential vaccine coverage expansion. Chapter 7 describes, in detail, the computational methodologies employed to model carbohydrate antigens and the valuable role that molecular simulations can play in furthering our understanding of carbohydrate immunogenicity and cross-protection. The authors outline molecular modeling case studies of polysaccharide antigens for meningococcal serogroups Y and W and pneumococcal serogroups 6, 19 and 23, as well O-antigens of *Salmonella enterica* and *Shigella flexneri*.

Synthetically derived saccharides have garnered significant interest due to their potential use as antigens to generate prophylactic (*Neisseria meningitidis* serogroup A, *Shigella flexneri*, etc.) vaccines, with positive results. Significant research efforts were aimed toward therapeutic (tumor associate carbohydrate antigens, TACA) glycoconjugate vaccines. Chapter 8 reviews the recent advances in oligosaccharide synthesis with desired stereo- and regio-selectivity, necessary to access high value targets. However, creating a synthesis platform that accommodates the large variation in oligosaccharide conformation and connectivity is a daunting task. Several challenges remain including poor immunogenicity, cost of goods (due to multi-step synthesis) and commercial viability that currently eludes their vaccine potential, especially for antigens with complex structures in their repeat units. For multi-antigen vaccines, this complexity is amplified. Nonetheless, over the last few decades researchers have made significant great progress toward addressing these shortcomings.

Formulation is the key final step to required produce a safe, stable and efficacious vaccine drug product, often involving an adjuvant to boost immunogenicity of the antigen components. Among the currently licensed vaccines, the production of multi-valent glycoconjugate vaccines are the most complex. Besides the complexity in manufacturing, there are several other considerations such as facilities, shipping, storage, and shelf-life considerations that need to be addressed. Chapter 9 discusses strategies involved toward the development, production and transfer of robust processes to developing countries.

Post conjugation with the carrier protein, the resulting glycoconjugate vaccine must retain the biological appearance of the pathogen's saccharide antigen to specifically direct an immune response that can recognize and facilitate killing of the pathogen. A significant structural diversity exists in saccharide antigens which vary across serotypes and even among strains within a serotype. Therefore, there is not a single platform conjugation approach that can be applied to all polysaccharide conjugate vaccines, and the ability of candidate vaccines to elicit functional immune responses is often empirically determined. Chapter 10 describes several glycoconjugate features/epitopes that are critical for eliciting functional immune responses. The chapter identifies and discusses important considerations for pre-clinical glycoconjugate vaccine evaluation including *in vivo* models to measure immunogenicity and *in vitro* assays that measure immune responses that facilitate killing of the pathogen.

Campylobacter jejuni is one of the most common causes of human diarrheal disease worldwide. In developed countries Campylobacteriosis cases are rare, but in developing countries, the incidence is estimated to be at least 10 times higher, which poses a life-threatening risk toward the pediatric population. Chapter 11 describes the development aspects of a multi-valent *Campylobacter jejuni* capsule polysaccharide conjugate candidate vaccine prepared for human clinical trials.

Quality considerations for consistency in the production of glycoconjugate vaccines is a major area that needs to be addressed by all developers and manufacturers during clinical trials as well as commercial supply. Chapter 12 provides a structured guide to the manufacturing processes, regulatory requirements and analytical methods applied to current products and for future candidates in development stage.

Despite the availability of significant information in public domain such as literature and patents, these data provide only a partial glimpse of best practices applicable toward the design and manufacture new vaccines. Therefore, the risk of technical failure resulting in poorly immunogenic products remains high for new entrants to the field. Multinational corporate vaccine players with decades of experience, who went through various clinical phases to licensure, have typically made substantial investments in complex glycoconjugate vaccine development and commercialization. Finally, tying all the chapter pieces together, in Chapter 13 several experts in the field, with decades of experience in vaccine industry, academia, government and non-profit organizations, outline the key concepts that define successful design and manufacture of new glycoconjugate vaccines. The chapter provides general guidance from "lessons learned" with the intent of improving the CMC and chances of success for the glycoconjugate vaccines, still in development.

Building on this extensive knowledge base from pneumococcal and meningococcal vaccines, glycoconjugates developers are now on the verge of extending this success to other encapsulated bacteria such as *Salmonella typhi, Staphylococcus aureus, Shigella* and *Escherichia coli*. Conjugate vaccine candidates, comprising Group B Streptococcus polysaccharide antigens, are currently undergoing clinical trials and hold significant promise to address morbidity and mortality in neonates.

While the chapters were designed with the intent to cover all major aspects of glycoconjugate vaccine development, a book of this size cannot capture information on all supporting topics and allied fields in infinite detail. We hope that this book will serve as a useful reference providing valuable insights into the design, development, manufacture and licensure of these complex products.

Acknowledgments

The editor thanks all the speakers at the ACS symposium and especially those who contributed chapters to the book. The authors fully recognize and acknowledge, with a deep sense of gratitude the importance, complexity and sensitive nature of the materials required to assemble the chapters to be shared with a vast audience. The authors and the editor gratefully acknowledge these efforts and the support of their respective managements which culminated in the successful completion of the book.

The following organizations, whose support enabled this collaborative project to be realized, are gratefully acknowledged: Pfizer, GSK Vaccines, Sanofi Pasteur, Janssen Vaccines, PATH, University of Georgia (Athens, GA), University of Toledo (Ohio), University of Cape Town (South Africa) and University of Guelph (Canada).

The editor thanks Drs. Kathrin Jansen, Annaliesa Anderson, Wendy Watson and Mark Ruppen for their overall support and valuable advice on the chapters contributed by Pfizer colleagues. The editor is grateful to Drs. Emilio Emini and Peter Paradiso for their support and to all the reviewers for their efforts to ensure the high scientific quality of the chapters in this book.

Chapter 2

Glycoconjugate Vaccines: The Clinical Journey

Stephen P. Lockhart,[*,1] Daniel A. Scott,[2] Kathrin U. Jansen,[3] Annaliesa S. Anderson,[3] and William C. Gruber[3]

[1]Pfizer Vaccines Clinical Research and Development, Horizon Building, Honey Lane, Hurley SL6 6RJ, United Kingdom
[2]Pfizer Vaccine Clinical Research and Development, 500 Arcola Rd., Collegeville, Pennsylvania 19426, United States
[3]Pfizer Vaccines Research and Development, 401 N. Middletown Rd., Pearl River, New York 10965-1299, United States
[*]E-mail: stephen.lockhart@pfizer.com.

We review the clinical journey of glyconjugate vaccines from their inception in the 1980s to contemporary vaccines and their positive public health impact. We focus on novel and ground breaking clinical trials, particularly those leading to marketing approvals for new categories of vaccines. Glyconjugate vaccines based on capsular polysaccharides have been highly successful against *Haemophilus influenzae* type b, a growing number of *Streptococcus pneumoniae* serotypes and *Neisseria meningitidis* serogroups A, C, W and Y. Glycoconjugate vaccines may extend this success to other encapsulated bacteria such as *Salmonella typhi,* Group B Streptococcus and *Staphylococcus aureus* as well as Shigella and *Escherichia coli*.

Introduction

Many important human bacterial pathogens are carried as colonizing organisms and only occasionally become pathogens. These pathogens have a polysaccharide capsule which is a key element in the close relationship between them and the host. The capsule paradoxically protects the organisms and yet also presents a key target for antibody-mediated host defense.

On the basis of experiments started in the 1920s (*1*), the potential of *Streptococcus pneumoniae* (also known as pneumococcus) capsular polysaccharide conjugates to induce protective immune responses in animal models was identified nearly 80 years ago (*2*). A number of bacterial vaccines based on plain, unconjugated capsular polysaccharides were developed in the second half of the twentieth century but limitations were noted, such as lack of immunogenicity in infants and failure to prime for immunological memory. It was not until the late 1970s that these shortcomings led to the development of capsular polysaccharide conjugate vaccines (*3*).

Although each vaccine is unique, there have been some common steps in the clinical journey for conjugate vaccines, which will be illustrated with specific examples in this chapter. Of note, the development pathway for *Haemophilus influenzae* type b (Hib) conjugate vaccines set a strong precedent for all the following conjugate vaccines, particularly with regard to characterization of vaccine-induced protective immune responses using appropriate serological assays. *Haemophilus influenzae* invasive disease in infants and young children presented a good first target for Hib conjugate vaccines as a single serotype was responsible for almost all disease. Several generalizations can be made for polysaccharide conjugate vaccines:

- Each type of conjugate vaccine currently marketed was preceded by plain polysaccharide (PS) vaccines. Plain PS vaccines induce protective antibodies through a T-cell independent mechanism, which induces very poor immune responses in infants, does not induce immune memory and often induces hyporesponsiveness to further vaccine doses.
- The early conjugate vaccines against some pathogens were approved on the basis of clinical efficacy studies, but vaccines for some pathogens and derivative vaccines were approved on the basis of immunological assessments based on experience with polysaccharide vaccine and only subsequently shown to be effective after implementation for widespread use. Immunological assessments for regulatory approval not only include achieving a concentration of antibody known or expected to be protective (also known as a correlate of protection), but also evidence of functional antibody being produced that kills the pathogen, induction of immune memory demonstrated by an enhanced response to a subsequent polysaccharide or conjugate vaccine dose challenge and evidence that the strength of antibody binding to its target (avidity) is enhanced.
- Development of standardized, sensitive, reproducible, high-throughput assays for measuring antibody responses to vaccines has been critical.

- In addition to direct protection of immunized subjects, reduced colonization produces "herd immunity," which adds to protection of immunized subjects as well as unimmunized populations.
- Antimicrobial resistance (AMR) has been reduced by conjugate vaccines, due to reductions in carriage of potentially antibiotic resistant isolates and reductions in antibiotic use for treatment of common bacterial diseases, such as otitis media.
- Development of multicomponent vaccines or combinations with other vaccines has been important, though not always straightforward.
- Use in lower and middle income countries (LMIC) generally requires specific clinical trial evidence, tailored vaccine presentations such as multidose vials (MDV) and sometime specific vaccine combinations.

Notes:

Except where specified, quoted efficacy is against first episode of an endpoint and calculated as intention to treat (ITT), primarily meaning that all subjects who received at least one dose are included from that dose onwards, whereas per protocol (PP) efficacy is generally calculated on first episodes occurring at least 2-4 weeks following completion of a full primary series of vaccination.

The term efficacy is sometimes reserved for vaccine studies with randomly allocated treatment and control populations using an endpoint that combines a clinical and a microbiological component (e.g. *Haemophilus influenzae* (Hib) invasive disease). In this review we have also used the term efficacy where subjects are randomly allocated to a treatment or control treatment, but only a clinical endpoint can be ascertained (e.g. radiologically confirmed community acquired pneumonia (CAP)), which would sometimes be referred to as effectiveness trials. We have reserved the term effectiveness studies for observational studies where treatment is not randomly allocated.

A range of numbers in parentheses following a point estimate in percent represents a 95% confidence interval, for example 5% (2, 17).

Haemophilus influenzae Type b (Hib)

Early Clinical Experience with Polysaccharide Hib Vaccines

Plain capsular polysaccharide vaccines for Hib, consisting of polyribosylribitol phosphate (PRP), were introduced in 1985 in USA for use in children 24 months to 6 years of age (*4*), based largely on clinical experience in Finland (*5*). Efficacy was limited to children over 18 months of age, whereas the peak of invasive disease was in younger infants (*5*). There was also inconsistency in measured effectiveness after introduction in the USA (*6*, *7*). These issues were to an extent resolved by approval of the first Hib conjugate vaccine in 1985 (*4*).

Serological Assays for Hib Vaccines

The development of practical, sensitive and specific assays for human antibody responses to Hib vaccines was essential for clinical development of both plain polysaccharide and conjugate Hib vaccines. These assays fall into two main groups, immunological assays for quantifying PRP-binding immunoglobulin or immunoglobulin classes, and functional assays such as serum bactericidal assays (SBA) that measure the actual killing of the pathogen. Consensus was that an immunological binding assay such as radioimmunosassay (RIA) (*8*) or enzyme-linked immunosorbent assay (ELISA) (*9*) was the primary method (*10*), as these are easier to perform, more reproducible, and more suitable for standardization (*9*). A functional assay such as an SBA is required to confirm that the binding antibodies are biologically active (*11*). Trials in Finland with the plain polysaccharide vaccine had suggested that Hib specific antibody concentrations of at least 1.0 μg/mL would be protective in older children (*12*).

Early Experience with Hib Conjugate Vaccines

The first Hib conjugate constructs were published in 1980 (*3*), following work that had started in 1968 (*13*). By 1984 PRP-D (PRP conjugated to diphtheria toxoid (DT)) had been shown to be safe and more immunogenic than PRP in adults (*14*). In children 15-24 months of age PRP-D was more immunogenic than PRP (*15*) and on this basis was approved by FDA for use in children 18 months and older in 1987 (*4*).

However, PRP-D efficacy studies in infants yielded conflicting results (Table 1), with good efficacy in Finnish infants (*16*) at 83% (26, 96) but no significant efficacy in Alaskan infants (*17*) at 35% (-57, 73). After three primary doses the Geometric Mean Concentration (GMC) was only 0.18 μg/mL in Alaskan infants, with only 48% achieving a concentration of 0.1 μg/mL (*17*). Even in the Finnish infants the GMC after three doses was only 0.42 μg/mL, with only 62% achieving a concentration of 0.15 μg/mL and 34% 1.0 μg/mL. PRP-D was approved and used in infants in Switzerland (*18*) but was never approved for use in infants in the USA, as other conjugate vaccines were soon shown to be more immunogenic and efficacious in infants.

HbOC (Hib capsular oligosaccharide conjugated to CRM_{197} , a genetically detoxified diphtheria toxin) (*19*) and PRP-OMP (PRP conjugated to meningococcal outer membrane protein, a highly immunogenic protein extracted from serotype 2 *Neisseria meningitidis*) (*20*) entered clinical trials soon after PRP-D. A series of studies on HbOC allowed selection of an optimal process and conjugate design, and showed safety, immunogenicity and priming in adults and infants over 12 months of age (*21*), progressing to demonstration of immunogenicity with three doses in infants at 2, 4 and 6 months of age (*22*). Priming for immune memory was shown in infants using a PRP dose at 12 months of age (*22*). A similar series of studies with PRP-OMP also showed the vaccine to be safe in adults (*23*) before demonstrating safety and immunogenicity (*23*) and priming (*24*) in infants down to 2 months of age. PRP-OMP was immunogenic in infants and induced priming with two doses (*23*, *24*). On the basis of immunogenicity being superior to that of the PRP vaccine in 24-26 month old children, HbOC was approved for use in toddlers in the USA in 1988 (*4*) and PRP-OMP a year later (*4*).

The induction of immune responses and priming in young infants led to efficacy studies in infants (Table 1). Excellent efficacy was shown in trials within the US for HbOC (*25*) and PRP-OMP (*26*). As a result, HbOC was the first conjugate vaccine approved for use in US in young infants, starting at 2 months of age, in 1990 (*4*), followed by PRP-OMP a few weeks later (*4*).

PRP conjugated to tetanus toxoid (TT, PRP-T) was included in head-to-head immunogenicity studies comparing Hib conjugate vaccines in infants (*27*, *28*). PRP-D was confirmed to be substantially less immunogenic than PRP-T, HbOC or PRP-OMP (*28*). After 3 doses these three vaccines were all similarly immunogenic (*27*). However, PRP-OMP showed an earlier response, with substantial immunogenicity after one dose, plateauing after two doses, whereas HbOC and PRP-T showed a substantial rise between 2 and 3 doses (*14*, *27*, *28*). The avidity of antibody after infant vaccination was, on the other hand, highest for HbOC, intermediate for PRP-T and lower for PRP-OMP (*11*).

PRP-T was subsequently approved in the US on the basis of immunogenicity for US infants in 1993, after randomized efficacy studies were interrupted by the approval of Hib conjugate vaccines for infants (*4*). However, staged introduction in different areas prior to a national introduction in the UK confirmed that PRP-T was highly efficacious in infants (Table 1) (*29*). PRP-T was subsequently used in an efficacy trial in the Gambia (Table 1) (*30*). This trial confirmed that Hib was as efficacious in an LMIC setting as in a developed country. Excellent efficacy was shown against confirmed Hib pneumonia as well as all Hib invasive disease (*30*). The vaccine was efficacious against all radiological pneumonia (21.1%, 95% CI 4.6-34.9) (*30*). Defining the etiological agent responsible for pediatric pneumonia can be challenging as good quality sputum is very difficult to obtain and most children will not have a positive blood culture. This study therefore indicates in an indirect fashion that Hib was responsible for a substantial proportion of radiological pneumonia in infants in the Gambia (*30*). This study provided strong support for extending use of Hib vaccine to infants around the world.

In Europe, the EMEA (European Medicines Evaluation Agency, now EMA) was not founded until 1995, so the regulatory roll-out of standalone Hib conjugate vaccines proceeded on a national basis in Europe. Nonetheless, deployment was rapid with substantial uptake in most of Western Europe and other developed countries (*31*).

Table 1. *Haemophilus influenzae* type b conjugate vaccine efficacy studies

Abbreviation (Conjugate protein) First approved monovalent vaccine by proprietary name	*First approval for use in infants in USA*	*Efficacy against invasive Hib disease % (95% CI)*	*Efficacy study schedule (months of age) Location*	*Efficacy study design*
PRP-D (Diphtheria toxoid) ProHIBit®	Only ≥18 months	83 (26, 96) (*16*)	3, 4, 6, 14 Finland	Treatment/ no treatment by date of birth
		35 (-57, 73) (*17*)	2, 4, 6 Alaska	RDBPCT
HbOC (CRM_{197}) HibTITER®	1990	100 (64-100) (*25*)	2, 4, 6 Kaiser Permanente Northern California (KPNC)	Treatment/ no treatment by date of birth
PRP-OMP (Meningococcal OMP) PedvaxHIB®	1990	93 (53-98) (*26*)	1-3, 3-5[a] Navajo Reservation	RDBPCT
PRP-T (Tetanus toxoid) ActHIB ®	1993	100 (80-100) (*29*)	2, 3, 4 UK	Staged introduction by area
		95 (67-100) (*30*)	2, 3, 4 Gambia	RDBPCT

[a] 2 doses at 42-90 days and 70-146 days, separated by at least 28 days. RDBPCT: Randomized, double-blind, placebo-controlled trial.

Correlates of Protection for Hib Conjugate Vaccines

A precise correlate of protection has been difficult to establish. After introduction of plain PRP vaccines ≥1.0 μg/mL was first used as a correlate of protection. However subsequently, with introduction of conjugate vaccines, a lower estimate of ≥0.15 μg/mL achieved within a month of completing a primary series or primary plus booster series of vaccination was used to assess response to conjugate vaccines in infants (*32*). The difference in these correlate values likely reflects qualitative differences between antibodies produced in response to plain polysaccharide vaccines rather than polysaccharide conjugate vaccine, such as differences in avidity maturation or development of immunological memory (*32*). Functional antibody titers determined in assays measuring complement mediated bactericidal killing were shown to correlate with binding antibody titers after Hib conjugate vaccine immunization (*11*). The WHO provides guidance on the clinical evaluation of Hib conjugate vaccines (*10*), which has been pivotal to developing combination vaccines and to validating new manufacturing sites on the basis of immunogenicity. Key elements to be described include;

- Percentage of subjects achieving Hib antibody concentrations ≥0.15 µg/mL and ≥1.0 µg/mL one month after completion of a primary series and one month after a booster dose
- Functional activity of antibodies as assessed by serum bactericidal assays
- Persistence of antibodies to 4 years of age
- Immune priming demonstrated by a booster response (initially to plain PS vaccine but now usually to a dose of Hib conjugate)
- Immune priming demonstrated by avidity maturation

Protein Carrier-Mediated Immune Differences

The question of the choice of carrier proteins in conjugate vaccines (e.g. CRM_{197}, DT, TT, OMP or others) has been of interest from the very beginning of conjugate vaccine trials. It is possible that the response to the conjugate could be enhanced or reduced by the choice of carrier proteins. In addition, there is the possibility of enhancing or inhibiting the vaccine response in relation to other vaccines that are concomitantly given with the conjugate vaccine. The sequence of administration could also be relevant, demonstrated by the fact that administration of pediatric diphtheria and tetanus toxoid vaccines before and after conjugate vaccines did have disparate impacts (*33*). This topic has continued to be of interest as co-administration of and creation of combinations of vaccines has grown. For Hib conjugate vaccines the topic has been well reviewed (*33*, *34*). In general, the standalone Hib conjugate vaccines had no clinically significant issues but there have been some interactions within combinations, particularly where multiple conjugates are combined, which will be touched upon below.

Combination Vaccines

The need for multiple injections poses a practical barrier to the introduction of new vaccines, particularly in infants and children. A logical next step was therefore the development of combinations including effective Hib conjugate vaccines with other primary infant vaccines.

The first such vaccine combined HbOC with the components of a DTPw vaccine (diphtheria and tetanus toxoids with whole cell pertussis vaccine). In this combination the antibody responses to PRP, diphtheria toxoid and pertussis agglutinins were enhanced by the combination, although this was not necessarily of clinical significance (*35*). This DTPwHib combination was approved in US in 1993 as Tetramune® and was successfully marketed in a number of other countries. As whole-cell pertussis vaccine (Pw) has remained important in LMIC, pentavalent combinations of DTPw with Hib and hepatitis B vaccine (HepB) have been a cornerstone for improving Hib uptake in these areas (*36*). The first such pentavalent combination gained WHO prequalification in 2006 (*37*) and, following substantial effort to transfer Hib conjugate technology (*38*), there are eight prequalified manufacturers, including companies in India and Indonesia as well as South Korea and Europe (*37*).

In the 1990s a number of acellular pertussis vaccine (Pa) combinations were being developed for infants in wealthier countries and combinations of DTPa and Hib were soon developed. At least in part, this was in response to falling pertussis vaccine coverage due to public concerns about the safety and tolerability of whole-cell pertussis vaccines. Unfortunately, in 1996 a combined DTPa-PRP-T was found to produce lower Hib immune responses than PRP-T administered separately (*39*). Assessments of the quality of the immune response suggested that the response remained likely to be protective (*40*). Subsequently however, effectiveness of the Hib vaccine in the UK declined in children born around 2000, coinciding with a period when DTPa-Hib combinations were first used (*41*). It was noted that the UK was unusual in not including a Hib toddler booster dose (*42*). Use of a catch-up program in children up to 5 years of age followed by inclusion of a booster dose in the schedule led to resumed protection (*42*, *43*), suggesting that sustained protection after a primary series with a conjugate vaccine cannot be assumed on the basis of an ability of infection to induce a memory response in individuals primed by conjugate vaccination; factors such as herd immunity due to reduced carriage in older subjects and sustained circulating antibody are also required.

In more economically advantaged countries, Hib conjugate vaccines for infant doses and booster doses up to 5 years of age are now largely included in combinations of five (pentavalent) or six (hexavalent) pediatric vaccines, also including DTPa and inactivated polio vaccine (IPV) with or without hepatitis B vaccine (HepB). Two hexavalent combination vaccines were approved in the EU in 2000 (*44*, *45*), although one was later withdrawn due to concerns about the level of HepB response (*45*), illustrating the difficulties of producing pediatric combination vaccines. Further hexavalent combinations are now approved in Europe and elsewhere (*46*, *47*). No hexavalent combination has yet been approved by the FDA, although one pentavalent combination including a Hib vaccine is approved (*48*).

Two additional types of combination Hib vaccines were developed.

In 1996 the FDA approved a combined PRP-OMP and hepatitis B vaccine (COMVAX®) (*49*). Before DTPa vaccines were combined with HepB this allowed a reduction in the number of injections. However, with the development of larger combinations including hepatitis B and Hib vaccines this product was no longer required and was withdrawn in 2014.

PRP-T has also been combined with meningococcal conjugate vaccines.

A combined PRP-T and meningococcal group C conjugate vaccine (Menitorix®) is available in Europe and is used as a toddler dose in the UK routine schedule (*50*). A combined PRP-T, meningococcal group C and Y vaccine (Menhibrix®) was approved by the FDA in 2012, but was never recommended for routine use and has now been withdrawn by the manufacturer due to the resulting very limited demand (*51*).

Impact

The impact of Hib vaccine was rapid and dramatic after its introduction in developed markets (*31*, *43*). For example, by 1997 the number of reported Hib cases in children less than 5 years of age in USA had declined by 99%, making the disease a rarity even to hospital-based pediatric infectious disease specialists (*52*). In the UK, the incidence in children less than 5 years of age was over 20/100,000 before vaccination was introduced and fell below 1.0/100,000 within 3 years (*41*). In Finland similar dramatic impacts were noted, falling from 30 cases/year in the Helsinki area in 1986 when the program started to 0 cases in 1991 (*53*). The introduction of vaccine had a similar impact in LMIC such as the Gambia, where PRP-T was introduced as part of the Gambian expanded program on immunization (EPI) in May 1997, following extensive involvement in Hib efficacy trials in 1993-1996 (*54*). In Western Gambia where a surveillance program was conducted rates of invasive Hib disease fell from over 200/100,000 in children < 1 year of age in 1990-1993 to 0/100,000 by 2002 (*54*).

However, continuing surveillance is critical as some reemergence of disease, albeit modest compared to the pre-vaccination incidence, was noted in UK (*41*) and Gambia (*55*). In the UK steps were taken to optimize schedules, as discussed above, leading to reversal of the resurgence (*42*, *43*); by 2012-2016 the incidence across the whole population was <0.05/100,000 compared to around 2.5/100,000 before the vaccine was introduced and over 0.5/100,000 at the peak of the recurrence (*56*).

Although global estimates of the incidence of Hib disease are challenging in the absence of a microbiological diagnosis in most cases, the Global Burden of Disease Survey 2015 estimated that the fraction of lower respiratory tract infection deaths in children due to Hib decreased 37·8% between 2005 and 2015 (*57*). By the end of 2014, all Gavi-supported countries had introduced Hib vaccine as part of the pentavalent vaccine (*58*). By the end of 2016 Hib vaccine had been introduced in 191 countries and global coverage with 3 doses of Hib vaccine is estimated at 70% (*59*). However, there is still room for improvement. In the WHO Region of the Americas, coverage is estimated at 90%, while it is only 28% in the WHO Western Pacific Region (*59*). With newly available pentavalent vaccines, the WHO South-East Asia Region raised coverage from 56% in 2015 to 80% in 2016 (*60*).

Streptococcus pneumoniae

Background to Formulation of Pneumococcal Conjugate Vaccines (PCV)

Pneumococcal polysaccharide vaccines (PPV) containing capsular polysaccharide from 14 serotypes were first approved in US and Canada in 1977 and subsequently vaccines including 23 serotypes (PPV23) were approved in 1983, followed by a series of national approvals in European markets and Australia (*61*). Polysaccharide vaccines are still used in adults but have limited immunogenicity in infants and young children (*62*, *63*), in whom the highest incidence of invasive pneumococcal disease was observed before conjugate vaccines were introduced (*64*). As a result this organism was an early target for conjugate vaccine development, particularly in view of excellent early experience with Hib conjugate vaccines.

However, unlike invasive *Haemophilus influenzae* disease, which was almost exclusively due to a single serotype, type b, it was known that many serotypes cause invasive pneumococcal disease (IPD) (*64, 65*). It appeared that a combination of at least 7 serotypes (4, 6B, 9V, 14, 18C, 19F, 23F) would be required to provide sufficient coverage for a vaccine to have impact in North America against IPD in infants and young children (*64*). It was estimated that such a PCV7 would cover around 70%–88% of IPD in young children in the United States and Canada, Oceania, Africa, and Europe and more than 60% in Latin America and Asia (*66*).

Measuring the Immune Response to Pneumococcal Vaccines

It is important to be able to assess the antibody response to a vaccine in relation to vaccine efficacy, using high throughput, sensitive and reproducible methods. This allows comparison of the vaccine across different populations and circumstances without the need to repeat efficacy studies, which would clearly be impossible given the size of the efficacy studies used to gain market approval for PCV7. Moreover, it is only by assessing antibody response that subsequent related vaccines can be introduced, as placebo-controlled efficacy studies cannot be performed once the vaccine class is shown to be efficacious. Following the success of immunoassays for measuring antibodies to Hib polysaccharide, standardized and optimized ELISA methods were developed for measuring human antibodies to pneumococcal polysaccharides (*67*). Specificity has been enhanced from the first assays by pre-adsorption of sera with pneumococcal C-polysaccharide and 22F polysaccharide (*67*) and this "third generation" assay is the basis of recommendations from WHO on clinical assessment of pneumococcal conjugate vaccines (*68*). Subsequently a multiplexed binding assay based upon the ELISA experience has been developed allowing much higher throughput (*69*).

As with Hib vaccines, it is important to assess not only antigen binding antibody, but also to confirm that the antibody is functional. Moreover, in older children and adults functional assays are the only method by which immune responses can be assessed, as binding antibodies are induced throughout life due to natural exposure during nasopharyngeal carriage and so protective binding antibody levels have not been determined. For pneumococcal vaccines the functional method is an opsonophagocytic killing activity (OPA) assay. These are complex assays with multiple biological components and reagents that need to be strictly controlled and qualified. Laboratory comparisons have been undertaken to ensure a degree of comparability (*70*). Assignment of agreed opsonic values to existing standard sera is a further step towards inter-laboratory comparability (*71*).

Early Clinical Experience with Experimental Pneumococcal Conjugate Vaccines

Manufacturers largely stayed with the conjugate carrier proteins that had been used for Hib conjugate vaccines, but explored increasingly large multivalent combinations. In the mid-1990s a bivalent conjugate vaccine was studied in children 18-30 months of age to select the most effective CRM_{197} conjugate constructs (*72*). A 5-component CRM_{197} conjugate vaccine (*73, 74*) and 4-component meningococcal outer membrane protein (OMP) (*75*), diphtheria toxoid (*76*) and tetanus toxoid (*76*) conjugate vaccines were studied in infants, demonstrating good immunogenicity and tolerance. Immunogenicity results were encouraging, leading to the development of formulations with an increased number of serotypes.

Clinical Experience with PCV7-CRM_{197} Leading to the First Marketing Approvals

Pivotal Efficacy against IPD

In 1998 the first study looking at PCV7-CRM_{197} (4, 6B, 9V, 14, 18C, 19F, 23F, see Figure 1) in infants administered alongside routine vaccines in a 2, 4, 6 and 12-15 month schedule was published (*77*). Consistent with previous studies, types 6B and 23F showed limited immunogenicity after 2 doses, but had a good response to a third dose and all serotypes had a good response after the toddler booster dose, indicating priming by the infant doses. A second phase 2 study in the same schedule was performed by the group running the pivotal efficacy study (*78*) and the results were very consistent between the studies.

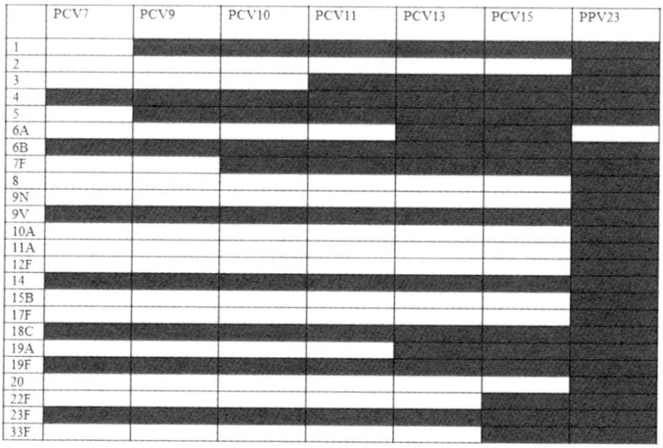

Figure 1. Serotypes covered by multivalent pneumococcal vaccines.

A pivotal efficacy study was conducted in the Kaiser Permanente North California (KPNC) (*79*). (Table 2). From October 1995, infants were randomized to receive PCV7-CRM_{197} or an active control, meningococcal group C CRM_{197} conjugate vaccine (MnC-CRM_{197}), at 2, 4, 6 and 12-15 months of age, alongside routine vaccination (which changed from whole-cell pertussis combinations to acellular pertussis combinations and from oral polio vaccine to inactivated polio vaccine in the course of the study). Cases of invasive pneumococcal disease (IPD) were identified from linked clinical and laboratory databases, and defined as positive cultures of *Streptococcus pneumoniae* from a normally sterile body fluid obtained from children presenting with an acute illness compatible with pneumococcal disease.

The study design included case-driven pre-planned analyses. By August 1998 17 cases of per-protocol vaccine type (VT) IPD and 22 cases of intention to treat (ITT, including partially vaccinated subjects) VT IPD had been identified in the study population. Astonishingly, all 22 ITT cases were in control subjects (*79*). Blinded case ascertainment was continued until April 1999. At this point 40 per-protocol cases of VT-IPD had been identified, with only one in a subject who received PCV7-CRM_{197}, giving an efficacy of 97.4% (82.7, 99.9) (*79*). An additional 12 cases (10 controls, 2 in PCV7-CRM_{197} group) were identified that were outside the per protocol case definition, mostly because subjects were only partially vaccinated, giving an intention-to-treat (ITT) efficacy against VT-IPD of 93.9% (79.6, 98.5) (*79*). A total of 61 cases of IPD were identified including vaccine and non-vaccine serotypes and the ITT efficacy against all IPD was 89.1% (73.7, 95.8) (*79*).

Efficacy against Pneumococcal otitis

At the time that the follow up for IPD was completed, otitis media efficacy was also assessed (Table 3), being 6.4% (3.9, 8.7) against clinical acute otitis media (AOM) episodes and 20.3% (3.6, 34.1) against the need for placement of tympanostomy tubes (*79*). Although AOM episodes are less severe than IPD cases, the number of AOM cases is much larger; compared to 61 cases of IPD, 52,789 AOM episodes occurred during the trial, and 432 children required tympanostomy tubes (*79*).

The study population was followed up further to 42 months of age for otitis media events (*80*). Overall PCV7-CRM_{197} had a 7.0% (4.7, 9.1) efficacy against otitis media visits, associated with a 5.4% (4.0, 6.7) reduction in antibiotic prescriptions, from first dose to completion of follow-up (*80*). There was a reduction of 23.2% (11.3, 33.5) in first tympanostomy tube placement during the same period (*80*). Frequent otitis media visits were reduced, with the highest frequency seeing the greatest reductions (*80*).

A very different efficacy study was conducted in Europe, investigating the effect of PCV7-CRM_{197} on pneumococcal otitis media (Table 3) (*81*). FinOM, which started in December 1995, randomized 1662 infants to receive PCV7-CRM_{197} or hepatitis B vaccine concomitantly with whole-cell DTP combined with Hib vaccine (half HbOC, half PRP-T) at 2, 4, 6 and 12 months of age. The subjects received all primary care at a trial clinic; whenever a case of acute otitis media (AOM) was identified, tympanostomy was performed and a sample of middle-ear fluid removed for microbiology (*81*). In the study population there were 2596 episodes, in which 685 were culture confirmed as pneumococcal (*81*). The vaccine was 57% (44, 67) efficacious against vaccine-type pneumococcal AOM, 51% (27, 67) against potentially cross-reactive types and 34% (21, 45) against all pneumococcal AOM, due to partial replacement with non-vaccine types (NVT) (*81*). Amongst potentially cross-reactive types, 6A showed significant efficacy from the 6B component in the vaccine (*81*). Overall there was non-significant 6% efficacy against all AOM (-4, 116) (*81*). Interestingly the point estimate for all AOM was similar to that in the KPNC study but, as the study was much smaller due to the intensity of follow-up, it did not reach statistical significance. Vaccine immunogenicity was consistent with that observed in other studies (*81*).

Efficacy against Pneumonia

The KPNC population was also investigated for episodes of all cause radiographic pneumonia (Table 3). An initial analysis using radiographic assessments in routine clinical practice showed 17.7% (4.8, 28.9) reduction in clinical pneumonia with a positive radiograph from the first dose of vaccine onwards (*82*). Interpretation of pediatric chest radiographs is subject to substantial inter-observer variability and, under the aegis of the WHO, much effort has been invested into standardizing methodologies for interpretation to improve specificity (*83*). With this improved methodology the *post hoc* calculated efficacy increased to 25.5% (6.5, 40.7) (*84*).

Initial Regulatory Approvals in USA and Europe

On the basis of the safety and efficacy data, PCV7-CRM_{197} was granted approval for marketing in the USA in February 2000, under the name Prevnar® (*85*).

EU marketing approval was granted in February 2001, under the name Prevenar® (*86*), the name used in most countries outside US. The file was supported by a number of studies looking at concomitant use with a range of European vaccination schedules. These included more accelerated schedules with a 2, 3, 4 primary series alongside a whole-cell pertussis vaccine combination and HbOC in UK (*87*), a whole-cell pertussis vaccine combination including PRP-T in France (*88*) and a pentavalent acellular pertussis combination including PRP-T in Germany (*89*). The German study also included concomitant use as a booster dose (*89*). A subsequent study in Germany also looked at concomitant use of PCV7-CRM_{197} and a hexavalent acellular pertussis combination including PRP-T and IPV (*90*). In general PCV7-CRM_{197} was well tolerated and was suitable for co-administration with routinely used vaccines and schedules using a three dose primary series and booster (3+1 schedule) in Europe.

A more pressing issue in Europe was that some countries, including Italy and some Scandinavian countries, routinely used only two vaccine doses as a primary series, with a booster dose from 11 months of age or later (2+1 schedule). Studies in Europe confirmed that types 6B and 23F showed lower immunogenicity than other types after the first two doses, but that after the third dose responses were good, indicating that the first two primary doses had induced immunological priming (*91*, *92*). On this basis, the 2+1 schedule was included as an option in the EMA product label and many countries launched with this schedule.

Studies on Combinations of Different CRM_{197} Conjugate Vaccines

Pneumococcal conjugate vaccines are already complex combinations and so adding other components was challenging. Nonetheless, three studies were performed examining combinations of PCV7-CRM_{197} with other CRM_{197} conjugate vaccines. Addition of further pneumococcal CRM_{197} types to produce extended pneumococcal conjugate vaccines will be discussed in subsequent sections.

In one study PCV7-CRM_{197} was mixed with HbOC before injection and pneumococcal IgG responses were significantly lower than with separate administration for 5 out of 7 serotypes after the primary series (*87*).

A combined PCV9-CRM_{197} and MnC-CRM_{197} vaccine was formulated, but found to induce much lower MnC antibody levels than MnC-CRM_{197} alone, both by SBA Geometric Mean Titer (GMT) 179 (133, 243) versus 808 (630, 1037) and MnC IgG ELISA GMCs 4.88 µg/mL (4.19, 5.70) versus 23.27 µg/mL (20.18, 26.84) (*93*).

It is possible that these results were a reflection of carrier-mediated immune response suppression as discussed above in the section on Hib conjugate vaccines. Neither combination, however, was further developed.

A third study randomized Gambian infants to receive three doses of Tetramune (DTPw-HbOC), Tetramune and PCV9-CRM_{197} administered at separate sites or Tetramune and PCV9-CRM_{197} mixed immediately prior to administration (*94*). Despite some minor statistical differences between groups, antibody responses to all antigens were good and similar across groups (*35*, *94*). The mixed product was used in a subsequent efficacy study (*95*) but was not further developed.

Table 2. Pneumococcal conjugate vaccines used in infant efficacy studies against IPD

Abbreviation *Conjugate protein* *Common proprietary names*	*First approval for use in infants*	*ITT Efficacy against VT IPD* *%* *(95% CI)*	*ITT Efficacy against all IPD* *%* *(95% CI)*	*Efficacy study schedule*	*Efficacy study design (active control vaccine)*	*Ref* *Location*
PCV7-CRM$_{197}$ Prevnar® Prevenar®	2000	93.9% (79.6, 98.5)	89.1% (73.7, 95.9)	2, 4, 6, 12-15 months	RDBACT MnC-CRM$_{197}$	*(79)* KPNC, US
		82·6% (21·4, 96·1)	46.3% (-16,5, 75.3)	2, 4, 6, 12-15 months and catch up	RDBACT Group randomized MnC-CRM$_{197}$	*(96)* Navajo, US
PCV9-CRM$_{197}$	-	72% (46, 87)	50% (23, 68)	6, 10, 14 weeks	RDBPCT	*(97)* Soweto, South Africa
		71% (46, 86)	45% (19, 62)	3 doses[a]	RDBPCT	*(95)* Gambia

Continued on next page.

Table 2. (Continued). Pneumococcal conjugate vaccines used in infant efficacy studies against IPD

Abbreviation *Conjugate protein* *Common proprietary names*	*First approval for use in infants*	*ITT Efficacy against VT IPD* *%* *(95% CI)*	*ITT Efficacy against all IPD* *%* *(95% CI)*	*Efficacy study schedule*	*Efficacy study design (active control vaccine)*	*Ref* *Location*
PCV10-HiD HiD for 8 types, D and T each 1 type Synflorix®	2009 (not based on efficacy data)	3+1 group 100% (83,100) 2+1 group 92% (58, 100)	All subjects 93% (75, 99)	<7m 3+1 7-11m 2+1 ≥12m 2 doses	RDBACT Cluster randomized HepB 3+1 HepA 2+1	*(98)* Finland
		100% (77.3, 100)	66.7% (21.8, 85.9)	2, 4, 6, 15-18 months	RDBACT HepB infant HepA booster	*(99)* Latin America

[a] First dose 6-51 weeks, subsequent doses a least 25 days apart. RDBACT: Randomized, double-blind, active-controlled trial. RDBPCT: Randomized, double-blind, placebo-controlled trial. HepA: Hepatitis A vaccine.

Subsequent Efficacy Studies with PCV-CRM_{197} and a Correlate of Protection for IPD

Three additional major efficacy studies against IPD in infants and young children were conducted with CRM_{197} based PCV (Table 2).

The first was a study with PCV7-CRM_{197} in a Native American population, which had a high burden of pneumococcal disease (Table 2) (*96*). The design was a group randomized study with an active control vaccine, MnC-CRM_{197}. A total of 8292 subjects were enrolled within 38 randomization groups, mostly infants receiving their primary vaccination but also subjects up to 23 months of age who received a catch-up schedule (*96*). The study was planned to be case-driven, but was stopped early due to the approval of PCV7-CRM_{197} combined with demonstration of a benefit to recipients of PCV7-CRM_{197} in the trial. Despite the early closure efficacy was demonstrated against IPD in subjects given a primary vaccination series, with ITT efficacy 82·6% (21·4, 96·1) (*96*).

The two further efficacy studies used a nine component vaccine, PCV9-CRM_{197}, which contained additional serotypes 1 and 5 (Figure 1). The use of 9 components was based on evidence that adding types 1 and 5 would bring serotype coverage in Africa close to that in USA and Europe (*66*).

The first of these PCV9-CRM_{197} trials was a double-blind, placebo-controlled, individually randomized study in South African infants at the EPI (Expanded Program on Immunization) schedule of 6, 10 and 14 weeks of age, alongside DTPw and HbOC (*97*). (Table 2) Almost 40,000 subjects were enrolled. HIV status was assessed in the event of hospitalization, but not at study entry. The study endpoints were IPD and radiologically confirmed pneumonia defined by WHO criteria (*83*). Against VT-IPD, ITT efficacy was 83% (32, 97) for non-HIV infected, fully immunized children, but was still significant in HIV infected children at 65% (24, 86) (*97*). For all cause pneumonia there was significant efficacy in non-HIV infected children 20% (2, 35), but not in HIV infected children at 13% (-7, 29) (*97*).

The second PCV9-CRM_{197} study was also a double-blind, placebo-controlled, individually randomized study in a LMIC population, but in the Gambia, with a much lower incidence of HIV infection (Table 2) (*95*). Over 17,000 subjects were randomized. ITT efficacy was 71% (46, 86) against VT IPD and 45% (19, 62) against all IPD (*95*). In addition 13% (6, 19) efficacy was found against all-cause admissions and 14% (2, 24) against mortality (*95*). This represents an astounding reduction in the absolute death rate of 4.1 deaths per 1000 child-years (0.4, 7.7) which reinforces the substantial potential public health benefit of pneumococcal conjugate vaccination (*95*).

The antibody correlate of protection for prevention of invasive pneumococcal disease (IPD) was based on relating the antigen-binding antibody concentration one month after completion of vaccination in populations in immunogenicity studies to the measured efficacy in different populations in efficacy studies (Table 2) (*100*). The method was first established based on the KPNC efficacy study and immunogenicity in a subset (*79*) and recommended a threshold of 0.2 μg/mL (*100*). The method was later refined (*101*) to include additional data from the efficacy studies in South African infants (excluding HIV infected subjects) (*97*) and Native Americans, to provide a correlate of protection more appropriate for a global population at ≥0.35 μg/mL. Although initial immunogenicity studies had been conducted before the ELISA assay was modified to include 22F absorption, it was found that the impact in infant sera was quite minor, as this absorption step was primarily intended to remove non-serotype specific pneumococcal antibodies in adult sera. The revised value would change only slightly to ≥0.32 μg/mL, which was insignificant given the uncertainties in the methodology for calculating a single threshold to be used for all serotypes (*101*). The threshold of ≥0.35 μg/mL was adopted as a recommendation by the WHO (*68*) as an appropriate basis of comparison (along with demonstration of serotype specific functional antibody by opsonophagocytic activity assay) for licensure of successor vaccines to PCV7-CRM_{197}.

Clinical Studies with PCV-HiD Constructs

Clinical trials with PCV based on non-typable Haemophilus influenzae D protein (HiD) as a carrier protein led to marketing approval of PCV10-HiD vaccine, Synflorix®, in 2009 (Figure 1) (*102*). Earlier studies had been performed using an 11-valent vaccine (Figure 1) with all polysaccharides conjugated to HiD, whereas PCV10-HiD was modified to remove serotype 3 conjugate and used HiD as a carrier protein on 8 serotypes (1, 4, 5, 6B, 7F, 9V, 14, 23F) with tetanus and diphtheria toxoids each used for a single serotype (18C, 19F respectively). Before this change was made, an efficacy study against otitis media with tympanostomy for children meeting certain criteria was performed, with subjects randomized to receive PCV11-HiD or hepatitis A vaccine at 3, 4, 5 and 12-15 months (Table 3) (*103*). In an ITT analysis efficacy for VT AOM was 52·6% (95% 36·1 to 64·9), and for non-typable Hi was 32·7% (0·77 to 54·3). Serotype 3 conjugate appeared to lack efficacy.

A clear review of clinical trials leading to approval of PCV10-HiD has been published (*104*). These included an extensive range of safety and immunogenicity studies confirming the ability of the vaccine to be administered safely and effectively in a range of schedules and with a range of co-administered vaccines in Europe and Asia (*104*).

In the absence of efficacy data, PCV10-HiD was approved in the European Union for prevention of invasive disease, pneumonia and acute otitis media caused by *Streptococcus pneumoniae* in infants and children from 6 weeks up to 5 years of age based on an immunogenicity non-inferiority comparison with the already approved PCV7-CRM for common serotypes (*105*), using the criteria established by WHO (*68*). The comparison was performed using a threshold of ≥0.2 μg/mL which was stated to be equivalent to ≥0.35 μg/mL in the analysis performed to support the WHO recommendation for a correlate of protection (*105*, *106*). Non-inferiority could not be demonstrated for types 6B and 23F after primary vaccination, although the choice of threshold was not relevant as similar differences were seen at the higher threshold. Nonetheless, a *post hoc* analysis suggested comparable OPA titers. Moreover, the same dose and construct for 6B and 23F had been included in the PCV11-HiD used in an otitis media efficacy study and had provided statistically significant serotype specific efficacy against

AOM (*103*). Overall then, the totality of the immunogenicity and efficacy profile supported licensure in Europe.

Subsequently a Latin American and a European PCV10-HiD efficacy study were published (Tables 2 and 3).

The COMPAS study examined the efficacy of PCV10-HiD against a range of pneumococcal diseases in Latin American infants, following the initial marketing approval (Tables 2 and 3) (*98*, *99*). Infants were randomized to receive PCV10-HiD or control vaccine (hepatitis B for infant doses, hepatitis A for booster dose) at 2, 4, 6 and 15-18 months. The primary objective was to study efficacy against likely bacterial CAP (B-CAP, defined as radiologically confirmed CAP by WHO criteria (*83*) with either alveolar consolidation/pleural effusion or non-alveolar infiltrates and CRP ≥40 mg/mL). After a mean follow-up of 23 months intent-to-treat efficacy was 18.2% (5.5, 29.1) against B-CAP and 23.4% (8.8, 35.7) against consolidated CAP, with per-protocol efficacy 22.0% (7.7, 34.2) against B-CAP and 25.7% (8.4, 39.6) against consolidated CAP (*99*).

Other endpoints were also studied in COMPAS including clinical AOM, other AOM endpoints and IPD (Table 2 and 3) (*99*). After a mean follow-up of 28–30 months per-protocol efficacy was 16.1% (21.1, 30.4) against clinical AOM, 67.1% (17.0, 86.9) against vaccine serotype clinical AOM, 100% (74.3, 100) against vaccine serotype IPD, and 65.0% (11.1, 86.2) against any IPD (Table 2) (*99*).

FinIP was an efficacy study in Finland focused on IPD in clusters randomized to receive PCV10-HiD or hepatitis vaccine in a 2:1 ratio (Table 2) (*98*). Subjects up to 19 months of age were enrolled. In half of the clusters infants under 7 months of age received in a 3+1 schedule and half in a 2+1 schedule. The older infants received a catch up schedule. Efficacy against IPD was 100% (83, 100) following a 3+1 schedule and 92% (58, 100) following a 2+1 schedule (*98*).

A nested study within FinIP examined effect upon AOM and nasopharyngeal carriage (Table 3) (*107*). ITT Efficacy against first AOM was not significant at 6.1% (-2.7, 14.1) for the 3+1 group and 7.4% (-2.8, 16.6) for the 2+1 group. Impact on NP carriage was noted to be higher for the 3+1 schedule and to be greatest following a booster dose (*107*).

Clinical Efficacy Studies with PCV Using Other Carrier Proteins

The efficacy of an 11-valent vaccine containing 7 tetanus toxoid conjugates and 4 diphtheria toxoid conjugates (Figure 1) was studied against pediatric pneumonia in the Philippines (Table 3) (*108*). An ITT analysis of radiologically confirmed pneumonia showed a trend towards efficacy of 16.0 % (-7.3, 34.2) (*108*). In the PP analysis this was almost significant 22.9 (-1.1, 41.2) and was significant in the PP analysis for children 3-11 months of age at 34.0% (4.8, 54.3) (*108*). In a nested immunogenicity study, one month after completing vaccination 93% to 100% of PCV11-DT recipients achieved an IgG concentration of ≥0.35 μg/mL for 9 serotypes, 76% for 6B, and 87% for 23F (*108*). However, this vaccine did not progress to the market.

The FinOM efficacy study was designed to study the efficacy of several conjugate vaccines (Table 3). A single control group receiving hepatitis B vaccine was to be used to compare to PCV groups. The PCV7-CRM_{197} and control groups proceeded as planned and a comparison of these groups yielding efficacy data for PCV7-CRM$_{197}$ was published separately, as described above (*81*). A third group of 835 subjects was randomized to receive PCV7-OMP at 2, 4, 6 and 12 months of age, but for the booster dose 187 of these received PPV-23 instead of PCV7-OMP (*109*). In the published PP analysis, efficacy against VT-AOM was 56% (44–66), which was similar to the outcome with PCV7-CRM$_{197}$, but there was no significant efficacy against potentially cross-reactive serotypes at -5% (-47, 25), which was different (*109*). There was a trend to negative efficacy for other serotypes -27 (-70, 6) (*109*). Overall efficacy to pneumococcal AOM was 25% (11, 37) with no overall efficacy against all AOM -1% (-12, 10) (*109*). This vaccine also did not proceed to the market.

Table 3. Clinical trials examining other pneumococcal disease end-points

Abbreviation Conjugate protein *Proprietary names*	*Main AOM endpoints (number of subjects)*	*Main pneumonia endpoints (number of subjects)*	*Efficacy study schedule* *Location*	*Efficacy study design (active control vaccine)*
PCV7-CRM$_{197}$ Prevnar® Prevenar®	AOM (*79, 80*) (37868)	R-CAP (*82, 84*) (37868)	2, 4, 6, 12-15 months KPNC	RDBACCT MnC-CRM
	C-AOM (*81*) M-AOM (1662)	-	2, 4, 6, 12 months Finland	RDBACCT HepB
PCV7-OMP				
PCV9-CRM$_{197}$	-	R-CAP (*97*) (39836)	6, 10, 14 weeks South Africa	RDBPCT
	-	R-CAP (*95*) (17437)	3 doses[a] Gambia	RDBPCT
PCV10-HiD HiD for 8 types, D and T each 1 type Synflorix ®	C-AOM M-AOM (*99*) (7214)	Likely bacterial CAP (*99*) (23579)	2, 4, 6, 15-18 months Latin America	RDBACT (HepB infant) (HepA booster)
	C-AOM (*107*) (6178)	-	<7m 3+1 7-11m 2+1 ≥12m 2 doses Finland	RDBACT (HepB 3+1) (HepA 2+1) Cluster randomized

Continued on next page.

Table 3. (Continued). Clinical trials examining other pneumococcal disease end-points

Abbreviation Conjugate protein *Proprietary names*	*Main AOM endpoints (number of subjects)*	*Main pneumonia endpoints (number of subjects)*	*Efficacy study schedule* *Location*	*Efficacy study design (active control vaccine)*
PCV11-HiD	C-AOM M-AOM (*103*) (4968)	-	3, 4, 5, 12-15 months Czechia	RDBACT (HepA)
PCV11-TD	-	R-CAP (*108*) (12031)	6, 10, 14 weeks Philippines	RDBPCT

[a] First dose 6-51 weeks, subsequent doses a least 25 days apart. RDBACT: Randomized, double-blind, active-controlled trial. RDBPCT: Randomized, double-blind, placebo-controlled trial. R-CAP: Radiologically confirmed community acquired pneumonia. C-AOM: Clinically confirmed AOM. M-AOM: Microbiologically characterized AOM on tympanosotomy.

Development and Use of PCV13-CRM_{197} in Infants and Children

Route to Initial Licensure for Use in Infants

PCV13-CRM_{197} contains CRM_{197} conjugates to types 1, 3, 5, 6A, 7F and 19A in addition to the 7 types in PCV7-CRM_{197} (Figure 1). Phase 1 trials comparing PCV13-CRM to PPV23 in US adults and in Japanese adults allowed progression into pediatric populations (*110, 111*).

Subsequent safety and immunogenicity studies were conducted in infants and toddlers in USA (*112–114*), Canada (*115*), Germany (*116*), UK (*117*), Spain (*118, 119*), Poland (*120*) and Italy (*121*) to support initial licensure. Overall safety and response to concomitant standard vaccination was unremarkable.

Two of these studies were pivotal for approval of PCV13-CRM_{197} based on noninferiority of the 7 shared types with PCV7-CRM_{197} and comparison of the immunogenicity of the 6 new types with the behavior of the existing 7 types, based on WHO criteria (*68*). A US study was pivotal for the US approval (*113*) whereas the EU approval was based on the US study (*113*) and an additional German study (*116*). The regulatory rationales for the approval are in the public domain (*122, 123*).

For the 7 shared types the first round of the non-inferiority assessment was to demonstrate if the lower confidence interval on the difference between the percentage achieving ≥0.35% µg/mL between PCV7-CRM_{197} and PCV13-CRM_{197} was less than -10% following the infant primary series. Type 6B narrowly missed this, the differences being -5.5% (-10.9. 0.1) (*113*) and -9.6 (-16.0, -3.3) (*116*) respectively in US and German studies. In the US study 9V also narrowly missed this at -7.9% (-12.4, -4.0). However, these types produced OPA responses comparable to other serotypes and produced good booster responses, including functional OPA responses (*113*, *116*). Following booster doses the shared serotypes also were non-inferior as assessed by IgG GMC ratio having a lower confidence interval ≥0.5 (*113*).

For the six new serotypes, non-inferiority comparisons were made to the lowest response rate observed among the shared serotypes in PCV13-CRM recipients, which was 6B. In the US study type 3 did not meet non-inferiority criteria after the third or the fourth dose (*122*). Type 3 met non-inferiority in the German study after the primary series, but narrowly missed after the booster dose, with a difference of -6% and a lower confidence interval of -10.2% (*123*). However, type 3 does show functional OPA responses (*113*, *116*) and regulatory approval was given based on the totality of the immune response (*122*, *123*).

In addition to these pivotal assessments, certain features of PCV13-CRM_{197} and PCV7-CRM_{197} immunogenicity were noted within these infant studies in US and Europe;

- Additional serotypes in PCV13-CRM_{197} show antibody responses after a primary series and toddler dose. The weakest response is for serotype 3 but there is a significant response compared to control subjects who received PCV7-CRM_{197} after primary and booster doses, with the percentage achieving at least 0.35 μg/mL by ELISA after a primary series of PCV13-CRM ranging from 63.5% to 98.9%.
- Serotypes 6A and 19A are described as cross-reactive types as even PCV7-CRM_{197} recipients develop IgG to these types, presumed to be due to cross-reactivity with PCV7-CRM_{197} 6B and 19F respectively. However, PCV7-CRM_{197} induces a much lower OPA GMT to these types than PCV13-CRM_{197}. This is consistent with the observation that type 19A was the most prominent non-vaccine type to increase after the introduction of PCV7-CRM_{197}, indicating that the antibody response induced by 19F in PCV7-CRM_{197} was not protective against 19A. In contrast, when 6A and 19A conjugates are included in PCV13-CRM_{197} they both induce protective IgG and OPA responses.
- In a 2-dose infant schedule, IgG responses to serotypes 6B and 23F in both PCV7-CRM_{197} and PCV13-CRM_{197} were lower than to other serotypes, but were similar to other serotypes after a toddler booster dose.

PCV13-CRM_{197} was approved for use in infants and children by FDA in February 2010 as Prevnar13® (*122*, *124*) and in the EU in December 2009 as Prevenar13® (*123*, *125*).

While many further countries approved PCV13-CRM_{197} on the studies already described, clinical trials in infants in Japan (*126*, *127*), Taiwan (*128*), Brazil (*129*), India (*130*), Mexico (*131*), Korea (*132*) and China (*133*) supported approval or introduction in those countries for use in infants.

With the widespread use of PCV in infants and acceptance of WHO criteria for extrapolating efficacy against IPD based on immunogenicity (*68*), it was not possible to perform placebo-controlled efficacy studies of PCV13-CRM_{197} against disease end-points. Comparative studies against PCV7-CRM_{197} would also have been impossible, based on the huge number of subjects that would have been required and the long duration of such studies beyond regulatory approval of PCV13-CRM_{197}. However, it was possible to perform a comparative study in nasopharyngeal (NP) carriage, which demonstrates a biological effect one step beyond immunogenicity in humans. A study was undertaken in 1866 healthy infants in Israel randomized to receive PCV13-CRM_{197} or PCV7-CRM_{197} at 2, 4, 6 and 12 months of age and followed to 24 months, with NP samples at 7, 12, 13, 18 and 24 months of age (*134*). A PP analysis was appropriate as the study was looking for a biological activity rather than a clinical endpoint. This study showed a significant effect of PCV13-CRM_{197} against the additional serotypes 1, 6A, 7F and 19A (*134*). There was no difference between the groups for common serotypes, except for a significant further reduction of 19F NP carriage with PCV13-CRM_{197}, possibly related to the greater antibody response to the toddler booster dose for 19F with PCV13-CRM_{197} than PCV7-CRM_{197} (*134*). There was activity against type 6C, which can be considered cross-reactive with vaccine type 6A (*134*). There was no activity in this study against carriage of type 3 (*134*), although a subsequent study in USA showed significant effectiveness against IPD due to serotype 3 (*135*). There were insufficient numbers of type 5 cases for an effect to be estimated.

Special Considerations for Lower and Middle Income Countries

Safety and immunogenicity of PCV13-CRM_{197} in single-dose presentations was studied in infants in lower and middle income countries (LMIC) (*129–131, 133*), but a multi-dose vial (MDV) has logistic advantages, particularly in the very poorest countries, due to reduced transport and storage costs. A 4-dose MDV was developed, with 2-phenoxyethanol selected as the preservative to prevent contamination after the first dose is removed, through a series of *in vitro* antimicrobial effectiveness studies and vaccine stability studies (*136*). A study comparing the safety and immunogenicity of the MDV was performed against a single-dose syringe (SDS) in infants in Gambia (*137*). This study showed non-inferiority of the MDV product against the SDS product. PCV13-CRM_{197} as a single-dose vial (SDV) was prequalified by WHO in 2010 (*37*); prequalification of the 4-dose MDV in 2016 now supports expanded use PCV13-CRM in GAVI and other LMIC countries (*37*). PCV10-HiD was also prequalified as an SDV in 2009, as a 2-dose MDV in 2010 and as a 4-dose MDV in 2017 (*37*).

By the end of 2016 PCV had been introduced in 134 countries but only 42% of infants received the vaccine (*60*). Pneumococcal pneumonia remains the predominant cause of lower respiratory mortality in children under 5 years of age (*57*), emphasizing the importance of expanded use of PCV in LMIC.

Development and Use of PCV13-CRM in Adults

Immunogenicity in Adults

The wider serotype coverage of PCV13 led to an interest in use of the vaccine in the elderly. In adults immune responses must be assessed by changes in functional opsonophagocytic activity (OPA) as capsular polysaccharide binding antibody does not generally reflect functional activity in adults (*71*). Early studies with PCV7-CRM_{197} showed that in adults 70 years of age or older conjugate vaccine induced higher OPA GMT than PPV23 for 5 of the shared types and similar titers for the other 2 types (*138*, *139*).

Immune measurements used to establish non-inferiority of pneumococcal vaccines in infants are not appropriate in adults, as no correlate of protection has been established. Instead a two-fold criterion of the OPA GMT has been used. With this criterion the lower 95% confidence interval of the ratio of the GMT induced by the new vaccine compared to the GMT induced by the reference vaccine should exceed 0.5 (*140*).

Two pivotal clinical trials demonstrated non-inferiority of the 12 common serotypes shared with PPV23.

- The first study was in adults 60-64 years of age, with an open cohort study of PCV13-CRM_{197} in adults 50-59 years of age (*140*). None of the subjects had received a prior dose of any pneumococcal vaccine. In this study, while all shared types were non-inferior, the OPA GMT was actually significantly higher for 8 of these types, as well as for type 6A which is not included in PPV23 (*140*). Responses to PCV13-CRM_{197} in the 50-59 year old were non-inferior to the responses induced by PCV13-CRM_{197} in 60-64 year olds, and superior for 9 types.
- The second was in adults 70 years of age and older who had received PPV23 at least 5 years earlier (*141*). In this study OPA GMTs were non-inferior after 13vPnC for all shared serotypes and 6A, and superior for 10 of the shared types and 6A.

On the basis of these non-inferiority studies PCV13-CRM_{197} was approved for use in adults $\geq$ 50 years of age in the USA (*142*) and EU in 2011 (*143*).

As PPV23 is widely used in older adults, it was important to examine the sequence of vaccination with PPV23 and PCV13-CRM in these age groups. A follow up within the study mentioned above in 50-64 year olds (*140*) looked at dosing four years later (*144*). In this follow-up study 60-64 year olds who had received PCV13-CRM were randomized to receive a dose of PCV13-CRM or PPV23, 60-64 year olds who had received PPV23 received another dose of PPV23 and all 50-59 year olds received a second dose of PCV13-CRM. A second dose of PCV13-CRM achieved GMTs that were superior to the first dose for 7 serotypes in the older group and 6 serotypes in the younger group (*144*). This was different to the response to a second dose of PPV23, which induced significantly lower GMTs for 9 of the 13 assessed types (*144*). This was also divergent from the response to PPV23 following PCV13-CRM, which was higher than the response to PPV23 alone for 10 of the 13 assessed types (*144*). In summary then, after a 4 year interval PPV23 may impair the response to some serotypes after a subsequent dose of PPV23, whereas PCV13-CRM_{197} does not impair and may even enhance the subsequent response to some serotypes in PPV23 and PCV13-CRM_{197}.

Another study compared the sequence of vaccinations in 60-64 year olds naïve to pneumococcal vaccine (*145*). Three sequences were studied with a one year interval, PCV13-CRM_{197} followed by PPV23 (PCV13/PPV), two doses of PCV13-CRM_{197} (PCV13/PCV13) or PPV23 followed by PCV13-CRM_{197} (PPV/PCV13). OPA GMTs to PCV13/PPV were statistically superior for 11 of the 12 shared types compared to when given in the opposite sequence, PPV/PCV13 (*145*). This study strongly suggests that if both PCV13-CRM_{197} and PPV23 are to be administered, the conjugate vaccine should be given first.

Use in age groups between children and older adults were subsequently added to the labels based on safety and immunogenicity studies in the intermediate age groups (*146–148*). Studies in older children and adults were performed in a number of countries to support local approvals, including Japan (*149*, *150*), India (*151*) and Mexico (*152*).

Efficacy in Adults

While the non-inferiority to PPV23 supported the use of PCV13-CRM_{197} to prevent IPD in older adults, it was unclear whether it could prevent the more common condition of pneumococcal community acquired pneumonia (CAP) particularly non-bacteremic pneumococcal CAP, which represents a large unmet medical need (*153*). To demonstrate PCV13-CRM_{197} efficacy against pneumococcal CAP, the Community-Acquired Pneumonia Immunization Trial in Adults (CAPiTA) study, a randomized, placebo-controlled clinical trial was conducted to assess whether PCV13-CRM_{197} could prevent vaccine-type CAP in adults 65 years of age or older (*153*, *154*). Such a study needed to be very large and several innovative features were used to make such a study feasible (*153*). First of all, a serotype-specific urinary antigen detection assay (SSUAD) needed to be developed to allow detection of non-bacteremic VT CAP, as many more subjects would have to be enrolled if the study would only be based on assessing vaccine impact on bacteremic CAP, diagnosed by blood culture and typing of the isolated pneumococcal strain (*153*). Given the large number of pneumococcal serotypes, it was critical that the SSUAD was also very specific (*155*, *156*). Subjects were recruited within catchment areas of specific hospitals and cases could be identified on hospitalization without being followed within the community (*153*).

Over 84,000 subjects were enrolled between September 2008 and January 2010 and follow up for cases continued through August 2013, based on case accrual. Per protocol (PP) efficacy against a first episode of VT CAP was 45.6% (21.8, 62.5) with intention to treat (ITT) efficacy 37.7% (14.3, 55.1) (*154*). Importantly, efficacy was also significant against VT non-bacteremic/non-invasive CAP at 45.0% (14.2, 65.3) PP, 41.1% (12.7, 60.7) ITT (*154*). Efficacy against VT IPD, which had been extrapolated from non-inferiority to shared serotypes in PPV23 during the regulatory process was confirmed at 75.0% (41.4, 90.8) PP and 75.8% (46.5, 90.3) ITT (*154*). Other key findings were significant efficacy against all pneumococcal CAP and all IPD (*154*).

Concomitant Vaccination in Adults

As annual influenza vaccination provides an opportunity to deliver pneumococcal vaccine to high risk groups, including older adults, clinical studies were performed to assess co-administration of PCV13-CRM_{197} and influenza vaccine. In a US clinical study subjects aged 50 to 59 years of age were randomized to receive PCV13-CRM_{197} or placebo with trivalent inactivated influenza vaccine (TIV), with placebo subjects receiving PCV13-CRM_{197} one month later (*157*). Although IgG GMCs to pneumococcal vaccine types were generally lower with co-administration, all types had a lower 95% confidence interval of the GMC ratio $\geq$ 0.5, the criterion for non-inferiority (*157*). Some types did not meet this criterion for OPA GMTs (*157*). Responses to TIV measured by haemagglutination inhibition (HAI) were similar when co-administered with PCV13-CRM_{197} or given separately (*157*).

Subjects in this study were followed for another 5 years to examine pneumococcal antibody persistence and the response to revaccination (*158*). OPA and IgG antibody levels remained above baseline throughout the follow up and were similar in both groups (*158*). OPA GMT and IgG GMC responses to revaccination with PCV13-CRM_{197} 5 years later were similar, and for some serotypes greater, than the response to initial vaccination (*158*). Moreover there were no differences between the groups who received the first dose with or without TIV (*158*).

Similar results were seen in a European study in adults $\geq$ 65 years of age, with HAI responses to TIV being similar in both groups (*159*). Pneumococcal IgG GMCs were generally lower with co-administration, though only one type, 19F, very narrowly missed the non-inferiority criterion, with the ratio of GMTs having a lower 95% confidence interval of 0.49 (*159*).

As quadrivalent inactivated influenza vaccine (QIV) is increasingly replacing TIV, co-administration of PCV13-CRM_{197} and QIV has been examined in a US study, currently only published as an abstract (*160*). In this study subjects were $\geq$ 50 years of age and had previously received PPV23. OPA GMTs tended to be lower with co-administration, but all types met non-inferiority criteria and responses to QIV types were similar with and without co-administration (*160*).

Use in Immunocompromising Conditions

Studies have also been performed to assess use of PCV in high-risk subjects with immunocompromising conditions. Even before development of PCV13-CRM_{197}, a study of two doses of PCV7-CRM in adolescents and adults $\geq$15 years of age with HIV infection in Malawi showed 74% (30, 90) efficacy of PCV7-CRM$_{197}$ against IPD due to vaccine types plus type 6A, a potentially cross-reactive type (*161*).

Studies of the safety and immunogenicity of PCV13-CRM$_{197}$ have been conducted in HIV infected subjects (CD4 cell counts ≥200 cells/mm^3 and viral loads <50,000 copies/mL) naïve to pneumococcal vaccine aged ≥ 6 years in South Africa (*162*) and in adults previously given PPV23 in the USA (*163*). Immunogenicity was demonstrated after a single dose with only modest increases in response after a second or third dose (*162, 163*). In a European study subjects ≥2 years of age who had undergone an allogeneic hemopoietic stem cell transplant were given three doses of PCV13-CRM$_{197}$ at one month intervals, a fourth dose 6 months later and one month after that a dose of PPV23 (*164*). Pneumococcal GMCs showed a rise over the first three doses of PCV13-CRM$_{197}$, with boosting after the fourth dose (*164*). The dose of PPV23 had no additional immunogenicity.

Impact of PCV Immunization

There is an extensive literature on the impact of PCV immunization, which is beyond the scope of this chapter, being primarily concerned with clinical trials. However, a few studies are noteworthy and show both primary impact in vaccinated subjects and secondary effects due to herd immunity, particularly extending early impact of PCV7-CRM$_{197}$ following the introduction of subsequent PCV vaccines.

An open label pre-licensure trial of PCV13-CRM$_{197}$ in Alaskan Native children in the Yukon Kuskokwim Delta, where rates of IPD were exceptionally high, was followed by continued post-licensure surveillance (*165*). During 2005 to 2008 the rate of IPD was almost 400/100,000 and fell to just over 100/100,000 from the introduction of PCV13-CRM$_{197}$ in 2009 to 2011, with no cases of VT IPD in immunized children, indicating a very rapid impact in a high risk population (*165*).

A larger UK study of IPD showed that following introduction of infant vaccination with PCV13, impact on PCV7 types has continued, while additional impact on the new types has been substantial in children, although serotype specific effectiveness was not shown for type 3 (incidence rate ratio post PCV13 compared to pre-PCV13 0.32 [0.11, 0.94]), and there were insufficient cases of type 5 to analyse (*166*). In a US case-control study modest effectiveness against serotype 3 IPD in children was demonstrated at 79·5% (30·3, 94·8) (*135*).

Of note, in the UK study there was a secondary impact on all new types (except type 5 with insufficient cases to study) in 5-64 year olds and ≥ 65 year olds, including serotype 3 (*166*). Increases in 19A and 7F that were noted in children and older populations following introduction of PCV7 were largely reversed following introduction of PCV13 (*166*). Nonetheless, increases in non-PCV13 types were noted, showing that continued vaccine development will be important (*166*). This is supported by a careful analysis of emerging non-vaccine types in UK, which provides useful guidance on which serotypes should be the focus of future efforts (*167*).

A US study demonstrated substantial impact on pediatric pneumococcal pneumonia comparing the PCV13 period to the PCV7 period, with rates falling from 53.6 to 23.3 per 100 000 admissions (*168*). However types 3, 19A, 7F and 1 continue to be responsible around 80% of cases of pneumococcal pneumonia; as only 86% of children had received even a single dose of PCV in the later period, this suggests that efforts to enhance vaccine coverage remain important. Again, a particular impact was noted on type 19A (*168*), which had increased in USA following introduction of PCV7 (*169*).

Impact of PCV13 on pediatric pneumonia has also been noted in the Gambia, with modest reductions in radiological pneumonia but more substantial impact on hypoxic pneumonia (*170*). Thus health benefits may be anticipated in LMIC as well as developed countries.

In a recent commentary it was noted that extended PCV immunization has had "a stunning and unprecedented" public health impact, but that continued observational studies of effectiveness and impact are essential as the disease background is highly dynamic and subject to change driven at least in part by the vaccines themselves (*167*, *171*).

Future Developments

Extended PCV Vaccines

There is interest from manufacturers in development of other PCV13 vaccines or expanding the number of serotypes in PCV to cover existing and emerging non-vaccine types (*167*). (Table 4)

Table 4. Extended PCV vaccines with clinical data or in clinical trials

Manufacturer	*Number of serotypes*	*Information source*
GSK	11, 12	Reference (*172*)
Merck	15	Reference (*173–175*) NCT02987972 NCT02573181
Walvax	13	Reference (*176*)
LG Life Sciences	13, 14	NCT02892812
SK Chemicals	13	NCT02927444 NCT02201030
Beijing Minhai Biotechnology	13	NCT02494999
Pfizer	20	NCT02955160 NCT03313037 NCT03313050

GSK has studied addition of type 19A and type 19A plus 6A conjugates to the other components in PCV10, to make 11 and 12-valent experimental vaccines but these components were not non-inferior to PCV13 (*172*). A number of manufacturers in Asia are working on pneumococcal conjugate vaccines containing the same types or the same number of types as PCV13 (Table 4).

Merck has been developing a PCV15, which contains the same types as PCV13 plus 22F and 33F (Figure 1) (*173*, *174*). However, types 6A and 19A were not non-inferior to PCV13 in infants who received PCV15 vaccines in two formulations (*175*). Since then PCV15 is in a further phase 2 study in infants and a phase 2 study in adults ≥ 65 years of age who previously received PPV23 (Table 4).

Pfizer has active phase 2 trials with a 20-valent PCV based on PCV13-CRM_{197} (Table 4).

Prevention of Early Onset Infant Disease

Early onset pneumococcal disease causes a substantial proportion of neonatal sepsis and meningitis in developing countries. Before PCV was available to protect infants there was an interest in maternal immunization with PPV23 (*177*).

A clinical trial in progress in the Gambia is examining the impact on early NP carriage of maternal or neonatal immunization with PCV13-CRM_{197} compared to the standard EPI schedule (Protecting from Pneumococcus in Early Life [PROPEL] NCT02628886).

Modified Vaccine Schedules

Noting the substantial direct impact of PCV13 on immunized children and to a lesser extent the indirect effect on other age groups in the UK following high levels of vaccine uptake (*166*), it has been questioned whether a reduced number of doses could sustain the level of protection. In a randomized clinical trial the immunogenicity of two doses of PCV13 at 3 and 12 months of age (1+1 schedule) was compared to the immunogenicity of three doses at 2, 4 and 12 months of age (2+1 schedule) (*178*). Unsurprisingly, the immunogenicity after the primary series was better for 12 of the 13 types after 2 primary doses than after a single primary dose. For types 6B and 23F, which were already known to have lower immunogenicity after 2 primary doses than after 3 primary doses, only 1.0% (0, 5.3) and 5.9% (2.2-12.4) respectively achieved antibody concentrations of at least 0.35 µg/mL after the single primary dose. After a booster dose percentages achieving at least 0.35 µg/mL were similar for both schedules although responses to types 6B and 23F continued to show some differences. If a 1+1 schedule is implemented in the UK it will be important to determine whether the immunogenicity reduction for certain types leads to recurrence of disease due to these types. In addition, any results in the UK may not be generalizable to other situations with lower vaccine uptake (*179*).

Neisseria meningitidis

As an encapsulated organism that often colonizes the upper respiratory tract but sometimes causes invasive disease with fatal or permanently disabling outcomes, *Neisseria meningitidis* is an obvious target for conjugate vaccines.

In the 1960s a series of studies showed that serum bactericidal antibodies (SBA) protect from meningococcal disease (*180*), high molecular weight meningococcal serogroup A and C capsular polysaccharides induce SBA activity (*181*) and that the group C polysaccharide could prevent acquisition of group C carriage in military recruits (*182*). In studies with military recruits, efficacy of group C polysaccharide vaccine was confirmed (*183*). There was also evidence in a randomized controlled trial in Finland that group A plain polysaccharide vaccine was efficacious in infants (*184*). Group W and Y polysaccharide vaccines were developed based on the ability to induce SBA activity and combinations including quadrivalent ACWY vaccines were developed. Polysaccharide vaccines proved useful in prevention of short term risk, such as in military recruits or in epidemics, but had marginal impact in younger infants, variable and short-term impact on carriage and group C polysaccharide can induce hyporesponsiveness to subsequent doses (*185*).

Group B polysaccharide (polysialic acid) is poorly immunogenic in humans (*186*) as a result of cross-reactivity to similar structures on embryonic N-CAM (neural cell adhesion molecules) (*187*) and so was never developed as a vaccine, either as a polysaccharide or as a conjugate vaccine. Instead vaccines have been developed that do not depend on polysaccharide antigens.

Meningococcal group C and A conjugate vaccines were developed and studied in clinical trials in the 1990s. Two manufacturers used CRM_{197} as a conjugate protein (Table 5). One produced a bivalent AC conjugate vaccine (*188*, *189*) that was taken into trials in Africa (*189*) and the other a monovalent C conjugate that was studied as a control vaccine in trials of PCV7-CRM_{197} (*77–79*). A third manufacturer developed a monovalent C TT conjugate using O-deacetylated capsular polysaccharide (*190*).

In 1994 the UK government noted a significant and rising incidence of meningococcal group C disease with a bimodal distribution peaking in infants and adolescents. In response to this public health emergency, manufacturers were invited to collaborate in bringing monovalent meningococcal group C conjugate vaccines to the market.

A series of clinical trials were conducted that complemented manufacturer sponsored trials to support marketing applications for use from infants through to adults (*191–194*). Key to the program was evidence from use of meningococcal polysaccharide vaccines that SBA activity was an adequate correlate of protection. The planning, implementation and early impact of this program has been well summarized (*195*). The marketing applications were made as national applications in the UK followed by rounds of mutual recognition in other EU countries. The first approval was for MnC-CRM_{197} (Wyeth, now part of Pfizer, see Table 5) in late 1999. The timing was important as the disease incidence increased in winter months. The vaccination programme was launched almost immediately at the beginning of November 1999 in the highest risk groups, infants and 15-17 year olds, with other groups receiving catch-up immunization through 2000, with MnC-CRM_{197} (Wyeth), MnC-CRM_{197} (Chiron, now part of GSK) and MnC-TT (Table 5). An almost immediate impact was noted in immunized age groups, with an 81% reduction of cases in the population <18 years of age in the period between July 2000 and April 2001 compared to the same period in 1988-1999 (*195*). Subsequently, effectiveness follow up showed rapid waning of protection after an infant primary series, demonstrating the importance of herd protection by immunization of older children and adolescents, as well as the need for a booster dose following an infant series (*196*), similar to that earlier seen with Hib conjugate vaccines (*42*, *43*). Monovalent meningococcal C conjugate vaccines were subsequently approved in other European countries by the Mutual Recognition process and also approved in a number of other countries including Canada and Australia.

However, changes to meningococcal epidemiology including increasing rates of W and Y serogroup disease as well as a risk of group A disease in travelers to sub-Saharan Africa and to the Hajj led to an interest in development of quadrivalent combinations of conjugate vaccines covering groups A, C, W and Y (Table 5). The monovalent group C conjugate vaccines were never approved in the US as the epidemiology drove interest to the quadrivalent vaccines. A large number of clinical trials supported approval and subsequent variations for these combination vaccines, which were approved on the basis of safety and immunogenicity without efficacy data. Because these vaccines are used in a wide range of age groups extensive information on safety and immunogenicity in different schedules, concomitant use with other vaccines and on persistence of antibodies was required for regulatory programs for each product and is readily available on the EMA (*197*, *198*) and FDA (*199*, *200*) websites, as well as review articles (*201–203*). A quadrivalent vaccine based on diphtheria toxoid (DT) as a conjugate vaccine (Mn4C-DT) was the first quadrivalent combination approved, but has not been approved for use below 9 months of age due to limitations of immunogenicity in infants (*204*). This was followed by a quadrivalent vaccine with CRM_{197} as a conjugate protein (Mn4C-CRM_{197}) and subsequently a quadrivalent vaccine with TT as a conjugate protein (Mn4C-TT). To date Mn4C-TT has not been marketed in the USA and Mn4C-DT has not been marketed in the EU, whereas Mn4C-CRM_{197} is marketed in the USA and EU.

Interest is now turning to combining quadrivalent conjugate vaccines with non-conjugate meningococcal vaccines, to produce vaccines that might be effective against A, B, C, W and Y disease. Both GSK (*205*, *206*) and Pfizer (NCT03135834) have experimental pentavalent meningococcal vaccines in the clinic.

Table 5. Marketed Meningococcal conjugate vaccines (excludes combinations with Hib vaccines)

Serogroup *Carrier protein*	*Abbreviation*	*Common proprietary name*	*First appro-val*	*Original and current manufacturers*
C CRM_{197}	MnC-CRM_{197}	Meningitec®	1999	Wyeth Nuron
C CRM_{197}	MnC-CRM_{197}	Menjugate®	2000	Chiron GSK
C Tetanus Toxoid (TT)	MnC-TT	NeissVacC®	2000	Baxter Pfizer
ACWY Diphtheria toxoid (DT)	Mn4C-DT	MenActra®	2005	Sanofi Pasteur
A Tetanus toxoid (TT)	MnA-TT	MenAfriVac®	2009	Serum Institute of India
ACWY CRM_{197}	Mn4C-CRM_{197}	MenVeo®	2010	Novartis GSK
ACWY Tetanus toxoid (TT)	Mn4C-TT	Nimenrix®	2012	GSK Pfizer

Parts of Africa have experienced epidemics of meningococcal disease in recent decades, largely due to group A strains. Polysaccharide vaccines had been used in outbreak control, but provided short duration of protection and had no effect on carriage so provided no herd immunity (*207*). The Meningitis Vaccine Project, a Bill and Melinda Gates Foundation funded partnership between WHO and PATH, was created in June 2001 to develop new Group A meningococcal conjugate vaccines specifically for control of epidemic disease in Africa. The group worked with Serum Institute of India (SII) to develop a low cost group A meningococcal vaccine (Table 5). After a phase 1 clinical trial in India (*208*), a series of safety and immunogenicity studies were conducted in the target population aged 1-29 years in Africa (*209*). The vaccine was approved in India in 2009 (*210*) and prequalified by WHO in 2010 (*37*) on the basis of safety and immunogenicity. The effectiveness in prevention of disease and carriage was established during a phased introduction starting in 2010 (*211*). The vaccine has been highly effective in reducing disease across the African meningitis belt as witnessed by a reduction in the percentage of CSF samples positive for group A *Neisseria meningitidis* from 24.9% in 2010 to 0.1% in 2011, following the introduction of MenAfriVac® in December 2010 (*211*). In Chad a 10 day campaign to administer a single dose to subjects 1-29 years of age in December 2011 effectively eliminated group A meningitis, despite continued cases in non-vaccinated areas (*212*). Total meningitis cases were 94% lower in immunized areas and group A carriage fell from 0.7% 2-4 months before immunization to 0.02% 2-4 months after immunization (*212*). Coverage will now be maintained by immunization of infants at the EPI vaccine 9 month visit concomitant with measles vaccine (*210*). Due to changes in the epidemiology of meningococcal disease in Africa, work has now started on development of a low cost ACWXY meningococcal combination vaccine (*210*).

Staphylococcus aureus

In the early 1990s conjugates were made with staphylococcal capsular types 5 and 8, the predominant types, conjugated to *Pseudomonas aeruginosa* recombinant exoprotein A (rEPA) and combined in a candidate vaccine, Staphvax®. After a first dose of type 5 and 8 conjugates, binding antibodies were observed but there was no increase after subsequent doses (*213*).This conjugate vaccine was administered in two efficacy studies against *S. aureus* bacteremia in patients undergoing hemodialysis (*214*, *215*). Efficacy endpoints were not met in either study. However, in the first study efficacy of 57% (10, 81) was noted up to week 40 in a *post hoc* analysis, despite non-significant efficacy of 27% at the 54 week primary time point (*214*). In a second study a booster dose was given at week 35 but no efficacy was observed (*215*). There is no clear explanation for the failure or the discrepant results, but possible factors include the conjugation chemistry strategy and quality of the CP conjugates, the lack of clear evidence of functional immune responses in large number of vaccinees, and preclinical evidence that multiple antigen targets are likely needed to prevent infection with this complex pathogen (*215*).

More recently type 5 and 8 conjugates have been included in multi-antigen vaccines. In one vaccine, CP5 and CP8 conjugated to tetanus toxoid were combined with genetically detoxified α-toxin and recombinant mutated clumping factor A (ClfA) (*216*). Four formulations were studied, with 5μg or 10μg of each conjugate, with or without $AS03_B$ adjuvant. In a phase I clinical trial, both types elicited an antibody response to a first dose but no additional response with further doses, as assessed in a multiplex antigen-binding assay. There were no differences between the four formulations.

A prototype three component *Staphylococcus aureus* vaccine (SA3Ag), from Wyeth/Pfizer, contained type 5 and 8 conjugates with CRM_{197} and a recombinant mutated ClfA. The vaccine was studied without adjuvant with polysaccharide dose levels of 10 μg, 30 μg and 100 μg paired with 30 μg, 60 μg and 200 μg does levels of ClfA respectively (*217*). All dose levels induced a rapid antibody response, peaking at 11-15 days following immunization as assessed by competitive Luminex assays (cLIA) and by opsonophagocytic activity assays, consistent with a memory response after priming by natural exposure. Similar responses were seen in healthy 18-24 year old and 50-85 year old adults. There was a small increase in cLIA responses from the lower to the middle dose of conjugate so a dose of 30μg/conjugate was used in further development. Antibody responses persisted well up to 6 months and a booster dose at that point had little effect (*218*). Incorporation of a fourth antigen, MntC, to make a 4-component *Staphylococcus aureus* vaccine (SA4Ag), by Wyeth/Pfizer, left responses to the CP5 and CP8 conjugates unchanged in 18-64 year old (*219*) and 65-85 year (*220*) old adults. The vaccine induced responses including bacterial killing activity that were substantial and their rapid rise and persistence supported advancing SA4Ag to a population with high risk for *S. aureus* disease (*219*, *220*). SA4Ag is now being evaluated, by Pfizer, in an efficacy study in subjects undergoing multilevel instrumented spinal surgery (*221*).

Group B Streptococcus

Group B Streptococcus (GBS) is the most common cause of neonatal sepsis and meningitis. It commonly colonizes the female genitourinary tract and is transmitted to the neonate during childbirth. Group B streptococci have polysaccharide capsules that define the serotype. In the 1970s a pivotal study by Baker and colleagues showed that the absence of serotype-specific antibody to GBS in maternal serum is related to the risk of neonatal invasive GBS and that serotype specific antibodies are placentally transferred to infants (*222*). This raised the possibility that maternal immunization might protect infants from invasive GBS disease. Although intrapartum antibiotic prophylaxis (AIP) has been used, few countries have clear policies and policies have limited penetration, showing the potential value of maternal immunization (*223*). A small study with type III polysaccharide during pregnancy confirmed transplacental transfer of vaccine-induced antibody, with opsonophagocytic activity in infants and antibody exceeding the putative protective concentration of 2μg/mL (*224*). However, responses were not seen in all mothers, suggesting that conjugate vaccines might be of value.

Studies in healthy adults have been conducted with TT conjugates of type 3 (*225*), type 2 (*226*), type Ia and Ib polysaccharide (*227*) and type V (*228*) as well as CRM_{197} type V conjugate (*228*), demonstrating excellent functional immunogenicity and safety. A bivalent type II and III conjugate retained the activity of the monovalent products (*229*), indicating that multivalent vaccines are feasible. This is important as the range of different GBS serotypes means that a multivalent product will be essential to have a useful impact on disease. Serotypes Ia, Ib, II, III, and V cover a large majority of strains causing neonatal disease, although type IV has also been emerging as a small but significant cause of neonatal disease in the US (*230*).

A dose ranging study of a trivalent CRM_{197} conjugate for types Ia, Ib and III by Novartis (now GSK) in South Africa showed good safety and immunogenicity in healthy women and subsequently in both mothers and infants following maternal administration at 28-35 weeks of pregnancy (*231*). A study in Canada and Belgium confirmed a similar profile using the selected dose level, 5 μg of each polysaccharide (*232*). Another study at the selected dose level in Malawi and South Africa showed lower responses in HIV-infected mothers (*233*), although antibodies still rose substantially from pre-vaccination concentrations and infants born to mothers with HIV infection appeared to have higher antibody concentrations than placebo recipients in the dose-ranging study (*231*, *233*). GSK are noted to be following on from the trivalent candidate vaccine with a pentavalent vaccine covering serotypes Ia, Ib, II, III and V (*234*). This level of coverage would enable a substantial impact on global disease if the vaccine can be developed successfully. Pfizer initiated a phase 1 study of a multivalent GBS vaccine in May 2017 (NCT03170609).

Typhoid

Salmonella typhi is the pathogenic bacterium in typhoid disease and expresses the capsular Vi polysaccharide (ViPS). As a vaccine ViPS is immunogenic and efficacious in adults and children over 2 years of age (*235*, *236*), but responses are not boostable and antibodies wane over time (*237*). ViPS was reported to be poorly immunogenic in infants (*238*). Even in children 2-5 years of age antibody responses are lower than in older children, with concentrations returning to baseline levels within 2 years (*239*).

As a result preparation of Vi conjugate vaccines became a priority. An early study looked at conjugates with rEPA (Vi-EPA) (*240*). After a single dose administered to adult volunteers, Vi-EPA conjugate produced a higher antibody level than unconjugated Vi polysaccharide and the response persisted longer (*240*). In a study in Vietnam, an endemic area where an efficacy trial was possible, an immunogenicity study allowed selection of the preferred conjugate linkage structure (*241*). Two doses in 2-4 year olds induced better responses than ViPS in older children.[241] In a randomized, placebo-controlled efficacy study in Vietnam using two doses at a 6 week interval in children 2-5 years of age the ITT efficacy was 91.1% (78.6–96.5) (*242*) and persisted over a 4 year period (*243*). However, to date this vaccine has not been marketed.

Antibody responses to typhoid vaccines have primarily been assessed by IgG ELISA (*244*). Efforts have been made to develop a reference standard (*244*). Using the Vi-EPA studies antibody concentrations of 4.3 μg/ml appear to be protective, although a defined correlate of protection has not been fully established.

Recognizing the complexity of new vaccine development, the WHO published guidelines in 2013 that broadly recommend approval based on immunogenicity using total Vi binding IgG as an endpoint (*245*). As no correlate has been firmly established, comparisons to the immunogenicity of unconjugated Vi vaccine may serve as a basis for approval of conjugate vaccines (*245*).

Following these guidelines, two typhoid conjugate vaccines have been approved and marketed in India based upon immunogenicity and safety studies. Both of these are tetanus toxoid conjugates (Vi-TT). A single dose of Vi-TT from Bharat Biotech International, Typbar-TCV®, has superior immunogenicity to ViPS in 24 month to 45 year olds (*246*). In 6-11 and 12-23 month infants and toddlers it produces responses similar to those produced by ViPS in older children and adults (*246*). Although antibodies wane with time, a second dose after 2 years produces a booster response (*246*). The vaccine has been studied in single-dose vials and multi-dose vials (*246*). As large scale efficacy studies would take many years to complete, efficacy has been assessed in a controlled human infection model (CHIM) to support WHO prequalification, which was primarily based on immunogenicity. In the CHIM study efficacy compared to a control group was 54·6% (26·8, 71·8) for Vi-TT and 52·0% (23·2, 70·0) for ViPS (*247*).

Due to the constraints of the model the estimate of efficacy may be lower than true field efficacy, but importantly the similarity between the efficacy of ViPS, which has established field efficacy and Vi-TT which has been approved based only on immunogenicity supports the use of the WHO guidelines for approval. This vaccine was prequalified in December 2017 (*37*).

Another Vi-TT vaccine, from BIO MED Pvt. Ltd., PedaTyph® was approved in India in 2008 based on safety and immunogenicity studies (*248*, *249*). A small cluster randomized study was conducted and shows evidence of efficacy, although there were limitations to the study design and power (*250*).

A CRM_{197} conjugate (Vi-CRM) has been studied across age groups, including infants receiving concomitant routine infant vaccination at an EPI schedule. Immunogenicity and persistence was good in all age groups and in young infants in the Philippines, but not in young infants in Pakistan (*251*). It may be that in future Vi conjugates would best be delivered at the EPI measles visit than in the youngest infants.

Lipopolysaccharide Conjugates To Prevent Gram Negative Infections

In 1992 it was proposed that antibodies to the O-specific polysaccharide of shigella and non-typhoidal salmonella lipopolysaccharide might be protective, and that the weak immunogenicity of the O-specific PS (O-PS) might be enhanced by conjugation to a carrier protein (*252*).

In a study in Israeli soldiers subjects were randomized to receive a *Shigella sonnei* O-PS rEPA conjugate (O-PS-EPA), a live-attenuated shigella vaccine or a meningococcal vaccine. Subjects were posted to a range of remote postings and in those sites with outbreaks of *Shigella sonnei* disease 70–155 days after vaccination, conjugate vaccine efficacy was 74% (28–100) (*253*).

In an efficacy study in Israeli children, subjects 1-4 years of age were randomized to receive O-PS-EPA conjugates of *Shigella sonnei* or *Shigella flexneri* 2a. In 3-4 year olds efficacy of the *Shigella sonnei* conjugate was 71.1% (-4.4, 92.0 and p=0.043). The point estimate trended lower with age and the efficacy was not significant in younger age groups. There were insufficient cases of *Shigella flexneri* 2a to estimate efficacy of this conjugate (*254*).

A phase 1 study was completed on an rEPA conjugate of the polysaccharide component of the O1 lipopolysaccharide of *Shigella dysenteriae* (*255*). This conjugate was produced in a modified *E. coli* rather than by chemical conjugation, a process termed bioconjugation. The clinical trial examined two dose levels with and without aluminum adjuvant and showed the vaccine to be adequately tolerated and immunogenic, producing anti-O1 IgG with evidence of functional antibody. The adjuvant was not required (*255*).

A similar rEPA bioconjugate to the polysaccharide component of *Shigella flexneri* 2a lipopolysaccharide was examined in a phase 1 study (*256*). This vaccine was also adequately tolerated and immunogenic, inducing functional antibodies without a need for aluminum adjuvant (*256*). A CHIM study appears to have been completed but not yet reported (NCT02646371).

A combination of conjugate vaccines against *Shigella sonnei* and *Shigella flexneri* serotypes 2a, 3a and 6 would cover around 88% of *Shigella* isolates from children aged <60 months with moderate-to-severe diarrheal disease at sites in Africa and Asia (*257*). This has led to renewed interest in development of a multivalent conjugate vaccine against *Shigella* (*258*).

A polyvalent *Escherichia coli (E. coli)* vaccine was constructed in the early 1990s in which the polysaccharide of 12 O-serotypes was conjugated to rEPA. In 88 volunteers the vaccine was immunogenic by ELISA and opsonophagocytic killing assays (*259*). However, this vaccine was not further developed.

Nearly 20 years later a candidate *E. coli* vaccine was produced using an *in vivo* bioconjugation approach, containing rEPA conjugates to the polysaccharide of 4 O-serotypes, ExPEC4V. Due to the large number of O-serotypes this is unlikely to be a practical candidate. A total of 188 women with recurrent urinary tract infection were randomly allocated to receive a single dose of ExPEC4V or placebo (*260*). In a 9 month follow up there was no reduction of UTI due to vaccine type *E. coli* at cut-offs of 10^3 or 10^5 cfu/mL although there was some reduction in UTI due to any *E. coli* in the vaccine group.

Conclusion

Since the 1980s a considerable body of clinical trial results has supported introduction of several conjugate vaccines and related combination vaccines into routine use across the world, with dramatic public health impact. Selected key events are shown in Table 6. However, work continues to improve the impact of many of these vaccines and to bring the technology to new infectious diseases due to organisms with a polysaccharide capsule as well as those with variable polysaccharide components to surface lipopolysaccharides. Development of novel trial designs as well as enhancement of standardized, high-throughput antibody assays have been critical within this work.

Table 6. Chronology of some key events

1980	First Hib conjugate constructs published
1981	
1982	
1983	
1984	
1985	First Hib conjugate vaccine approved for toddlers (PRP-D in US)
1986	
1987	
1988	HbOC approved for use in toddlers in US
1989	PRP-OMP approved for use in toddlers in US

Continued on next page.

Table 6. (Continued). Chronology of some key events

1990	HbOC and PRP-OMP approved in US for infants
1991	
1992	
1993	
1994	UK call for collaborative meningococcal C conjugate development
1995	First PCV7-CRM efficacy study starts
1996	
1997	First Shigella conjugate efficacy demonstrated
1998	First PCV7-CRM efficacy study completed
1999	First meningococcal conjugate approved (MnC-CRM in UK)
2000	First pneumococcal conjugate approved (PCV7-CRM in US)
2001	Meningitis Vaccine Project created. First typhoid conjugate efficacy published
2002	
2003	
2004	
2005	First meningococcal ACWY conjugate approval
2006	
2007	
2008	First typhoid conjugate Vi-TT vaccine approved. CAPiTA study starts
2009	First PCV10-HiD and PCV13-CRM approvals. MenAfriVac approval.
2010	Introduction of MenAfriVac®
2011	PCV13-CRM first approval in adults ≥ 50 years of age
2012	
2013	
2014	
2015	CAPiTA study shows efficacy of PCV13-CRM against adult pneumonia
2016	
2017	Typhoid conjugate Vi-TT CHIM study published and vaccine prequalified
2018	

References

1. Avery, O. T.; Goebel, W. F. *J.Exp. Med.* **1929**, *50*, 533–550.
2. Goebel, W. F. *J. Exp. Med.* **1940**, *72*, 33–48.
3. Schneerson, R.; Barrera, O.; Sutton, A.; Robbins, J. B. *J. Exp. Med* **1980**, *152*, 361–376.
4. Frasch, C. In *Development and Clinical Uses of Haemophilus b Conjugate Vaccines*; Marcell Dekker, Inc: New York, 1994; 435–453.
5. Peltola, H.; Käythy, H.; Sivonen, A.; Mäkelä, P. H. *Pediatrics* **1977**, *60*, 730–737.
6. Granoff, D.; Osterholm, M. *Pediatrics* **1987**, *80*, 590–592.
7. Osterholm, M. T.; Rambeck, J. H.; White, K. E.; Jacobs, J. L.; Pierson, L. M.; Neaton, J. D.; Hedberg, C. W.; MacDonald, K. L.; Granoff, D. M. *JAMA, J. Am. Med. Assoc.* **1988**, *260*, 1423–1428.
8. Robbins, J. B.; Parke, J. C.; Schneerson, R.; Whisnant, J. K. *Pediatr. Res.* **1973**, *7*, 103–110.
9. Madore, D. V.; Anderson, P.; Baxter, B. D.; Carlone, G. M.; Edwards, K. M.; Hamilton, R. G.; Holder, P.; Kayhty, H.; Phipps, D. C.; Peeters, C. C.; Schneerson, R.; Siber, G. R.; Ward, J. I.; Frasch, C. E. *Clin. Diagn. Lab. Immunol.* **1996**, *3*, 84–88.
10. World Health Organization. *Expert Committee on Biological Standardization: Forty-Ninth Report*; Geneva, 1998.
11. Schlesinger, Y.; Granoff, D. M. *JAMA, J. Am. Med. Assoc.* **1992**, *267*, 1489–1494.
12. Käyhty, R.; Peltola, H.; Karanko, V.; Mäkelä, P. H. *J. Infect. Dis* **1983**, *147*, 1100–1100.
13. Robbins, J.; Schneerson, R.; Anderson, P.; Smith, D. *JAMA, J. Am. Med. Assoc.* **1996**, *276*, 1181–1185.
14. Granoff, D. M.; Boies, E. G.; Munson, R. S. *J. Pediatr* **1984**, *105*, 22–27.
15. Berkowitz, C. D.; Ward, J. I.; Meier, K.; Hendley, J. O.; Brunell, P. A.; Barkin, R. A.; Zahradnik, J. M.; Samuelson, J.; Gordon, L. *J. Pediatr* **1987**, *110*, 509–514.
16. Eskola, J.; Peltola, H.; Takala, A. K.; Käyhty, H.; Hakulinen, M.; Karanko, V.; Kela, E.; Rekola, P.; Rönnberg, P.-R.; Samuelson, J. S.; Gordon, L. K.; Mäkelä, P. H. *N. Engl. J. Med* **1987**, *317*, 717–722.
17. Ward, J.; Brenneman, G.; Letson, G.; Heyward, W. *N. Engl. J. Med* **1990**, *323*, 1393–1401.
18. Desgrandchamps, D.; Schmid, R.; Zimmermann, H. P.; Imahorn, P.; Kabus, K.; Schubiger, G. *Schweiz. Med. Wochenschr.* **1994**, *124*, 575–582.
19. Malito, E.; Bursulaya, B.; Chen, C.; Surdo, P. L.; Picchianti, M.; Balducci, E.; Biancucci, M.; Brock, A.; Berti, F.; Bottomley, M. J.; Nissum, M.; Costantino, P.; Rappuoli, R.; Spraggon, G. *Proc. Natl. Acad. Sci.* **2012**, *109*, 5229–5234.
20. Frasch, C. E.; Robbins, J. D. *J. Exp. Med* **1978**, *147*, 629–644.
21. Anderson, P.; Pichichero, M. E.; Insel, R. A. *J. Clin. Invest.* **1985**, *76*, 52–59.
22. Anderson, P.; Pichichero, M.; Edwards, K.; Porch, C. R.; Insel, R. *J. Pediatr.* **1987**, *111*, 644–650.

23. Einhorn, M.; Anderson, E.; Weinberg, G.; Granoff, P.; Granoff, D. *Lancet* **1986**, *328*, 299–302.
24. Weinberg, G. A.; Einhorn, M. S.; Lenoir, A. A.; Granoff, P. D.; Granoff, D. M. *J. Pediatr.* **1987**, *111*, 22–27.
25. Black, S. B.; Shinefield, H. R.; Fireman, B.; Hiatt, R.; Polen, M.; Vittinghoff, E. *Pediatr. Infect. Dis. J.* **1991**, *10*, 97–103.
26. Santosham, M.; Wolff, M.; Reid, R.; Hohenboken, M.; Bateman, M.; Goepp, J.; Cortese, M.; Sack, D.; Hill, J.; Newcomer, W.; Capriotti, L.; Smith, J.; Owen, M.; Gahagan, S.; Hu, D.; Kling, R.; Lukacs, L.; Ellis, R. W.; Vella, P. P.; Calandra, G.; Matthews, H.; Ahonkhai, V. *N. Engl. J. Med.* **1991**, *324*, 1767–1772.
27. Granoff, D. M.; Anderson, E. L.; Osterholm, M. T.; Holmes, S. J.; McHugh, J. E.; Belshe, R. B.; Medley, F.; Murphy, T. V. *J. Pediatr.* **1992**, *121*, 187–194.
28. Decker, M. D.; Edwards, K. M.; Bradley, R.; Palmer, P. *J. Pediatr.* **1992**, *120*, 184–189.
29. Booy, R.; Hodgson, S.; Moxon, E.; Mayon-White, R.; Slack, M.; Carpenter, L.; Macfarlane, J.; Haworth, E.; Kiddle, M.; Shribman, S. *Lancet* **1994**, *344*, 362–366.
30. Mulholland, K.; Hilton, S.; Adegbola, R.; Usen, S.; Oparaugo, A.; Omosigho, C.; Weber, M.; Palmer, A.; Schneider, G.; Jobe, K.; Lahai, G.; Jaffar, S.; Secka, O.; Lin, K.; Ethevenaux, C.; Greenwood, B. *Lancet* **1997**, *349*, 1191–1197.
31. Peltola, H. *Clin. Microbiol. Rev.* **2000**, *13*, 302–17.
32. Granoff, D. M.; Lucas, A. H. *Ann. N. Y. Acad. Sci.* **1995**, *754*, 278–288.
33. Daum, R.; Granoff, D. In *Development and Clinical Uses of Haemophilus b Conjugate Vaccines*; Marcell Dekker, Inc: New York, 1994; pp 291–312.
34. Dagan, R.; Poolman, J.; Siegrist, C.-A. *Vaccine* **2010**, *28*, 5513–5523.
35. Paradiso, P. R.; Hogerman, D. A.; Madore, D. V.; Keyserling, H.; King, J.; Reisinger, K. S.; Blatter, M. M.; Rothstein, E.; Bernstein, H. H.; Hackell, J. *Pediatrics* **1993**, *92*, 827–832.
36. World Health Organization. Progress Introducing *Haemophilus influenzae* Type b Vaccine in Low-Income Countries, 2004–2008. *Wkly. Epidemiol. Rec.* **2008**, *83*, 62–67.
37. World Health Organization. Prequalified Vaccines. https://extranet.who.int/gavi/PQ_Web/ (accessed January 9, 2018).
38. Beurret, M.; Hamidi, A.; Kreeftenberg, H. *Vaccine* **2012**, *30*, 4897–4906.
39. Eskola, J.; Ölander, R.-M.; Hovi, T.; Litmanen, L.; Peltola, S.; Käyhty, H. *Lancet* **1996**, *348*, 1688–1692.
40. Poolman, J.; Kaufhold, A.; De Grave, D.; Goldblatt, D. *Vaccine* **2001**, *19*, 2280–2285.
41. Ramsay, M. E.; McVernon, J.; Andrews, N. J.; Heath, P. T.; Slack, M. P. *J. Infect. Dis.* **2003**, *188*, 481–485.
42. McVernon, J.; Ramsay, M.; McLean, A. *Epidemiol. Infect.* **2008**, *136*, 800–812.
43. Morris, S. K.; Moss, W. J.; Halsey, N. *Lancet. Infect. Dis.* **2008**, *8*, 435–443.
44. Zepp, F.; Schmitt, H.-J.; Cleerbout, J.; Verstraeten, T.; Schuerman, L.; Jacquet, J.-M. *Expert Rev. Vaccines* **2009**, *8*, 663–678.

45. Zanetti, A.; Desole, M. G.; Romanò, L.; d'Alessandro, A.; Conversano, M.; Ferrera, G.; Panico, M. G.; Tomasi, A.; Zoppi, G.; Zuliani, M.; Thomas, S.; Soubeyrand, B.; Eymin, C.; Lockhart, S. *Vaccine* **2017**, *35*, 4034–4040.
46. Lee, A. W.; Jordanov, E.; Boisnard, F.; Marshall, G. S. *Expert Rev. Vaccines* **2017**, *16*, 85–92.
47. Lyseng-Williamson, K. A.; McCormack, P. L. *Drugs Ther. Perspect.* **2013**, *29*, 329–335.
48. FDA. Pentacel Prescribing Information. https://www.fda.gov/downloads/BiologicsBloodVaccines/Vaccines/ApprovedProducts/UCM109810.pdf (accessed February 8. 2018).
49. Weniger, B. G. *Pediatrics* **1999**, *103*, 1064–1077.
50. Pace, D.; Khatami, A.; McKenna, J.; Campbell, D.; Attard-Montalto, S.; Birks, J.; Voysey, M.; White, C.; Finn, A.; Macloed, E.; Faust, S. N.; Kent, A. L.; Heath, P. T.; Borrow, R.; Snape, M. D.; Pollard, A. J. *Br. Med. J.* **2015**, *350*, h1554.
51. GSK. https://www.gskdirect.com/medias/Menhibrix-Discontinuation-Notice.pdf?context=bWFzdGVyfHJvb3R8NTIyNzQxfGFwcGxpY2F0aW9uL3BkZnxoZTMvaDg1Lzg4NTg3MTMwNjM0NTQucGRmfGYxMjAyMDFmMmU2YjAwYzYyZmUxNWQyNTdjYmUzNjFjMTQ2MDUyODIyMjg4YWIzMTIzNjJhZTZjMWRlNjBlMzg (accessed January 5, 2018).
52. CDC. *MMWR Morb. Mortal. Wkly. Rep.*; 1998, 47, 993–988.
53. Peltola, H.; Kilpi, T.; Anttila, M.; Peltola, H.; Kilpi, T. *Lancet* **1992**, *340*, 592–594.
54. Adegbola, R. A.; Secka, O.; Lahai, G.; Lloyd-Evans, N.; Njie, A.; Usen, S.; Oluwalana, C.; Obaro, S.; Weber, M.; Corrah, T.; Mulholland, K.; McAdam, K.; Greenwood, B.; Milligan, P. J. M. *Lancet* **2005**, *366*, 144–150.
55. Howie, S. R. C.; Antonio, M.; Akisanya, A.; Sambou, S.; Hakeem, I.; Secka, O.; Adegbola, R. A. *Vaccine* **2007**, *25*, 6305–6309.
56. Collins, S.; Litt, D.; Almond, R.; Findlow, J.; Linley, E.; Ramsay, M.; Borrow, R.; Ladhani, S. *J. Infect.* **2018**, *76*, 335–341.
57. Troeger, C.; Forouzanfar, M.; Rao, P. C.; Khalil, I.; Brown, A.; Swartz, S.; Fullman, N.; Mosser, J.; Thompson, R. L.; Reiner, R. C., Jr.; Abajobir, A.; Alam, N.; Alemayohu, M. A.; Amare, A. T.; Antonio, C. A.; Asayesh, H.; Avokpaho, E.; Barac, A.; Beshir, M. A.; Boneya, D. J.; Brauer, M.; Dandona, L.; Dandona, R.; Fitchett, J. R. A.; Gebrehiwot, T. T.; Hailu, G. B.; Hotez, P. J.; Kasaeian, A.; Khoja, T.; Kissoon, N.; Knibbs, L.; Kumar, G. A.; Rai, R. K.; El Razek, H. M. A.; Mohammed, M. S. K.; Nielson, K.; Oren, E.; Osman, A.; Patton, G.; Qorbani, M.; Roba, H. S.; Sartorius, B.; Savic, M.; Shigematsu, M.; Sykes, B.; Swaminathan, S.; Topor-Madry, R.; Ukwaja, K.; Werdecker, A.; Yonemoto, N.; El Sayed Zaki, M.; Lim, S. S.; Naghavi, M.; Vos, T.; Hay, S. I.; Murray, C. J. L.; Mokdad, A. H. *Lancet. Infect. Dis.* **2017**, *17*, 1133–1161.
58. GAVI. http://www.gavi.org/support/nvs/hib/ (accessed October 24, 2017).
59. Feldstein, L. R. M.; Stephanie; Gagic-Dobo, M.; Diallo, M. S.; Conklin, L. M.; Wallace, A. S. *Wkly. Epidemiol. Rec.* **2017**, *92*, 701–707.
60. World Health Organization. http://www.who.int/mediacentre/factsheets/fs378/en/ (accessed October 25, 2017).

61. Mäkelä, P. H.; Butler, J. C. In *Pneumococcal vaccines*, American Society of Microbiology: Washington, D.C., 2008; pp 19–29.
62. Borgoño, J. M.; McLean, A. A.; Vella, P. P.; Woodhour, A. F.; Canepa, I.; Davidson, W. L.; Hilleman, M. R. *Proc. Soc. Exp. Biol. Med.* **1978**, *157*, 148–154.
63. Douglas, R. M.; Paton, J. C.; Duncan, S. J.; Hansman, D. J. *J. Infect. Dis.* **1983**, *148*, 131–137.
64. Robinson, K. A.; Baughman, W.; Rothrock, G.; Barrett, N. L.; Pass, M.; Lexau, C.; Damaske, B.; Stefonek, K.; Barnes, B.; Patterson, J. *JAMA, J. Am. Med. Assoc.* **2001**, *285*, 1729–1735.
65. Hausdorff, W. P.; Feikin, D. R.; Klugman, K. P. *Lancet Infect. Dis.* **2005**, *5*, 83–93.
66. Hausdorff, W. P.; Bryant, J.; Paradiso, P. R.; Siber, G. R. *Clin. Infect. Dis.* **2000**, *30*, 100–121.
67. Wernette, C. M.; Frasch, C. E.; Madore, D.; Carlone, G.; Goldblatt, D.; Plikaytis, B.; Benjamin, W.; Quataert, S. A.; Hildreth, S.; Sikkema, D. J.; Kayhty, H.; Jonsdottir, I.; Nahm, M. H. *Clin. Diagn. Lab. Immunol.* **2003**, *10*, 514–519.
68. World Health Organization. *Expert Committee on Biological Standardization: Fifty-Fourth Report*; Geneva, 2005.
69. Pickering, J. W.; Martins, T. B.; Greer, R. W.; Schroder, M. C.; Astill, M. E.; Litwin, C. M.; Hildreth, S. W.; Hill, H. R. *Am. J. Clin. Pathol.* **2002**, *117*, 589–596.
70. Rose, C. E.; Romero-Steiner, S.; Burton, R. L.; Carlone, G. M.; Goldblatt, D.; Nahm, M. H.; Ashton, L.; Haston, M.; Ekstrom, N.; Haikala, R.; Kayhty, H.; Henckaerts, I.; Durant, N.; Poolman, J. T.; Fernsten, P.; Yu, X.; Hu, B. T.; Jansen, K. U.; Blake, M.; Simonetti, E. R.; Hermans, P. W.; Plikaytis, B. D. *Clin. Vaccine Immunol.* **2011**, *18*, 135–142.
71. Burton, R. L.; Antonello, J.; Cooper, D.; Goldblatt, D.; Kim, K. H.; Plikaytis, B. D.; Roalfe, L.; Wauters, D.; Williams, F.; Xie, G. L.; Nahm, M. H.; Akkoyunlu, M. *Clin. Vaccine Immunol.* **2017**, *24*in press.
72. Steinhoff, M.; Edwards, K.; Keyserling, H.; Thomas, M.; Johnson, C.; Madore, D.; Hogerman, D. *Pediatr. Infect. Dis. J.* **1994**, *13*, 368–372.
73. Åhman, H.; Käyhty, H.; Tamminen, P.; Vuorela, A.; Malinoski, F.; Eskola, J. *Pediatr. Infect. Dis. J.* **1996**, *15*, 134–139.
74. Leach, A.; Ceesay, S. J.; Banya, W. A.; Greenwood, B. M. *Pediatr. Infect. Dis. J.* **1996**, *15*, 333–339.
75. Käyhty, H.; Åhman, H.; Rönnberg, P.-R.; Tillikainen, R.; Eskola, J. *J. Infect. Dis.* **1995**, *172*, 1273–1278.
76. Dagan, R.; Melamed, R.; Zamir, O.; Leroy, O. *Pediatr. Infect. Dis. J.* **1997**, *16*, 1053–1059.
77. Rennels, M. B.; Edwards, K. M.; Keyserling, H. L.; Reisinger, K. S.; Hogerman, D. A.; Madore, D. V.; Chang, I.; Paradiso, P. R.; Malinoski, F. J.; Kimura, A. *Pediatrics* **1998**, *101*, 604–611.
78. Shinefield, H. R.; Black, S.; Ray, P.; Chang, I.; Lewis, N.; Fireman, B.; Hackell, J.; Paradiso, P. R.; Siber, G.; Kohberger, R. *Pediatr. Infect. Dis. J.* **1999**, *18*, 757–763.

79. Black, S.; Shinefield, H.; Fireman, B.; Lewis, E.; Ray, P.; Hansen, J. R.; Elvin, L.; Ensor, K. M.; Hackell, J.; Siber, G. *Pediatr. Infect. Dis. J.* **2000**, *19*, 187–195.
80. Fireman, B.; Black, S. B.; Shinefield, H. R.; Lee, J.; Lewis, E.; Ray, P. *Pediatr. Infect. Dis. J.* **2003**, *22*, 10–16.
81. Eskola, J.; Kilpi, T.; Palmu, A.; Jokinen, J.; Eerola, M.; Haapakoski, J.; Herva, E.; Takala, A.; Käyhty, H.; Karma, P.; Lockhart, S. *N. Engl. J. Med.* **2001**, *344*, 403–409.
82. Black, S. B.; Shinefield, H. R.; Ling, S.; Hansen, J.; Fireman, B.; Spring, D.; Noyes, J.; Lewis, E.; Ray, P.; Lee, J. *Pediatr. Infect. Dis. J.* **2002**, *21*, 810–815.
83. World Health Organization. *Pneumonia Vaccine Trial Investigators' Group*; Geneva, 2001; document WHO/V&B/01.35.
84. Hansen, J.; Black, S.; Shinefield, H.; Cherian, T.; Benson, J.; Fireman, B.; Lewis, E.; Ray, P.; Lee, J. *Pediatr. Infect. Dis. J.* **2006**, *25*, 779–781.
85. FDA. *February 17, 2000 Approval Letter - Prevnar*. http://wayback.archive-it.org/7993/20170723031136/https://www.fda.gov/BiologicsBloodVaccines/Vaccines/ApprovedProducts/ucm137057.htm (accessed January 8. 2018).
86. EMA. Prevenar EPAR. http://www.ema.europa.eu/docs/en_GB/document_library/EPAR_-_Summary_for_the_public/human/000323/WC500041558.pdf (accessed January 8, 2018).
87. Choo, S.; Seymour, L.; Morris, R.; Quataert, S.; Lockhart, S.; Cartwright, K.; Finn, A. *Pediatr. Infect. Dis. J.* **2000**, *19*, 854–62.
88. Reinert, P.; Guy, M.; Girier, B.; Szelechowski, B.; Baudoin, B.; Deberdt, P.; Wollner, A.; Kemeny, G.; Amzallag, M.; Moat, C. *Arch. Pediatr.* **2003**, *10*, 1048–1055.
89. Schmitt, H.; Faber, J.; Lorenz, I.; Schmöle-Thoma, B.; Ahlers, N. *Vaccine* **2003**, *21*, 3653–3662.
90. Knuf, M.; Habermehl, P.; Cimino, C.; Petersen, G.; Schmitt, H.-J. *Vaccine* **2006**, *24*, 4727–4736.
91. Käyhty, H.; Åhman, H.; Eriksson, K.; Sörberg, M.; Nilsson, L. *Pediatr. Infect. Dis. J.* **2005**, *24*, 108–114.
92. Esposito, S.; Pugni, L.; Bosis, S.; Proto, A.; Cesati, L.; Bianchi, C.; Cimino, C.; Mosca, F.; Principi, N. *Vaccine* **2005**, *23*, 1703–1708.
93. Buttery, J. P.; Riddell, A.; McVernon, J.; Chantler, T.; Lane, L.; Bowen-Morris, J.; Diggle, L.; Morris, R.; Harnden, A.; Lockhart, S.; Pollard, A. J.; Cartwright, K.; Moxon, E. R. *JAMA, J. Am. Med. Assoc.* **2005**, *293*, 1751–1758.
94. Obaro, S. K.; Enwere, G. C.; Deloria, M.; Jaffar, S.; Goldblatt, D.; Brainsby, K.; Hallander, H.; McInnes, P.; Greenwood, B. M.; Mcadam, K. P. *Pediatr. Infect. Dis. J.* **2002**, *21*, 940–946.
95. Cutts, F. T.; Zaman, S. M. A.; Enwere, G.; Jaffar, S.; Levine, O. S.; Okoko, J. B.; Oluwalana, C.; Vaughan, A.; Obaro, S. K.; Leach, A.; McAdam, K. P.; Biney, E.; Saaka, M.; Onwuchekwa, U.; Yallop, F.; Pierce, N. F.; Greenwood, B. M.; Adegbola, R. A. *Lancet* **2005**, *365*, 1139–1146.
96. O'Brien, K. L.; Moulton, L. H.; Reid, R.; Weatherholtz, R.; Oski, J.; Brown, L.; Kumar, G.; Parkinson, A.; Hu, D.; Hackell, J.; Chang, I.; Kohberger, R.; Siber, G.; Santosham, M. *Lancet* **2003**, *362*, 355–361.

97. Klugman, K. P.; Madhi, S. A.; Huebner, R. E.; Kohberger, R.; Mbelle, N.; Pierce, N. *N. Engl. J. Med.* **2003**, *349*, 1341–1348.
98. Palmu, A. A.; Jokinen, J.; Borys, D.; Nieminen, H.; Ruokokoski, E.; Siira, L.; Puumalainen, T.; Lommel, P.; Hezareh, M.; Moreira, M.; Schuerman, L.; Kilpi, T. M. *Lancet* **2013**, *381*, 214–222.
99. Tregnaghi, M. W.; Sáez-Llorens, X.; López, P.; Abate, H.; Smith, E.; Pósleman, A.; Calvo, A.; Wong, D.; Cortes-Barbosa, C.; Ceballos, A.; Tregnaghi, M.; Sierra, A.; Rodriguez, M.; Troitiño, M.; Carabajal, C.; Falaschi, A.; Leandro, A.; Castrejón, M. M.; Lepetic, A.; Lommel, P.; Hausdorff, W. P.; Borys, D.; Guiñazú, J. R.; Ortega-Barría, E.; Yarzábal, J. P.; Schuerman, L. *PLoS Med.* **2014**, *11*, e1001657.
100. Jodar, L.; Butler, J.; Carlone, G.; Dagan, R.; Goldblatt, D.; Kayhty, H.; Klugman, K.; Plikaytis, B.; Siber, G.; Kohberger, R.; Chang, I.; Cherian, T. *Vaccine* **2003**, *21*, 3265–3272.
101. Siber, G. R.; Chang, I.; Baker, S.; Fernsten, P.; O'Brien, K. L.; Santosham, M.; Klugman, K. P.; Madhi, S. A.; Paradiso, P.; Kohberger, R. *Vaccine* **2007**, *25*, 3816–3826.
102. EMA. Synflorix Summary of Product Characteristics. http://www.ema.europa.eu/docs/en_GB/document_library/EPAR_-_Product_Information/human/000973/WC500054346.pdf (accessed October 26, 2017).
103. Prymula, R.; Peeters, P.; Chrobok, V.; Kriz, P.; Novakova, E.; Kaliskova, E.; Kohl, I.; Lommel, P.; Poolman, J.; Prieels, J.-P. *Lancet* **2006**, *367*, 740–748.
104. Prymula, R.; Schuerman, L. *Expert Rev. Vaccines* **2009**, *8*, 1479–1500.
105. Vesikari, T.; Wysocki, J.; Chevallier, B.; Karvonen, A.; Czajka, H.; Arsène, J.-P.; Lommel, P.; Dieussaert, I.; Schuerman, L. *Pediatr. Infect. Dis. J.* **2009**, *28*, S66–S76.
106. Henckaerts, I.; Goldblatt, D.; Ashton, L.; Poolman, J. *Clin. Vaccine Immunol.* **2006**, *13*, 356–360.
107. Vesikari, T.; Forsten, A.; Seppä, I.; Kaijalainen, T.; Puumalainen, T.; Soininen, A.; Traskine, M.; Lommel, P.; Schoonbroodt, S.; Hezareh, M.; Moreira, M.; Borys, D.; Schuerman, L. *J. Pediatric Infect. Dis. Soc.* **2016**, *5*, 237–248.
108. Lucero, M. G.; Nohynek, H.; Williams, G.; Tallo, V.; Simões, E. A.; Lupisan, S.; Sanvictores, D.; Forsyth, S.; Puumalainen, T.; Ugpo, J. *Pediatr. Infect. Dis. J.* **2009**, *28*, 455–462.
109. Kilpi, T.; Åhman, H.; Jokinen, J.; Lankinen, K. S.; Palmu, A.; Savolainen, H.; Grönholm, M.; Leinonen, M.; Hovi, T.; Eskola, J. *Clin. Infect. Dis.* **2003**, *37*, 1155–1164.
110. Scott, D. A.; Komjathy, S. F.; Hu, B. T.; Baker, S.; Supan, L. A.; Monahan, C. A.; Gruber, W.; Siber, G. R.; Lockhart, S. P. *Vaccine* **2007**, *25*, 6164–6166.
111. Scott, D.; Ruckle, J.; Dar, M.; Baker, S.; Kondoh, H.; Lockhart, S. *Pediatr. Int.* **2008**, *50*, 295–299.
112. Bryant, K. A.; Block, S. L.; Baker, S. A.; Gruber, W. C.; Scott, D. A.; Group, P. I. S. *Pediatrics* **2010**, *125*, 866–875.

113. Yeh, S. H.; Gurtman, A.; Hurley, D. C.; Block, S. L.; Schwartz, R. H.; Patterson, S.; Jansen, K. U.; Love, J.; Gruber, W. C.; Emini, E. A. *Pediatrics* **2010**, *126*, e493–e505.
114. Payton, T.; Girgenti, D.; Frenck, R. W.; Patterson, S.; Love, J.; Razmpour, A.; Sidhu, M. S.; Emini, E. A.; Gruber, W. C.; Scott, D. A. *Pediatr. Infect. Dis. J.* **2013**, *32*, 871–880.
115. Vanderkooi, O. G.; Scheifele, D. W.; Girgenti, D.; Halperin, S. A.; Patterson, S. D.; Gruber, W. C.; Emini, E. A.; Scott, D. A.; Kellner, J. D.; Group, C. P. S. *Pediatr. Infect. Dis. J.* **2012**, *31*, 72–77.
116. Kieninger, D. M.; Kueper, K.; Steul, K.; Juergens, C.; Ahlers, N.; Baker, S.; Jansen, K. U.; Devlin, C.; Gruber, W. C.; Emini, E. A. *Vaccine* **2010**, *28*, 4192–4203.
117. Snape, M. D.; Klinger, C. L.; Daniels, E. D.; John, T. M.; Layton, H.; Rollinson, L.; Pestridge, S.; Dymond, S.; Galiza, E.; Tansey, S. *Pediatr. Infect. Dis. J.* **2010**, *29*, e80–e90.
118. Martinón-Torres, F.; Gimenez-Sanchez, F.; Gurtman, A.; Bernaola, E.; Diez-Domingo, J.; Carmona, A.; Sidhu, M.; Sarkozy, D. A.; Gruber, W. C.; Emini, E. A. *Pediatr. Infect. Dis. J.* **2012**, *31*, 392–399.
119. Diez-Domingo, J.; Gurtman, A.; Bernaola, E.; Gimenez-Sanchez, F.; Martinon-Torres, F.; Pineda-Solas, V.; Delgado, A.; Infante-Marquez, P.; Liang, J. Z.; Giardina, P. C. *Vaccine* **2013**, *31*, 5486–5494.
120. Gadzinowski, J.; Albrecht, P.; Hasiec, B.; Konior, R.; Dziduch, J.; Witor, A.; Mellelieu, T.; Tansey, S. P.; Jones, T.; Sarkozy, D.; Emini, E. A.; Gruber, W. C.; Scott, D. A. *Vaccine* **2011**, *29*, 2947–2955.
121. Esposito, S.; Tansey, S.; Thompson, A.; Razmpour, A.; Liang, J.; Jones, T. R.; Ferrera, G.; Maida, A.; Bona, G.; Sabatini, C. *Clin. Vaccine Immunol.* **2010**, *17*, 1017–1026.
122. FDA. Summary Basis for Regulatory Action Prevnar 13. http://wayback.archive-it.org/7993/20170723031324/https://www.fda.gov/downloads/BiologicsBloodVaccines/Vaccines/ApprovedProducts/UCM206140.pdf (accessed January 18, 2018).
123. EMA. Prevenar 13 Public Assessment Report EMA/798877/2009. http://www.ema.europa.eu/docs/en_GB/document_library/EPAR_-_Public_assessment_report/human/001104/WC500057250.pdf (accessed January 18, 2018).
124. FDA. February 24, 2010 Approval Letter - Prevnar 13. http://wayback.archive-it.org/7993/20170723031322/https://www.fda.gov/BiologicsBloodVaccines/Vaccines/ApprovedProducts/ucm201741.htm (accessed January 18, 2018).
125. EMA. Prevenar 13 EPAR. http://www.ema.europa.eu/docs/en_GB/document_library/EPAR_-_Product_Information/human/001104/WC500057247.pdf (accessed January 18, 2018).
126. Togashi, T.; Yamaji, M.; Thompson, A.; Giardina, P. C.; Aizawa, M.; Patterson, S.; Gruber, W. C.; Scott, D. A.; Group, S. *Pediatr. Infect. Dis. J.* **2013**, *32*, 984–989.

127. Togashi, T.; Okada, K.; Yamaji, M.; Thompson, A.; Gurtman, A.; Cutler, M.; Aizawa, M.; Gruber, W. C.; Scott, D. A. *Pediatr. Infect. Dis. J.* **2015**, *34*, 1096–1104.
128. Huang, L.-M.; Lin, T.-Y.; Juergens, C. *Vaccine* **2012**, *30*, 2054–2059.
129. Weckx, L. Y.; Thompson, A.; Berezin, E. N.; Faria, S. M..; Cunha, C. A.; Pride, M.; Patterson, S.; Gruber, W. C.; Emini, E. A.; Scott, D. A. *Vaccine* **2012**, *30*, 7566–7572.
130. Amdekar, Y. K.; Lalwani, S. K.; Bavdekar, A.; Balasubramanian, S.; Chhatwal, J.; Bhat, S. R.; Verghese, V. P.; Tansey, S. P.; Gadgil, D.; Jiang, Q. *Pediatr. Infect. Dis. J.* **2013**, *32*, 509–516.
131. Gutiérrez Brito, M.; Thompson, A.; Girgenti, D.; Giardina, P. C.; Sarkozy, D. A.; Gruber, W. C.; Emini, E. A.; Scott, D. A. *Pan Am. J. Public Health* **2013**, *33*, 414–421.
132. Kim, D. S.; Shin, S. H.; Lee, H. J.; Hong, Y. J.; Lee, S. Y.; Choi, K. M.; Oh, C. E.; Kim, K. H.; Juergens, C.; Patterson, S. *Pediatr. Infect. Dis. J.* **2013**, *32*, 266–273.
133. Zhu, F.; Hu, Y.; Li, J.; Ye, Q.; Young, M. M.; Zhou, X.; Chen, Z.; Yan, B.; Liang, J. Z.; Gruber, W. C. *Pediatr. Infect. Dis. J.* **2016**, *35*, 999–1010.
134. Dagan, R.; Patterson, S.; Juergens, C.; Greenberg, D.; Givon-Lavi, N.; Porat, N.; Gurtman, A.; Gruber, W. C.; Scott, D. A. *Clin. Infect. Dis.* **2013**, *57*, 952–962.
135. Moore, M. R.; Link-Gelles, R.; Schaffner, W.; Lynfield, R.; Holtzman, C.; Harrison, L. H.; Zansky, S. M.; Rosen, J. B.; Reingold, A.; Scherzinger, K. *Lancet Respir. Med.* **2016**, *4*, 399–406.
136. Khandke, L.; Yang, C.; Krylova, K.; Jansen, K. U.; Rashidbaigi, A. *Vaccine* **2011**, *29*, 7144–7153.
137. Idoko, O. T.; Mboizi, R. B.; Okoye, M.; Laudat, F.; Ceesay, B.; Liang, J. Z.; Le Dren-Narayanin, N.; Jansen, K. U.; Gurtman, A.; Center, K. J. *Vaccine* **2017**, *35*, 3256–3263.
138. Lode, H.; Schmoele-Thoma, B.; Gruber, W.; Ahlers, N.; Fernsten, P.; Baker, S.; Razmpour, A.; Siber, G.; Hackell, J.; Lockhart, S.; Burkhardt, O.; Welte, T.; de Roux, A. *Vaccine* **2011**, *29*, 4940–4946.
139. de Roux, A.; Schmole-Thoma, B.; Siber, G. R.; Hackell, J. G.; Kuhnke, A.; Ahlers, N.; Baker, S. A.; Razmpour, A.; Emini, E. A.; Fernsten, P. D.; Gruber, W. C.; Lockhart, S.; Burkhardt, O.; Welte, T.; Lode, H. M. *Clin. Infect. Dis.* **2008**, *46*, 1015–1023.
140. Jackson, L. A.; Gurtman, A.; van Cleeff, M.; Jansen, K. U.; Jayawardene, D.; Devlin, C.; Scott, D. A.; Emini, E. A.; Gruber, W. C.; Schmoele-Thoma, B. *Vaccine* **2013**, *31*, 3577–3584.
141. Jackson, L. A.; Gurtman, A.; Rice, K.; Pauksens, K.; Greenberg, R. N.; Jones, T. R.; Scott, D. A.; Emini, E. A.; Gruber, W. C.; Schmoele-Thoma, B. *Vaccine* **2013**, *31*, 3585–3593.
142. FDA. December 29, 2011 Summary Basis for Regulatory Action - Prevnar 13. http://wayback.archive-it.org/7993/20170723031315/https:/www.fda.gov/downloads/BiologicsBloodVaccines/Vaccines/ApprovedProducts/UCM287412.pdf (accessed November 7, 2017).

143. EMA. CHMP Variation Assessment Report. http://www.ema.europa.eu/docs/en_GB/document_library/EPAR_-_Assessment_Report_-_Variation/human/001104/WC500119784.pdf (accessed November 7, 2017).
144. Jackson, L. A.; Gurtman, A.; van Cleeff, M.; Frenck, R. W.; Treanor, J.; Jansen, K. U.; Scott, D. A.; Emini, E. A.; Gruber, W. C.; Schmoele-Thoma, B. *Vaccine* **2013**, *31*, 3594–602.
145. Greenberg, R. N.; Gurtman, A.; Frenck, R. W.; Strout, C.; Jansen, K. U.; Trammel, J.; Scott, D. A.; Emini, E. A.; Gruber, W. C.; Schmoele-Thoma, B. *Vaccine* **2014**, *32*, 2364–2374.
146. Frenck, R., Jr.; Thompson, A.; Senders, S.; Harris-Ford, L.; Sperling, M.; Patterson, S.; Devlin, C.; Jansen, K. U.; Gruber, W. C.; Emini, E. A.; Scott, D. A.; Gurtman, A. *Pediatr. Infect. Dis. J.* **2014**, *33*, 183–189.
147. Bryant, K. A.; Frenck, R.; Gurtman, A.; Rubino, J.; Treanor, J.; Thompson, A.; Jones, T. R.; Sundaraiyer, V.; Baxter, L. M.; Gruber, W. C.; Emini, E. A.; Scott, D. A.; Schmoele-Thoma, B. *Vaccine* **2015**, *33*, 5854–3860.
148. Frenck, R., Jr.; Thompson, A.; Yeh, S. H.; London, A.; Sidhu, M. S.; Patterson, S.; Gruber, W. C.; Emini, E. A.; Scott, D. A.; Gurtman, A. *Pediatr. Infect. Dis. J.* **2011**, *30*, 1086–91.
149. Shiramoto, M.; Irie, S.; Juergens, C.; Yamaji, M.; Tamai, S.; Aizawa, M.; Belanger, T.; Gruber, W. C.; Scott, D. A.; Schmoele-Thoma, B. *Hum. Vaccines Immunother.* **2014**, *10*, 1850–1858.
150. Shiramoto, M.; Hanada, R.; Juergens, C.; Shoji, Y.; Yoshida, M.; Ballan, B.; Cooper, D.; Gruber, W. C.; Scott, D. A.; Schmoele-Thoma, B. *Hum. Vaccines Immunother.* **2015**, *11*, 2198–2206.
151. Solanki, B. B.; Juergens, C.; Chopada, M. B.; Supe, P.; Sundaraiyer, V.; Le Dren-Narayanin, N.; Cutler, M. W.; Gruber, W. C.; Scott, D. A.; Schmoele-Thoma, B. *Hum. Vaccines Immunother.* **2017**, *13*, 2065–2071.
152. Tinoco, J. C.; Juergens, C.; Ruiz Palacios, G. M.; Vazquez-Narvaez, J.; Enkerlin-Pauwells, H. L.; Sundaraiyer, V.; Pathirana, S.; Kalinina, E.; Gruber, W. C.; Scott, D. A.; Schmoele-Thoma, B. *Clin. Vaccine Immunol.* **2015**, *22*, 185–192.
153. Hak, E.; Grobbee, D.; Sanders, E.; Verheij, T.; Bolkenbaas, M.; Huijts, S.; Gruber, W.; Tansey, S.; McDonough, A.; Thoma, B. *Neth. J. Med.* **2008**, *66*, 378–383.
154. Bonten, M. J.; Huijts, S. M.; Bolkenbaas, M.; Webber, C.; Patterson, S.; Gault, S.; van Werkhoven, C. H.; van Deursen, A. M.; Sanders, E. A.; Verheij, T. J.; Patton, M.; McDonough, A.; Moradoghli-Haftvani, A.; Smith, H.; Mellelieu, T.; Pride, M. W.; Crowther, G.; Schmoele-Thoma, B.; Scott, D. A.; Jansen, K. U.; Lobatto, R.; Oosterman, B.; Visser, N.; Caspers, E.; Smorenburg, A.; Emini, E. A.; Gruber, W. C.; Grobbee, D. E. *N. Engl. J. Med.* **2015**, *372*, 1114–1125.
155. Pride, M. W.; Huijts, S. M.; Wu, K.; Souza, V.; Passador, S.; Tinder, C.; Song, E.; Elfassy, A.; McNeil, L.; Menton, R. *Clin. Vaccine Immunol.* **2012**, *19*, 1131–1141.
156. Huijts, S. M.; Pride, M. W.; Vos, J. M.; Jansen, K. U.; Webber, C.; Gruber, W.; Boersma, W. G.; Snijders, D.; Kluytmans, J. A.; van der Lee, I. *Eur. Respir. J.* **2013**, *42*, 1283–1290.

157. Frenck, R. W., Jr.; Gurtman, A.; Rubino, J.; Smith, W.; van Cleeff, M.; Jayawardene, D.; Giardina, P. C.; Emini, E. A.; Gruber, W. C.; Scott, D. A.; Schmole-Thoma, B. *Clin. Vaccine Immunol.* **2012**, *19*, 1296–1303.
158. Frenck, R. W.; Fiquet, A.; Gurtman, A.; van Cleeff, M.; Davis, M.; Rubino, J.; Smith, W.; Sundaraiyer, V.; Sidhu, M.; Emini, E. A. *Vaccine* **2016**, *34*, 3454–3462.
159. Schwarz, T. F.; Flamaing, J.; Rumke, H. C.; Penzes, J.; Juergens, C.; Wenz, A.; Jayawardene, D.; Giardina, P.; Emini, E. A.; Gruber, W. C.; Schmoele-Thoma, B. *Vaccine* **2011**, *29*, 5195–5202.
160. Baxter, R.; Downey, H. J.; Patterson, S. D.; Sundaraiyer, V.; Watson, W.; Clarke, K.; Gruber, W. C.; Scott, D. A.; Thompson, A.; Schmoele-Thoma, B. *Open Forum Infect. Dis.* **2015**, *2*, 67–70.
161. French, N.; Gordon, S. B.; Mwalukomo, T.; White, S. A.; Mwafulirwa, G.; Longwe, H.; Mwaiponya, M.; Zijlstra, E. E.; Molyneux, M. E.; Gilks, C. F. *N. Engl. J. Med* **2010**, *362*, 812–822.
162. Bhorat, A. E.; Madhi, S. A.; Laudat, F.; Sundaraiyer, V.; Gurtman, A.; Jansen, K. U.; Scott, D. A.; Emini, E. A.; Gruber, W. C.; Schmoele-Thoma, B. *AIDS* **2015**, *29*, 1345–1354.
163. Glesby, M. J.; Watson, W.; Brinson, C.; Greenberg, R. N.; Lalezari, J. P.; Skiest, D.; Sundaraiyer, V.; Natuk, R.; Gurtman, A.; Scott, D. A.; Emini, E. A.; Gruber, W. C.; Schmoele-Thoma, B. *J. Infect. Dis.* **2015**, *212*, 18–27.
164. Cordonnier, C.; Ljungman, P.; Juergens, C.; Maertens, J.; Selleslag, D.; Sundaraiyer, V.; Giardina, P. C.; Clarke, K.; Gruber, W. C.; Scott, D. A.; Schmoele-Thoma, B.; Study, G. *Clin. Infect. Dis.* **2015**, *61*, 313–323.
165. Singleton, R.; Wenger, J.; Klejka, J. A.; Bulkow, L. R.; Thompson, A.; Sarkozy, D.; Emini, E. A.; Gruber, W. C.; Scott, D. A. *Pediatr. Infect. Dis. J.* **2013**, *32*, 257–263.
166. Waight, P. A.; Andrews, N. J.; Ladhani, S. N.; Sheppard, C. L.; Slack, M. P. E.; Miller, E. *Lancet Infect. Dis.* **2015**, *15*, 535–543.
167. Ladhani, S. N.; Collins, S.; Djennad, A.; Sheppard, C. L.; Borrow, R.; Fry, N. K.; Andrews, N. J.; Miller, E.; Ramsay, M. E. *Lancet Infect. Dis.* **2018**, *14*, 441–451.
168. Olarte, L.; Barson, W. J.; Barson, R. M.; Romero, J. R.; Bradley, J. S.; Tan, T. Q.; Givner, L. B.; Hoffman, J. A.; Lin, P. L.; Hultén, K. G.; Mason, E. O.; Kaplan, S. L. *Clin. Infect. Dis.* **2017**, *64*, 1699–1704.
169. Hulten, K. G.; Kaplan, S. L.; Lamberth, L. B.; Barson, W. J.; Romero, J. R.; Lin, P. L.; Bradley, J. S.; Givner, L. B.; Tan, T. Q.; Hoffman, J. A.; Mason, E. O. *J. Clin. Microbiol.* **2013**, *51*, 1294–1297.
170. Mackenzie, G. A.; Hill, P. C.; Sahito, S. M.; Jeffries, D. J.; Hossain, I.; Bottomley, C.; Uchendu, U.; Ameh, D.; Ndiaye, M.; Osuorah, C. D.; Adeyemi, O.; Pathirana, J.; Olatunji, Y.; Abatan, B.; Ahameefula, E.; Muhammad, B. S.; Fombah, A. E.; Saha, D.; Mackenzie, R.; Plumb, I.; Akano, A.; Ebruke, B.; Ideh, R. C.; Kuti, B.; Githua, P.; Olutunde, E.; Ofordile, O.; Green, E.; Usuf, E.; Badji, H.; Ikumapayi, U. N. A.; Manjang, A.; Salaudeen, R.; Nsekpong, E. D.; Jarju, S.; Antonio, M.; Sambou, S.; Ceesay, L.; Lowe-Jallow, Y.; Sowe, D.; Jasseh, M.; Mulholland, K.; Knoll, M.;

Levine, O. S.; Howie, S. R.; Adegbola, R. A.; Greenwood, B. M.; Corrah, T. *Lancet Infect. Dis.* **2017**, *17*, 965–973.

171. O'Brien, K. L. *Lancet Respir. Med.* **2016**, *4*, 343–345.
172. Martinez, A. Abstract P-196 and poster; *9th International Symposium on Pneumococci and Pneumococcal Diseases*, Hyderabad, India, 2014.
173. McFetridge, R.; Sobanjo-ter Meulen, A.; Folkerth, S. D.; Hoekstra, J. A.; Dallas, M.; Hoover, P. A.; Marchese, R. D.; Zacholski, D. M.; Watson, W. J.; Stek, J. E. *Vaccine* **2015**, *33*, 2793–2799.
174. Sobanjo-ter Meulen, A.; Vesikari, T.; Malacaman, E. A.; Shapiro, S. A.; Dallas, M. J.; Hoover, P. A.; McFetridge, R.; Stek, J. E.; Marchese, R. D.; Hartzel, J. *Pediatr. Infect. Dis. J.* **2015**, *34*, 186–194.
175. Greenberg, D. Abstract 0197; *9th International Symposium on Pneumococci and Pneumococcal Diseases*, Hyderabad, India, 2014.
176. Chen, J. J.; Yuan, L.; Huang, Z.; Shi, N. M.; Zhao, Y. L.; Xia, S. L.; Li, G. H.; Li, R. C.; Li, Y. P.; Yang, S. Y. *BMJ Open* **2016**, *6*, e012488.
177. Shahid, N. S.; Hoque, S. S.; Begum, T.; Steinhoff, M. C.; Thompson, C.; Siber, G. R. *Lancet* **1995**, *346*, 1252–1257.
178. Goldblatt, D.; Southern, J.; Andrews, N. J.; Burbidge, P.; Partington, J.; Roalfe, L.; Valente Pinto, M.; Thalasselis, V.; Plested, E.; Richardson, H.; Snape, M. D.; Miller, E. *Lancet Infect. Dis.* **2018**, *18*, 171–179.
179. O'Brien, K. L. *Lancet Infect. Dis.* **2018**, *18*, 127–128.
180. Goldschneider, I.; Gotschlich, E. C.; Artenstein, M. S. *J. Exp. Med.* **1969**, *129*, 1307–1326.
181. Gotschlich, E. C.; Goldschneider, I.; Artenstein, M. S. *J. Exp. Med.* **1969**, *129*, 1367–1384.
182. Gotschlich, E. C.; Goldschneider, I.; Artenstein, M. S. *J. Exp. Med.* **1969**, *129*, 1385–1395.
183. Gold, R.; Artenstein, M. S. *Bull. W.H.O.* **1971**, *45*, 279.
184. Peltola, H.; Mäkelä, P. H.; Käyhty, H.; Jousimies, H.; Herva, E.; Hällström, K.; Sivonen, A.; Renkonen, O.-V.; Pettay, O.; Karanko, V.; Ahvonen, P.; Sarna, S. *N. Engl. J. Med* **1977**, *297*, 686–691.
185. Granoff, D. M.; Pelton, S.; Harrison, L. H. Meningococcal Vaccines. In *Vaccines*, 6th ed.; Plotkin, S., Orenstein, W., Offit, P., Eds.; Elsevier Saunders: London, 2012; pp 388–418.
186. Wyle, F. A.; Artenstein, M. S.; Brandt, B. L.; Tramont, E. C.; Kasper, D. L.; Altieri, P. L.; Berman, S. L.; Lowenthal, J. P. *J. Infect. Dis.* **1972**, *126*, 514–522.
187. Rougon, G.; Dubois, C.; Buckley, N.; Magnani, J. L.; Zollinger, W. *J. Cell Biol.* **1986**, *103*, 2429–2437.
188. Costantino, P.; Viti, S.; Podda, A.; Velmonte, M. A.; Nencioni, L.; Rappuoli, R. *Vaccine* **1992**, *10*, 691–698.
189. Leach, A.; Twumasi, P. A.; Kumah, S.; Banya, W. S.; Jaffar, S.; Forrest, B. D.; Granoff, D. M.; LiButti, D. E.; Carlone, G. M.; Pais, L. B. *J. Infect. Dis.* **1997**, *175*, 200–204.
190. Richmond, P.; Goldblatt, D.; Fusco, P. C.; Fusco, J. D. S.; Heron, I.; Clark, S.; Borrow, R.; Michon, F. *Vaccine* **1999**, *18*, 641–646.

191. Richmond, P.; Borrow, R.; Goldblatt, D.; Findlow, J.; Martin, S.; Morris, R.; Cartwright, K.; Miller, E. *J. Infect. Dis.* **2001**, *183*, 160–163.
192. Burrage, M.; Robinson, A.; Borrow, R.; Andrews, N.; Southern, J.; Findlow, J.; Martin, S.; Thornton, C.; Goldblatt, D.; Corbel, M.; Sesardic, D.; Cartwight, K.; Richmond, P.; Miller, E. *Infect. Immun.* **2002**, *70*, 4946–4954.
193. Richmond, P.; Kaczmarski, E.; Borrow, R.; Findlow, J.; Clark, S.; McCann, R.; Hill, J.; Barker, M.; Miller, E. *J. Infect. Dis.* **2000**, *181*, 761–764.
194. Borrow, R.; Southern, J.; Andrews, N.; Peake, N.; Rahim, R.; Acuna, M.; Martin, S.; Miller, E.; Kaczmarski, E. *Vaccine* **2001**, *19*, 3043–3050.
195. Miller, E.; Salisbury, D.; Ramsay, M. *Vaccine* **2001**, *20*, S58–S67.
196. Trotter, C. L.; Andrews, N. J.; Kaczmarski, E. B.; Miller, E.; Ramsay, M. E. *Lancet* **2004**, *364*, 365–367.
197. EMA. Nimenrix Assessment History. http://www.ema.europa.eu/ema/index.jsp?curl=pages/medicines/human/medicines/002226/human_med_001548.jsp&mid=WC0b01ac058001d124 (accessed January 11, 2018).
198. EMA. MenVeo Assessment History. http://www.ema.europa.eu/ema/index.jsp?curl=pages/medicines/human/medicines/001095/human_med_001323.jsp&mid=WC0b01ac058001d124 (accessed January 11, 2018).
199. FDA. Menactra https://www.fda.gov/BiologicsBloodVaccines/Vaccines/ApprovedProducts/ucm176044.htm (accessed January 11, 2018).
200. FDA. MenVeo. https://www.fda.gov/biologicsbloodvaccines/vaccines/approvedproducts/ucm201342.htm (accessed January 11, 2018).
201. Assaf-Casals, A.; Dbaibo, G. *Hum. Vaccines Immunother.* **2016**, *12*, 1825–1837.
202. Dhillon, S.; Pace, D. *Drugs* **2017**, *77*, 1881–1896.
203. Cooper, B.; DeTora, L.; Stoddard, J. *Expert Rev. Vaccines* **2011**, *10*, 21–33.
204. Rennels, M.; King, J., Jr; Ryall, R.; Papa, T.; Froeschle, J. *Pediatr. Infect. Dis. J.* **2004**, *23*, 429–435.
205. Block, S. L.; Szenborn, L.; Daly, W.; Jackowska, T.; D'Agostino, D.; Han, L.; Dull, P. M.; Smolenov, I. *Vaccine* **2015**, *33*, 2500–2510.
206. Welsch, J. A.; Senders, S.; Middleton, R.; Pedotti, P.; Han, L.; Dull, P.; Smolenov, I. *Open Forum Infect. Dis.* **2016**, *3*, S136.
207. LaForce, F. M.; Konde, K.; Viviani, S.; Préziosi, M.-P. *Vaccine* **2007**, *25*, A97–A100.
208. Kshirsagar, N.; Mur, N.; Thatte, U.; Gogtay, N.; Viviani, S.; Préziosi, M.-P.; Elie, C.; Findlow, H.; Carlone, G.; Borrow, R. *Vaccine* **2007**, *25*, A101–A107.
209. Sow, S. O.; Okoko, B. J.; Diallo, A.; Viviani, S.; Borrow, R.; Carlone, G.; Tapia, M.; Akinsola, A. K.; Arduin, P.; Findlow, H.; Elie, C.; Haidara, F. C.; Adegbola, R. A.; Diop, D.; Parulekar, V.; Chaumont, J.; Martellet, L.; Diallo, F.; Idoko, O. T.; Tang, Y.; Plikaytis, B. D.; Kulkarni, P. S.; Marchetti, E.; LaForce, F. M.; Preziosi, M.-P. *N. Engl. J. Med.* **2011**, *364*, 2293–2304.
210. LaForce, F. M.; Djingarey, M.; Viviani, S.; Preziosi, M. P. *Hum. Vaccines Immunother.* **2017**in press.

211. Trotter, C. L.; Lingani, C.; Fernandez, K.; Cooper, L. V.; Bita, A.; Tevi-Benissan, C.; Ronveaux, O.; Préziosi, M.-P.; Stuart, J. M. *Lancet Infect. Dis.* **2017**, *17*, 867–872.
212. Daugla, D.; Gami, J.; Gamougam, K.; Naibei, N.; Mbainadji, L.; Narbé, M.; Toralta, J.; Kodbesse, B.; Ngadoua, C.; Coldiron, M. *Lancet* **2014**, *383*, 40–47.
213. Fattom, A.; Schneerson, R.; Watson, D.; Karakawa, W.; Fitzgerald, D.; Pastan, I.; Li, X.; Shiloach, J.; Bryla, D.; Robbins, J. *Infect. Immun.* **1993**, *61*, 1023–1032.
214. Shinefield, H.; Black, S.; Fattom, A.; Horwith, G.; Rasgon, S.; Ordonez, J.; Yeoh, H.; Law, D.; Robbins, J. B.; Schneerson, R.; Muenz, L.; Fuller, S.; Johnson, J.; Fireman, B.; Alcorn, H.; Naso, R. *N. Engl. J. Med.* **2002**, *346*, 491–496.
215. Fattom, A.; Matalon, A.; Buerkert, J.; Taylor, K.; Damaso, S.; Boutriau, D. *Hum. Vaccines Immunother.* **2015**, *11*, 632–641.
216. Levy, J.; Licini, L.; Haelterman, E.; Moris, P.; Lestrate, P.; Damaso, S.; Van Belle, P.; Boutriau, D. *Hum. Vaccines Immunother.* **2015**, *11*, 620–631.
217. Nissen, M.; Marshall, H.; Richmond, P.; Shakib, S.; Jiang, Q.; Cooper, D.; Rill, D.; Baber, J.; Eiden, J.; Gruber, W.; Jansen, K. U.; Emini, E. A.; Anderson, A. S.; Zito, E. T.; Girgenti, D. *Vaccine* **2015**, *33*, 1846–54.
218. Marshall, H.; Nissen, M.; Richmond, P.; Shakib, S.; Jiang, Q.; Cooper, D.; Rill, D.; Baber, J.; Eiden, J.; Gruber, W. C.; Jansen, K. U.; Anderson, A. S.; Zito, E. T.; Girgenti, D. *J. Infect.* **2016**, *73*, 437–454.
219. Frenck, R. W.; Creech, C. B.; Sheldon, E. A.; Seiden, D. J.; Kankam, M. K.; Baber, J.; Zito, E.; Hubler, R.; Eiden, J.; Severs, J. M. *Vaccine* **2017**, *35*, 375–384.
220. Creech, C. B.; Frenck, R. W.; Sheldon, E. A.; Seiden, D. J.; Kankam, M. K.; Zito, E. T.; Girgenti, D.; Severs, J. M.; Immermann, F. W.; McNeil, L. K. *Vaccine* **2017**, *35*, 385–394.
221. Wang, M. Y.; Dayan, G.; Hassanzadeh, H.; Huec, J.-C. L.; Patel, V. V.; Liljenqvist, U.; Vaidya, R.; Noriega, D. C.; Maseres, J. M.; Fehlings, M. G.; Sabharwal, C.; Begier, E.; Baber, J.; Jaques, A.; Gruber, W.; Eiden, J.; Gurtman, A. *Spine J.* **2017**, *17*, S206–S207.
222. Baker, C. J.; Kasper, D. L. *N. Engl. J. Med.* **1976**, *294*, 753–756.
223. Le Doare, K.; O'Driscoll, M.; Turner, K.; Seedat, F.; Russell, N. J.; Seale, A. C.; Heath, P. T.; Lawn, J. E.; Baker, C. J.; Bartlett, L.; Cutland, C.; Gravett, M. G.; Ip, M.; Madhi, S. A.; Rubens, C. E.; Saha, S. K.; Schrag, S.; Sobanjo-ter Meulen, A.; Vekemans, J.; Kampmann, B. *Clin. Infect. Dis.* **2017**, *65*, S143–S151.
224. Baker, C. J.; Rench, M. A.; Edwards, M. S.; Carpenter, R. J.; Hays, B. M.; Kasper, D. L. *N. Engl. J. Med* **1988**, *319*, 1180–1185.
225. Kasper, D. L.; Paoletti, L. C.; Wessels, M. R.; Guttormsen, H. K.; Carey, V. J.; Jennings, H. J.; Baker, C. J. *J. Clin. Invest.* **1996**, *98*, 2308–2314.
226. Baker, C. J.; Paoletti, L. C.; Rench, M. A.; Guttormsen, H.-K.; Carey, V. J.; Hickman, M. E.; Kasper, D. L. *J. Infect. Dis.* **2000**, *182*, 1129–1138.
227. Brigtsen, A. K.; Kasper, D. L.; Baker, C. J.; Jennings, H. J.; Guttormsen, H.-K. *J. Infect. Dis.* **2002**, *185*, 1277–1284.

228. Baker, C. J.; Paoletti, L. C.; Rench, M. A.; Guttormsen, H.-K.; Edwards, M. S.; Kasper, D. L. *J. Infect. Dis.* **2004**, *189*, 1103–1112.
229. Baker, C. J.; Rench, M. A.; Fernandez, M.; Paoletti, L. C.; Kasper, D. L.; Edwards, M. S. *J. Infect. Dis.* **2003**, *188*, 66–73.
230. Ferrieri, P.; Lynfield, R.; Creti, R.; Flores, A. E. *Emerging Infect. Dis.* **2013**, *19*, 551–558.
231. Madhi, S. A.; Cutland, C. L.; Jose, L.; Koen, A.; Govender, N.; Wittke, F.; Olugbosi, M.; Sobanjo-ter Meulen, A.; Baker, S.; Dull, P. M. *Lancet Infect. Dis.* **2016**, *16*, 923–934.
232. Donders, G. G. G.; Halperin, S. A.; Devlieger, R.; Baker, S.; Forte, P.; Wittke, F.; Slobod, K. S.; Dull, P. M. *Obstet. Gynecol. (Philadelphia, PA, U.S.)* **2016**, *127*, 213–221.
233. Heyderman, R. S.; Madhi, S. A.; French, N.; Cutland, C.; Ngwira, B.; Kayambo, D.; Mboizi, R.; Koen, A.; Jose, L.; Olugbosi, M. *Lancet Infect. Dis.* **2016**, *16*, 546–555.
234. Kobayashi, M.; Schrag, S. J.; Alderson, M. R.; Madhi, S. A.; Baker, C. J.; Sobanjo-ter Meulen, A.; Kaslow, D. C.; Smith, P. G.; Moorthy, V. S.; Vekemans, J. *Vaccine* **2016**in press.
235. Plotkin, S. A.; Cam, N. *Arch. Intern. Med.* **1995**, *155*, 2293–2299.
236. Sur, D.; Ochiai, R. L.; Bhattacharya, S. K.; Ganguly, N. K.; Ali, M.; Manna, B.; Dutta, S.; Donner, A.; Kanungo, S.; Park, J. K.; Puri, M. K.; Kim, D. R.; Dutta, D.; Bhaduri, B.; Acosta, C. J.; Clemens, J. D. *N. Engl. J. Med.* **2009**, *361*, 335–344.
237. Keitel, W. A.; Bond, N. L.; Zahradnik, J. M.; Cramton, T. A.; Robbins, J. B. *Vaccine* **1994**, *12*, 195–199.
238. Cadoz, M. *Vaccine* **1998**, *16*, 1391–1395.
239. Ochiai, R. L.; Khan, M. I.; Soofi, S. B.; Sur, D.; Kanungo, S.; You, Y. A.; Habib, M. A.; Sahito, S. M.; Manna, B.; Dutta, S. *Clin. Vaccine Immunol.* **2014**, *21*, 661–666.
240. Szu, S. C.; Taylor, D. N.; Trofa, A. C.; Clements, J. D.; Shiloach, J.; Sadoff, J. C.; Bryla, D. A.; Robbins, J. B. *Infect. Immun.* **1994**, *62*, 4440–4444.
241. Kossaczka, Z.; Lin, F.-Y. C.; Ho, V. A.; Thuy, N. T. T.; Van Bay, P.; Thanh, T. C.; Khiem, H. B.; Trach, D. D.; Karpas, A.; Hunt, S. *Infect. Immun.* **1999**, *67*, 5806–5810.
242. Lin, F. Y. C.; Ho, V. A.; Khiem, H. B.; Trach, D. D.; Bay, P. V.; Thanh, T. C.; Kossaczka, Z.; Bryla, D. A.; Shiloach, J.; Robbins, J. B. *N. Engl. J. Med.* **2001**, *344*, 1263–1269.
243. Lanh, M. N.; Van Bay, P.; Ho, V. A.; Thanh, T. C.; Lin, F. Y. C.; Bryla, D. A.; Chu, C.; Schiloach, J.; Robbins, J. B.; Schneerson, R. *N. Engl. J. Med* **2003**, *349*, 1390–1391.
244. Szu, S. C.; Hunt, S.; Xie, G.; Robbins, J. B.; Schneerson, R.; Gupta, R. K.; Zhao, Z.; Tan, X. *Vaccine* **2013**, *31*, 1970–1974.
245. World Health Organization. *Expert Committee on Biological Standardization: Sixty-Fourth Report*; Geneva, 2014.
246. Mohan, V. K.; Varanasi, V.; Singh, A.; Pasetti, M. F.; Levine, M. M.; Venkatesan, R.; Ella, K. M. *Clin. Infect. Dis.* **2015**, *61*, 393–402.

247. Jin, C.; Gibani, M. M.; Moore, M.; Juel, H. B.; Jones, E.; Meiring, J.; Harris, V.; Gardner, J.; Nebykova, A.; Kerridge, S. A.; Hill, J.; Thomaides-Brears, H.; Blohmke, C. J.; Yu, L.-M.; Angus, B.; Pollard, A. J. *Lancet* **2017**, *390*, 2472–2480.
248. Chinnasami, B.; Mangayarkarasi, V.; Prema, A.; Sadasivam, K.; Davis, M. J. *Int. J. Sci. Res. Publ. IJSRP* **2013**, *3*, 1–5.
249. Chinnasami, B.; Sadasivam, K.; Vivekanandhan, A.; Arunachalam, P.; Pasupathy, S. *J. Clin. Diagn. Res.* **2015**, *9*, SC01–C03.
250. Mitra, M.; Shah, N.; Ghosh, A.; Chatterjee, S.; Kaur, I.; Bhattacharya, N.; Basu, S. *Hum. Vaccines Immunother.* **2016**, *12*, 939–945.
251. Bhutta, Z. A.; Capeding, M. R.; Bavdekar, A.; Marchetti, E.; Ariff, S.; Soofi, S. B.; Anemona, A.; Habib, M. A.; Alberto, E.; Juvekar, S. *Lancet Infect. Dis.* **2014**, *14*, 119–129.
252. Robbins, J. B.; Chu, C.; Schneerson, R. *Clin. Infect. Dis.* **1992**, *15*, 346–361.
253. Cohen, D.; Ashkenazi, S.; Green, M. S.; Gdalevich, M.; Robin, G.; Slepon, R.; Yavzori, M.; Orr, N.; Block, C.; Ashkenazi, I.; Shemer, J.; Taylor, D. N.; Hale, T. L.; Sadoff, J. C.; Pavliakova, D.; Schneerson, R.; Robbins, J. B. *Lancet* **1997**, *349*, 155–159.
254. Passwell, J. H.; Ashkenzi, S.; Banet-Levi, Y.; Ramon-Saraf, R.; Farzam, N.; Lerner-Geva, L.; Even-Nir, H.; Yerushalmi, B.; Chu, C.; Shiloach, J. *Vaccine* **2010**, *28*, 2231–2235.
255. Hatz, C. F.; Bally, B.; Rohrer, S.; Steffen, R.; Kramme, S.; Siegrist, C. A.; Wacker, M.; Alaimo, C.; Fonck, V. G. *Vaccine* **2015**, *33*, 4594–4601.
256. Riddle, M. S.; Kaminski, R. W.; Di Paolo, C.; Porter, C. K.; Gutierrez, R. L.; Clarkson, K. A.; Weerts, H. E.; Duplessis, C.; Castellano, A.; Alaimo, C. *Clin. Vaccine Immunol.* **2016**, *23*, 908–917.
257. Livio, S.; Strockbine, N. A.; Panchalingam, S.; Tennant, S. M.; Barry, E. M.; Marohn, M. E.; Antonio, M.; Hossain, A.; Mandomando, I.; Ochieng, J. B.; Oundo, J. O.; Qureshi, S.; Ramamurthy, T.; Tamboura, B.; Adegbola, R. A.; Hossain, M. J.; Saha, D.; Sen, S.; Faruque, A. S. G.; Alonso, P. L.; Breiman, R. F.; Zaidi, A. K. M.; Sur, D.; Sow, S. O.; Berkeley, L. Y.; O'Reilly, C. E.; Mintz, E. D.; Biswas, K.; Cohen, D.; Farag, T. H.; Nasrin, D.; Wu, Y.; Blackwelder, W. C.; Kotloff, K. L.; Nataro, J. P.; Levine, M. M. *Clin. Infect. Dis.* **2014**, *59*, 933–941.
258. Hosangadi, D.; Smith, P. G.; Kaslow, D. C.; Giersing, B. K. WHO consultation on ETEC and Shigella burden of disease, Geneva, 6-7th April 2017: Meeting report. *Vaccine* **2018**in press.
259. Cross, A.; Artenstein, A.; Que, J.; Fredeking, T.; Furer, E.; Sadoff, J. C.; J. Cryz Jr, S. *J. Infect. Dis.* **1994**, *170*, 834–840.
260. Huttner, A.; Hatz, C.; van den Dobbelsteen, G.; Abbanat, D.; Hornacek, A.; Frölich, R.; Dreyer, A. M.; Martin, P.; Davies, T.; Fae, K.; van den Nieuwenhof, I.; Thoelen, S.; de Vallière, S.; Kuhn, A.; Bernasconi, E.; Viereck, V.; Kavvadias, T.; Kling, K.; Ryu, G.; Hülder, T.; Gröger, S.; Scheiner, D.; Alaimo, C.; Harbarth, S.; Poolman, J.; Fonck, V. G. *Lancet Infect. Dis.* **2017**, *17*, 528–537.

Chapter 3

Immunological Mechanisms of Glycoconjugate Vaccines

Jeremy A. Duke and Fikri Y. Avci*

Department of Biochemistry and Molecular Biology, Center for Molecular Medicine and Complex Carbohydrate Research Center, University of Georgia, Athens, Georgia 30602, United States
***E-mail: avci@uga.edu.**

On the surface of many infectious pathogens, a dense array of complex carbohydrate structures are presented. These glycans can be seen as possible targets for the body's immune system to recognize and elicit protection against infection. However, vaccination with a sole polysaccharide antigen does not confer long-lasting protective immunity, as most carbohydrates are T-cell independent antigens. Glycoconjugate vaccines, in which the polysaccharide is covalently attached to a protein, allows for activation of the adaptive immune system and recognition of the antigen in a T-cell dependent manner. In this chapter, an overview of carbohydrate vaccinology will be discussed, with an emphasis on T-cell immunity elicited by glycoconjugate vaccines.

Introduction

The first point of interaction between cells, viruses, and bacteria is a coating of carbohydrates composed of the many glycans, branching out from membrane-bound structures (*i.e.,* proteins, lipids, peptidoglycans). These carbohydrates serve a variety of functions, from working as a cellular receptor for cytokines and other signaling molecules in mammalian cells, to a providing a dense coating to mediate bacterial adhesion (*1*). The fact that glycans are ubiquitous, but still heterogeneous in terms of function, means the unique carbohydrate signature of certain bacteria, viruses, fungi, and even malformed mammalian cells, such as those of oncogenic or mutagenic origin, express a distinct glycan signature that can act as tentative markers for identity by the immune system (*2–5*). A drawback to this is the immune response induced by entirely carbohydrate antigens acts in a T cell-independent manner outside of special circumstances, which does not provide memory cells for induction of a booster response following re-exposure, and is not inducible in infants. This has since been overcome, with early accounts noting protective immunity can be elicited against these T cell-independent polysaccharide antigens when they are first covalently conjugated to an immunogenic protein (*6*).

The beginning of immunity lies in the mechanical barrier protecting one from infectious agents, for instance the human skin. A pathogen that has breached the physical barrier or has otherwise gained access to more sensitive epithelial tissues will then initiate the development of immune response in two parts: innate and adaptive immunity.

Innate immunity involves the initial resistance the body mounts in response to an invasive pathogen. Recognition events occur in a broad manner by macrophages and neutrophils that can recognize and bind conserved constituents on the surface of pathogens, triggering phagocytic induction and the release of chemokines and cytokines that act as the herald for the immune response. Their release triggers a cascade of events, notably the induction of inflammation and increase of blood and lymph flow to the spot of inflammation, further agitating the release of phagocytes to the site of infection for additional pathogen uptake and cytokine release. The cytokines allow the liaison between innate and adaptive immunity, the professional antigen presenting cells (APCs), to arrive and internalize the offensive microbial product for intracellular degradation and presentation to T lymphocytes. These cells, upon uptake of a pathogen, activate, maturing in the lymph. The activation catalyst, presentation of antigenic molecules on Major Histocompatibility Complex (MHC) molecules, acts as the trigger for T cell activation upon return to secondary lymph tissues by a recognition event between the mature APC and a T cell receptor (TCR) found on the naïve T cell. This process will be more thoroughly explained in context to glycoconjugates in the current chapter, with an introduction to antibody classes and class switching, followed by insights into carbohydrate antigen processing, presentation, and memory T cell formation.

Antibody Classes: Class Switching

Antibodies (Ab), or proteins in the Immunoglobulin family, occur in different classes and subtypes that are categorized on the basis of structural differences in their heavy chain domains, or more accurately the fragment crystallizable (Fc) region. Their production occurs through a special type of lymphocyte known as a B cell and are the keystone of immunity, with antibody matching to antigen to form a unique complex that initiates the destruction of the offending molecule. A thorough analysis of structure and function of all members of the immunoglobulin family is beyond the scope of this chapter and has been more thoroughly reviewed elsewhere, but a brief overview and the aspects important for carbohydrate antigen binding and vaccine efficacy will be covered herein (*7*).

There are five isotypes of antibodies: IgA, IgD, IgE, IgG, and IgM. In humans, both the IgA and IgG classes are further split into IgA1 and IgA2, and IgG1, IgG2, IgG3 and IgG4 subclasses, respectively, on a basis of differing biological function. In total, nearly a quarter of all circulating proteins in the plasma of any single individual is comprised of immunoglobulins of these various kinds, with an estimated diversity in binding specificities greater than ten million ($>10^7$) unique targets (*8*).

The IgA class is found throughout various bodily secretions and contains two subtypes, with IgA1 found predominantly in serum and IgA2 in mucosal membranes, such as through the gastrointestinal tract and tear ducts. IgA2 is known to be resistant to proteolytic degradation, taking residence and preventing the colonization of invasive pathogens. These work together to prevent infection through the entrances of the body. IgD immunoglobulins have not been thoroughly studied, but are known to occur in low levels on the surface of naïve B cells, with a putative role in fine-tuning specificity of the BCR, in concert with IgM, as well as stabilizing the complex (*9*). The IgE class has a known role in allergic response: binding to allergens, especially parasites, and triggering the release of histamine and most potent inflammatory response.

IgG is the most predominant class of antibody by far, expressed on the surface of plasma B cells and found circulating in serum. Importantly, IgG antibodies are the only class capable of transferring passive immunity from mother to placenta through a neonatal IgG receptor. While each subtype of IgG presents a distinct biological function, the IgG isotype majorly represents the predominant secondary immune response. IgG1 responses are normally the most abundant, reported as more than 50% of the total IgG class and is widespread against proteinaceous antigens. IgG2 is the prevalent immunoglobulin response against polysaccharide antigens, with nearly all bacterial carbohydrate antigens isolated to this class. The paradigm that carbohydrate-specific antibodies solely derived from the IgG2 subtype was maintained for many years, but this tenet has shifted more recently. A study on a *Streptococcus pneumoniae* serotype 6B glycoconjugate revealed a carbohydrate-recognition repertoire of immunoglobulins that contained more than the IgG2 subclass (*10*). More recently, the human anti-polysaccharide IgG population was evaluated by glycan array. These results revealed that of the 610 glycans immobilized, 104 targets could still be recognized in IgG2-depletion analyses, further indicating a wider carbohydrate-recognition repertoire (*11*).

IgG3 is a potent effector-mediating antibody, binding to the antigen before binding to the Fc receptor on dendritic cells for uptake and neutralization (*12*). IgG4 is the least populous of the Immunoglobulin G proteins, forming after long-term exposure and unable to elicit effector response, with their physiological role in immune response not greatly understood.

Finally, IgM is the largest class of antibody and earliest to express, frequently on the surface of B cells, with secretory IgM arranging as a dense pentamer through crosslinking by disulfide bonds. Primary immune response can be characterized by degree of IgM response, and their multimeric form allows for a very high avidity bond, especially against antigens of polymeric nature as is seen frequently in capsular polysaccharides.

In all immunoglobulins, biological activity is heavily supported by the presence of post-translational modifications on the antibodies themselves, namely glycosylation. In the context of immunology, Ab glycans have a known role in the interaction of the antibody with the Fc receptor on cell surfaces through which phagocytosis and cell-mediated cytotoxicity mechanisms occur (*13*). While much of this topic is still unknown, many different glycoforms of N-linked glycosylation has been shown to occur, and the lack of which is implicated in some autoimmune diseases (*14*, *15*).

The phenomenon of isotype class switching is what allows the immune response to more efficiently circulate and diversify effector function of generated immunoglobulins. When immunized with a purely carbohydrate antigen, such as a *Neisseria meningitidis* capsular polysaccharide, an almost exclusively IgM response is seen (*16*). While this does elicit short-term protection, there is no memory effect or booster response, and this is not feasible for conferring immunity to infants (*17*). However, through the chemical attachment to an immunogenic protein, B cell responses can then present epitopes generated by antigen-processing and harness helper T cells (*18*). This interaction results in the secretion of critical stimulatory molecules that drive the switch in production of avidity-driven IgM, to high affinity IgG specific to the carbohydrate antigen that has been presented. This is due to the fact the variable region of the antibody is not what is being altered. Instead, a process of genetic alteration occurs to modify the heavy chain of newly produced antibodies, resulting in a differentiation of effector function. This is performed by the enzyme Activation-Induced cytidine Deaminase (AID), which is expressed by B cells after maturation (*19*). AID initiates DNA double strand breaks along the Ig locus in areas corresponding to the heavy chain, triggering the class switching event by inserting the heavy chain regions corresponding to the framework of an alternate isotype (*20*). These newly expressed IgG can now circulate and offer protective immunity against the epitope they have been encoded to recognize. In the case of a glycoconjugate, it is the polysaccharide that was conjugated.

MHC Presentation

Antigens that are processed by the immune system subscribe to one of two systems within the adaptive immune response, dependent upon the origin of the antigen and the manner in which the degraded antigen is presented. Two classes of the cell surface MHC molecules, MHC-I and MHC-II, are responsible for the presentation of antigens, and each subscribes to its own branch of differentiated lymphocyte. In the case of MHC-I, recognition occurs through Cluster of Differentiation 8 (CD8+) expressing T cells, whereas MHC-II subscribes to the use of Cluster of Differentiation 4 (CD4+) T cells. A further parallel between the two pathways is if the antigen being loaded is of a cytosolic/endogenous origin as per MHC-I, or of exogenous nature as seen in MHC-II presentation (*21*).

MHC-I

Class 1 MHC processing and presentation harnesses molecules that are endogenous to the cell and have been marked for degradation by the proteasome through ubiquitination on a lysine residue (*22*). Examples of endogenous proteins include misfolded and those marked for turnover, but can include viral proteins in the case of an infected cell as well. The proteasome will degrade the protein into peptides of about nine residues in length, which will then undergo translocation to the ER by the Transporter associated with Antigen Processing (TAP), a heterodimer composed of the proteins TAP-1 and TAP-2 (*23*). One of the two ER associated aminopeptidases (ERAP-1 or ERAP-2) will trim the peptide to a length usable by the MHC-I if this is an issue (*24*). TAP is involved in a peptide loading complex (PLC) that involves multiple parts; tapasin is a lumenal ER membrane glycoprotein that acts as a bridge between TAP and MHC-I through a supposed concert action with ERp57, and calreticulin assists in tapasin stability as well as the PLC stability as a whole, through interaction with the conserved N-linked glycosylation sites found on tapasin and MHC-I molecules respectively (*25–27*). PLC binds the peptide onto the MHC-I molecule, which is then transported from the ER to the membrane and ultimately the surface of the APC.

Recognition of the MHC-I presenting APC is performed by the CD8+ T cell, by interaction of the TCR. This activates the CD8+ T cell, becoming a CD8+ cytotoxic T cell, with the capacity to now recognize the epitope presented by MHC-I and destroy the cell by release of effector proteins contained within a lytic granule (*28*). These effector proteins include perforin, which opens pores in the infected cell and upsets osmolarity and allows entry of special cytosolic serine proteases known as granzymes. It is not believed that the MHC-I pathway plays a role in current glycoconjugates, although a previous study has illustrated the capacity of MHC-I to present a novel glycoconjugate (*29*). Moreover, various natural and synthetic glycopeptides have been shown to be presented by MHC-I pathway for cytotoxic T cell recognition (*30–33*).

MHC-II

Class 2 MHC processing and presentation instead uses molecules that are exogenous and have been introduced to the cell by a phagocytic event. B cells are produced in the bone marrow, derived from hematopoietic stem cells. Upon exit of the primary lymph tissue, they begin to express IgM and to a lesser degree IgD molecules that will arrange on the surface of the B cell to act as the B Cell Receptor (BCR). B cells then patrol through the circulatory system, attempting to bind offending residues that will be endocytosed into the cell. Once internalized, low pH vesicles containing proteases, phosphatases, and hydrolases fuse to form an endosome that works to degrade the internalized molecule into fragments that can be loaded onto the MHC-II molecule (*34*). The binding of antigen to BCR stimulates the production of CCR7 and chemotactic homing to secondary lymph tissue as the maturation process of MHC+antigen complex is presented on the surface of the B cell (*35*). The activated B cell, referred to as a follicular B cell and now presenting the epitope on its MHC-II molecule, is susceptible to recognition by CD4+ follicular helper T cells (T_{FH}) that contain TCR specific to the antigenic epitope presented. Engagement of other costimulatory factors, including the T cell expressed CD28 and CD40 ligand, which will interact with the B cell expressed CD80/86 and CD40, respectively, further activates the B cell (*36*, *37*). The released cytokines Interleukin 2 and 4 (IL-2 and IL-4) act in this context as a positive feedback loop for TFH development and B cell differentiation (*38*).

A subset of the B cells can begin to secrete antibodies and will cause the dynamic formation of a germinal center in the secondary lymph, where the remaining B cells will move to for further differentiation. The germinal center is divided into two sites: the dark zone and the light zone, which were named on a basis of histological staining patterns (*39*). The dark zone is where three major events: B cell proliferation, IgM-to-IgG class switching, and somatic hypermutation of the genes in the variable region of immunoglobulin take place. The introduction of the somatic hypermutations is part of a process known as affinity maturation, which allows for a selected repertoire of B cells to develop a higher affinity IgG against the antigen.

Affinity maturation is a lengthy process in which the body introduces variation within the binding region of the immunoglobulins of the B cell, with the ultimate goal being to find a set of mutations that allows for a more efficacious antibody (Figure 1). Within the light zone, surface immunoglobulin is produced and iterative selection steps with T_{FH} occur to ensure antigen binding is still potent. A specialized follicular dendritic cell will display native antigen allowing B cells to compete, with only those with high affinity binding capable of acquiring antigen for presentation to T_{FH} (*40*). Most of the cells generated do not introduce beneficial mutations and, thus, are fated for apoptosis upon failure to tightly bind the presented antigen (*41*). Surviving B cells that have exhibited high affinity binding and survive the selection will then continue differentiating to form the IgG antibody secreting plasma cells or the long-lived memory cells that have the epigenetic coding to rapidly differentiate to antibody secreting cells upon re-exposure of the antigen.

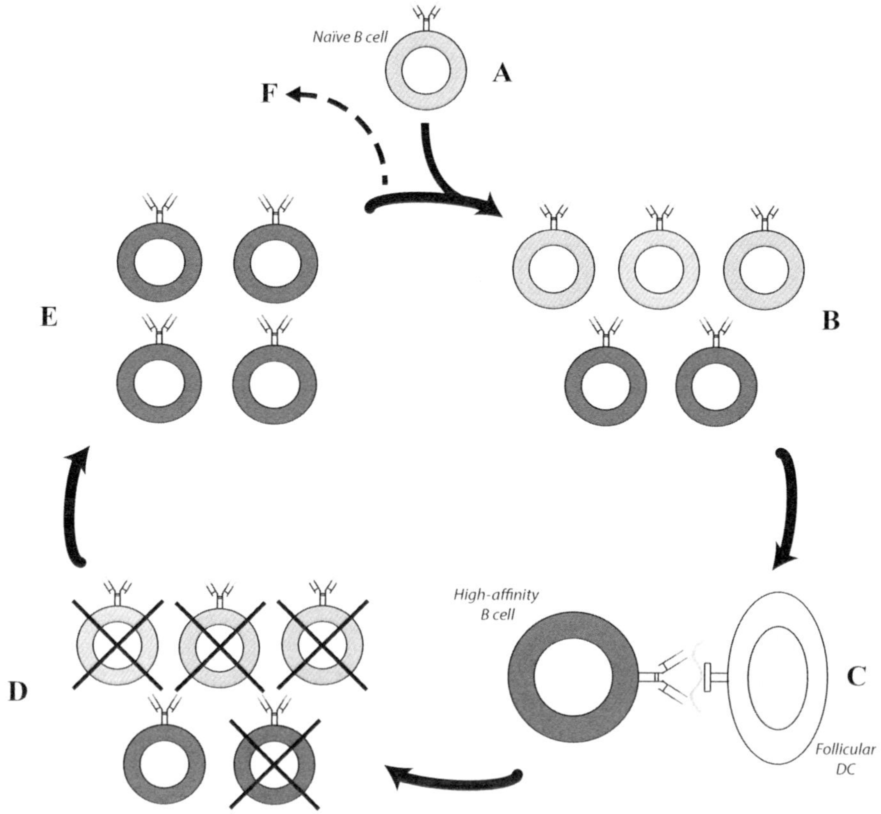

Figure 1. Affinity Maturation as it occurs in the germinal center. (A) Initiating B cells with affinity to the target arrive at the germinal center. (B) The B cells undergo clonal expansion with somatic hypermutation within the antigen recognizing portions of the immunoglobulin in order to diversify binding capability. (C) Selection against the target. Follicular dendritic cells present the antigen, allowing only the tight binding B cells to outcompete and acquire antigen for presentation to T_{FH} cells. (D) Most mutations do not generate a more efficient antibody and are removed by apoptosis after failure to bind antigen on follicular dendritic cells. (E) The B cells with most potent binding survive and proliferate. This pool undergoes iterative mutation and selection pressures until high-affinity exhibiting B cells are produced. (F) After the affinity maturation process has undergone multiple rounds, the winners of selection move on to become the antibody-secreting plasma cell, or memory cells.

Zwitterionic Polysaccharides

In the case of nearly all complex carbohydrates, presentation by MHC is not possible due to incapacity to bind to the cleft of MHC without attachment to a protein. However, one particular class of polysaccharide, the Zwitterionic Polysaccharides (ZPSs), have been shown to bind to MHC-II and be presented by APCs due to the conserved motif of alternating positive and negative residues within each repeat unit, allowing for electrostatic interactions with MHC-II that are not possible with most other polysaccharides. This ZPS motif is known to occur in multiple pathogens, as the polysaccharide A of *Bacteroides fragilis* and type 1 serotype of *Streptococcus pneumoniae* two examples of such (*42*, *43*). The ability for T cell activation with this class of polysaccharide has been thoroughly explored, showing it is in fact MHC-II presentation and CD4+ T cells acting as the immunological mechanism (*44–47*). The use of ZPSs as a potential carrier in an entirely carbohydrate glycoconjugate vaccine is an idea that has gained traction with the synthesis of the ZPS bound tumor associated carbohydrate antigen Sialyl-Tn (*48*). In an investigation by Shi et al., high levels of antigen specific IgGs were generated in a T cell-dependent manner, with *in vitro* tumor cell specific recognition and cytolysis responses seen (*49*).

Glycoconjugate Processing

In the case of most polysaccharides, the ZPS motif does not occur and secreted B cells do not receive any sort of antigen-directed T cell assistance after induction of the polysaccharide by BCR. Instead, plasma cell differentiation occurs without affinity maturation, which results in limited amounts of low affinity immunoglobulin response consisting almost entirely of IgM, and there are no memory B cells that arise to provide protection upon re-exposure (*50*). However, by covalently attaching the T cell-independent responding polysaccharide onto a known proteinaceous immunogen, T cell help can be promoted against the polysaccharide in a phenomenon that was first reported in 1929 (Figure 2) (*6*). These glycoconjugates enable isotype class switching and memory B cell formation, and have seen enormous biomedical interest. The first licensed glycoconjugate vaccine was introduced to the market in 1987 against the pathogen *Haemophilus influenza* and saw great success (*51*, *52*). Since then, glycoconjugates have played a major role in lowering numbers of infection to other polysaccharide encapsulated bacteria including *Streptococcus pneumoniae* and *Neisseria meningitidis*. In these conjugate vaccines, the MHC-II pathway is harnessed in a mechanism that allows for the generation of now carbohydrate-specific CD4+ T cells, Tcarbs (*53–56*).

Initial hypotheses attributed peptide presentation to the immunomodulating abilities of glycoconjugate vaccines (*18*). However, the use of different carrier proteins between primary and secondary vaccination still saw an antigen-primed booster response, suggesting a carbohydrate recognizing population of memory cells only possible through presentation to T cells (*47*, *53*, *57*). Upon recognition of the glycoconjugate by the BCR, endocytosis and endosomal degradation occurs. Within the endosome, processing of the proteinaceous features can occur by proteases and hydrolases, but due to the lack of glycosidases, polysaccharide trimming is believed to occur through the generation of reactive oxygen and/or nitrogen species (ROS/RNS) (*58*, *59*). The strong covalent bond between protein and carbohydrate is unlikely to be broken in the endosome, forming glycopeptides. Thus, when the MHC-II complex is loaded with peptide, it has a trimmed polysaccharide component still covalently attached, allowing for the presentation of a carbohydrate to T cells. This has been proven through microscopy studies in which the fate of the glycoconjugate is followed through the antigen processing pathway, with the resultant glycopeptide colocalizing with MHC-II, whereas unconjugated polysaccharide does not (*56*). As described above, the MHC-II/glycopeptide complex interacts with the TCR, in concert with the CD40 ligand (CD40L) and CD28 of T cells associating with CD40 and CD80 or CD86 of B cells, respectively. The interaction of CD80/CD86 with CD28 results in the secretion of the cytokine IL-2 by the T cell, sending a signal for T cell proliferation by positive feedback with its IL-2 receptor. B cell proliferation occurs synchronously, as CD40 binding to its ligand stimulates the release of IL-4 by the T cell, which interacts on the IL-4 receptor on B cells. These proliferation events are what drive the formation of a germinal center, affinity maturation, and ultimately the formation of memory cells against the antigen. The presence and isolation of Tcarbs has been demonstrated against the type III polysaccharide of both group B *Streptococcus* as well as the type 3 serotype of *Streptococcus pneumoniae* (*53–55*). Additional studies have suggested the presence of Tcarbs, in recognition events against serotypes 3 and 14 of *S. pneumoniae* by way of virus-like particle glycopeptide presentation, as well as a group C meningococcal glycoconjugate (*60, 61*).

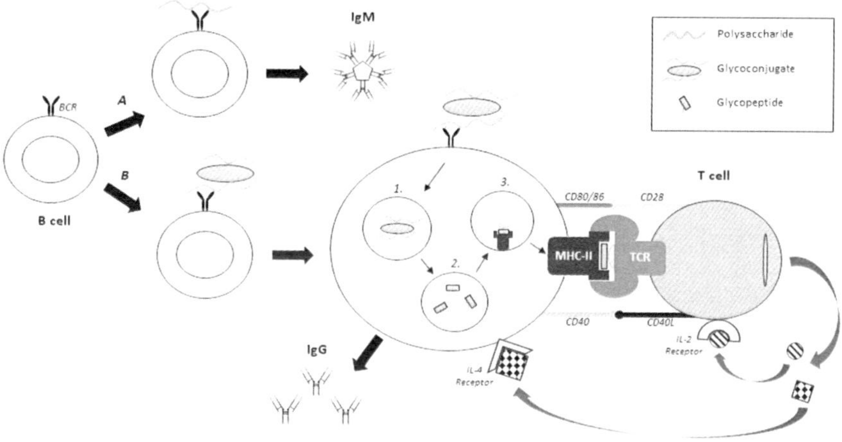

Figure 2. Mechanism of immune activation by carbohydrate vaccines. (A) In purely polysaccharide containing vaccines, B cell uptake is not presented to T cells, thus immune response is limited to an entirely IgM secreting response. (B) Glycoconjugate vaccination allows for T cell assistance, starting with the carbohydrate component binding to the BCR. 1. The glycoconjugate is endocytosed into an endolysosome 2. The protein portion of the glycoconjugate is degraded by hydrolases and proteases, whereas reactive oxygen and/or nitrogen species degraded the polysaccharide to form glycopeptides. 3. These glycopeptides are now able to be bound to MHC-II, where the glycopeptide-MHC complex will travel to the surface and become presented to CD4+ T cells. T cell activation by MHC-II, along with co-stimulation by CD molecules releases cytokines that induce B cell maturation and resultant anti-carbohydrate IgG formation.

Methods of Analysis

Vaccine efficiency in terms of protective capacity is quantified by analyzing correlates of protection for the pathogen being immunized against (*62*). The standard correlate in glycoconjugate vaccines are the antibodies that are generated against bacterial polysaccharide, ideally by their opsonophagocytic capabilities, but binding capacity can be used as an alternative (*63*). The Opsonophagocytic killing assay (OPA) or Serum Bactericidal assay (SBA), act as measures to determine if the antibodies produced through vaccination are of a biologically relevant nature (*64*, *65*). Are they robust enough to elicit complement-mediated destruction of the pathogen? This requires an understanding of the synergy between the classic complement system and immunoglobulins in the ability to eliminate pathogens. IgGs that are bound to antigen will interact, through their Fc portion, with what is known as the C1 complex of the complement. The C1 complex is a large molecule comprised of six C1q proteins, which act as the receptor for the IgG Fc, and two serine proteases, C1s and C1r. Upon activation by the IgG, C1s can release and work downstream on the proteins C2 and C4, cleaving them into their biologically active form as the noncovalently linked C3 convertase. This complex will then cleave an additional complement protein, C3, into C3a and C3b. While C3a acts as an anaphylatoxin and signals for chemokine and phagocyte recruitment, C3b will bind to the surface of the antigen and tag it for destruction, in what is known as opsonization. Phagocytes use receptors to recognize these C3b tagged antigens and ingest them, thus inducing destruction by opsanophagocytosis (*66*). Unencapsulated pathogens are much more susceptible to opsanophagocytosis. The problem originates in capsular polysaccharides acting as a virulence factor, thus deterring immune recognition and complement activation (*67*, *68*). Without immune memory to the particular capsular polysaccharides, a poor immunoglobulin recognition response and limited C3b deposition is seen.

OPA or SBA is performed by first incubating various dilutions of serum that contains the putative antibodies of interest with the pathogen that they were generated against (for instance, a particular serotype of a bacteria). Afterwards, effector cells and complement are added to the serum-pathogen mixture to allow for protective killing to occur. Samples are then cultured to determine a measure of activity as compared to non-serum containing controls by way of colonization (*69–72*).

While serum-mediated protection is the current gold standard in determining efficacy of a vaccine, as it allows *in vitro* responses emulous to an *in vivo* system, Enzyme-Linked Immunosorbent Assay (ELISA) can be used to verify antibody-target binding (*63*). In ELISA, either the pathogen or the structure in which vaccination has been directed against is used. The antigen is immobilized and incubated with serum containing the antibodies, allowing them a chance to bind. Non-binding immunoglobulins are washed off, and a secondary antibody that recognizes the structural framework of the first is introduced. After rinsing these non-binders away, a read-out of the amount of immunoglobulins contained within the sera can give an idea for immune response.

Conclusion

For many years, glycoconjugate vaccine formation was an unsystematic process of attaching immune-eliciting protein to the carbohydrate one wished to generate a response against. Various issues have been shown to arise in glycoconjugate vaccine development: batch variation, reproducibility, and heterogeneous presentation due to ill-defined chemistries (*73–75*). By understanding the mechanism and process glycoconjugate vaccines undergo to elicit an immune response, a forward-thinking approach to future vaccines can begin. One such example for knowledge-based vaccine design and synthesis through exploitation of antigen processing and presentation mechanisms elicited by glycoconjugate vaccines came from one of our studies, in which a glycoconjugate vaccine designed to enrich for MHC-II glycopeptide epitopes elicited up to two orders of magnitude higher IgG titers compared to a traditional glycoconjugate preparation (*53*).

By marrying the mechanistic insights on glycoconjugate immunology with more recent conjugation technologies, the field of glycoimmunology can overcome problems previously associated with glycoconjugates. One such example uses strings of MHC-II presenting glycan-coupled peptide epitopes to present four different serotypes of *Neisseria meningitidis*, with results of immunization outperforming comparable carrier protein vaccinations (*76*). Another harnesses the bacterial machinery of *E. coli* to attach N-linked glycans to protein, or *E. coli* itself, as a glycoconjugate (*73*, *77*, *78*). Several advances have been made such that one can envision application in the development of future glycoconjugates: from site-selective click chemistries, in which azido-modified sugars are incorporated onto organisms and react, to *in vitro* glycoengineering, where the cellular glycosylation machinery can build glycans on a cell-free substrate (*79–81*). With this in mind, a shift from qualitative to a quantitative and knowledge-based direction for future glycoconjugate vaccines can be expected.

Acknowledgments

We thank Ms. Cathryn Quinn for her excellent editorial assistance. This work was supported by funding from the National Institute of Health grant: 1R01AI123383.

References

1. Ohtsubo, K.; Marth, J. D. *Cell* **2006**, 855–867.
2. Comstock, L. E.; Kasper, D. L. *Cell* **2006**, 847–850.
3. Takahashi, M.; Kizuka, Y.; Ohtsubo, K.; Gu, J.; Taniguchi, N. *Mol. Aspects Med.* **2016**, 56–70.
4. An, H. J.; Kronewitter, S. R.; de Leoz, M. L.; Lebrilla, C. B. *Curr. Opin. Chem. Biol.* **2009**, 601–607.
5. Raman, R.; Tharakaraman, K.; Sasisekharan, V.; Sasisekharan, R. *Curr. Opin. Struct. Biol.* **2016**, 153–162.
6. Avery, O. T.; Goebel, W. F. *J. Exp. Med.* **1929**, 533–550.

7. Schroeder, H. W.; Cavacini, L. *J. Allergy Clin. Immunol.* **2010**, S41–S52.
8. Alberts, B.; Johnson, A.; Lewis, J.; Raff, M.; Roberts, K.; Walter, P. *Molecular Biology of the Cell*, 4th ed.; Garland Science: New York, 2002.
9. Geisberger, R.; Lamers, M.; Achatz, G. *Immunology* **2006**, 429–437.
10. Vidarsson, G.; Sigurdardottir, S. T.; Gudnason, T.; Kjartansson, S.; Kristinsson, K. G.; Ingolfsdottir, G.; Jonsson, S.; Valdimarsson, H.; Schiffman, G.; Schneerson, R.; I, J. *Infect. Immun.* **1998**, 2866–2870.
11. Schneider, C.; Smith, D. F.; Cummings, R. D.; Boligan, K. F.; Hamilton, R. G.; Bochner, B. S.; Miescher, S.; Simon, H. U.; Pashov, A.; Vassilev, T.; von Gunten, S. *Sci. Transl. Med.* **2015**, 269ra261–269ra261.
12. Scharf, O.; Golding, H.; King, L. R.; Eller, N.; Frazier, D.; Golding, B.; Scott, D. E. *J. Virol.* **2001**, 6558–6565.
13. Krapp, S.; Mimura, Y.; Jefferis, R.; Huber, R.; Sondermann, P. *J. Mol. Biol.* **2003**, 979–989.
14. Seeling, M.; Bruckner, C.; Nimmerjahn, F. *Nat. Rev. Rheumatol.* **2017**, 621–630.
15. Anthony, R. M.; Wermeling, F.; Ravetch, J. V. *Ann. N.Y. Acad. Sci.* **2012**, 170–180.
16. Andersen, J.; Berthelsen, L.; Lind, I. *Clin. Diagn. Lab. Immunol.* **1997**, 345–351.
17. Rijkers, G.; Sanders, E. A. M.; Breukels, M. A.; Zegers, B. J. M. *Vaccine* **1998**, 1396–1400.
18. Avci, F. Y.; Kasper, D. L. *Annu. Rev. Immunol.* **2009**, 107–130.
19. Kumar, R.; DiMenna, L.; Chaudhuri, J.; Evans, T. *Biomed. J.* **2014**, 269–269.
20. Stavnezer, J.; Schrader, C. E. *J. Immunol.* **2014**, 5370–5378.
21. Rock, K. L.; Reits, E.; Neefjes, J. *Trends Immunol.* **2016**, 724–737.
22. Lecker, S. H.; Goldberg, A. L.; Mitch, W. E. *J. Am. Soc. Nephrol.* **2006**, 1807–1819.
23. Spiliotis, E. T.; Osorio, M.; Zuniga, M. C.; Edidin, M. *Immunity* **2000**, 841–851.
24. York, I. A.; Brehm, M. A.; Zendzian, S.; Towne, C. F.; Rock, K. L. *Proc. Natl. Acad. Sci. U.S.A.* **2006**, 9202–9207.
25. Blees, A.; Januliene, D.; Hofmann, T.; Koller, N.; Schmidt, C.; Trowitzsch, S.; Moeller, A.; Tampe, R. *Nature* **2017**, 525–528.
26. Williams, A. P.; Peh, C. A.; Purcell, A. W.; McCluskey, J.; Elliott, T. *Immunity* **2002**, 509–520.
27. Zhang, Y. N.; Baig, E.; Williams, D. B. *J. Biol. Chem.* **2006**, 14622–14631.
28. Peters, P. J.; Borst, J.; Oorschot, V.; Fukuda, M.; Krahenbuhl, O.; Tschopp, J.; Slot, J. W.; Geuze, H. J. *J. Exp. Med.* **1991**, 1099–1109.
29. Singh, S. K.; Streng-Ouwehand, I.; Litjens, M.; Kalay, H.; Burgdorf, S.; Saeland, E.; Kurts, C.; Unger, W. W.; van Kooyk, Y. *Eur. J. Immunol.* **2011**, 916–925.
30. Deck, M. B.; Sjölin, P.; Unanue, E. R.; Kihlberg, J. *J. Immunol.* **1999**, 4740–4744.
31. Speir, J. A.; Abdel-Motal, U. M.; Jondal, M.; Wilson, I. A. *Immunity* **1999**, 51–61.
32. Apostolopoulos, V.; Yuriev, E.; Ramsland, P. A.; Halton, J.; Osinski, C.; Li, W.; Plebanski, M.; Paulsen, H.; McKenzie, I. F. *Proc. Natl. Acad. Sci. U.S.A.* **2003**, 15029–15034.

33. Haurum, J. S.; Høier, I. B.; Arsequell, G.; Neisig, A.; Valencia, G.; Zeuthen, J.; Neefjes, J.; Elliott, T. *J. Exp. Med.* **1999**, 145–150.
34. West, M. A.; Lucocq, J. M.; Watts, C. *Nature* **1994**, 147–151.
35. Dijkstra, I. M.; de Haas, A. H.; Brouwer, N.; Boddeke, H. W. G. M.; Biber, K. *Glia* **2006**, 861–872.
36. Linsley, P. S.; Clark, E. A.; Ledbetter, J. A. *Proc. Natl. Acad. Sci. U.S.A.* **1990**, 5031–5035.
37. Noelle, R. J.; Ledbetter, J. A.; Aruffo, A. *Immunol. Today* **1992**, 431–433.
38. Valle, A.; Aubry, J. P.; Durand, I.; Banchereau, J. *Int. Immunol.* **1991**, 229–235.
39. Thorbecke, G. J.; Amin, A. R.; Tsiagbe, V. K. *FASEB J.* **1994**, 832–840.
40. Oropallo, M. A.; Cerutti, A. *Trends Immunol.* **2014**, 287–289.
41. Tarlinton, D. M.; Smith, K. G. *Immunol. Today* **2000**, 436–441.
42. Lindberg, B.; Lindqvist, B.; Lonngren, J.; Powell, D. A. *Carbohydr. Res.* **1980**, 111–117.
43. Baumann, H.; Tzianabos, A. O.; Brisson, J. R.; Kasper, D. L.; Jennings, H. J. *Biochemistry* **1992**, 4081–4089.
44. Kalka-Moll, W. M.; Tzianabos, A. O.; Bryant, P. W.; Niemeyer, M.; Ploegh, H. L.; Kasper, D. L. *J. Immunol.* **2002**, 6149–6153.
45. Johnson, J. L.; Jones, M. B.; Cobb, B. A. *J. Biol. Chem.* **2015**, 5007–5014.
46. Cobb, B. A.; Wang, Q.; Tzianabos, A. O.; Kasper, D. L. *Cell* **2004**, 677–687.
47. Avci, F. Y.; Li, X. M.; Tsuji, M.; Kasper, D. L. *Semin. Immunol.* **2013**, 146–151.
48. De Silva, R. A.; Wang, Q.; Chidley, T.; Appulage, D. K.; Andreana, P. R. *J. Am. Chem. Soc.* **2009**, 9622–9623.
49. Shi, M.; Kleski, K. A.; Trabbic, K. R.; Bourgault, J.-P.; Andreana, P. R. *J. Am. Chem. Soc.* **2016**, 14264–14272.
50. Stein, K. E. *J. Infect. Dis.* **1992**, S49–S52.
51. Eskola, J.; Kayhty, H.; Takala, A. K.; Peltola, H.; Ronnberg, P. R.; Kela, E.; Pekkanen, E.; Mcverry, P. H.; Makela, P. H. *New Engl. J. Med.* **1990**, 1381–1387.
52. Schneerson, R.; Barrera, O.; Sutton, A.; Robbins, J. B. *J. Exp. Med.* **1980**, 361–376.
53. Avci, F. Y.; Li, X. M.; Tsuji, M.; Kasper, D. L. *Nat. Med.* **2011**, 1602–1609.
54. Avci, F. Y.; Li, X. M.; Tsuji, M.; Kasper, D. L. *Nat. Protoc.* **2012**, 2180–2192.
55. Middleton, D. R.; Sun, L.; Paschall, A. V.; Avci, F. Y. *J. Immunol.* **2017**, 598–603.
56. Lai, Z. Z.; Schreiber, J. R. *Vaccine* **2009**, 3137–3144.
57. Greenberg, D. P.; Lieberman, J. M.; Marcy, S. M.; Wong, V. K.; Partridge, S.; Chang, S. J.; Chiu, C. Y.; Ward, J. I. *J. Pediatr.* **1995**, 206–211.
58. Duan, J. Y.; Kasper, D. L. *Glycobiology* **2011**, 401–409.
59. Duan, J. Y.; Avci, F. Y.; Kasper, D. L. *Proc. Natl. Acad. Sci. U.S.A.* **2008**, 5183–5188.
60. Polonskaya, Z.; Deng, S.; Sarkar, A.; Kain, L.; Comellas-Aragones, M.; McKay, C. S.; Kaczanowska, K.; Holt, M.; McBride, R.; Palomo, V.; Self, K. M.; Taylor, S.; Irimia, A.; Mehta, S. R.; Dan, J. M.; Brigger, M.; Crotty, S.; Schoenberger, S. P.; Paulson, J. C.; Wilson, I. A.; Savage, P. B.; Finn, M. G.; Teyton, L. *J. Clin. Invest.* **2017**, 1491–1504.
61. Muthukkumar, S.; Stein, K. E. *Vaccine* **2004**, 1290–1299.
62. Plotkin, S. A. *Clin. Vaccine Immunol.* **2010**, 1055–1065.

63. Vernacchio, L.; Romero-Steiner, S.; Martinez, J. E.; MacDonald, K.; Barnard, S.; Pilishvili, T.; Carlone, G. M.; Ambrosino, D. M.; Molrine, D. C. *J. Infect. Dis.* **2000**, 1162–1166.
64. Romero-Steiner, S.; Libutti, D.; Pais, L. B.; Dykes, J.; Anderson, P.; Whitin, J. C.; Keyserling, H. L.; Carlone, G. M. *Clin. Diagn. Lab. Immunol.* **1997**, 415–422.
65. Stratton, C. W. *Clin. Microbiol. Rev.* **1988**, 19–26.
66. Merle, N. S.; Church, S. E.; Fremeaux-Bacchi, V.; Roumenina, L. T. *Front. Immunol.* **2015**, 262.
67. Hyams, C.; Camberlein, E.; Cohen, J. M.; Bax, K.; Brown, J. S. *Infect. Immun.* **2010**, 704–715.
68. Nanra, J. S.; Buitrago, S. M.; Crawford, S.; Ng, J.; Fink, P. S.; Hawkins, J.; Scully, I. L.; McNeil, L. K.; Aste-Amezaga, J. M.; Cooper, D.; Jansen, K. U.; Anderson, A. S. *Hum. Vaccin. Immunother.* **2013**, 480–487.
69. Burton, R. L.; Nahm, M. H. *Clin. Vaccine Immunol.* **2006**, 1004–1009.
70. Romero-Steiner, S.; Frasch, C. E.; Carlone, G.; Fleck, R. A.; Goldblatt, D.; Nahm, M. H. *Clin. Vaccine Immunol.* **2006**, 165–169.
71. McIntosh, E.; Bröker, M.; Wassil, J.; Welsch, J.; Borrow, R. *Vaccine* **2015**, 4414–4421.
72. Bash, M. C.; Lynn, F.; Mocca, B.; Borrow, R.; Findlow, H.; Hassan-King, M.; Preziosi, M.-P.; Idoko, O.; Sow, S.; Kulkarni, P. *Clin. Vaccine Immunol.* **2014**, 755–761.
73. Feldman, M. F.; Wacker, M.; Hernandez, M.; Hitchen, P. G.; Marolda, C. L.; Kowarik, M.; Morris, H. R.; Dell, A.; Valvano, M. A.; Aebi, M. *Proc. Natl. Acad. Sci. U.S.A.* **2005**, 3016–3021.
74. Seppälä, I.; Mäkelä, O. *J. Immunol.* **1989**, 1259–1264.
75. Carmenate, T.; Canaán, L.; Álvarez, A.; Delgado, M.; González, S.; Menéndez, T.; Rodés, L.; Guillén, G. *FEMS Immunol. Med. Microbiol.* **2004**, 193–199.
76. Baraldo, K.; Mori, E.; Bartoloni, A.; Norelli, F.; Grandi, G.; Rappuoli, R.; Finco, O.; Del Giudice, G. *Infect. Immun.* **2005**, 5835–5841.
77. Nothaft, H.; Davis, B.; Lock, Y. Y.; Perez-Munoz, M. E.; Vinogradov, E.; Walter, J.; Coros, C.; Szymanski, C. M. *Sci. Rep.* **2016**, 26511.
78. Wacker, M.; Wang, L.; Kowarik, M.; Dowd, M.; Lipowsky, G.; Faridmoayer, A.; Shields, K.; Park, S.; Alaimo, C.; Kelley, K. A. *J. Infect. Dis.* **2013**, 1551–1561.
79. Schoborg, J. A.; Hershewe, J. M.; Stark, J. C.; Kightlinger, W.; Kath, J. E.; Jaroentomeechai, T.; Natarajan, A.; DeLisa, M. P.; Jewett, M. C. *Biotechnol. Bioeng.* **2017**, 739–750.
80. Saxon, E.; Bertozzi, C. R. *Science* **2000**, 2007–2010.
81. Hong, V.; Steinmetz, N. F.; Manchester, M.; Finn, M. *Bioconjugate Chem.* **2010**, 1912–1916.

Chapter 4

Design and Development of Glycoconjugate Vaccines

A. Krishna Prasad,* Jin-hwan Kim, and Jianxin Gu

Pfizer Vaccines Research and Development, 401 N. Middletown Rd., Pearl River, New York 10965, United States
***E-mail: krishna.prasad1@pfizer.com.**

An optimal glyconjugate vaccine stimulates a potent and specific immune response to a saccharide antigen, while irreversibly bound to the carrier protein. In order to elicit a robust and potent immune response, it is important that the critical immunogenic epitopes in the saccharide antigen, are preserved. In addition, the overall conjugate structure should be of an optimal size and the carrier protein T-cell epitopes are also largely preserved. As a result of the need to maintain these functional and structural requirements, several key elements need to be considered during the design and development of conjugate vaccines. These elements include (a) preservation of critical immunogenic saccharide epitopes, including the key non-saccharide substituents; length; (b) minimal antigen modification; (c) optimal chain (d) efficient conjugation chemistry; (e) selection of an effective and well tolerated carrier protein; and (f) stability of the conjugate in the drug substance as well as drug product. In this chapter, we review the key design features necessary to develop the "optimal conjugate vaccine construct" and the "toolbox" to achieve these ends.

Introduction

The first demonstration of the basic principle that polysaccharide-protein conjugates have the ability to increase the immunogenicity of bacterial capsular polysaccharides was provided by Goebel and Avery in 1929. The field has been further expanded considerably with the introduction of a number of glycoconjugate vaccines (*1–3*). The clinical success based on the work related to the monovalent *Haemophilus infleuenzae* type b (Hib) polysaccharide conjugate vaccines (*4–6*) paved the way for the subsequent development of several monovalent and multivalent glycoconjugate vaccines. These vaccines have been developed primarily to target other bacterial diseases using pneumococcal, meningococcal, staphylococcal and group B streptococcal antigens. The pioneering studies on Hib oligosaccharide-CRM_{197} conjugate (HbOC) allowed the selection of a suitable conjugate design, which resulted in the demonstration of safety and immunogenicity in adults and infants.

The selection of an appropriate chemistry route to produce a glycoconjugate from a specific saccharide antigen and carrier protein, however, fulfils only the first minimal step towards the generation of an "optimal glycoconjugate vaccine construct". In order to produce a commercially viable vaccine suitable for subsequent licensure, many design elements need to be considered. This, in turn, involves a significant research and development effort. The early development studies should take into consideration several factors such as stability, consistency in quality and manufacturability, from the very outset. Several of these key elements that play a key role during the design and development of an "optimal conjugate construct" will be discussed in the following sections.

Even though a number of routes are available for choosing the conjugation chemistry, to bind the saccharide antigen to the carrier protein, the unique structural features of the polysaccharide often dictate the choice. There are several key elements, spanning multiple disciplines, that need to be considered during the design of an "optimal construct." A customized design approach for each saccharide antigen, based on the specific structural requirement, is particularly important for the development of multi-valent vaccines. It is important that the immunogenic epitopes are available in abundance, without interference from other constituent serotypes, for successful antigen-antibody interactions in the complex mixtures of multi-valent vaccines. A common "platform chemistry and process" approach, uniformly applicable to all the constituent serotypes, has the advantage of simplifying development effort and manufacturing process. However, this could potentially undermine the specific structural needs, such as stability and/or immunogenicity, for some selected constituent antigens. The unique design and development needs of glycoconjugate vaccines, due to specific structural and stability requirements sets them apart from other purely protein-based vaccines. The "one size fits all" platform process approach for all saccharide antigens, for example in a multivalent conjugate vaccine, without customization that takes into account the specific structural features of each constituent, is likely to pose significant challenges. These challenges include quality, consistency and stability in order to generate a robust Chemistry, Manufacturing and Control (CMC) package important to develope a commercially viable process and product, in particular.

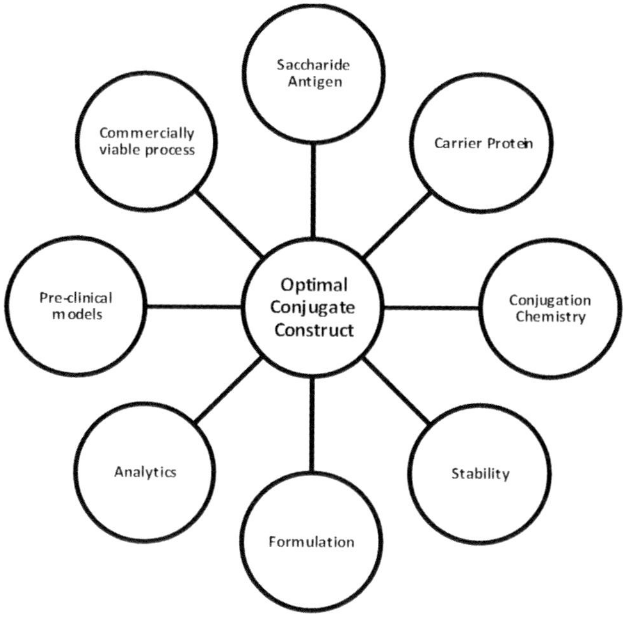

Figure 1. Design of an optimal conjugate vaccine construct: key elements.

The "multi-disciplinary constellation" of these, often interdependent, key control elements that have significant impact towards the design and development of an optimal conjugate construct, is illustrated in Figure 1. Even though there are several choices available for the conjugation chemistry, it is important to select a route such that (a) the immune response is directed preferentially towards the saccharide antigen; (b) the critical immunogenic saccharide epitopes of the antigen are preserved, with minimal modification; (c) the non-specific immune responses towards linkers and other potential neo-epitopes, introduced as a result of conjugation are minimized; (d) key non-saccharide substituents, functional groups such as O-acetyl, phosphodiester, glycerol phosphate, pyruvyl etc., are preserved, depending on the bacterial antigen; and (e) the stability of the conjugate is maintained, for maximum shelf-life. Keeping in mind the stringent human safety considerations required for prophylactic vaccines, administered to a healthy human population, the choice of the carrier protein (and any associated chemical linkers) is often limited to a few options. The following carrier proteins have been used, so far, in licensed conjugate vaccines: a genetically modified cross-reacting material (CRM_{197}) of diphtheria toxin, tetanus toxoid, meningococcal outer membrane protein complex; diphtheria toxoid and *H. influenzae* protein D. Many of these clinically evaluated carrier proteins have been effective in enhancing vaccine immunogenicity. However, each of these carrier proteins differ in the quantity and avidity of antibody they elicit, ability to carry multiple saccharide antigens in the same vaccine and to be administered concurrently with other vaccines. The multiple considerations that play a key role during the selection of a suitable carrier protein have been described in earlier reviews (*7–10*). The key factors that play a role in the selection of the conjugation chemistry and the optimal attributes of the saccharide antigen, such as saccharide chain length and degree of activation are discussed, in detail, later in this chapter.

The careful utilization and interpretation of immunogenicity and animal models, towards the design and development is described in Chapter 10 (*11*). Several key elements required towards the development of a successful and commercially viable product includes the definition of a process control strategy. This strategy typically includes extensive use of statistical process control using Design of Experiments (DOE), during early development stage for a full process understanding. A good understanding of the critical and key process parameters that ultimately define critical quality attributes is a vital part of the control strategy, which in turn, helps identify the risks for failure of the manufacturing process. Several of these key elements related to scale-up and process development of glycoconjugate vaccines are discussed in further detail in Chapter 5 (*12*). The role of adjuvants are covered in Chapter 9 (*13*). Key analytical characterization and stability considerations are covered in Chapter 12 (*14*). The role of carrier proteins and additional development considerations, based on historical experience, are discussed in further detail in Chapter 13 (*15*).

In this chapter, our discussion focusses on the primary considerations for selection of (a) appropriate conjugation chemistry strategies, including the deployment of linkers; (b) associated activation/conjugation reaction parameters; (c) optimal attributes of the polysaccharide antigens including length, degree of antigen activation; (d) impact of non-saccharide substituents and (e) degree of carrier protein modification. The primary drivers for the ultimate selection from the above menu are (a) the unique polysaccharide structural attributes, which typically vary from serotype to serotype often by a single linkage site difference; (b) the stability of the polysaccharide as well as the activated polysaccharide, prior to conjugation and stability of the (c) conjugate at the drug substance as well as the drug product stage.

The Conjugate Toolbox

The production of a glycoconjugate vaccine requires an irreversible, and preferably a covalent linkage between the saccharide antigen and the carrier protein. The attachment of the saccharide antigens to the carrier proteins could be accomplished using one of the many conjugation routes, described earlier in a number of reviews (*16–20*). Conjugation could be achieved by the linking of saccharide antigen using either (a) direct route or (b) *via* using a linker/spacer. The direct conjugation route is widely used where the antigen is a long chain length polysaccharide, having an excess of saccharide repeat units (RU), typically 15 or more. The use of a linker/spacer is preferred for small chain length oligosaccharide antigens (< 15 RUs). Jennings (*21*) noted that the antigenicity and immunogenicity of saccharide antigens are length-dependent and antigenicity typically is associated with oligosaccharide epitopes no larger than an antibody site. In order to generate a robust immune response, an optimal conjugate construct requires multiple epitopes, typically present only in long chain length polysaccharides. This particularly plays an important role in multivalent or multicomponent vaccines where the potential for interference, due to competition and/or steric factors, becomes an important hurdle to overcome. Linkers, however can also be used to facilitate the conjugation of long saccharide chains to overcome any potential electrostatic repulsion between polysaccharide and the carrier protein.

Cyanylation Chemistry

Cyanylation Using Cyanogen Bromide

Cyanogen bromide has been widely used to produce several conjugate vaccines such as *Hemophilus influenzae* b (Hib) *via* cyanylation of the hydroxyl groups of the polysaccharides, using linkers such as adipic dihydrazide (ADH) (*22*).

Activation Using CDAP Chemistry

CNBr has been largely replaced by the organic cyanylating reagent, 1-cyano-4-dimethylamminopyridinium tetrafluoroborate (CDAP) which activates the polysaccharides under mild conditions. CDAP has been used to couple carrier proteins with or without linkers (*23*). This approach has been applied for the production of multivalent pneumococcal conjugate vaccine (Synflorix®) and meningococcal conjugate vaccine (Nimenrix®).

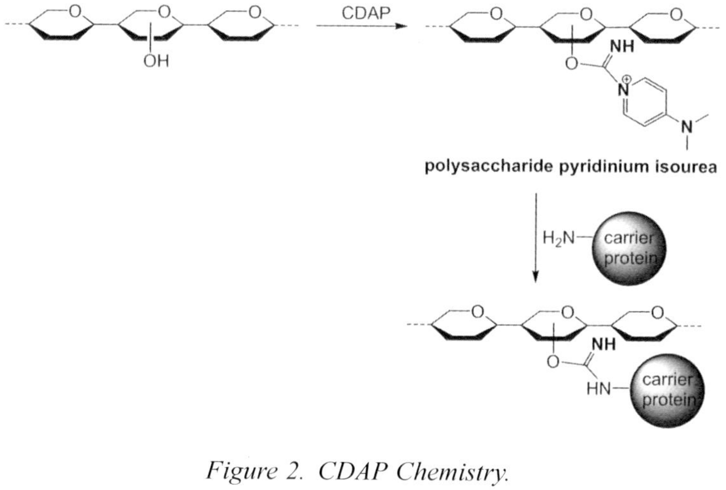

Figure 2. CDAP Chemistry.

The CDAP activation of polysaccharides is performed under basic conditions, generating a pyridinium isourea intermediate. Addition of the carrier protein results in nucleophilic displacement of dimethylaminopyridine (DMAP) and formation of an isourea linkage with lysines of the protein (*24*) (Figure 2).

Reductive Amination Chemistry (RAC)

Reductive amination is a commercially proven technology that has been successfully and consistently used in the manufacturing of glycoconjugate vaccines (*17*), such as Hib (HbOC), meningococcal and pneumococcal conjugate (Prevnar13®) vaccines.

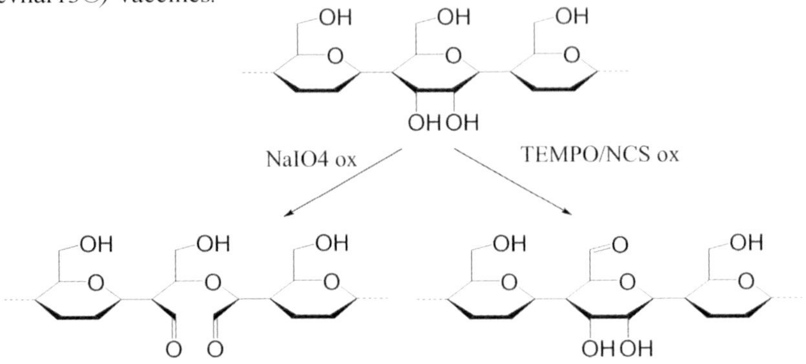

Figure 3. Oxidation of saccharide chains using periodate or TEMPO/NCS.

The first step requires the activation of the saccharide antigen (Figure 3). Itinvolves the periodate oxidation (PO) of vicinal diols on the repeat units of polysaccharide, which typically results in the formation of aldehydes. An alternative reagent is TEMPO (2,2,6,6-tetramethylpiperidine 1-oxyl radical) –N-chlorosuccinimide (NCS) combination, which results in the formation of aldehydes from primary alcohol groups. In aqueous media, TEMPO is oxidized by the stoichiometric oxidant to generate a nitrosonium cation, which is the actual oxidant of the alcohol.

Periodate oxidation of polysaccharide transforms the unreactive hydroxyls of sugar residues into amine-reactive aldehydes.

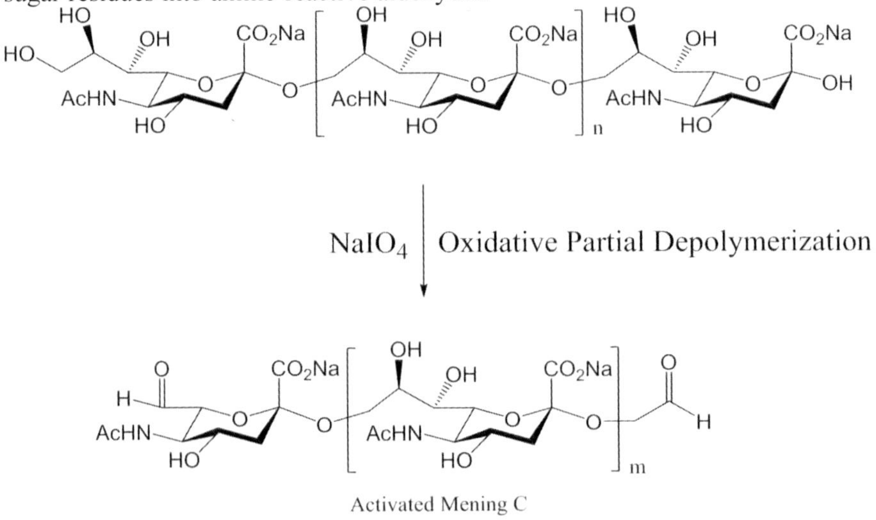

Figure 4. Oxidation of Meningococcal C polysaccharide.

A unique case of periodate activation of polysaccharide is the preparation of meningococcal serotype C conjugate vaccine (Meningitec®). Meningococcal serogroup C, which has repeating sialic acid units, generates two aldehyde groups at both ends with partially depolymerized polysaccharide chain (Figure 4). This depolymerization phenomenon also occurs during the oxidation of Hib polysaccharide.

When vicinal diols are not available, chemoselective oxidation of C-6 primary alcohol to aldehyde using TEMPO/NCS is another practical way to activate the polysaccharide. This approach was applied, by our group, to oxidize *S. pneumoniae* polysaccharide serotypes 3, 10A,12F and 33F and subsequently conjugated to CRM_{197} using RAC approach (*25*).

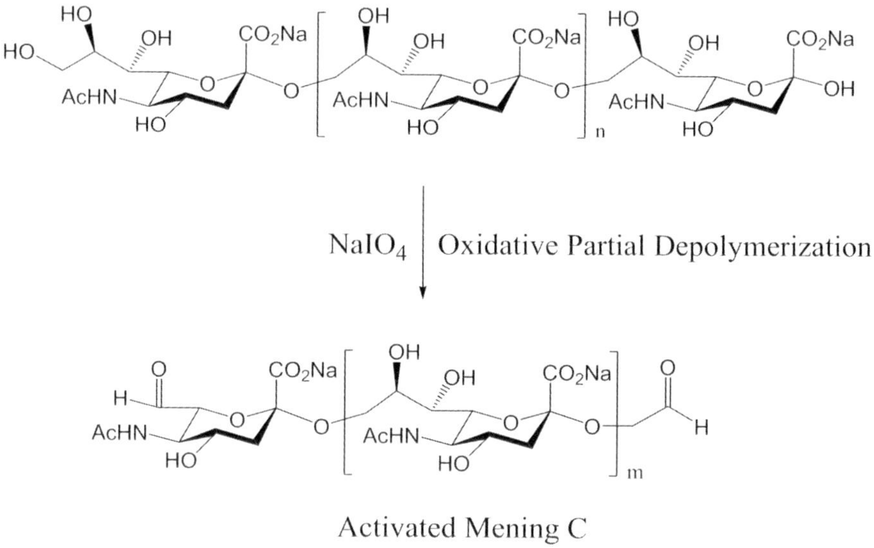

Figure 5. Conjugation step via Reductive Amination Chemistry (RAC).

Following the generation of aldehydes on oligo or polysaccharide, the activated saccharide is subsequently reacted with lysines on carrier protein to form Schiff base linkages. These Schiff bases are then reduced in the presence of sodium cyanoborohydride, to create stable covalent linkage between carbohydrate and protein (Figure 5). Conjugation reactions can be carried out either in a buffered aqueous medium or an aprotic solvent, such as DMSO.

The residual unreacted aldehyde functional groups of the activated saccharides are typically "capped" using a strong reducing agent such as sodium borohydride. However, for some polysaccharides containing keto sugars, such as pneumococcal serotype 5 (Figure 6), sodium borohydride has the potential to reduce the keto functionalities, potentially interfering with the key immunogenic epitopes. The "capping" step for these polysaccharides, containing sugars that are sensitive to borohydride, involves a second incubation step using sodium cyanoborohydride.

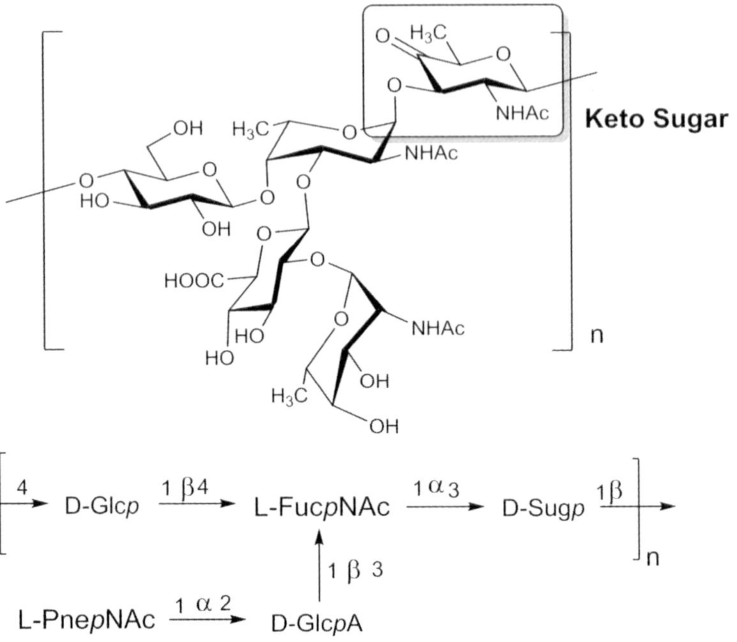

D-Sugp = 2-acetamido-2,6-dideoxy-D-*xylo*-hexopyranos-4-ulose
L-PnepNAc = 2-acetamido-2,6-dideoxy-L-talopyranose

Figure 6. Structure of pneumococcal polysaccharide serotypes 5.

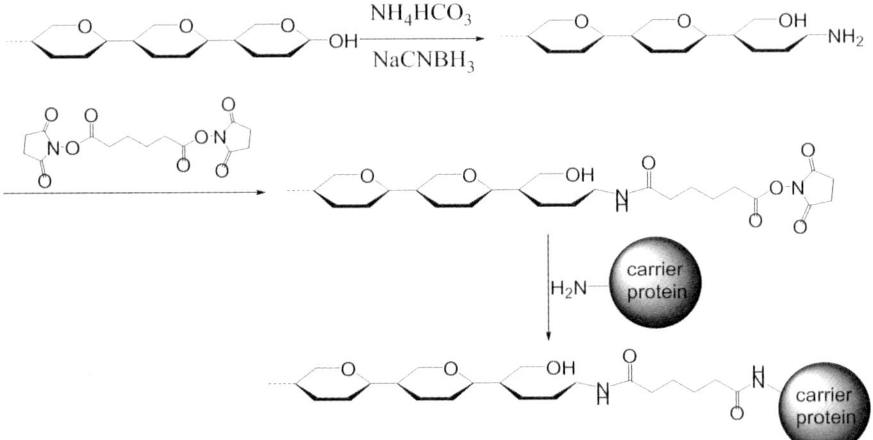

Figure 7. Conjugation via reducing end.

Conjugation through the reducing end (Figure 7) is another practical way to link protein by reductive amination process. This approach is preferably used for oligosaccharide antigens, the reducing end is converted to amine derivative by treating oligosaccharide with ammonium bicarbonate and sodium cyanoborohydride. This newly created amine reactive group then offers a number of chemistry options for conjugation, such as nucleophilic substitution of N-hydroxysuccinimide (NHS) ester.

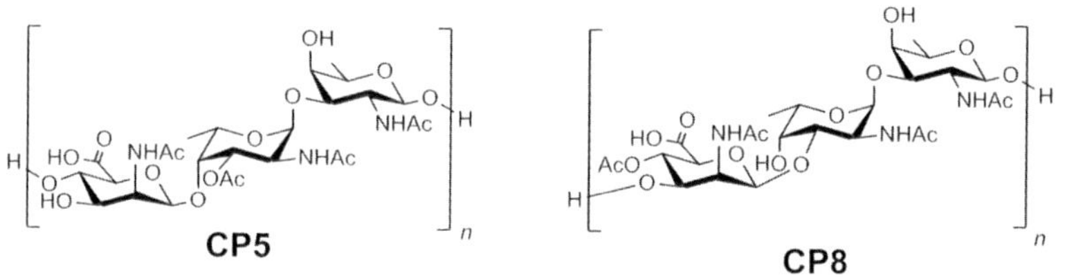

Figure 8. Structures of staphylococcal polysaccharide serotypes 5 and 8.

In order to produce conjugates using polysaccharide antigens that do not contain vicinal hydroxyl groups, e.g. staphylococcal polysaccharides (Figure 8), alternate conjugation routes need to be evaluated.

Alternative Conjugation Chemistry Strategies in Clinical Development

Conjugates have been successfully generated using various cross-linking or coupling reagents, such as homobifunctional, heterobifunctional, or zero-length crosslinkers. Many methods are currently available for coupling immunogenic molecules, such as saccharides, proteins, and peptides, to peptide or protein carriers. Most methods create amine, amide, urethane, isothiourea, or disulfide bonds, or in some cases thioethers.

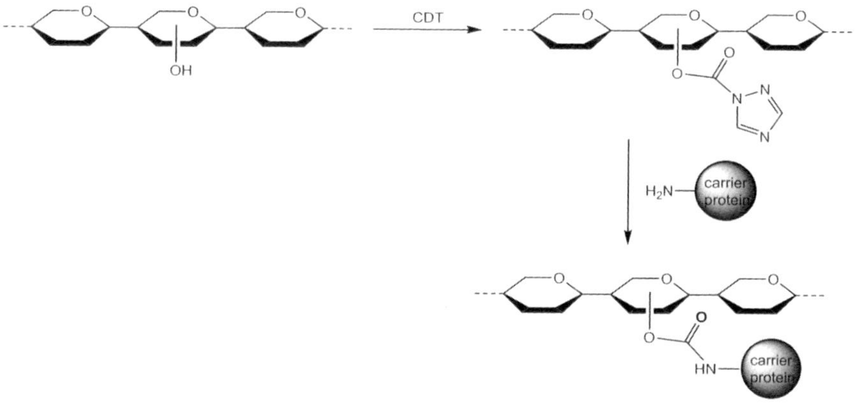

Figure 9. Conjugation using CDT (or CDI).

The conjugation route to couple saccharide antigens to carrier proteins could be accomplished by using bis electrophilic reagents (*26*) such as carbonyldiimidazole (CDI) or carbonylditriazole (CDT) in aprotic solvents such as DMF or DMSO *via* a direct route (*27*) (Figure 9) or using bigeneric linkers (*28*).

CDI/CDT chemistry was found by us to be particularly suitable for polysaccharides that do not contain vicinal hydroxyl groups, e.g. staphylococcal polysaccharides (Figure 8), to produce a multicomponent conjugate vaccine which is currently in clinical trial (*27*, *29*, *30*).

A novel method of making glycoconjugates comprising a saccharide covalently conjugated to a carrier protein through a bivalent, heterobifunctional spacer (2-((2-oxoethyl)thio)ethyl)carbamate (eTEC) was recently developed by our group (*31*).

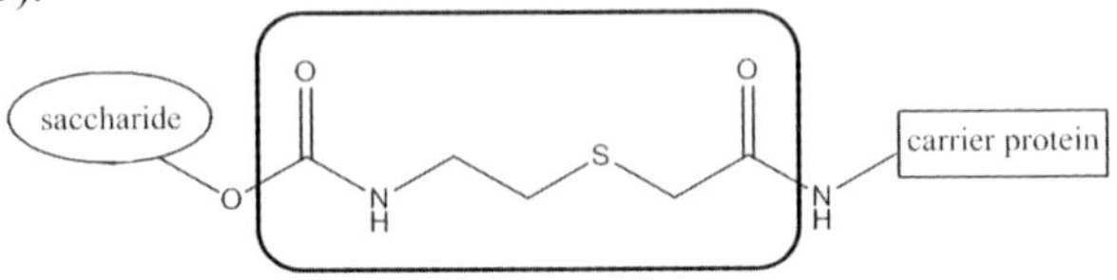

Figure 10. Bifunctional (2-((2-oxoethyl)thio)ethyl)carbamate (eTEC) spacer.

The eTEC spacer includes seven linear atoms ($-C(O)NH(CH_2)_2SCH_2C(O)-$) and provides stable thioether, carbamate and amide bonds between the saccharide and carrier protein (Figure 10).

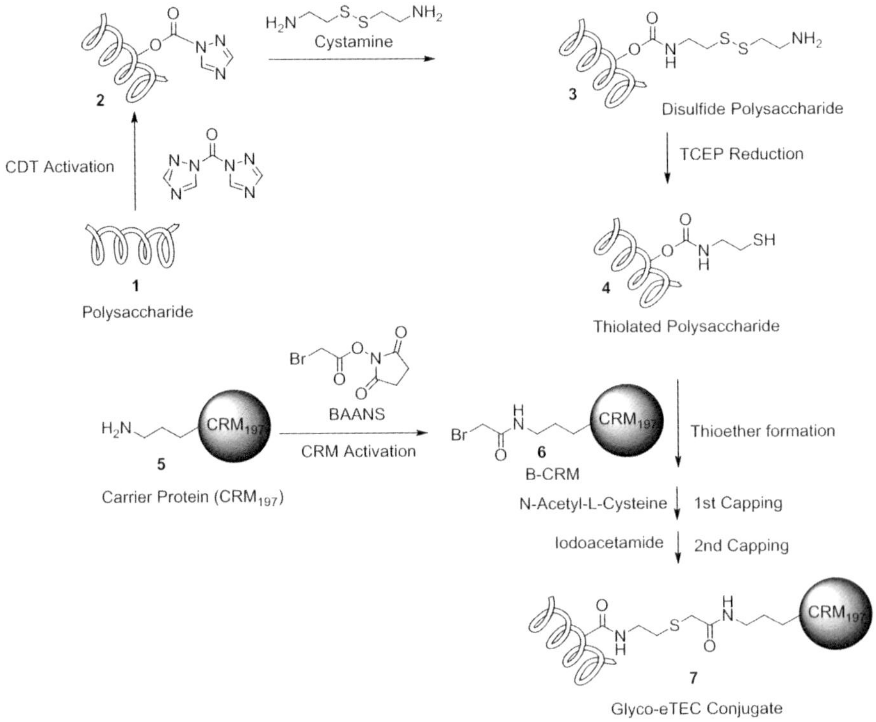

Figure 11. Thioether glycoconjugation via eTEC spacer.

This novel approach using thioether chemistry *via* the eTEC spacer (Figure 11) has been successfully used by our group to produce stable immunogenic conjugates (*31*) for pneumococcal serotypes 10A, 11A, 22F and 33F.

Additional Conjugation Chemistry Applications in Early Development

The following alternative routes have been explored, by various groups, for the generation of immunogenic conjugates. However, these approaches have not yet been widely applied, past the pre-clinical/clinical arena towards vaccine licensure.

In Vacuo Conjugation Approach

Pneumococcal polysaccharide-CRM_{197} protein conjugates were produced using a cyanide-free, *in vacuo* glycation conjugation approach. *In vitro* characterization, followed by *in vivo* immunogenicity screening by animal experiments confirmed the generation of higher molecular weight conjugates (*32*).

Generation of Glycoconjugates Using Click Chemistry

Lipinski and Bundle (*33*) describe a strategy that permits three sequential orthogonal conjugation reactions to prepare glycoconjugates, using click chemistry, targeted for uptake by DCs. Stefanetti et al. (*34*), describe a solid-phase conjugation method to generate an O-antigen based glycoconjugate vaccine against *Salmonella typhimurium*, with CRM_{197} as the carrier protein. Copper-free click chemistry was used for derivatizing the saccharide antigen and the protein components with alkyne and azido linkers, respectively.

Squaric Acid Linker Based Conjugation

Saccharide antigens from *Vibrio cholera* O1 lipopolysaccharide (LPS) and synthetic analogs were coupled to protein carriers through squaric acid chemistry to form conjugates suitable for vaccine candidates (*35*).

Even though several linker-based conjugation approaches have been evaluated, at the research and pre-clinical stages, ADH remains the widely used spacer implemented in the licensed vaccines. The flexible nature of the aliphatic chain linkers, such as ADH, provide a major advantage to direct the immune response specifically to the saccharide antigen and without potential immune interference. Linkers containing bulky groups, such as cyclic derivatives, have the potential disadvantage of directing the immune response also towards the linker instead of the target antigen. Buskas et al., for example, found that the rigid cyclohexyl maleimide linker often employed in conjugation chemistry, because of its rapid and selective reaction with thiol-derivatives at near neutral pH, dramatically reduced the immune response towards the Le^y antigen (*36*).

Bioconjugation Approach

Feldman et al., describe an approach using engineered *E. coli* cells to exploit the bacterial N-glycosylation machinery to enable PglB, the key enzyme of the *Campylobacter jejuni*, to transfer the oligosaccharide O-antigen *in situ* from a prenyl lipid carrier directly to the carrier protein (*37*). The safety and immunogenicity of a bioconjugate vaccine, containing the O-antigens of four *E coli* serotypes produced using this technology, was evaluated in a recent clinical trial (*38*). Another bioconjugate vaccine candidate prepared using the same technology was shown to be safe, immunogenic and elicit functional antibodies against *Shigella flexneri* 2a (*39*).

Non-Covalent Based Vaccine Approaches

A multiple antigen-presenting system (MAPS) based approach was developed using avidin-biotin based macromolecular complex, integrating various vaccine antigen components, including polysaccharides and proteins, in the same construct (*40*). Using antigens from various pathogens (*Streptococcus pneumoniae, Salmonella typhimurium* and *Mycobacterium tuberculosis*), the utility of the MAPS system and its feasibility for the design was assessed in pre-clinical models, using subunit vaccines to exploit multiple immune mechanisms.

The type of chemistry used to generate conjugate vaccine constructs is one of the parameters that can affect the immunogenicity. Stefanetti et al., investigated the impact of this variable on the immunogenicity of *Salmonella typhimurium* saccharide antigen-CRM_{197} conjugate vaccines in mice (*41*). The study examined random derivatization along the O-antigen chain and compared with site-selective activation of the terminal KDO sugar residue of the core oligosaccharide. It was observed that random conjugates elicited antibodies with greater bactericidal activity than selective ones, and an inverse correlation was found between degree of O-antigen modification and antibody functional activity.

Saccharide Antigen: Key Features

Saccharide Size

Many of the licensed first generation conjugate vaccines incorporating *Hemophilus influenzae* b and Meningococcal serogroup C antigens, involved oligosaccharides or small size polysaccharides. Initial binding studies involving antibody to polysaccharide antigen showed that the basic recognition of relatively simple oligosaccharide epitopes requires about six or seven sugar residues (*42*). These conjugates achieved significant success in the clinic.

However, it was later recognized that saccharide conformational epitopes may play an important role in order to generate a robust immune response. The affinity of antibody binding to saccharide antigen increases as the chain length increases and conformational epitopes are typically fully expressed only in the high-molecular weight form of the saccharide. Therefore, the chain length of saccharides may have a significant impact on the design of glycoconjugate vaccines. A specific and robust immune response for each of the saccharide antigens plays a particularly important role for multi-antigen vaccines.

Pillai *et al.,* studied the role of chain length of oligosaccharide antigens required for antibody binding, by using polysaccharide from *Haemophilus influenzae* type b or oligosaccharides derived from it (*43*). It was noted that the oligosaccharide-protein conjugate binds antibody better than the free oligosaccharides do. However, no difference in binding was observed between the polysaccharide and a polysaccharide-protein conjugate. These data indicated that identical epitopes are expressed by oligomeric and polymeric forms of the antigen and that a particularly more stable conformation in polysaccharides is preferred by antibodies.

Lafarriere *et al.*, established, both immunologically and antigenically, that at least four RU of pneumococcal serotype 14 polysaccharide were required to form an extended conformational epitope and that approximately 22 RU of the polysaccharide were required to duplicate the same epitope on the same saccharide chain (*44*). The conformational epitope was found to be critical for the induction of antibodies with high opsonophagocytic activity and the augmentation of antigenicity/immunogenicity was also dependent on further chain extension. The significance of conformational epitopes of saccharide antigens is discussed in further detail in Chapter 7 (*45*).

Daum *et al.*, evaluated the role of saccharide chain length and saccharide quantity, in a clinical trial involving 400 infants, using pneumococcal serotypes 6B, 14, 18C, 19F and 23F. Conjugates produced using polysaccharides were generally demonstrated to be more immunogenic than their oligosaccharide counterparts (*46*). It was concluded that polysaccharide based pneumococcal conjugate vaccines, for these pneumococcal serotypes were more immunogenic than oligosaccharide based vaccines.

In general, the molecular weight of the conjugates generated using polysaccharides are likely to be higher compared to their oligosaccharide counterparts, produced using identical conjugation reaction conditions. Therefore, the contribution of conjugate molecular weight may play a significant role in determining the immunogenicity, in tandem with the saccharide chain length.

A polysaccharide can become slightly reduced in size during normal downstream purification procedures. Alternatively, polysaccharides can be subjected to sizing techniques before activation step, prior to conjugation. A number of procedures are available for the control of the saccharide chain length. The most widely used methods currently applied for licensed vaccines and clinical candidates are (i) mechanical sizing using high pressure homogenization and (ii) chemical hydrolysis.

For mechanical sizing of polysaccharides, using high pressure homogenization, high shear rates are applied by pumping the process stream through a flow path with sufficiently small dimensions. The high pressure homogenization process is particularly suitable for reducing the size of polysaccharides containing non-saccharide substituents, such as O-acetyl, glycerol phosphate, pyruvyl, etc. commonly present in pneumococcal, meningococcal, staphylococcal, group B streptococcus capsular polysaccharides. Chemical hydrolysis is more suitable for polysaccharides which do not contain these non-saccharide substituents, such as pneumococcal serotypes 3, 6A, 8, 12F, etc. However, for polysaccharides that are susceptible to cleavage, resulting in the loss of side chain sugars from the backbone, the preservation of integrity of the repeat units need to be monitored closely.

Saccharide Antigen Modification (Degree of Activation)

In order to generate a stable conjugate, certain minimal level of saccharide antigen modification is required. However, the degree of saccharide modification plays a significant role in preserving the structural integrity of the critical immunogenic epitopes of the antigen.

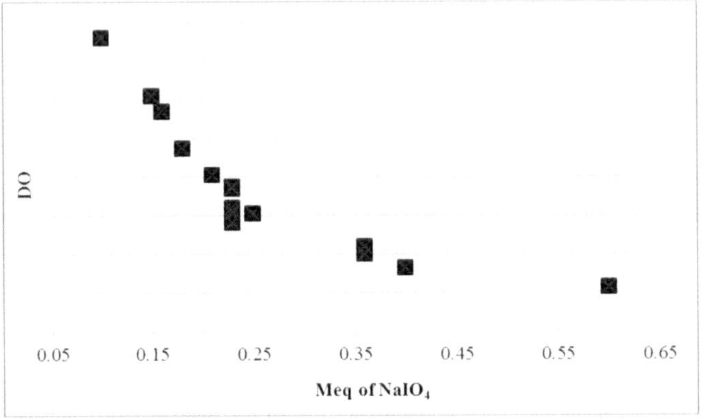

Figure 12. Degree of Oxidation.

Varying the molar equivalents (meq) of sodium periodate relative to polysaccharide repeat unit and temperature during oxidation results in varying levels of degree of oxidation (DO). Typically a curved response of DO is observed with varying levels of $NaIO_4$ (Figure 12). The degree of oxidation (DO) is defined as moles of sugar repeat unit / moles of aldehyde) of the activated polysaccharide. The saccharide and aldehyde concentrations are typically determined by colorimetric assays.

The target degree of oxidation, in turn, is a major contributor towards the determination of (a) saccharide modification, (b) molecular weight (MW) of the activated polysaccharide (c) MW of the conjugate; (d) output saccharide/protein ratio of the conjugate; (e) free saccharide levels; (f) conjugation efficiency; (g) stability and/or (h) preservation non-saccharide substituents, such as glycerol phosphate.

The degree of oxidation for the TEMPO/NCS based polysaccharide activation, exemplified for pneumococcal serotype 12F, is controlled by the meq levels of NCS (Figure 13).

Kim *et al.*, determined the location and order of activation for several representative saccharide antigens (*47*). Pneumococcal serotypes 7F (Pn 7F) and 18C (Pn 18C) were evaluated as polysaccharides containing multiple potential sites for activation. Monitoring for the site and degree of activation by sodium periodate oxidation and linkage/methylation analysis and gas chromatography-mass spectrometry (GC-MS). Sialyllactose was used as a model oligosaccharide compound to evaluate oxidation of terminally linked sialic acids and reducing sugar residues.

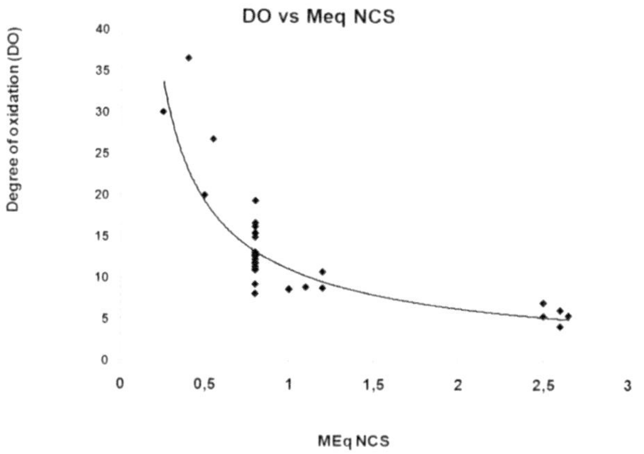

Figure 13. Degree of Oxidation (using TEMPO/NCS).

Table 1. Sites of activation: pneumococcal serotypes 7F, 18C and 3′-sialyllactose

Sample	*Order of periodate oxditaion*
Pn 7F	t-Gal>2,3-Gal>t-GlcNAc, 2-Rha>4-Glc
Pn 18C	Glycerol>t-Glc>4-Glc (2)
3'-sialyllactose	C9>Glc-ol>C8>4-Glc

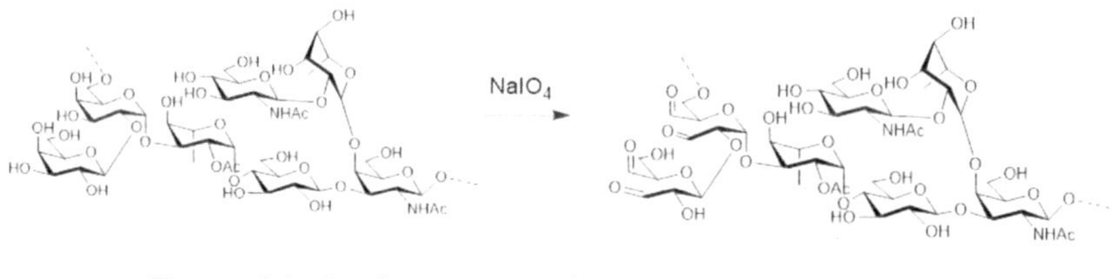

Figure 14. Oxidation sites of pneumococcal serotype 7F.

In the case of serotype 7F, using higher meq of oxidant, the three residues GlcNAc, Rha and Glc containing trans diols were found to be susceptible for oxidation (Table 1, Figure 14) . However, the Gal residues containing cis diols (C3-C4) were the only residues to be oxidized at low meq levels, with a preference for terminally linked Gal (t-Gal).

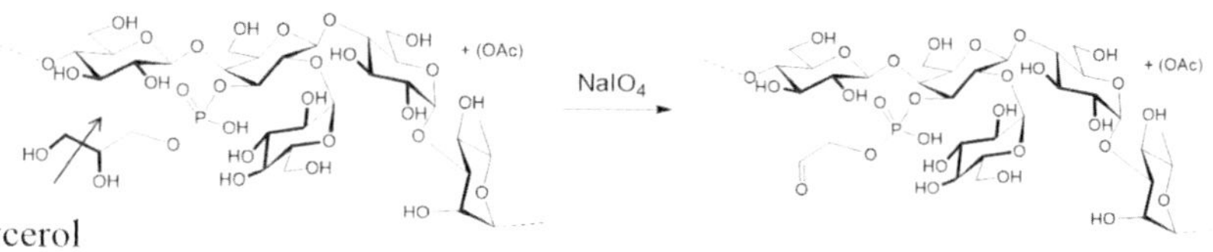

glycerol

Figure 15. Oxidation sites of pneumococcal serotype 18C.

In the case of serotype 18C, the glycerol phosphate residue was preferentially modified at low oxidation levels (Table 1, Figure 15).

Periodate oxidation results in random activation. However, a customized approach of controlled modification of the polysaccharide could be used, depending on the structure of the antigen, to target specific sugars such as t-Gal (exemplified by Pn 7F oxidation), or non-saccharide substituents (exemplified by glycerol phosphate, by Pn 18C oxidation). The activation and conjugation parameters, therefore, need to be tailor-made depending on the structural motif of the specific saccharide antigen target.

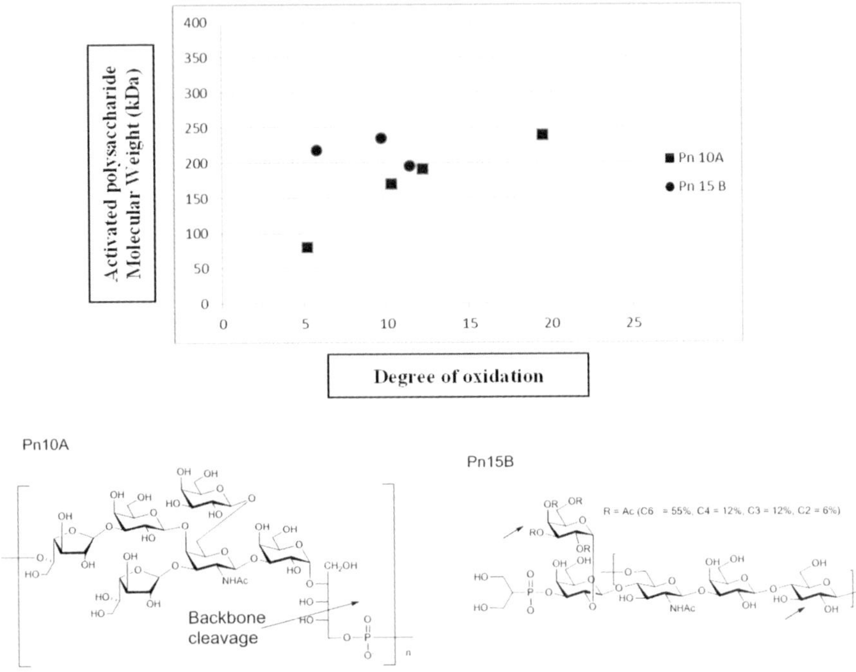

Figure 16. Impact of degree of oxidation on MW of activated saccharide.

The degree of oxidation can have a significant impact on the molecular weight (MW) of the activated polysaccharide. In the case of pneumococcal serotype 10A (Pn 10A), we have observed that excessive oxidation results in significant reduction in the MW of the activated polysaccharide (Figure 16) due to cleavage of the polysaccharide backbone (ribitol) (*48*). In contrast, for pneumococcal serotype 15B (Pn 15B), no significant change in the MW of the activated polysaccharides is observed as a function of degree of oxidation, since the cleavage of backbone repeat unit does not occur in this polysaccharide (*48*).

Degree of oxidation may have a significant impact on the degree of conjugation (Figure 17), as measured by the number of lysine residues modified (as determined by amino acid analysis). The distinct differences in the structural patterns of the two polysaccharides serotypes results in varying pattern in oxidation sites, and the lysines in the carrier protein react accordingly during conjugation (*48*). The degree of conjugation does not vary significantly, as a function of degree of oxidation for Pn 15B. In contrast, the highly oxidized activated polysaccharide for Pn 10A results in significantly higher lysine (more than 3 fold, compared to Pn 15B) modification, in CRM_{197}.

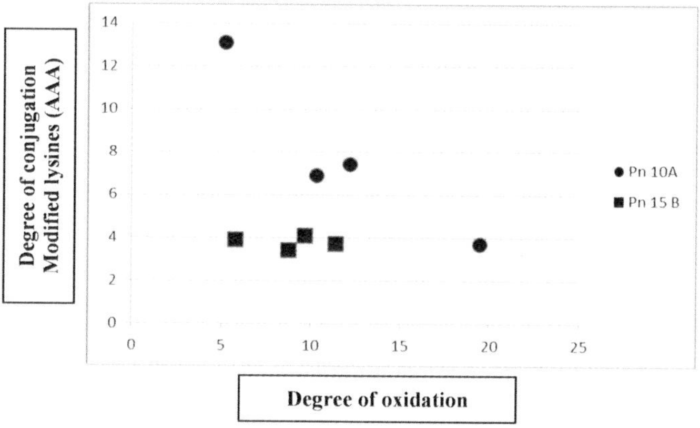

Figure 17. Impact of degree of oxidation on the degree of conjugation.

Table 2. Pneumococcal serotype 15B-CRM_{197} conjugates produced in Aqueous and DMSO media

	DMSO	*Aqueous Buffer*
Activated Polysaccharide MW (kDa)	235K	270K
Degree of Oxidation (DO)	9.7	8.8
Conjugate MW (kDa)	7937	1029
% of O-Acetyl Retained	99.5%	67%

We have observed that a notable advantage of performing conjugation reactions in DMSO instead of aqueous solvents is that it can preserve base sensitive functional groups, such as O-Acetyl of the polysaccharide becaue the reactions can be carried out under mild pH and temperature conditions, close to 23 °C (Table 2) (*48*). Conjugation in DMSO is faster and more efficient, typically yielding glycoconjugates with higher molecular weight containing more cross links. Conjugation in DMSO can dramatically improve the reaction efficiency with regard to conjugation yield and filterability. The formation of imine is kinetically more favored in DMSO than in an aqueous solution. We have successfully used conjugation in DMSO for the production of conjugates for pneumococcal serotypes 6A, 6B, 7F, 19A, 19F and 23F, for the licensed Prevnar13® conjugate vaccine (*49*), and subsequently for pneumococcal serotypes 8, 10A, 15B and 22F for the pneumococcal next generation clinical candidate vaccine (*48*).

A conjugation process that produces conjugates with lower levels of “free” (unreacted) polysaccharide is advantageous and preferable. It is well known that high levels of “free” (unreacted) polysaccharide may cause an excessive T-cell independent immune response. This may result in the dilution of the T-cell dependent response generated by the polysaccharide-protein conjugate, thereby lowering the immunogenic response generated by the conjugate.

We produced conjugates for group B streptoccus (GBS) serotypes Ia, Ib, II, III, IV and V polysaccharides, by systematically (i) varying periodate oxidation/reductive amination chemistry (PO/RAC) reaction parameters, (ii) conjugation solvent (aqueous versus DMSO medium), (iii) varying levels of sialic acid in the initial polysaccharide and (iv) degree of oxidation/saccharide epitope modification. In general, the PO/RAC conjugates produced using DMSO as the solvent were found to have lower levels of (free) polysaccharide, higher conjugate molecular weight, and higher saccharide/protein ratios than conjugates produced in aqueous medium (*50*).

Specific sugars in various polysaccharide antigens, such as sialic acid (N-acetylneuraminic acid) may define, in full or part, critical immunogenic epitopes. The modification of these sugars which are part of these critical immunogenic epitopes may have a significant on immunogenicity. Selected GBS polysaccharides were chemically desialylated, by our group, to generate conjugate variants to determine the impact of % desialylation on immunogenicity (*50*). Desialylation of more than about 40% (i.e. sialic acid levels less than about 60%) had a negative impact on immunogenicity, as observed from screening in animal models. In most cases, a degree of oxidation of less than about 5, or saccharide epitope modification greater than about 20%, had a negative impact on immunogenicity. Oxidation occurs through the sialic acid on the capsular polysaccharide and ultimately results in reduced immunogenicity (*50*).

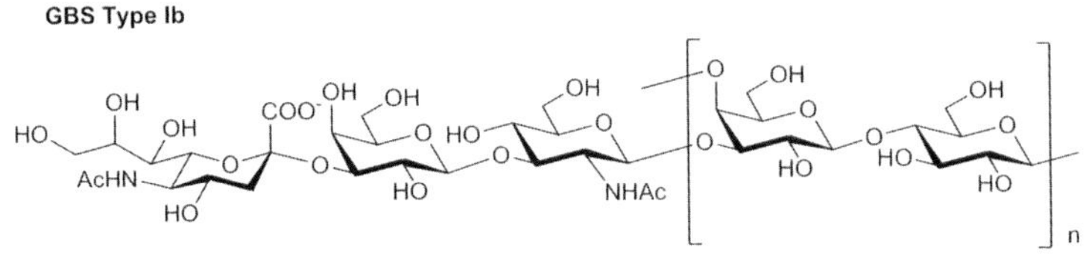

Figure 18. Structure of group B streptococcal polysaccharide serotype Ib.

The structure GBS polysaccharide serotype Ib is shown in Figure 18. A prominent feature of the GBS polysaccharides is the presence of terminal sialic acid (N-Acetylneuraminic acid) as part of the repeat unit.

GBS Serotype Ib Polysaccharide-CRM_{197} Conjugates

We generated conjugates, using PO/RAC and activated polysaccharides having a DO of 15.8 (approximately 6% saccharide epitope modification) in DMSO was demonstrated to be immunogenic in mice (*50*). The conjugate generated by PO/RAC in DMSO was slightly more immunogenic than the conjugate generated by PO/RAC in the aqueous medium when all other conjugate molecular attributes were similar (conjugates 1 and 3, respectively). However, using activated polysaccharides having a DO of 4.7 (approximately 21% saccharide epitope modification) had a negative impact on immunogenicity (conjugate 2). Immunogenicity was almost completely abolished, with very few responders, in the conjugate generated using PO/RAC and a 95% desialylated (5% sialic acid level) polysaccharide (conjugate 4). Results are summarized in Table 3.

Table 3. Effects of varying conjugation parameters: GBS Serotype Ib-CRM$_{197}$ conjugates

Conjugate	*1*	*2*	*3*	*4*
Solvent	DMSO	DMSO	Aqueous	DMSO
Poly MW (kDa)	120	120	120	120
%Sialic Acid in initial polysaccharide	>95	>95	>95	5
%Modification	6	21	6	9
Degree of Oxidation (DO)	15.8	4.7	15.8	11.7
Saccharide/Protein Ratio	1.1	1	2	1.1
% Free Saccharide	11	<5	33	7
MW by SEC-MALLS, kDa	2608	7302	381	8418
OPA Titer	417	159	278	62

The degree of saccharide antigen activation ultimately may have a significant impact on the immunogenicity of the conjugates (Table 4). We produced six conjugate variants, using three different conjugation chemistry routes, by varying the activation levels. The antibody titer data of the pneumococcal polysaccharide serotype 10A-CRM$_{197}$ conjugate variants, produced by deliberately varying the saccharide antigen activation levels, using three different conjugation chemistry routes were measured by opsonophygocytic assay (OPA). The antibody titers of the control unconjugated polysaccharide are low, as expected. In each case, high modification levels of the polysaccharide, due to higher degree of activation, result in lower anibody titers, also as expected. The lowest activation level is the preferred approach to minimize significant modification of the critical immunogenic epitopes. However, conjugates produced using the lowest level of polysaccharide modification may not necessarily possess optimal conjugate attributes such as (i) MW, (ii) low levels of unconjugated free saccharide, (iii) optimal saccharide/protein ratio, (iv) sufficient levels of linkages to the carrier protein modification, such as lysine modification (v) desired stability profile, etc. In addition to the degree of saccharide antigen modification, there are several other conjugation reaction process parameters that need to be tailor-made for each polysaccharide antigen, depending on the structure, to achieve the target attributes and stability.

Table 4. Impact of degree of activation on the conjugate immunogenicity (mice)

		OPA GMT (95% CI)		
	Conjugate Variant	*0.001 μg*	*0.01 μg*	*0.1 μg*
1	Medium Activation PO/RAC Chemistry	443 (270, 726)	906 (568, 1445)	2838 (1452, 5545)
2	Low Activation eTEC Chemistry	378 (176, 810)	590 (292, 1191)	4956 (2570, 9559)
3	Medium Activation eTEC Chemistry	248 (126, 489)	587 (361, 953)	3148 (1732, 5724)
4	High Activation eTEC Chemistry	329 (181, 600)	649 (325, 1294)	1454 (672, 3147)
5	Medium Activation TEMPO/RAC Chemistry	103 (42, 251)	582 (275, 1231)	3310 (1816, 6033)
6	High Activation TEMPO/RAC Chemistry	210 (94, 472)	518 (255, 1050)	1193 (500, 2843)
7	Unconjugated PS			602 (193, 1882)

Non-Saccharide Substituents (NSS)

Non-saccharide substituents of the polysaccharide antigens, such O-acetyl, glycerol phosphate, pyruvyl functional groups may play an important role in the immunogenicity of the conjugates. In contrast to the variations in sugar composition and linkage sites among the constituent saccharide units, NSS can introduce subtle antigen modiciations that nonetheless could potentially result in a significant impact on capsular shielding and recognition of the capsule by vaccine-elicited antibodies. Among these, O-acetyl groups are the most widely studied, in terms of understanding their role in immunogenicity of the vaccines, that constitute these antigens (*29*, *51–54*). Some examples among the pneumococcal polysaccharide serotypes, which contain O-acetyl functional groups, include 1, 7F, 9V, 11A, 15B, 18C, 22F and 33F.

Carrier Protein

A number of carrier proteins have been used for the production of glycoconjugate vaccines, as dicussed in Chapter 13 (*15*). The choice of an appropriate carrier may influence the immunogenic properties of the conjugate vaccines. This may be particularly more important for the production of multi-valent conjugate vaccines.

CRM_{197}

Shinefield provided a comprehensive review of the use of CRM_{197} as a carrier protein, contrasting it to other carrier proteins used in single-antigen pediatric vaccines as well as identifying areas for future study (*7*). The successful development of the first CRM_{197} conjugate vaccine developed to target *Hemophilus influenza* b, HbOC (HibTITER®), protected infants against Hib-related disease (*55, 56*), which resulted in the expansion to other bacterial pathogen targets, namely meningococcal, pneumococcal, staphylococcal and group B streptococcus (*57, 58*).

In order to precisely understand the function of the CRM_{197} in the vaccine, Bixler *et al*., mapped the T-cell epitopes of the protein. Peptide sequences which represent a segment of the primary sequence of the carrier protein were found to stimulate diphtheria toxoid or CRM_{197}-primed T-lymphocytes (*59, 60*). In particular, the peptide region 357-383 was found to be recognized by both CRM_{197} and diphtheria toxin (DT)-primed T-cells. Synthetic peptides representing defined T-cell epitopes of CRM_{197} were functionally demonstrated based on their ability to act as carrier molecules in a PRP conjugate vaccine (*59*).

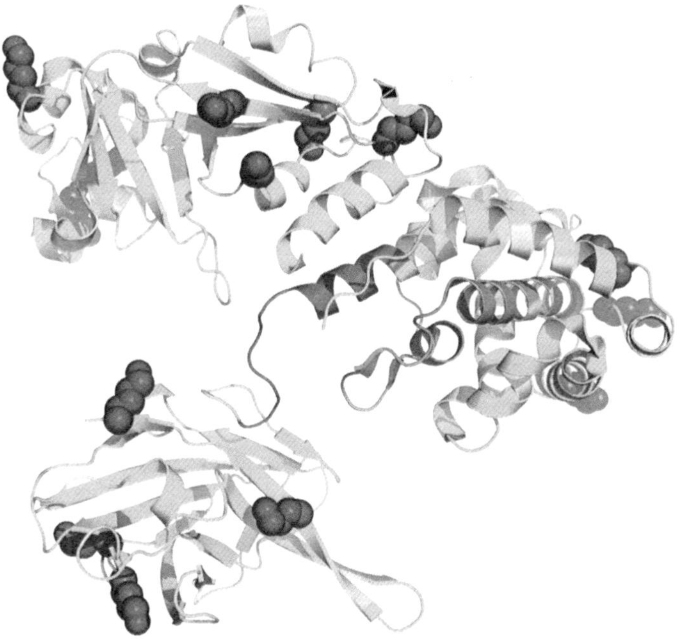

Figure 19. CRM_{197} structure showing T cell epitopes and lysine residues.

Figure 19 shows the crystal structure model of CRM_{197} (ribbon, PDB ID: 5I82) in light green. The two T-cell epitopes reported by Bixler *et al* (*59, 60*) are highlighted in dark red (amino acid residues 369-383) and orange (306-334) respectively. The lysine side chains are shown as blue spheres (Model structure, by courtesy of Rahel Rezaei-Araghi, Pfizer).

A significant observation of the study by Bixler *et al.*, is the lack of lysine residues in the T cell epitopes (*59*, *60*). For the vast majority of the licensed glycoconjugate vaccines, as well the clinical candidates, conjugation occurs via the modification of the lysine residues in CRM_{197}. Therefore, it is important to note that high levels of lysine modification by the activated saccharide antigens, in the conjugate vaccines, is less likely to directly interfere with the T cell functional aspects of the carrier protein. The lack of lysine residues, in the T cell epitopes, presents itself as a major advantage to consider CRM_{197} as the carrier protein of choice towards the design and development of conjugate vaccines.

Conclusions

The selection of an appropriate chemistry route to produce a glycoconjugate from a specific saccharide antigen and carrier protein, however, fulfils only the first minimal step towards the generation of an "optimal glycoconjugate vaccine construct". Due to the need to maintain all the functional and structural requirements of a saccharide antigen, several key elements need to be considered during the design and development of conjugate vaccines. The early development studies should take into consideration several factors such as stability, consistency in quality and manufacturability, from the very outset. A customized design approach, based on the specific structural features of the saccharide antigen, is particularly important for the development of multi-valent vaccines. It is important that the immunogenic epitopes are available in abundance, without interference from other constituent serotypes, for successful antigen-antibody interactions in the complex mixtures of multi-valent vaccines. A common "platform chemistry and process" approach, uniformly applicable to all the constituent serotypes, has the advantage of simplifying development effort and manufacturing process. However, this could potentially undermine the specific structural needs, such as stability, for some selected constituent antigens. The unique design and development needs of glycoconjugate vaccines, due to specific structural and stability requirements sets them apart from other purely protein-based vaccines. Due to the absence of reliable pre-clinical models that could translate fully into a favorable clinical outcome, the design and development of conjugate vaccines still heavily rely significantly on biochemical, biophysical and immunochemical attributes. Since the glycoconjugate vaccines are typically heterogeneous and complex in nature, in order to maintain product quality and consistency the design and process features developed during the very early clinical development stages essentially define the final product.

Acknowledgments

The authors acknowledge with gratitude Drs. Mark Ruppen, Suddham Singh, Profs. Neil Ravenscroft and Michelle Kuttel for reviewing the manuscript.

References

1. Avery, O. T.; Goebel, W. F. *J. Exp. Med.* **1929**, *50*, 533–550.
2. Vella, M.; Pace, D. *Expert Opin. Biol Ther.* **2014**, *15*, 529–546.
3. Ada, G.; Isaacs, D. *Clin. Microbiol. Infect.* **2003**, *9*, 79–85.
4. Frasch, C. E. Regulatory Perspectives in Vaccine Licensure. In *Development and Clinical Uses of Haemophilus b Conjugate Vaccines*; Ellis, R. W.; Granoff, D. M., Eds.; Marcel Dekker: New York, 1994; pp 435−435.
5. Anderson, P.; Pichichero, M. E.; Insel, R. A. *J. Clin. Invest.* **1985**, *76*, 52–59.
6. Anderson, P.; Pichichero, M.; Edwards, K.; Porch, C. R.; Insel, R. *J. Pediatr.* **1987**, *111*, 644–650.
7. Shinefield, H. R. *Vaccine* **2010**, *28*, 4335–4339.
8. Lockyer, K.; Gao, F.; Derrick, J. P.; Bolgiano, B. *Vaccine* **2015**, *33*, 1345–1352.
9. Anderson, P.; Treanor, J.; Porcelli, S.; Pichichero, M. *Vaccine* **2003**, *21*, 1554–1559.
10. Pichichero, M. E. *Human Vaccines Immunother.* **2013**, *9*, 2505–2523.
11. Scully, I. L.; Swanson, K. A.; Kanevsky, I.; Prasad, A. K.; Anderson, A. S. Pre-Clinical Assessment of Glycoconjugate Vaccines. In Prasad, A. K. (Ed) *Carbohydrate-Based Vaccines: From Concept to Clinic*; ACS Symposium Series 1290; American Chemical Society: Washington, DC, 2018; Chapter 10.
12. Prasad, A. K.; Murphy, K.; Kolodziej, S. A. Manufacturing Considerations for Glycoconjugate Vaccines. In *Carbohydrate-Based Vaccines: From Concept to Clinic*; Prasad, A. K., Ed.; ACS Symposium Series 1290; American Chemical Society: Washington, DC, 2018; Chapter 5.
13. Khandke, L.; Welsh, J. A.; Alderson, M. R. Formulation Development of Glycoconjugate Vaccines for Low- and Middle-Income Countries. In *Carbohydrate-Based Vaccines: From Concept to Clinic*; Prasad, A. K., Ed.; ACS Symposium Series 1290; American Chemical Society: Washington, DC, 2018; Chapter 9.
14. Jones, C. Chemistry Manufacturing, Control, and Licensure Criteria for Carbohydrate-Based Vaccines. In *Carbohydrate-Based Vaccines: From Concept to Clinic*; Prasad, A. K., Ed.; ACS Symposium Series 1290; American Chemical Society: Washington, DC, 2018; Chapter 12.
15. Hennessey, J. P., Jr.; Costantino, P.; Talaga, P.; Beurret, M.; Ravenscroft, N.; Alderson, M. R.; Zablackis, E.; Prasad, A. K.; Frasch, C. Lessons Learned and Future Challenges in the Design and Manufacture of Glycoconjugate Vaccines. In *Carbohydrate-Based Vaccines: From Concept to Clinic*; Prasad, A. K., Ed.; ACS Symposium Series 1290; American Chemical Society: Washington, DC, 2018; Chapter 13.
16. Hermanson, G. T., Ed.; *Bioconjugate Applications in Bioconjugate Techniques*; Academic Press: San Diego, CA, 2008; pp 745−857.
17. Dick, W. E.; Beurret, M. Conjugate Vaccines. In *Contributions to Microbiology and Immunology*; Cruse, J. M.; Lewis, R. E., Jr., Eds.; Basel Karger: Switzerland, 1989; Vol. 10, pp 48−114.

18. Jennings, H. J.; Sood, R. K. Synthetic Glycoconjugates As Human Vaccines. In *Neoglycoconjugates: Preparation and Applications*; Lee, Y. C.; Lee, R. T., Eds.; Academic Press: San Diego, CA, 1994; Vol. 10, pp 325–371.
19. Lees, A.; Puvanesarajah, V.; Frasch, C. E. Conjugation Chemistry in Siber. In *Pneumococcal Vaccines: The Impact of Conjugate Vaccines*; Siber, G. R.; Klugman, K. P.; Makela, P. H., Eds.; ASM Press: Washington, DC, 2008; pp 163–174.
20. Jennings, H. J; Pon, R. A. Bacterial Polysaccharide Vaccines: Glycoconjugates and Peptide-Mimetics. In *Microbial Glycobiology*; Holst, O.; Brennan, P. J.; von Itzstein, M., Eds.; Academic Press: San Diego, CA, 2009; pp 933–956.
21. Jennings, H . J. *J. Infect Dis.* **1992** (Suppl 1), S156–9.
22. Schneerson, R.; Barrera, O.; Suton, O.; Robbins, J. B. *J. Exp. Med.* **1980**, *152*, 361–376.
23. Shafer, D. E.; Toll, B.; Schuman, R. F; Nelson, B. L.; Mond, J. J.; Lees, A. *Vaccine* **2000**, *18*, 1273–1281.
24. Stollner, D.; Scheller, F. W.; Warsinke, A. *Anal. Biochem.* **2002**, *304*, 157–165.
25. Han, M.; Kainthan, R. K.; Kim, J. Prasad, A. K. PCT Patent Application, 2014/097099 A2; June 26, 2014.
26. Bethell, G. S.; Ayres, J. S.; Hancock, W. S.; Hearn, M. T. W. *J. Biol. Chem.* **1979**, *254*, 2572–2574.
27. Freese, S. J.; Anderson, A.; Pavliak, V.; Jansen, K. U.; Scully, I. L.,; Scott, T. D.; Nanra, J. S.; Prasad, A. K.; Green, B. A. U.S. Patent, 9,623,100, April 18, 2017.
28. Marburg, S.; Jorn, D.; Tolman, R. L.; Arison, B.; McCauley, J.; Kinskern, P. J.; Hagopian, A.; Vella, P. P. *J. Am. Chem. Soc.* **1986**, *108*, 5282–5287.
29. Scully, I. L.; Pavliak, V.; Timofeyeva; Liu, Y.; Singer, C.; Anderson, A. S. *Human Vaccines Immunother.* **2017**, *14*, 81–84.
30. Begier, E.; Seiden, D. J.; Patton, M.; Zito, E.; Severs, J.; Cooper, D.; Gruber, W. C.; Jansen, K. U.; Anderson, A. S.; Gurtman, A. *Vaccine* **2017**, *35*, 1132–39.
31. Gu, J.; Kim, J.; Prasad, A. K.; Yang, Y. U.S. Patent, 9,517,274, December 13, 2016.
32. Turner, A. E. B.; Gerson, J. E.; So, H. Y.; Krasznai, D. J.; St Hillaire, A. J.; Gerson, D. F. *Synthetic Systems Biotechnol.* **2017**, *2*, 49–58.
33. Lipinski, T.; Bundle, D. R. Temporary Conversion of Protein Amino Groups to Azides: A Synthetic Strategy for Glycoconjugate Vaccines. In *Carbohydrate-Based Vaccines; Methods and Protocols, Methods in Molecular Biology*; Lepenies, B. , Ed.; Humana Press: New York, 2015; Vol. 1331, pp 145–158.
34. Stefanetti, G.; Saul, A.; MacLennan, C. A.; Micoli, F. *Bioconjug. Chem.* **2015**, *26*, 2507–13.
35. Xu, P.; Kelly, M.; Vann, W. F.; Qadri, E. T.; Ryan, E. T.; Kovac, P. *ChemBioChem.* **2017**, *18*, 799–815.
36. Buskas, T.; Thompson, P.; Boons, G-J. *Chem. Commun. (Cambridge, U.K.)* **2009**, *36*, 5335–49.

37. Feldman, M. F.; Wacker, M.; Hernadez, M. *Proc. Natl. Acad. Sci. U.S.A.* **2005**, *102*, 3016–3021.
38. Huttner, A.; Hatz, C.; van den Dobbelstein, G.; Abbanat, D.; Hornacek, A.; Frolich, R.; Dreyer, A. M.; Martin, P.; Davies, T.; Fae, K.; van den Nieuwenhof, I.; Thoelen, S.; de Valliere, S.; Kuhn, A.; Bernasconi, E.; Viereck, V.; Kavvadias, T.; Kling, K.; Ryu, G.; Hulder, T.; Groger, S.; Scheiner, D.; Alaimo, C.; Harbrath, S.; Poolman, J.; Fonck, V. G. *Lancet Infect. Dis.* **2017**, *17*, 528–553.
39. Riddle, M. S.; Kaminski, R. W.; Di Paolo, C.; Porter, C. K.; Gitterze, R. L.; Clarkson, K. A.; Weerts, H. E.; Duplessis, C.; Castellano, A.; Alaimo, C.; Paolino, K. *Clin. Vaccine Immunol.* **2016**, *23*, 908–917.
40. Zhang, F.; Lu, Y. J.; Malley, R. *Proc. Natl. Acad. Sci. U.S.A.* **2013**, *110*, 13564–13569.
41. Stefanetti, G.; Rondino, S.; Lanzilao, L.; Saul, A.; MacLennan, C. A.; Micoli, F. *Vaccine* **2014**, *32*, 6122–6129.
42. Glaudemans, C. P. *J. Mol. Immunol.* **1987**, *24*, 371–377.
43. Pillai, S.; Cicirello, S.; Koster, M.; Eby, R. *Infect. Immun.* **1991**, *59*, 4371–4376.
44. Laferriere, A. C.; Sood, R. K.; de Muys, J-M.; Michon, F.; Jennings, H. J. *Infect. Immun.* **1998**, *66*, 2441–2446.
45. Kuttel, M. M.; Ravenscroft, N. The Role of Molecular Modeling in Predicting Carbohydrate Antigen Conformation and Understanding Vaccine Immunogenicity. In *Carbohydrate-Based Vaccines: From Concept to Clinic*; Prasad, A. K., Ed.; ACS Symposium Series 1290; American Chemical Society: Washington, DC, 2018; Chapter 7.
46. Daum, R. S.; Hogerman, D.; Rennels, M. B.; Bewley, K.; Mailoski, F.; Reisinger, K.; Block, S.; Keyserling, H.; Steinhoff, M. *J. Infect. Dis.* **1997**, *176*, 445–455.
47. Kim, J. S.; Laskowich, E. R.; Michon, F.; Kaiser, R. E.; Arumugham, R. G. *Anal. Biochem.* **2006**, *358*, 136–142.
48. Emini, E. A.; Watson, W. J.; Prasad, A. K.; Han, M.; Kim, J.; Gu, J.; Yang, Y.; Kainthan, R. K.; Cooper, D.; Pride, M. W; Jansen, K. U. U.S. Patent 9,492,559, November 15, 2016.
49. Prasad, A. K. U.S. Patent 7,955,605, June 7, 2011.
50. Anderson, A. S.; Bhalla, B. B.; Donald, R. G. K.; Gu, J.; Jansen, K. U.; Kainthan, R. K.; Khandke, L.; Kim, J.; Liberator, P.; Prasad, A. K.; Ruppen, M. E.; Scully, I. L.; Singh, S.; Yang, C. X. U.S. Patent Application, 2016/0324950, 2016.
51. Fusco, P. C.; Farley, E. K.; Huang, C. H.; Moore, S.; Michon, F. *Clin. Vaccin Immunol.* **2007**, *14*, 577–584.
52. Tielen, P.; Strathmann, M.; Jaeger, K. E.; Flemming, H. C.; Wingender *J. Microbiol. Res.* **2005**, *160*, 165–176.
53. Slauch, J. M.; Mahan, M. J.; Michetti, P.; Neutra, M. R.; Mekalanos, J. *J. Infect. Immun.* **1995**, *63*, 437–441.
54. Spencer, B. L.; Shenoy, A. T.; Orihuela, C. J.; Nahm, M. H. *Clin. Vaccine Immunol.* **2017**, *24*, 1–13.

55. Robbins, J.; Schneerson, R.; Anderson, P.; Smith, D. J. *JAMA, J. Am. Med. Assoc.* **1996**, *276*, 1181–1185.
56. Eby, R. Pneumococcal Conjugate Vaccines. In *Vaccine Design: The Subunit and Adjuvant Approach*; Powell, M. F., Newman, M. J., Eds.; Plenum Press: New York, 1995; pp 695–718.
57. Lockhart, S. P.; Scott, D. A.; Jansen, K. U.; Anderson, A. S.; Gruber, W. C. Glycoconjugate Vaccines: The Clinical Journey. In *Carbohydrate-Based Vaccines: From Concept to Clinic*; Prasad, A. K., Ed,; ACS Symposium Series 1290; American Chemical Society: Washington, DC, 2018; Chapter 2.
58. Berti, F. Multivalent Meningococcal Conjugate Vaccines: Chemical Conjugation Strategies Used for the Preparation of Vaccines Licensed or in Clinical Trials. In *Carbohydrate-Based Vaccines: From Concept to Clinic*; Prasad, A. K., Ed,; ACS Symposium Series 1290; American Chemical Society: Washington, DC, 2018; Chapter 6.
59. Bixler, G. S.; Eby, R.; Dermody, K. M.; Woods, R. M.; Seid, R. C.; Pillai, S. In *Immunobiology of Proteins and Peptides V*; Atassi, M. Z., Ed.; Springer Science + Buisness Media: New York, 1989; pp 175–180.
60. Bixler, G.; Pillai, S.; Insel, R. U.S. Patent 5,785,973, July 28, 1998.

Chapter 5

Manufacturing Considerations for Glycoconjugate Vaccines

A. Krishna Prasad,*,[1] Kent Murphy,[2] and Stephen A. Kolodziej[3]

[1]Pfizer Vaccines Research and Development, 401 N. Middletown Rd., Pearl River, New York 10965, United States
[2]Pfizer Global Supply, 4300 Oak Park, Sanford, North Carolina 27330, United States
[3]Pfizer Biotherapeutics and Pharmaceutical Sciences, 700 Chesterfiled Parkway W., Chesterfield, Missouri 63017, United States
*E-mail: krishna.prasad1@pfizer.com.

We review the process development and manufacturing considerations of glyconjugate vaccines. We focus on key elements required for process/product control strategy approaches during various clinical phases through licensure including statistical process control using Design of Experiments and performance qualification necessary for a full process understanding. Central to this approach is the development of a control strategy to demonstrate consistency, robustness testing and transfer of commercially viable process technologies into manufacturing facilities, along with the production of clinical trial material. In this chapter we will focus our discussion on the key elements that define the commercial process development of glycoconjugate vaccine. We also identify the key steps involved to define critical and key process parameters that, in turn, define and impact critical quality attributes and the risks for failure.

Introduction

Vaccine development is a complex and lengthy process typically spanning more than ten to twelve years from discovery through product licensure. The development of a glycoconjugate vaccine presents itself with an additional layer of complexity in order to define the manufacturing process since the process development, scale-up and large scale manufacturing is an area typically implemented within industry. Even though several key aspects of biopharmaceutical process development, such as therapeutic monoclonal antibodies, have been discussed earlier (*1*), information with respect to glycoconjugate vaccine development is much more complex and not readily available in the public domain. Approaches to early phase process development of antibody drug conjugates (ADCs) have some basic similarities with glycoconjugates (*2*), particularly with respect to aspects related to conjugation routes, linkers and the general process control elements that are required for consistency of manufacturing. Since prophylactic vaccines are administered to healthy people, the bar for regulatory hurdles is much higher when compared to therapeutic products such as ADCs which are typically given to sick patients. Secondly, due to the presence of multiple sites of attachment in the carbohydrate antigen as well as the carrier protein, the complexity of glycoconjugate vaccine product is also typically higher. Due to these extra layers of structural complexity associated with glycoconjugate vaccines, a significant amount of process development is required. Therefore, these multi-layered development efforts require a significant development effort, from the outset, in order to define a commercially viable manufacturing process.

Glycoconjugate Vaccine Development Paradigm

The primary goal during the early development stage is to develop an optimal glycoconjugate vaccine construct, based on lab scale development studies, typically in conjunction with some animal immunogenicity studies. This is followed by late phase process development, during which the final commercial manufacturing process, suitable for licensure, is defined. Equally important is the need to manufacture clinical supplies of the vaccine that will be representative of the final commercial process. After approval by the regulatory agencies, it is incumbent on the vaccine manufacturer to demonstrate that the commercial process remains in a state of control producing a safe and efficacious vaccine. A lifecycle approach to commercial glycoconjugate vaccine process development consists of three general phases: commercial process definition, process control, and process surveillance (Figure 1). Central to this approach is the development of a control strategy to demonstrate consistency, robustness testing and transfer of commercially viable process technologies into manufacturing facilities, along with the production of clinical trial material. In this chapter, we will focus our discussion on the late phase glycoconjugate commercial process development. Various related and preceding aspects pertaining to early phase glycoconjugate vaccine design and process development, in particular, have been covered in the other chapters of this book (*3–5*). The development and implementation of a robust control strategy, during all stages of development to licensure, is important since the opportunity to make significant process changes are minimal for glycoconjugate vaccines. This is due to their unique structural complexity which is in contrast to well characterized protein therapeutics, such as monoclonal antibodies.

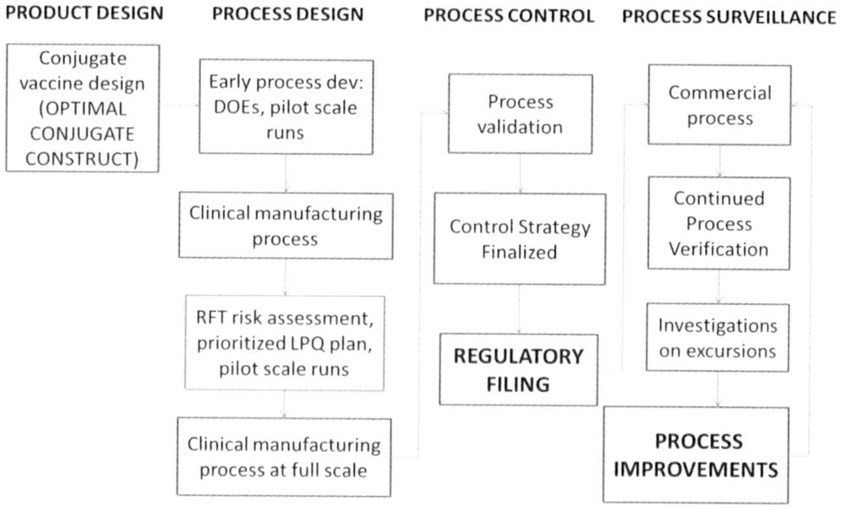

Figure 1. Lifecycle approach to glycoconjugate vaccine commercial process development.

A Deep Process Understanding Underlies Robust Process Design

A systematic approach to late phase process development for a typical monoclonal antibody (MAb) product has been discussed in detail earlier (*6*). The requisite process understanding was obtained through the application of a quality by design (QbD) approach focused on establishing a functional relationship between process parameters and product quality attributes (QAs). A similar strategy is applicable for glycoconjugate vaccines in general, but the approach is significantly more complex. Glycoconjugate vaccines are derived from two drug substance intermediates (DSIs), namely, a capsular polysaccharide (CPS) antigen and a carrier protein. Commercial processes must be designed and validated for both. An additional chemical activation step is typically required for the CPS prior to the actual conjugation process, and again commercial processes must be designed and validated for both. While the manufacturing processes for the DSIs, the CPS, activated CPS, and carrier protein, are similar to the MAb example cited above, process design and control for a glycoconjugate drug substance (DS) is more complex for a variety of reasons. Structural heterogeneity complicates the analytical characterization of final glycoconjugate DS due to several attributes. CPSs are biopolymers with variable size distribution, and subsequent controlled, partial activation results in apportioned functionality across the CPS chain. Heterogeneity also results from the manner in which the CPSs are conjugated to the carrier protein, similar to ADCs (*2*). Material attributes of the DSIs impact the QAs of the glycoconjugate DS and final drug product (DP), and batch to batch differences for DSIs must be evaluated throughout the product lifecycle. Furthermore, each serotype from the various microbial CPS groups, requires a unique CPS DSI and glycoconjugate DS, and the final DP is produced from multiple glycoconjugate DSs, unlike a MAb which typically consists of a single DS formulated into the final DP.

Therefore, a much more systematic, robust, and risk-based approach is necessary to generate process and product understanding for various glycoconjugate vaccine modalities based on lab scale multivariate experimentation using design of experiments (DOE), scale-up/scale-down strategies, and heightened analytical characterization of DSIs and DS to elucidate glycoconjugate vaccine structure.

Design space is the output of process understanding studies that are initiated during early phase process development and continued through the late phase (*6*). Design space is the summation of parameter ranges that yield a glycoconjugate with the desired QAs, referred to as the proven acceptable ranges (PARs) (Figure 2). Commercial processes are designed to operate within the design space/PARs for process parameters in the control space, and a control strategy is applied to ensure that the commercial process remains in a state of control. The process parameter ranges within the control space are known as normal operating ranges (NORs). PARs are determined during late phase process development through process characterization experiments referred to as lab pre-qualifications (LPQs), for example, under Pfizer's Right First Time (RFT) approach. To efficiently cover parameter space, LPQ studies are heavily dependent on DOEs with multivariate statistical analyses of the resulting data.

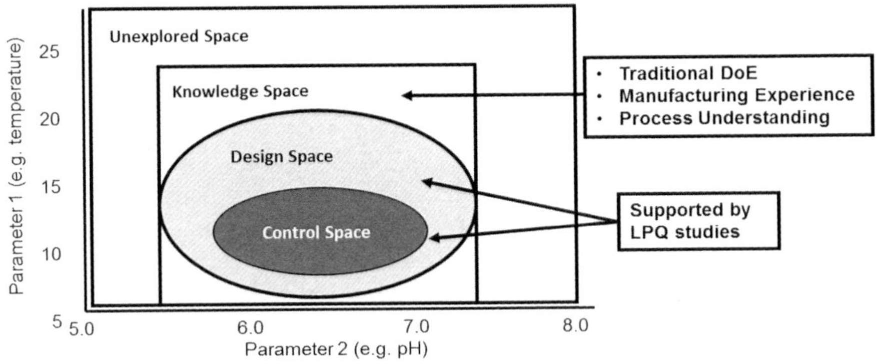

Figure 2. Relationship between design space and control space.

The LPQ experimental plans for glycoconjugate vaccines are prioritized during a series of risk assessments on the DSIs and the glycoconjugate DSs that comprise the final vaccine DP. Risk assessments are conducted by a multidisciplinary team consisting of representatives from early phase development, process development, analytical, manufacturing, regulatory and quality. Data and knowledge from various sources, including relevant process history, platform process understanding and literature are used to prioritize an experimental plan designed to characterize the process. Typical processes for CPS DSI and glycoconjugate DS each consist of 10-20 unit operations and these are divided into unique focus areas for the risk assessment based on equipment and process similarities. Process parameters (inputs) and QAs (outputs) are identified and scored using a predefined scale. The effect of each process parameter on relevant QAs is assessed for each focus area, generating a cause and effect matrix for each focus area. A cumulative score is calculated for each parameter and these are ranked to delineate the relative importance of parameters within each focus area and to prioritize experimental plans.

Prior to conducting lab scale experimentation, a scale-down strategy must be in place for all relevant unit operations. A concerted effort is necessary across research, development and commercial groups to standardize the lab-scale conjugation reaction design in order to reduce development and facilitate technology transfers. The approach includes modeling conjugation chemistry with well understood mechanisms and kinetics, establishing verification criteria and mapping lab-scale model reactors corresponding to pilot and commercial scale parameters to facilitate scale-up. A preferred outcome of this elaborate effort is the identification of a commercially available lab-scale reactor that has the capability of generating scalable data to existing pilot and commercial systems within the various manufacturing suites and facilities.

Once the key unit operations and all process steps, including the conjugation chemistry, have been identified, it is important to define the ranges of various process parameters using strategically planned and executed experiments, using statistical process control. These DOE studies are required to obtain a detailed process understanding leading to a comprehensive picture of the product with desired attributes of an "optimal glycoconjugate construct". For glycoconjugates, the effect on a response may be dependent on multiple parameters since the one-factor-at-a-time (OFAT) approach to process understanding is not only time and resource consuming but also inefficient when compared with changing factor levels simultaneously. These DOE studies typically answer the following questions:

(a) What are the key steps and factors that control the process?
(b) What are the process settings and ranges that control key and critical quality attributes (CQAs)?
(c) Are there any interactive effects and dependent parameters that control the process?

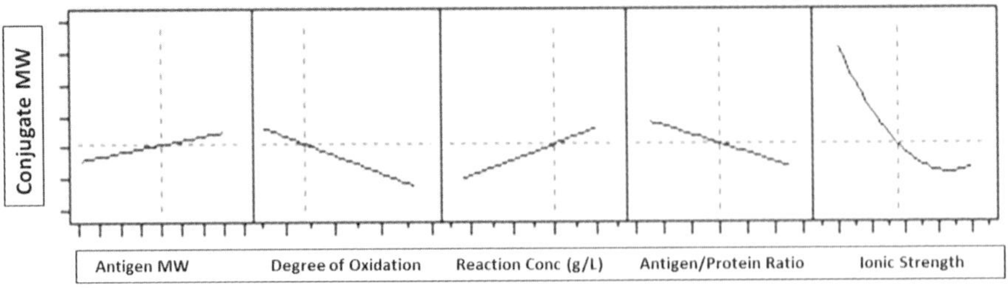

Figure 3. Conjugation parameters that impact glycoconjugate molecular weight (MW).

Figure 3 shows the various conjugation parameters that control a CQA, glycoconjugate molecular weight (MW), as an example. An increase in glycoconjugate MW is seen with an increase in the CPS antigen MW and conjugation reaction concentration. A decrease in glycoconjugate MW is seen with an increase in degree of CPS oxidation and CPS antigen/carrier protein ratio and ionic strength. Ionic Strength has the most impact in terms of significance with a curvature while all the other parameters have a direct linear relationship. Once the desired glycoconjugate MW range has been identified, based on previous historical knowledge and/or animal immunogenicity studies, the appropriate ranges for various process parameters could be delineated. If two process parameters have an opposite effect in terms of an impact on glycoconjugate MW, e.g. antigen MW and degree of oxidation, we found that DOE is particularly helpful in terms of finding a balance between these two parameters to obtain the desired glycoconjugate MW. Once the process parameter ranges have been identified, it is important to control these parameters in a tight window, during the routine manufacturing stage, to manage the QAs.

Control Strategy Considerations Supporting Product Development, Scale-Up, and Transfer

The application of a comprehensive process development control strategy is consistent with industry guidance for biopharmaceutical products (*7*). Glycoconjugate vaccines are produced through a complex continuum of processing steps comprising production of DSI/DS/DP and therefore present challenges consistent with other complex biopharmaceuticals. The application of a product development control strategy for glycoconjugate vaccines controls for challenges associated with the product development lifecycle, from early stage pilot scale clinical production through commercial scale production, which include:

- Scale of operations, processing equipment, and site of production,
- Change of conjugation or purification operating conditions that may potentially impact CQAs
- Change in critical reagents, buffers, and/or processing aides
- Change in storage conditions or container of DSI/DS/DP
- Change in analytical methods used to characterize CQA's

It is recommended to apply a risk management approach consistent with elements of the control strategy to:

- Rank critical process parameters (CPPs) and CQAs throughout the production of DSI/DS/DP
- Rank the priority of processing factors to study
- Determine the appropriate control strategy elements and associated acceptance criteria

A limited number of process changes may be introduced during various phases of clinical trial material production. Therefore, significant process changes should be discouraged that can alter the glycoconjugate attributes such as molecular size, saccharide/protein ratio, levels of free saccharide and patterns of non-saccharide substituents. If any changes to DS buffer matrix, excipients and final formulation and container closure are needed, it is recommended that these measures be initiated during the very early stages of process development.

To ensure consistent process performance and product quality a key element of a control strategy is to perform comparability assessments through controlled studies to connect clinical performance and manufacturing operations, CPPs to unit operation and product CQA's. Performing comparability assessments throughout the product lifecycle from early stage clinical production through scale-up to commercial scale production is a key element to demonstrate a continuum of process control and conservation of product QAs throughout the development lifecycle.

Control Strategy for a Well-Characterized Vaccine Process

Control strategy is defined by ICH Q10 as a planned set of controls, derived from current product and process understanding that ensures process performance and product quality. These controls can include parameters and attributes related to DSI/DS/DP materials and components, facility and equipment operating conditions, in-process controls, finished product specifications, and the associated methods and frequency of monitoring and control (*8–12*). Development of a comprehensive control strategy enables the design of the appropriate continued process verification (CPV) program for implementation during the product life cycle. CPV monitoring and trending may identify areas of improvement in the process, leading to the evolution of the control strategy.

It is important to note that a control strategy is not a new requirement by regulatory agencies and not a new paradigm within the biopharmaceutical industry. What is relatively new is the guidance for how to more effectively develop, implement and clearly communicate a given product's customized control strategy. The control strategy is not just the final deliverable, but is developed and refined throughout the product development lifecycle, based on a number of production runs at various scales.

A well-defined control strategy determines appropriate controls to consistently ensure product quality, safety, efficacy and potency. The control strategy should consider CQAs, QAs and process performance attributes (PPAs) through the application of process monitoring. The combined elements assure product quality (i.e., QAs within the demonstrated acceptable ranges) and mitigate risks of failures in process performance.

A well-defined control strategy should describe the design of the process to manufacture DSI/DS/DP with the desired QAs. Control strategy is comprised of three major components: Facilities, Materials, and Process and Product. The goal is to build knowledge of these components over the product lifecycle to maintain control of product quality and process performance to ensure consistency in commercial manufacture and to demonstrate to regulators an understanding of the CPPs to control and deliver a pure, safe, and efficacious drug to patients.

The overall control strategy is insufficient to ensure consistent DSI/DS/DP quality. CPPs and in-process controls (IPCs) for each process step are determined based on the CQAs. The absence of such an analysis of the whole process and each of the steps is considered a major deficiency in the process development effort.

What Is Required?

Notwithstanding regulatory requirements for a suitable control strategy, the overarching purpose of a control strategy is to ensure consistent process performance and quality of an approved product. The controls may include:

- Parameters and attributes related to DSI/DS/DP manufacturing
- Raw materials, excipients, stabilizers, adjuvants and process intermediates
- Facility and equipment operating conditions
- IPCs
- Specifications for DSI/DS/DP (that may contain excipients, stabilizers and adjuvants).
- Associated methods and frequency of monitoring, sampling and testing, including Process Analytical Technology.

A control strategy is communicated to the various boards of health (e.g., FDA) via a comprehensive document(s) (e.g., CTD - Common Technical Document) provided in a license application, e.g., Biologics License Application (BLA), within the 3.2.S.2.6 Manufacturing Process Development (DSI/DS) and 3.2.P.2.3 Manufacturing Process Development (DP) sections of the dossier. These documents provide clear discussion on how the control strategy was developed and where in the license each element of control is detailed.

The information provided on the control strategy should include detailed descriptions of the individual elements of the control strategy for each of the components, as appropriate (DSI/DS/DP) plus, a holistic overview of the overall control strategy for DSI/DS/DP consisting of all of the individual control components. The control strategy is based on the following information filed in the registration dossier:

- List of CQAs and non-CQAs from Section 3.2.S.2.6 (DSI/DS) and 3.2.P.2 (DP) for which control strategy is filed.
- Material attributes and process parameters, filed in Section 3.2.S.2.2 (DSI/DS) and Section 3.2.P.3.3 (DP), including acceptable limits/ranges for these controls that become the manufacturers commitment for commercial manufacture.
- In-process tests filed in Section 3.2.S.2.4 (DSI/DS) and Section 3.2.P.3.4 (DP), including acceptance criteria for these controls that become our commitment for commercial manufacture.
- Elements of process understanding that qualitatively link operation steps, process parameters etc. with product attributes, filed in Section 3.2.S.2.6 (DSI/DS) and 3.2.P.2.3 (DP).
- Raw materials, including cell banks, starting materials, intermediates and container closure system, for which a direct link to QAs is indicated in Section 3.2.S.2.3 (DSI/DS) and 3.2.P.2.4 (DP).
- Filed release and stability specifications (Section 3.2.S.4.1 and 3.2.P.5.1)
- Elements of product understanding (Section 3.2.S.3.1) that support control strategy for non-CQAs.

Control Strategy Development Methodology

In defining the product control strategy a product development control strategy team should be formed and composed of subject matter experts from relevant functions including analytical, formulation and process development, manufacturing, regulatory and quality.

A holistic product development lifecycle approach should be fit-for-purpose where at each stage of development control attributes are consistent with level of knowledge and clinical stage.

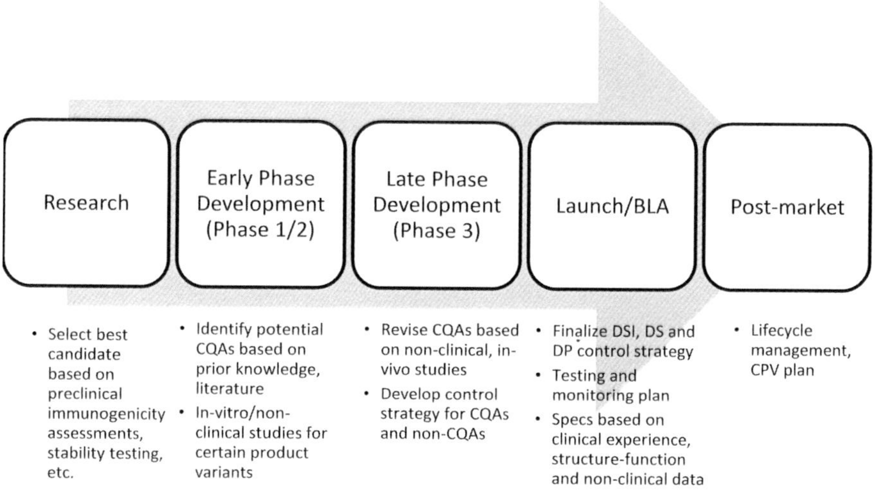

Figure 4. Holistic process and product control strategy.

A product's holistic control strategy (Figure 4) is likely to evolve with time and is developed as part of the development process and is built into plan milestones and deliverables.

To assist the product development team to examine the overall control of the process and product, certain elements of control should be considered for each product or process attribute to ensure a robust control strategy. Elements of the process and product control strategy that should be considered are outlined in Table 1. As part of a control strategy all product and process input and outputs are risk assessed, and the critical process controls that are linked to control of CQAs are categorized as established conditions. The BLA will reflect the most crucial controls and summarized as CQAs versus the relevant elements of controls. A Control Strategy Plan Document should capture the holistic control strategy of all attributes [CQA, QA, PPA (process performance attributes)] versus all eight elements of control listed in Table 1. A Control Strategy Plan Document should be the source document to develop a BLA control strategy document.

Capturing the elements of control, application of risk assessment and refinement of the control strategy are iterative processes. The control strategy for the DSI/DS/DP is developed using a holistic approach considering and assessing several elements of control. The control strategy for the DSI is linked to DS, and DS to DP. Prior to finalization of the control strategy, DSI/DS/DP elements are considered in totality to assure final DP quality is met throughout the product shelf life. Source information, which is assembled to enable the control strategy risk assessment, includes:

- Quality target product profile (QTPP)
- QA risk assessment
- Manufacturing process description
- Process understanding around the impact of process parameters (including operating ranges) and material attributes on the CQAs for each unit operation
- cGMPs (including facility, equipment and procedural elements)
- Failure Mode and Effects Analysis (FMEA) risk assessment for the DSI/DS/DP manufacturing process
- Pharmaceutical quality system

Elements of Process Control Strategy

The first stage of developing the control strategy entails systematically identifying all of the elements of control applied for the DSI/DS/DP. The proposed controls are identified in line with the elements of control as described in Table 1.

Table 1. Description of Elements of the Process and Product Control Strategy

Control Element	*Definition*	*Type of Control*
Element 1	In-process control of product QAs	Direct in-process tests, or surrogates, of product QAs and their control limits
		Product QA in-process control demonstrated through process validation (PV)
Element 2	Control of process parameters	Control implicit in the design of manufacturing process or unit operations
		Manufacturing process controls
		Process parameters that impact product quality or process performance attributes and control limits / acceptable ranges.
		Manufacturing hold time control limits
		Manufacturing process development and history for understanding and application of acceptable ranges
Element 3	Control of process performance	In-process tests of process performance and their control limits
Element 4	Non-routine product characterization testing	Elucidation of structure and other characteristics.
		Non-routine tests for characterization and demonstration of product comparability
		Characterization included in reference standard or reference material qualification
Element 5	Product control through QA testing	Routine release test and acceptance criteria in product specification
		Justification of specification
		Analytical procedure and its validation
		Product QA control demonstrated through PV

Continued on next page.

Table 1. (Continued). Description of Elements of the Process and Product Control Strategy

Control Element	*Definition*	*Type of Control*
Element 6	Product control through stability testing	Routine stability test and acceptance criteria in stability protocols
		Stability data and conclusions
Element 7	Control of materials	Specifications for raw materials
		Manufacture and testing of cell banks; cell bank controls
		Characteristics of incoming materials (such as raw materials, starting material, intermediates, primary packaging materials) that impact product QAs and their acceptable ranges
		Compatibility with container closure system; container closure controls
Element 8	Facility and equipment controls	cGMP and pharmaceutical quality systems for the manufacturing facilities
		Environmental and equipment controls directly impacting product QAs

Product Development Control Strategy Considerations

As noted due to the structural complexity of glycoconjugate vaccines it is important to establish a robust product development control strategy for all phases of development through licensure. Prior to each production phase (clinical phases) elements of control should be demonstrated and fit-for-purpose where control attributes are consistent with the level of knowledge. Outlined for each stage of development below are considerations for product development and commercialization of glycoconjugate vaccines.

Pre-Clinical (TOX)

- Research criteria are met to advance product candidate for further development through pre-Clinical phase.
- Process expectations include: Early clinical phase 1&2 readiness, manufacturing process development, and manufacturing platform fit assessment.
- Product expectations include: Initial QA assessment, cell bank generation, stability and formulation.
- Analytical expectations include method evaluation, toxicology specifications, release testing, and in-process monitoring, reference material, and critical reagent management.
- Manufacturing expectations include EHS assessment, site selection for Tox material manufacture.
- Knowledge management: Assay, process and formulation development reports, process description, technology transfer documents, testing and monitoring sampling plans, Tox manufacturing batch records, and Tox specifications documents.
- Evaluation of *in-vivo/in-vitro* model study design to support product performance, preliminary QA assessment aligned with clinical expectations, establish master and working cell banks (MCB & WCB), DSI/DS/DP/placebo stability, including process capability, defined DSI/DS/DP container closure systems, initial QTPP
- Regulatory: Pre-IND meetings, IND submission.

Early ClinicaL (PH 1-2)

- Defined process, formulation & analytical methods.
- Process expectations include: Late clinical phase 3 readiness, manufacturing process development, definition, and description.
- Product expectations include: Summary of quality characteristics of all DSI/DS/DP, control of impurities through release testing, initial raw materials assessment, demonstrated comparability of Tox and early clinical materials, dosage and administration instructions (DAI) study to support clinical study design, and identify primary packaging requirements.

- Analytical expectations include: Method development, qualification and transfer, clinical DSI/DS/DP specifications, release and in-process testing (IPT), excipient, safety testing, and critical reagent management.
- Manufacturing expectations include: EHS assessment, manufacturing site, GMP manufacturing controls, process and analytical technology transfer, and equipment/facility qualification.
- Knowledge management: Development report, clinical campaign summary, clinical specification report, potential CQAs designation and report QTPP, assay and process development report from Tox to clinical phase 1-2, assay qualification report, container closure report, process description, testing and monitoring sample plan, and key material/excipient list.
- Quality risk management: QTPP and CQA assessment.
- Regulatory: Regulatory strategy, IND submission

Late Clinical (PH 2B-3)

- Historical batch and stability data meet intended quality targets or clinical specifications. Draft product Control Strategy plan. Clinical manufacturing process steps and equipment considered representative of commercial manufacturing process. Process and analytical transfer completed to clinical manufacturing site. Process description, including Formulation & Packaging descriptions.
- Process expectations include: Process robustness assessment, product and process characterization, manufacturing process description, process parameter criticality and functional relationship with QAs, sampling and analytical testing requirements, characterize pre-validation studies.
- Product expectations include: CQAs assessment, structure-function studies aligned with clinical specifications, defined cell bank strategy, WCB qualification and release, DSI/DS/DP/placebo stability, including process stability for ICH materials, commercial formulation, defined DSI/DS/DP container closure and integrity system, revised QTPP as appropriate, developed and documented process capability for impurity removal, identified critical DSI/DS/DP raw materials and risk mitigation identified, analytical testing requirements, demonstrated comparability of early/late clinical materials, finalized commercial comparability plan, additional DAI study to support new regions, defined primary secondary packaging requirements, and assess impact of leachables and extractables from product contact parts.
- Analytical expectations include: Product specific method development, qualification, validation and transfer, revised clinical specifications, revised release and IPT excipient testing, safety testing, establish reference material for new processes, critical reagent management (re-supply).
- Manufacturing expectations include: EHS assessment, site selection for commercial manufacture, GMP manufacturing controls, process and

analytical technology transfer, equipment/facility qualification, product release schedule, cleaning procedures, reproducibility runs, define shipping configuration.

- Knowledge management: Process description, process characterization reports and other technical studies, initial experimental process plan or technology lifecycle plan, criticality designation (PP and QA) and report, commercial specification report, updated QTPP, container closure documents (extractable assessment), technology transfer report, comparability plan, analytical development history, assay qualification report.
- Quality risk management: Process risk assessment (cause and effect (C&E), FMEA), PP and CQA assessment, materials risk assessment, stability risk assessment, process fit and technology/analytical transfer assessment, extractables and leachables assessment.
- Regulatory: Revise regulatory strategy and initiate planning for BLA authoring/review and commercial supply.

Process Performance Qualification (PPQ)

- Draft Control Strategy plan. Process and analytical characterization. PP/CQA assessment. Stability characteristics for DSI/DS/DP including ICH plan. Defined primary and secondary packaging requirements (DSI/DS/DP). Comparability plan (process, product and stability), completed technology and analytical transfer activities, analytical method validation.
- Process expectations include: PV readiness, master manufacturing batch records, PV acceptance criteria, PV protocols and reports, sampling and analytical testing requirements for PV, characterization lifetime studies, validation plan including PPQ and CPV.
- Product expectations include: Revised CQA assessment, aligned with commercial specifications, DSI/DS/DP placebo stability, including in-process stability for ICH and PV materials, validate container closure system, revise/finalize QTPP as appropriate, define plan for validation of impurity removal and commercial testing, established raw material release specifications and analytical methods, comparability assessment (product, process, and stability), commercial dosage administration instructions, finalize leachables and extractables report from product contact parts.
- Analytical expectations include: Analytical methods PV risk assessment, validated analytical methods, commercial/validation specifications, validation testing (release, stability, and in-process), establish primary/commercial reference materials).
- Manufacturing expectations include: EHS assessment (safety), GMP manufacturing controls, validated process controls (PP/QA), equipment/facility qualification, production plan (PV lot manufacture and testing, product specific cleaning validation, trial runs (if required),

shipping qualification/validation, media hold/sterility studies, in-process holds/storage intermediates, qualified material vendors.

- Knowledge management: Validation documentation (PV master plan, validation protocols and reports), site control and monitoring plan, site CPV plan, commercial process understanding plan or technology lifecycle plan, risk assessment knowledge capture, site process description.
- Quality risk management: Process risk assessment, validation readiness assessment, extractables and leachables assessment, commercial manufacturing readiness.
- Regulatory: Revised regulatory strategy, BLA authoring/review.

Manufacturing (GMP)

- Commercial readiness activities. Final product Control Strategy plan. Plan for CPV. Commercial specifications for DSI/DS/DP. Validated stability properties for intermediates, bulk DS and DP. Site Controls and Monitoring plan.
- Process expectations include: Performance capability assessment, master manufacturing batch records, sampling requirements for commercial manufacturing, site control and monitoring plan, supplemental validation studies and reports, and lifecycle continuous process performance.
- Product expectation includes: Cell bank qualification and release as required, material/vendor qualification and release Bill of Materials.
- Analytical expectations include: Commercial specifications, commercial testing (release, stability, and in-process), and critical reagent management (lifecycle).
- Manufacturing expectations include: EHS assessment (safety), GMP manufacturing controls, lifecycle revisions to documented control strategy, commercial production plan, cleaning and lifecycle validation.
- Knowledge management: Site documentation Quality System, site data acquisition/analysis, GMP batch documentation, CPV report.
- Quality risk management: Process risk assessment, process capability analysis (e.g. PpK, robustness contours), manufacturing investigations, process risk evaluation (e.g. FMEA), continuous improvement and lean deployment.
- Regulatory: BLA/Market Authorization submission/questions, pre-approval inspection readiness (PAI), annual product quality review, compliance with regulatory/PAI commitments.

Alignment of Control Strategy with Regulatory Requirements for a Successful Process Performance Qualification

Where production of glycogonjugates is complex, involving DSI/DS/DP consisting of multiple unit operations and where structural heterogeneity of glycoconjugates complicates analytical characterization the approach to validation must be built upon a holistic product lifecycle development view that encompasses process understanding connecting the process to clinical and product QA performance.

As outlined in "Guidance for Industry. Process Validation: General Principles and Practices" (*13*), the application of a holistic control strategy aligns PV with a product lifecycle concept and with existing FDA guidance, where "successful validation is identified as depending upon information and knowledge from product and process development". Several industrial case studies, pertaining to process validation in manufacturing of biopharmaceuticals, have been provided by Rathore and Sofer (*14*). A holistic product development control strategy is "the basis for establishing an approach to control the manufacturing process that results in products with the desired quality attributes". It is through quality by design and the use of multivariate experimental DOE that we define functional relationships between process parameters and product QAs and by use of control elements in their totality that we understand sources of variation and our ability to detect the presence and degree of variation and their impact on the process and product QAs.

The application of the elements of process control throughout product development is the basis for identifying CPPs and IPCs that when successfully executed for each unit operation and the process overall, CQAs for DSI/DS/DP are assured. "The FDA expects controls to include both examination of material quality and equipment monitoring with special attention to control the process through operational limits and in-process monitoring when intermediates and products cannot be highly characterized and well-defined QAs cannot be identified". "A successful process performance qualification will confirm the process design and demonstrate that the commercial manufacturing process performs as expected".

Process qualification "must specify manufacturing conditions, controls, testing, and expected outcomes". "Guidance recommends that the protocol discuss the following elements:"

- The manufacturing conditions, e.g., operating parameters, processing limits, and component inputs
- The data to be collected and how it will be evaluated
- Tests to be performed (in-process, release, characterization) and acceptance criteria for each step
- The sampling plan, including sampling points, number of samples, and frequency of sampling for each unit operation and attribute
- Criteria and process performance indicators that allow for a science and risk based decision about the ability of the process to consistently produce quality products
- Design of the facility and qualification of utilities and equipment, personnel training and qualification, and verification of material sources
- Status of validation of analytical methods used in measuring process, in-process materials, and product
- Review and approval of the validation protocol by appropriate departments and the quality unit.

Scale-Up and Technology Transfer Considerations

Technology transfer organization is critical to facilitate planning, implementation and management of end-to-end product transfers. Consideration for scale-up and transfer of complex processes should be performed well in advance of supplying pivotal clinical material. Adequate planning in advance is critical to transfer and production success and should account for engineering/trial/demonstration batches, and cGMP clinical and validation production.

Key considerations needed to define and develop the transfer plan:

- Process robustness. Assess process robustness and capabilities to decide how to transfer, e.g., improve the process and then transfer or introduce improvements during the transfer.
- Perform a GAP assessment and documentation of changes. Identify changes associated with manufacturing and packaging the product. Each change should be documented with justification and comparability plan. The GAP assessment is the basis of the full regulatory strategy and subject to change management.
- Knowledge management. Documentation and tacit knowledge that support process transfer is captured.
- Once gap assessments are completed and changes are agreed, a full Regulatory Strategy Document is developed. This document will contain detailed regulatory requirements, a submission plan and risk registry assessment.

Conclusions

Even though several key aspects of biopharmaceutical process development, such as therapeutic monoclonal antibodies, have been discussed earlier, information with respect to glycoconjugate vaccine development is much more complex and not readily available in the public domain. Glycoconjugate vaccines are produced through a complex continuum of processing steps comprising production of drug substance intermediates/drug substance/drug product and therefore present challenges consistent with other complex biopharmaceuticals. Since prophylactic vaccines are administered to healthy people, the bar for regulatory hurdles is much higher when compared to biotherapeutics which are typically given to sick patients. Secondly, due to the presence of multiple sites of attachment in the carbohydrate antigen as well as the carrier protein, the complexity of glycoconjugate vaccine product is also typically higher. Due to these extra layers of structural complexity associated with glycoconjugate vaccines, a significant amount of process development is required. The development of a glycoconjugate vaccine presents itself with an additional layer of complexity in order to define the manufacturing process since a large amount of information regarding several key aspects of process development, scale-up and large scale manufacturing is typically reserved within the domain vaccine industry. The development and implementation of a robust control strategy, during all stages of development to licensure, is important since the opportunity to make significant process changes are minimal for glycoconjugate vaccines. This is due to their unique structural complexity which is in contrast to well characterized protein therapeutics, such as monoclonal antibodies. Therefore, these multi-layered development efforts require a significant development effort, from the outset, in order to define a commercially viable manufacturing process. In this chapter, we identify a lifecycle approach and some key control strategies including steps involved to define critical and key process parameters that, in turn, define and impact critical quality attributes and the risks for failure.

Control Strategy Definitions

Quality Attribute (QA):

An attribute that affects product quality such as the identity, strength, and purity

Critical Quality Attribute (CQA):

CQA is a physical, chemical, biological, or microbiological property or characteristic that should be within an appropriate limit, range, or distribution to ensure the desired product quality. The assignment of criticality to a QA is based upon the potential of the QA to impact patient safety or efficacy.

CQAs are distinguished from QA by an iterative process of quality risk management and experimentation that assesses the extent to which their variation can have an impact on the quality of the drug product

Process Parameter:

A process variable (e.g., temperature, compression force) that can be assigned values to be used as control levels or operating limits. Process parameters are categorized into CPP and non-CPPs.

Deviation during commercial manufacturing will result in a formal investigation, which will include a regulatory impact assessment as well as a product quality impact assessment. Batch disposition decision is dependent on this investigation outcome.

Critical Process Parameter (CPP):

A process parameter whose variability has an impact on a CQA and therefore should be monitored or controlled to ensure the process produces the desired quality.

Non-Critical Process Parameter (non-CPP):

All input parameters that fall outside of the definition for critical process parameters are non-critical process parameters.

Process parameter ranges can be within the normal operating range (NOR) and/or within the proven acceptable range (PAR):

Normal Operating Range (NOR):

A defined range specified in the manufacturing batch record instructions at which a process parameter is controlled

Proven Acceptable Range (PAR):

A characterized range of a process parameter for which operation within this range, while keeping other parameters constant, will result in a material meeting relevant quality criteria.

Process Performance Attribute (PPA):

An output variable or outcome that cannot be directly controlled but is an indicator that the process performed as expected (e.g. step yield, chromatographic profile)

In Process Tests (IPT):

In-process tests are tests that are either used to monitor or control the process at appropriate stages.

Control: The in-process test is used to control a QA/CQA to within a specified value so that it meets desired DSI/DS/DP quality.

Acknowledgements

The authors would like to thank Dr. Bo Arve for reviewing the chapter and provding useful insights.

References

1. Steinmeyer, D. E.; McCormick, E. L. *Drug Discovery Today* **2008**, *13*, 613–618.
2. Prashad, A.; Nolting, A.; Patel, V.; Xu, A.; Arve, B.; Letendre, L. *Org. Process Res. Dev.* **2017**, *21*, 590–600.
3. Prasad, A. K.: Kim, J.-h.; Gu, J. Design and Development of Glycoconjugate Vaccines. In *Carbohydrate-Based Vaccines: From Concept to Clinic*; Prasad, A. K., Ed.; ACS Symposium Series 1290; American Chemical Society: Washington, DC, 2018; Chapter 4.
4. Berti, F. Multivalent Meningococcal Conjugate Vaccines: Chemical Conjugation Strategies Used for the Preparation of Vaccines Licensed or in Clinical Trials. In *Carbohydrate-Based Vaccines: From Concept to Clinic*; Prasad, A. K., Ed.; ACS Symposium Series 1290; American Chemical Society: Washington, DC, 2018; Chapter 6.
5. Hennessey, J. P., Jr.; Costantino, P.; Talaga, P.; Beurret, M.; Ravenscroft, N.; Alderson, M. R.; Zablackis, E.; Prasad, A. K.; Frasch, C. Lessons Learned and Future Challenges in the Design and Manufacture of Glycoconjugate

Vaccines. In *Carbohydrate-based Vaccines: From Concept to Clinic*; Prasad, A. K., Ed.; ACS Symposium Series 1290; American Chemical Society: Washington, DC, 2018; Chapter 13.
6. Banerjee, A. *BioPharm. International* **2010**, *43*, 26–40.
7. *Handbook/Manual/Guide by International Society for Pharmaceutical Engineering*, 10/31/2013. ISPE Guide: Biopharmaceutical Process Development and Manufacturing. https://www.ispe.org/publications/guidance-documents/biopharmaceutical-process-development-manufacturing (accessed 5 May 2018).
8. *International Conference on Harmonisation of Technical Requirements for Registration of Pharmaceuticals for Human Use: ICH Harmonised Tripartite Guideline Pharmaceutical Development Q8(R2)*, Current Step 4 version dated August 2009. http://www.ich.org/fileadmin/Public_Web_Site/ICH_Products/Guidelines/Quality/Q8_R1/Step4/Q8_R2_Guideline.pdf (accessed 5 May 2018).
9. *International Conference on Harmonisation of Technical Requirements for Registration of Pharmaceuticals for Human Use: ICH Harmonised Tripartite Guideline Quality Risk Management Q9*, Current Step 4 version dated 9 November 2005. http://www.ich.org/fileadmin/Public_Web_Site/ICH_Products/Guidelines/Quality/Q9/Step4/Q9_Guideline.pdf (accessed 5 May 2018).
10. *International Conference on Harmonisation of Technical Requirements for Registration of Pharmaceuticals for Human Use: ICH Harmonised Tripartite Guideline Pharmaceutical Quality System Q10*, Current Step 4 version dated 4 June 2008. http://www.ich.org/fileadmin/Public_Web_Site/ICH_Products/Guidelines/Quality/Q10/Step4/Q10_Guideline.pdf (accessed 5 May 2018).
11. *International Conference on Harmonisation of Technical Requirements for Registration of Pharmaceuticals for Human Use: ICH Harmonised Tripartite Guideline Development and Manufacture of Drug Substances (Chemical Entities and Biotechnological/Biological Entities) Q11*, Current Step 4 version dated 1 May 2012. http://www.ich.org/fileadmin/Public_Web_Site/ICH_Products/Guidelines/Quality/Q11/Q11_Step_4.pdf (accessed 5 May 2018).
12. *Established Conditions: Reportable CMC Changes for Approved Drug and Biological Products, Guidance for Industry U.S. Department of Health and Human Services, Food and Drug Administration, Center for Drug Evaluation and Research (CDER), Center for Biologics Evaluation and Research (CBER)*, May 2015 Pharmaceutical Quality/CMC. https://www.fda.gov/downloads/drugs/guidancecomplianceregulatoryinformation/guidances/ucm448638.pdf (accessed 5 May 2018).
13. *Guidance for Industry Process Validation: General Principles and Practices U.S. Department of Health and Human Services Food and Drug Administration, Center for Drug Evaluation and Research (CDER), Center for Biologics Evaluation and Research (CBER), Center for Veterinary Medicine (CVM)*, January 2011 Current Good Manufacturing Practices

(CGMP) Revision 1. https://www.fda.gov/downloads/drugs/guidances/ucm070336.pdf (accessed 5 May 2018).
14. Rathore, A. S.; Sofer, G., Eds.; in *Process Validation in Manufacturing of Biopharmaceuticals: Guidelines, Current Practices and Industrial Case Studies*; Taylor & Francis, CRCPress, Boca Raton, FL, USA, 2005.

Chapter 6

Multivalent Meningococcal Conjugate Vaccines: Chemical Conjugation Strategies Used for the Preparation of Vaccines Licensed or in Clinical Trials

Francesco Berti*

Technical R&D, GSK Vaccines, Siena 53100, Italy
***E-mail: francesco.x.berti@gsk.com.**

Several multi-valent carbohydrate-based vaccines have been licensed or are currently in clinical development, so far, to prevent bacterial infection caused by several serogroup of *Neisseria meningitidis*. In this chapter, the conjugation strategies, including the chemical structures and functional groups involved in the covalent coupling of polysaccharides to carrier proteins for the preparation of tetravalent meningococcal serogroup A, C, W, Y glycoconjugates vaccines, have been described in depth.

Introduction

Meningococcal disease is caused by the gram-negative diplococcus *Neisseria meningitidis* (Nm). It was identified as one of the causative agents of bacterial meningitis by Weichselbaum in 1887. Reports of this disease date back to 1807 and nowadays meningococci are known as one of the major causes of bacterial meningitis. The incidence of invasive meningococcal disease (IMD) varies according to geographical location and time, and usually results in septicemia and/or meningitis, and can be complicated by purpuric rash, purpura fulminans, arthritis, myocarditis, pericarditis, endophthalmitis, or pneumonia. The symptoms of bacterial meningitis include headache, stiff neck, fever, chills, malaise, and prostration (*1*).

Meningococcal disease is predominantly a disease of infants and young children, although the disease does have a bimodal distribution with a second age peak in adolescents. Neonates are relatively protected against meningococcal disease as a result of passive acquisition of transplacental maternal antibodies (*2*).

With increasing age, there is a decreasing level of maternal antibodies and increasing susceptibility to meningococcal disease. The lowest antibody levels are found in infants between 6 months and 2 years of age after which natural immunity begins to appear.

The antigenic diversity is defined by chemically and immunologically distinctive capsular polysaccharides which led to serogroup definition, from which 12 different serogroups have been reported. The additional 13th serogroup D initially reported does not exist. The structure has not been described yet and genetic analyses revealed that the attributed serogroup D isolate was found to contain serogroup C capsule biosynthesis genes with internal stop codons in some genes resulting in an unencapsulated phenotype (*3*).

Multiple serogroups are associated with IMD (i.e. serogroup A, B, C, W and Y). However, recently, serogroup X has become important in some regions (*4*).

The meningococcal capsular polysaccharides have the following chemical structure (*5*, *6*):

NmA: →6)-α-D-Man*p*NAc(3/4OAc)-(1→OPO_3→
NmB: →8)-α-D-Neu5Ac-(2→
NmC: →9)-α-D-Neu5Ac(7/8OAc)-(2→
NmW: →6)-α-D-Gal*p*-(1→4)-α-D-Neu5Ac(7/9OAc)-(2→
NmY: →6)-α-D-Glc*p*-(1→4)-α-D-Neu5Ac(7/9OAc)-(2→
NmX: →4)-α-D-Glc*p*NAc-(1→OPO_3→

Vaccine formulations containing purified meningococcal polysaccharides (bivalent, serogroups A and C; trivalent, serogroups A, C and W; tetravalent, serogroups A, C, W and Y), lyophilized together with lactose excipient and to be reconstituted with diluent before use, have been available. Tetravalent serogroup A, C, W and Y polysaccharide vaccines (Mencevax®, GSK Vaccines; Menomune®, Sanofi Pasteur) were licensed in the 1980s (*7*).

Immunization with the meningococcal serogroup B polysaccharide antigen, which is composed of the →8)-α-D-Neu5Ac-(2→ repeating unit that mimics a protein-attached glycan expressed in the foetus meaning that immune tolerance exists against this structure (*8*), resulted in a very low antibody responses.

Molecular mimetics of meningococcal serogroup B polysaccharide epitopes have also been explored as potential vaccine candidates (*9, 10*). Using a N-propionyl-tetanus toxoid conjugate vaccine, Granoff et al. produced a panel of anti-N-propionyl meningococcal serogroup B polysaccharide monoclonal antibodies (mAbs) that were reactive with *N. meningitidis* serogroup B infection but were either not cross-reactive or minimally cross-reactive with purified polysaccharide or human polysialic acid antigens (*11*). Nevertheless, the manufacturers have been reluctant to test a system that may result in a breach of immune tolerance against a foetal glycan associated with neural development.

The development focus of vaccines against *N. meningitidis* serogroup B infection was later shifted towards protein-based antigens. Outer membrane vesicles (OMVs) containing outer membrane proteins (OMPs), chemically treated by detergent to reduce reactogenicity, and, more recently, a limited selection of recombinant OMPs selected by searching the genome have been used as antigens to prepare vaccine formulations with a wide strains coverage (*12–14*).

Vaccines containing both meningococcal polysaccharide of serogroups C and OMPs of serogroup B (VA-Mengoc-BC, Finley Institute), or containing recombinant OMPs only (Trumenba®, Pfizer), or a combination of recombinant OMPs and OMVs (Bexsero®, GSK Vaccines) have been developed and licensed (*15*).

Polysaccharide vaccines, which have been used for over 40 years, have proven immunogenicity and an acceptable tolerability and safety profiles, but they suffer fromsome limitations. As T-cell independent antigens, they are poorly immunogenic in individuals younger than 2 years of age, they cannot induce immune memory and may induce hyporesponsiveness.

To overcome these limitations, conjugate vaccines in which a capsular polysaccharide is chemically linked to a protein that is thus termed a 'carrier protein' have been successfully developed and used, as discussed in earlier chapters (see Lockhart et al. (*16*), Chapter 2; Duke and Avci (*17*), Chapter 3; and Prasad et al. (*18*), Chapter 4). In contrast to polysaccharide vaccines, the conjugate vaccines induce T cell-dependent responses and therefore, they induce immunological memory. They are also highly immunogenic, even when used in infants, and have a longer duration of protection than polysaccharide vaccines. The conjugate vaccines may induce a robust immune response that reduces nasopharyngeal carriage, and have the potential to provide herd immunity.[7]

Conjugate Vaccines

As the culmination of an intensive 5 year clinical trial research programme sponsored by the Department of Health (DH) for England, whith the objective to accelerate the availability a vaccine aganst meningococcal serogroup C disease, in November 1999 the UK became the first country implementing a large immunisation programme using protein conjugate vaccines. Three vaccines (Meningitec®, Nuron Biotech, formerly Pfizer; Menjugate®, GSK Vaccines, formerly Novartis Vaccines; NeisVac-C®, Pfizer, formerly Baxter) were developed, licensed and introduced as part of routine immunization of all children aged >4 months to <18 years (*19*). A dramatic reduction in the number of confirmed cases of NmC IMD in the immunized age groups in the UK was observed, with a vaccine effectiveness of 81% overall and 92-97% in teenagers; the number of deaths fell from 67 in 1999 to 5 in 2001 (*20*).

Despite the use of millions of doses of meningococcal polysaccharide vaccine, meningitis epidemics, largely due to serogroup A *Neisseria meningitidis*, continued to plague meningitis belt countries. To finally eliminate epidemic meningitis due to meningococcal serogroup A, reduce the rate of endemic meningitis up to 50% (the fraction of endemic meningitis due to this serogroup) and establish an herd immunity (control of this serogroup) in infants and persons above 30 years of age (*21*), a monovalent conjugate vaccine (MenAfriVac®, Serum Institute of India Ltd) was successfully developed, tested, licensed (*22*) and distributed as part of the Meningitis Vaccine Project (MVP), a Gates Foundation funded partnership between PATH and the World Health Organization (WHO), in the African 'meningitis belt,' where there is a high incidence of serogroup A IMD. Over 260 million Africans have received a dose of the vaccine in campaigns largely directed at 1-29-year olds. The impact was huge with dramatic elimination of serogroup A meningococcal infections wherever the vaccine has been used at public health scale. To ensure that African populations remain protected, LaForce et al. (*23*), suggested the implementation of an Expanded Program Immunization (EPI) program: birth cohorts to be immunized with a formulation of MenAfriVac given at 9 months with measles, and the populations born after the initial 1-29 year old campaigns, and consequently not yet immunized with the serogroup A conjugate vaccine, immunized in country-specific catch-up campaigns. Countries with poor EPI coverage (measles coverage <60%) will likely need quinquennial vaccination campaigns aimed at covering 1-4 year olds.

As summarized in Table 1, bivalent meningococcal A and C conjugate vaccines have been developed and licensed in China.

A bivalent serogroup C/Hib (Menitorix®, GSK Vaccines) and a trivalent CY/Hib (MenHibrix®, GSK Vaccines) conjugate vaccine have been licensed. While all these conjugate vaccines are marketed internationally, some conjugates have been specifically developed for national immunization programs (i.e. Menitorix® is mainly used in UK and Australia).

Three tetravalent serogroups ACWY conjugates, Menactra® (MenACWY-DT; Sanofi Pasteur), Menveo® (MenACWY-CRM; GSK Vaccines, formerly Novartis Vaccines) and Nimenrix® (MenACWY-TT; Pfizer, formerly GSK Vaccines) have been developed and are available for several markets.

Excluding the monovalent vaccines against meningococcal serogroup C (Meningitec® and NeisVac-C®), the bivalent combinations NmAC regionally distributed in China, and the tetravalent NmACWY Menactra® vaccine, the 'valency' of all the other vaccines (i.e. mono- or bi-valent one) have been extended up to four or five combinations. For instance, the tetravalent Menveo® vaccine has been developed by combining the serogroup C conjugate contained in Menjugate® vaccine with the serogroup AWY conjugates. The serogroup C and Y conjugates of Menitorix® and MenHibrix® are also contained in tetravalent Nimenrix® vaccine. As discussed later, the monovalent NmA MenAfriVac® vaccine is being extended to a pentavalent ACWYX vaccine, whereas licensed

ACWY conjugates have been proposed as pentavalent combinations with serogroup B antigens (Bexsero® and Trumenba®).

Table 1. Multivalent meningococcal conjugate vaccine licensed or in clinical trials. The multivalent conjugate vaccines containing at least two Nm serogroups are listed here.

Serogroup	*Commercial trade name/clinical trial phase*	*Manufacturer*
AC [a]	*MengNa Kang*	Beijing Luzhu (subsidiary of ChongquingZhifei)
	Nao Man King	Xiangrui (also known as Beijing Sanroad)
	Wo Er Kang	Yunnan Walvax
	MeningACon	BeijingZhifei Lvzhu Biopharmaceutical
	N/A	Royal Wuxi
CY/Hib	MenHibrix®	GSK Vaccines
ACWY	Menactra®	Sanofi Pasteur
	MenveoMenveo®	GSK Vaccines
	NimenrixNimenrix®	Pfizer
	MenQuadTT (Phase III)	Sanofi Pasteur
ACWYX	Phase I	Serum Institute of India/PATH
ABCWY [b]	Phase I	GSK Vaccines
	Phase I	Pfizer

[a] Vaccine for national immunization programs. [b] Multivalent conjugate vaccines also including non-conjugate antigens.

Conjugation Chemistry for Preparing Conjugate Vaccines

The main scope of this chapter is to describe the conjugation strategies/chemistries applied for preparing NmACWY conjugates contained in the tetra- or pentavalent vaccine combinations (Table 2). Details of chemical structures and functional groups involved in the covalent coupling of polysaccharides to carrier proteins are reported.

Three carrier proteins (TT, Tetanus Toxoid; DT, Diphtheria Toxoid; CRM_{197}, Cross Reacting Material 197) have been used for all meningococcal conjugate vaccines licensed or currently in clinical development. The toxoids DT and TT are obtained from the respective toxins by chemical detoxification with formaldehyde, while CRM_{197}, a 58 kDa nontoxic mutant of diphtheria toxin, is isolated from the supernatant of *Corynebacterium diphtheriae* C7(b197) tox(-) strain bacterial growths. Other technologies for producing recombinant CRM_{197} are currently under development (*25–27*).

As described in Table 2, for the first time two different carrier proteins have been used for meningococcal multivalent conjugate vaccine (NmACWYX combination; TT for NmA and NmX polysaccharides, and CRM_{197} for NmC, NmW, NmY polysaccharides) (*28*).

Table 2. Tetra- or pentavalent meningococcal conjugate vaccine licensed or in clinical trials with details on carrier proteins

Serogroup	*Commercial trade name/clinical trial phase*	*Carrier protein*
ACWY	Menactra®	DT
	Menveo®	CRM_{197}
	Nimenrix®	TT
	MenQuadTT (Phase III)	TT
ACWYX	Phase I	TT, CRM_{197}
ABCWY [b]	Phase I	CRM_{197}
	Phase I[a]	TT

[a] Clinical Study Identifier: NCT03135834 - A trial to describe the immunogenicity and safety of 2 doses of divalent rLP2086 (Trumenba) and a pentavalent meningococcal vaccine in healthy subjects >=10 to <26 years of age. This study is examining safety and immunogenicity of 2 doses of Trumenba® administered on a 0-, 6-month schedule. This clinical trial is also studying safety and immunogenicity of a meningococcal pentavalent vaccine (*24*).

The meningococcal conjugates have been prepared by different conjugation strategies, which result in either a cross-linked structure with the saccharide chains (long or size-reduced) randomly linked to the carrier protein (Figure 1a) or a well-defined structure with the size-reduced saccharide chains terminally linked to the carrier (Figure 1b).

As described in detail by Broker et al. (*29*), NmACWY antigens of Menveo are prepared using approach depicted in Figure 1b, starting with the controlled acidic hydrolysis of the polysaccharides chains. A polydispersion of oligosaccharides with defined length is then fractionated, activated *via* reductive amination followed by introduction of an adipic acid linker bearing a terminal active ester and finally conjugated to CRM_{197} via its primary amino groups (i.e. ε-NH_2 groups of Lys residues and protein N-terminus). Since the modification of saccharide chains is limited to the reducing ends, the conjugated oligosaccharides are all radially oriented relative to the carrier protein (Figure 2). This manufacturing process results in a very well defined product with excellent retention of the structural identity of the oligosaccharides following conjugation and a consistent saccharide to protein ratio (loading of saccharide chains on the protein) (*29–32*).

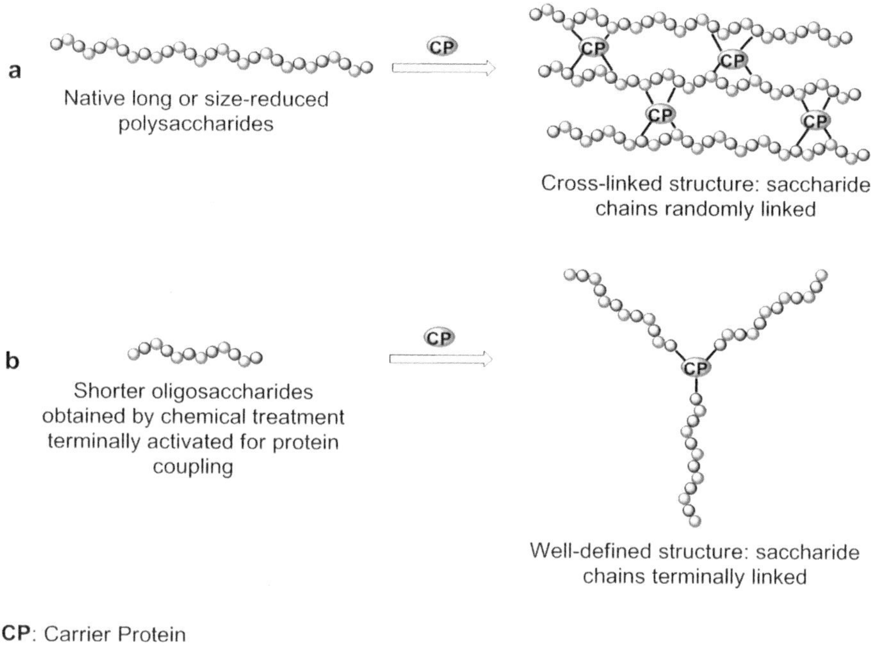

Figure 1. Schematic representation of NmACWY conjugate vaccines prepared by using different conjugation strategies. (see color insert)

A second example of conjugation chemistry has been proposed by Ryall et al.(Figure 3) (*33*). The meningococcal polysaccharides are first partially depolymerized by hydrogen peroxide treatment followed by an activation step to provide chemical groups capable of reacting with the carrier protein. In addition to analytical controls commonly used [i.e. ^{1}H and ^{13}C NMR spectroscopy, O-acetyl content (*34*), reducing activity (*35*)], the integrity of the reducing end group and the repeating sialic acid units of depolymerized serogroup C polysaccharide upon depolymerization using hydrogen peroxide was assessed by using a Liquid Chromatography/Mass Spectrometry method (*36*). This study confirms that hydrogen peroxide depolymerization of serogroup C polysaccharide occurs by cleavage at the α2-9 glycosidic linkage, yielding intact sialic acid units without loss of O-acetyl or carboxyl groups and results in a higher concentration of aldehyde groups at the reducing end terminus due the increase of shorter saccharide chains in solution.

The activation method involves treatment with adipic acid dihyrazide (ADH) in physiological saline at pH ~5.0 by applying carbodiimide crosslinker chemistry and/or the reductive amination with sodium cyanoborohydride ($NaBH_3CN$) to reduce Schiff's base, generated between an aldehyde and an amine, to form a stable covalent bond. EDC [1-ethyl-3-(3-dimethylaminopropyl)carbodiimide] is the most commonly used zero-length crosslinker; triazine activators such as DMTMM [4-(4,6-dimethoxy[1,3,5]triazin-2-yl)-4-methyl-morpholinium] can also be used. Once activated, the polysaccharides are conjugated to a carrier protein (i.e. DT) also using the EDC chemistry. An LC-MS/MS method was also developed for quantification of residual EDC in the presence of meningococcal polysaccharide-DT conjugate vaccine and other vaccine matrix compounds (*37*).

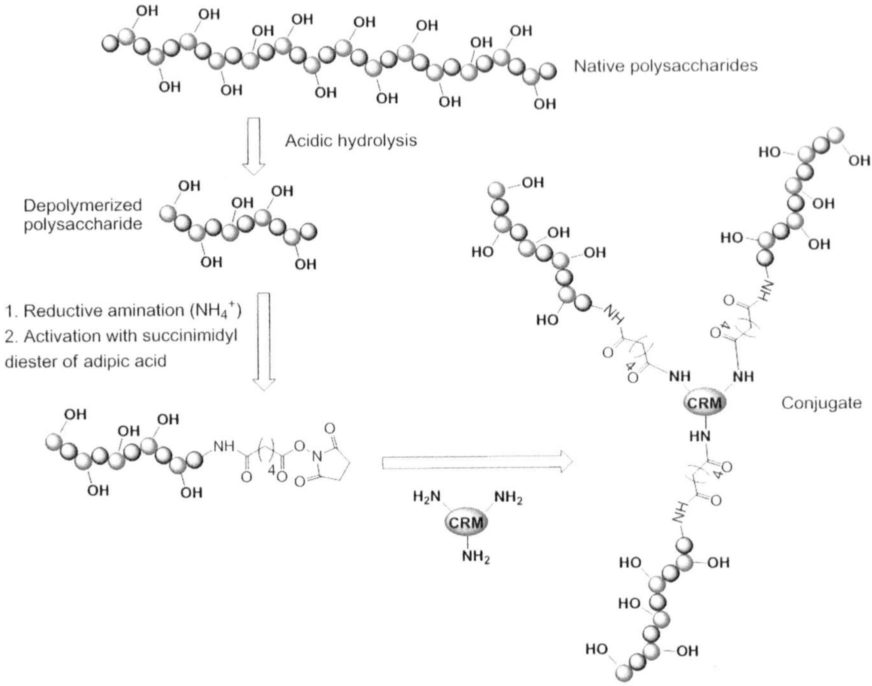

Figure 2. Schematic representation of NmACWY conjugation chemistry used for Menveo® antigens. (see color insert)

Another example of conjugation chemistry has been proposed by Knuf et al. (*38, 39*). Four ACWY-TT vaccine formulation vaccines were developed and manufactured first by GSK (Rixensart, Belgium) and currently by Pfizer, which differed in the conjugation approaches. The NmW and NmY polysaccharides were directly coupled to TT carrier protein following activation of their hydroxyl groups with CDAP reagent (1-cyano-4-dimethylaminopyridinium tetrafluoroborate). CDAP has generally replaced cyanogen bromide (CNBr) as the cyanylation reagent as activation with CDAP can be achieved at a lower pH.

The NmA and NmC polysaccharides were each coupled on a spacer molecule (ADH) by using CDAP chemistry and then coupled to TT by EDC chemistry (*40*). This approach was chosen following testing of a vaccine formulation in which only the NmA polysaccharide was coupled using the spacer molecule (*38*). In Figure 4, a schematic representation of conjugation strategies applied for NmACWY is reported.

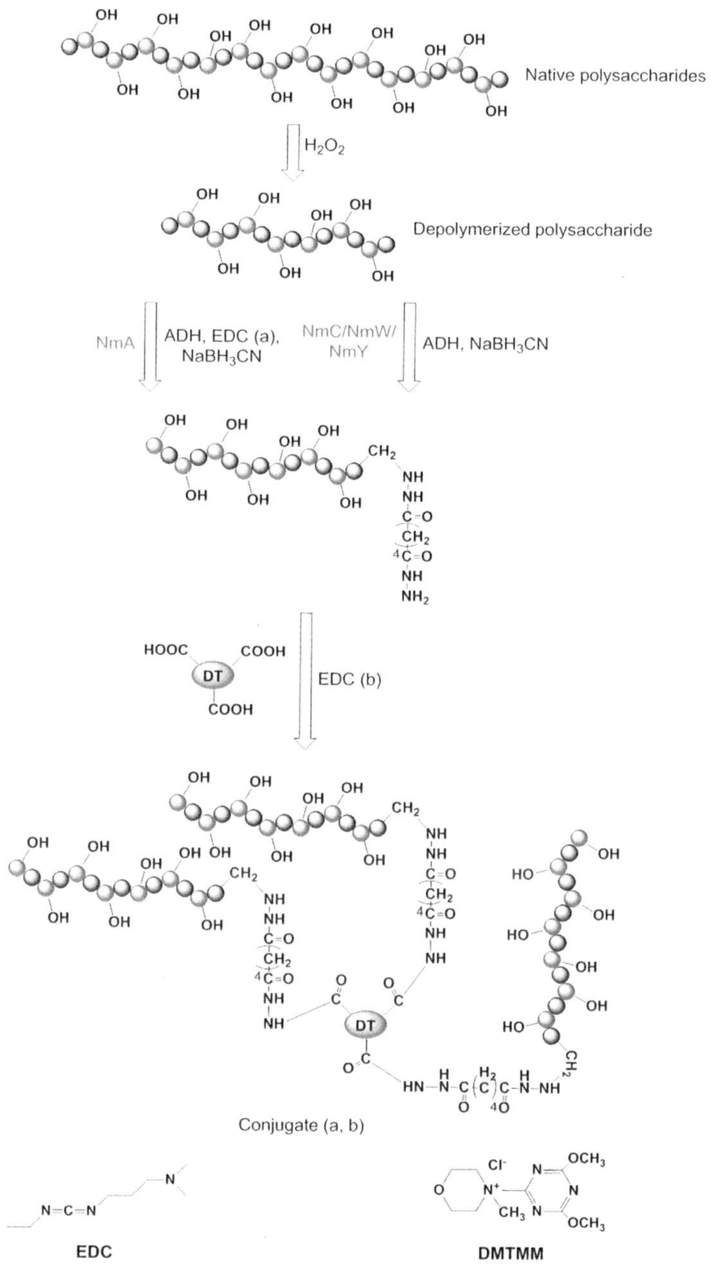

Figure 3. Schematic representation of NmACWY-DT conjugation chemistry using meningococcal polysaccharides firstly depolymerized by hydrogen peroxide treatment. The chemical structure of EDC and DMTMM zero-length crosslinker reagents are reported. (a) The use of EDC allows the NmA phosphodiester group coupling to the amine groups of ADH. (b) The use of EDC allows the NmA phosphodiester group and carboxyl groups of NmCWY coupling to amine groups of DT resulting in a cross-linked conjugate. (see color insert)

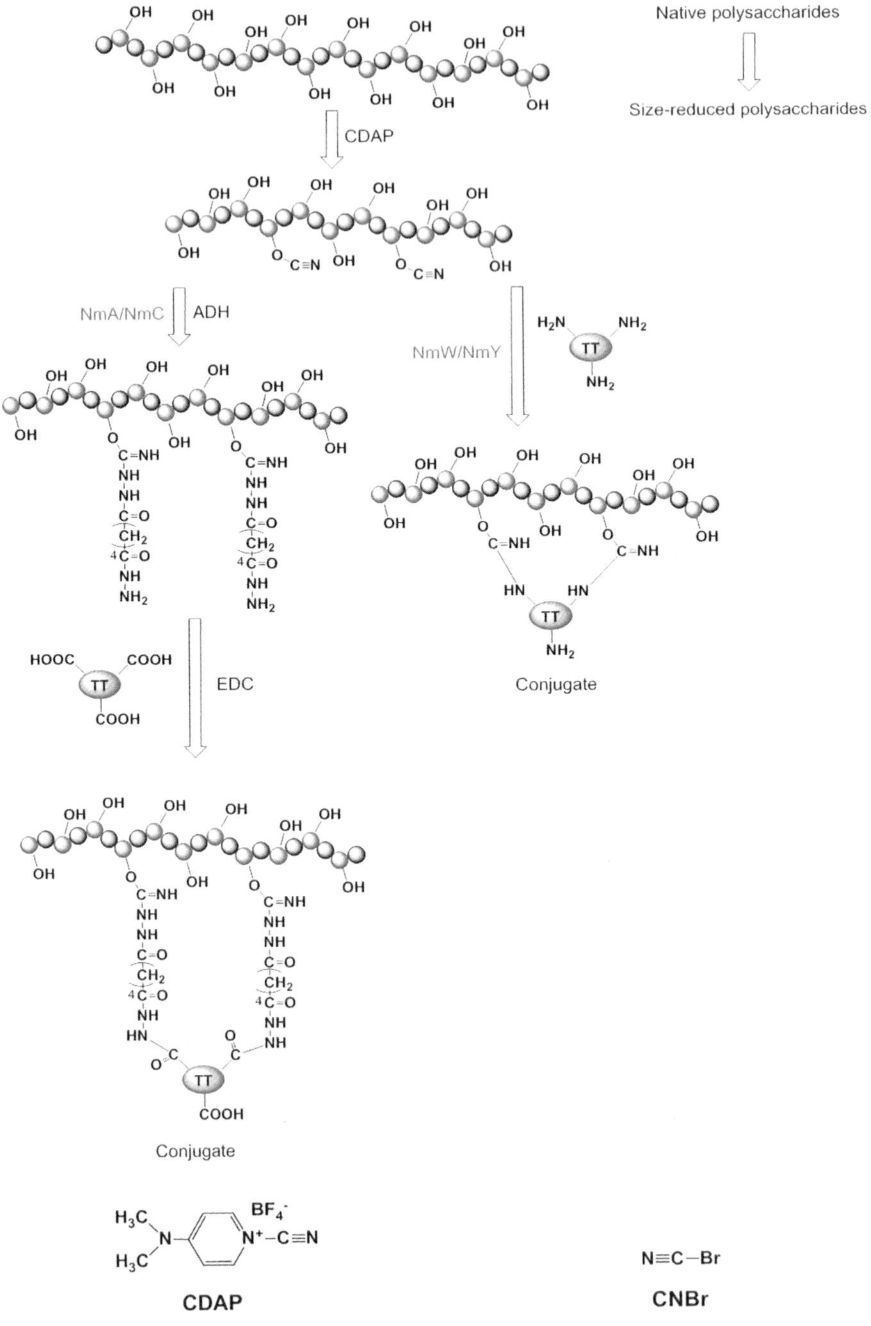

Figure 4. Schematic representation of NmACWY-TT conjugation chemistry using native/size-reduced polysaccharides. (see color insert)

A further example of conjugation chemistry has been proposed by Lee C-HR, et al. (*41–43*), (Figure 5), which has been applied for preparing the NmA conjugate of MenAfriVac®.

The carboxyl groups of aspartic and glutamic acid residues of TT carrier protein are activated by treatment with hydrazine (NH_2-NH_2) to formhydrazide groups in the presence of the zero-length coupling reagent EDC at acidic conditions. The NmA native polysaccharide, which contains some non O-acetylated repeating units →6)-α-D-Man*p*NAc-(1→OPO_3→ (5-30% randomly distributed), is oxidized at positions C_3 and C_4 to produce two aldehyde groups per residue. The activated protein and the activated NmA polysaccharide are mixed and react to form a Schiff's base which is finally reduced to a stable covalent linkage by treatment with sodium borohydride ($NaBH_4$). This reagent also acts as a quenching agent by reducing the possible residual aldehyde groups of NmA polysaccharide to secondary alcohol groups.

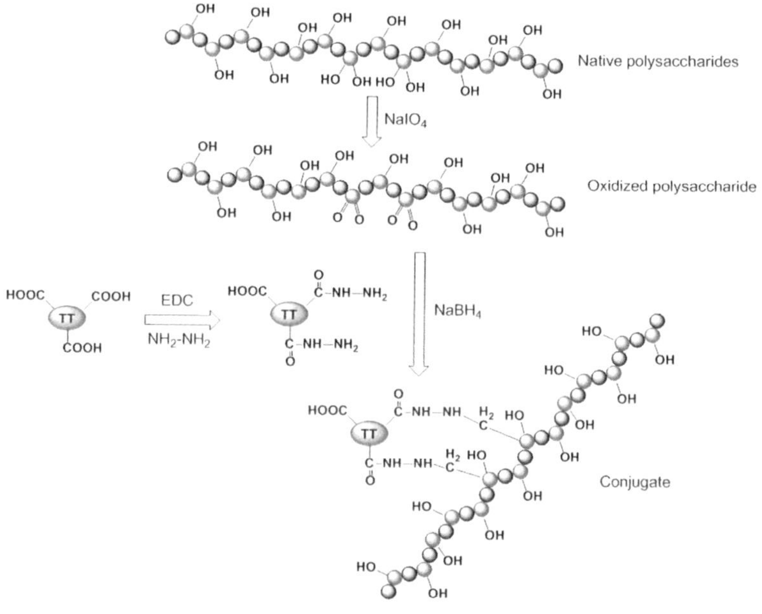

Figure 5. Schematic representation of NmA conjugation chemistry used for preparing NmA-TT conjugate of MenAfriVac®. (see color insert)

The first sodium periodate treatment used for NmA conjugate preparation, which oxidizes vicinal diols (two adjacent carbons bearing hydroxyl groups) into aldehydes and breaks the corresponding C-C bond, can be potentially applied to the other serogroups NmCWYX. This methodology has already been used for the preparation of the licensed Meningitec® vaccine. A controlled treatment of the NmC polysaccharide with sodium periodate generates two aldehydes in the non O-acetylated repeating units →9)-α-D-Neu5Ac-(2→ (5-10% randomly distributed), and oxidizes and cleaves the C_7-C_8 bond to generate shorter saccharide chains (a lower MW polydispersion is produced). The aldehyde groups are then readily conjugated to the amino groups of CRM_{197} by reductive amination.

For NmWY, two diols systems are potentially available for periodate oxidation if not O-acetylated at C_7 or C_9. The C_8-C_9 and C_7-C_8 of Neu5Ac moiety of their repeating units [→6)-α-D-Gal*p*-(1→4)-α-D-Neu5Ac(7/9OAc)-(2→ and →6)-α-D-Glc*p*-(1→4)-α-D-Neu5Ac(7/9OAc)-(2→ for NmW and NmY respectively] can be oxidized and activated without resulting in depolymerization. No diol system is available in the NmX repeating unit [→4)-α-D-Glc*p*NAc-(1→OPO_3→], however, reductive amination chemistry can be employed if the C_6 primary hydroxyl is selectively oxidized using TEMPO (2,2,6,6-tetramethylpiperidine 1-oxyl radical) –N-chlorosuccinimide (NCS) activation chemistry (see Prasad et al. (*18*), Chapter 4).

Several efficient conjugation methods have been developed with good conjugation yields for preparation of conjugate vaccines, as discussed earlier in Chapter 4. The reductive amination, cyanylation and carbodiimide chemistries have been applied, also using bifunctional linkers (e.g. ADH, succinimidyl diester of adipic acid, hydrazine) to increase the coupling reactivity between polysaccharide chains and carrier protein (e.g. by limiting the possible steric hindrance due to their large size, possible rigid conformations and relatively low mobility in solution). The nucleophilic amino groups (ε-NH_2 groups of Lys residues and protein N-terminus) or electrophilic carboxyl groups (β- and γ-COOH groups of Asp and Glu residues respectively and protein C-terminus) have been involved at the carrier protein moiety.

Meningococcal Vaccines in Development

To further identify optimal scheduling of tetravalent NmACWY meningococcal conjugates and then facilitate inclusion of these vaccines into national immunization programs, a second vaccine using TT (TetraMen-T; Sanofi Pasteur) instead of DT (Menactra®, Sanofi Pasteur) as carrier protein, is currently in clinical development (*44*).

Following the great success obtained by widespread introduction of MenAfriVac® with the support of UK Department for International Development (DFID), the Serum Institute of India and PATH, the development of a pentavalent meningococcal (NmACYWX) vaccine designed to meet further Africa's needs started few years ago. In addition to the emerging serogroup X related cases, other meningococcal serogroup strains continue to circulate and cause disease outbreaks (i.e. serogroup C strain has caused more than 14,000 cases in Nigeria and Niger in 2017) (*28*). NmW epidemics have emerged and spread across the Africa continent, further justifying the development of a pentavalent vaccine (*45*).

The Phase 1 clinical trial, conducted at the Center for Vaccine Development, University of Maryland, Baltimore (August 2016 to March 2017), on sixty adults randomized to receive a single dose of either NmACWYX (Alum-adjuvanted), NmACWYX (non-adjuvanted) or NmACWY Menactra®, confirmed a high safety profile of pentavalent vaccine, with no serious adverse effects recorded, and the solicited and unsolicited adverse events were few and mild.

High rSBA titers against all meningococcal serogroups were elicited by non-adjuvanted vaccine, which was found to be not inferior to the adjuvanted formulation. In addition to the strong NmX immunogenicity induced, the response against the other serogroups was not inferior in comparison to the control vaccine Menactra®.

A Phase 2 study in Malian 12-24 month olds will begin in late 2017 while a Phase 3 study in 18-75 year old Indian adults is targeted to commence in 2018 (*28*).

To simplify the immunization schedules, a single combination meningococcal vaccine that includes the antigens of the licensed NmACWY and NmB vaccines would be recommended. For this reason, a combination of NmACWY-CRM_{197} conjugate antigens, also contained in Menveo® vaccine, and the tetravalent Alum-adjuvanted Bexsero® vaccine (containing factor H-binding protein, fHbp; Neisseria adhesin A, NadA; Neisserial Heparin Binding Antigen, NHBA, outer membrane vesicle, OMV, from the New Zealand outbreak strain NZ98/254), recently approved in the EU, and several other countries for protection against serogroup B infection and used in national immunization programs (*46–48*).

The Trumemba® vaccine, which contains two recombinant lipidated fHbp variants, is being combined into corresponding pentavalent NmACWYB vaccine formulations and currently in clinical trials (*24*)

Conclusions

Conjugate vaccines are semi-synthetic biologicals with complex structures which differ from each other by several attributes (i.e. polysaccharide chain length, non-saccharide substituents such as O-acetylation, carrier protein, conjugation strategy/chemistry, conjugate size). A detailed description of the polysaccharide-protein coupling strategy/chemistry allows for better understanding of the structural attributes of the different conjugates and potentially provides insight into their immunological behavior in clinical trials.

References

1. Goldschneider, I.; Gotschlich, E. C.; Artenstein, M. S. *J. Exp. Med.* **1969**, *129*, 1307–1326.
2. Rosenstein, N. E.; Perkins, B. A.; Stephens, D. S.; Popovic, T.; Hughes, J. M. *N. Engl. J. Med.* **2001**, *344*, 1378–1388.
3. Harrison, O. B.; Claus, H.; Jiang, Y.; Bennett, J. S.; Bratcher, H. B.; Jolley, K. A.; Corton, C.; Care, R.; Poolman, J. T.; Zollinger, W. D.; Frasch, C. E.; Stephens, D. S.; Feavers, I.; Frosch, M.; Parkhill, J.; Vogel, U.; Quail, M. A.; Bentley, S. D.; Maiden, M. C. *Emerg. Infect. Dis.* **2013**, *19*, 566–573.
4. Bröker, M.; Berti, F.; Costantino, P. *Human Vaccines Immunother.* **2016**, *12*, 1808–1824.
5. Jones, C. *An. Acad. Bras. Cienc.* **2005**, *77*, 293–324.
6. Bundle, D. R.; Smith, I. C.; Jennings, H. J. *J. Biol. Chem.* **1974**, *249*, 2275–2281.

7. Granoff, D. M.; Feavers, I. M.; Borrow, R. Meningococcal Vaccines. In *Vaccines*, 4th ed.; Plotkin, S. A., Orenstein W., Offit, P. A., Eds.; Elsevier-Saunders: Philadelphia, PA, USA, 2004; pp 959–987.
8. Finne, J.; Leinonen, M.; Mäkelä, P. H. *Lancet* **1983**, *2*, 355–357.
9. Moe, G. R.; Tan, S.; Granoff, D. M. *FEMS Immunol. Med. Microbiol.* **1999**, *26*, 209–206.
10. Moe, G. R.; Bhandari, T. S.; Flitter, B. A. *J. Immunol.* **2009**, *182*, 6610–6617.
11. Moe, G. R.; Dave, A.; Granoff, D. M. *Infect. Immun.* **2005**, *73*, 2123–2128.
12. Holst, J.; Martin, D.; Arnold, R.; Huergo, C. C.; Oster, P.; O'Hallahan, J.; Rosenqvist, E. *Vaccine* **2009**, *27*, B3–B12.
13. Lewis, S.; Sadarangani, M.; Hoe, J. C.; Pollard, A. J. *Expert Rev. Vaccines* **2009**, *8*, 729–745.
14. Feavers, I. M.; Pizza, M. *Vaccine* **2009**, *27*, B42–B50.
15. Crum-Cianflone, N.; Sullivan, E. *Infect. Dis. Ther.* **2016**, *5*, 89–112.
16. Lockhart, S. P.; Scott, D. A.; Jansen, K. U.; Anderson, A. S.; Gruber, W. C. Glycoconjugate Vaccines: The Clinical Journey. In *Carbohydrate-Based Vaccines: From Concept to Clinic*; Prasad, A. K., Ed.; ACS Symposium Series 1290; American Chemical Society: Washington, DC, 2018; Chapter 2.
17. Duke, J. A.; Avci, F. Y. Immunological Mechanisms of Glycoconjugate Vaccines. In *Carbohydrate-Based Vaccines: From Concept to Clinic*; Prasad, A. K., Ed.; ACS Symposium Series 1290; American Chemical Society: Washington, DC, 2018; Chapter 3.
18. Prasad, A. K.; Kim, J.-h.; Gu, J. Design and Development of Glycoconjugate Vaccines. In *Carbohydrate-Based Vaccines: From Concept to Clinic*; Prasad, A. K., Ed.; ACS Symposium Series 1290; American Chemical Society: Washington, DC, 2018; Chapter 4.
19. Miller, E.; Salisbury, D.; Ramsay, M. *Vaccine* **2001**, *20*, S58–67.
20. Snape, M. D.; Pollard, A. J. *Lancet Infect. Dis.* **2005**, *5*, 21–30.
21. LaForce, M. F.; Ravenscroft, N.; Djingarey, M.; Viviani, S. *Vaccine* **2009**, *27*, B13–19.
22. Frasch, C. E.; Preziosi, M. P.; LaForce, F. M. *Hum. Vaccin. Immunother.* **2012**, *8*, 715–724.
23. LaForce, F. M.; Djingarey, M.; Viviani, S.; Preziosi, M. P. *Human Vaccines Immunother.* **2017**, 1–510.1080/21645515.2017.1378841.
24. *Pfizer, Find A Trial.* https://www.pfizer.com/science/find-a-trial/NCT03135834 (accessed 26 March 2018).
25. Goffin, P.; Dewerchin, M.; De Rop, P.; Blais, N.; Dehottay, P. *Biotechnol. J.* **2017**, *12*, 1–11.
26. *Pfenex, Products.* http://www.reagentproteins.com/product_detail/CRM197_Lyophilized.html (accessed 26 March 2018).
27. *ScaraGenomics, Products.* https://www.scarabgenomics.com/c-46-crm197.aspx (accessed 26 March 2018).
28. LaForce, M. In *Progress on MenACWYX Vaccine: Phase 1 Trial Results*, MRF Conference London, November 14–15, 2017.
29. Bröker, M.; Dull, P. M.; Rappuoli, R.; Costantino, P. *Vaccine* **2009**, *27*, 5574–5580.

30. Bardotti, A.; Averani, G.; Berti, F.; Berti, S.; Carinci, V.; D'Ascenzi, S.; Fabbri, B.; Giannini, S.; Giannozzi, A.; Magagnoli, C.; Proietti, D.; Norelli, F.; Rappuoli, R.; Ricci, S.; Costantino, P. *Vaccine* **2008**, *26*, 2284–2296.
31. Ravenscroft, N.; Averani, G.; Bartoloni, A.; Berti, S.; Bigio, M.; Carinci, V.; Costantino, P.; D'Ascenzi, S.; Giannozzi, A.; Norelli, F.; Pennatini, C.; Proietti, D.; Ceccarini, C.; Cescutti, P. *Vaccine* **1999**, *17*, 2802–2816.
32. Costantino, P.; Norelli, F.; Giannozzi, A.; D'Ascenzi, S.; Bartoloni, A.; Kaur, S.; Tang, D.; Seid, R.; Viti, S.; Paffetti, R.; Bigio, M.; Pennatini, C.; Averani, G.; Guarnieri, V.; Gallo, E.; Ravenscroft, N.; Lazzeroni, C.; Rappuoli, R.; Ceccarini, C. *Vaccine* **1999**, *17*, 1251–1263.
33. Ryall, R. P. A3 WO Patent 2004103400, 2002.
34. Hestrin, S. *J. Biol. Chem.* **1949**, *180*, 249–261.
35. Park, J. T.; Johnson, M. J. *J. Biol. Chem.* **1949**, *181*, 149–151.
36. Cai, X.; Lei, P. Q.; Lamb, D. H.; Shannon, A.; Jacoby, J.; Kruk, J.; Kensinger, R. D.; Ryall, R.; Zablackis, E.; Cash, P. *Anal. Chem.* **2004**, *76*, 7387–7390.
37. Lei, Q. P.; Lamb, D. H; Shannon, A. G.; Cai, X.; Heller, R. K.; Huang, M.; Zablackis, E.; Ryall, R.; Cash, P. *J. Chromatogr. B: Anal. Technol. Biomed. Life Sci.* **2004**, *813*, 103–112.
38. Knuf, M.; Kieninger-Baum, D.; Habermehl, P.; Muttonen, P.; Maurer, H.; Vink, P.; Poolman, J.; Boutriau, D. *Vaccine* **2010**, *28*, 744–753.
39. Ghanem, S.; Hassan, S.; Saad, R.; Dbaibo, G. S. *Expert Opin. Biol. Ther.* **2013**, *13*, 1197–1205.
40. Biemans, R. L.; Duvivier, P. A2 WO Patent 2007000343, 2008.
41. Lee, C.-H. R.; Pinto, V. B. B2 U.S. Patent 9427476, 2007.
42. Lee, C.-H. R.; Kuo, W. C.; Beri, S.; Kapre, S.; Joshi, J. S.; Bouveret, N.; LaForce, F. M.; Frasch, CE. *Vaccine* **2009**, *27*, 726–732.
43. Frasch, C. E.; Kapre, S. V.; Lee, C. H.; Préaud, J. M. *Clin. Infect. Dis.* **2015**, *61*, S404–409.
44. McVernon, J.; Nolan, T.; Richmond, P.; Reynolds, G.; Nissen, M.; Lambert, S. B.; Marshall, H.; Papa, T.; Rehm, C. *Pediatr. Infect. Dis. J.* **2012**, *31*, e15–23.
45. Mustapha, M. M.; Harrison, L. H. *Human Vaccin. Immunother.* **2017**, *0*, 1–9.
46. Saez-Llorens, X.; Aguilera Vaca, D. C.; Abarca, K.; Maho, E.; Han, L.; Smolenov, I.; Dull, P. *Pediatr. Infect. Dis. J.* **2015**, *34*, e264–278.
47. Saez-Llorens, X.; Aguilera Vaca, D. C.; Abarca, K.; Maho, E.; Graña, M. G.; Heijnen, E.; Smolenov, I.; Dull, P. M. *Human Vaccin. Immunother.* **2015**, *11*, 1507–1517.
48. Block, S. L.; Szenborn, L.; Daly, W.; Jackowska, T.; D'Agostino, D.; Han, L.; Dull, P. M.; Smolenov, I. *Vaccine* **2015**, *33*, 2500–2510.

Chapter 7

The Role of Molecular Modeling in Predicting Carbohydrate Antigen Conformation and Understanding Vaccine Immunogenicity

Michelle M. Kuttel[*,1] and Neil Ravenscroft[1]

[1]Department of Computer Science, University of Cape Town, Private Bag X3, Rondebosch 7701, Cape Town, South Africa
[2]Department of Chemistry, University of Cape Town, Private Bag X3, Rondebosch 7701, Cape Town, South Africa
[*]E-mail: mkuttel@cs.uct.ac.za.

Licensed conjugate vaccines have proven to be highly effective in preventing bacterial disease. Coverage of a vaccine is extended when specific antigens elicit immune responses that provide cross-protection against infection by closely related, non-vaccine strains. However, structural similarity between carbohydrate antigens has not proven to be a reliable predictor of cross-protection and the current understanding of the role of saccharide antigen conformation in immunogenicity is sparse. Identification of the conformational effect of specific structural changes in conjugate vaccine antigens may usefully inform the development of conjugate vaccines. The limited ability of experimental methods to establish saccharide conformation has led to the development of systematic molecular modeling protocols. Here we cover the computational methodologies employed to model carbohydrate antigens and demonstrate, through case studies, the valuable role that molecular simulations can play in furthering our understanding of carbohydrate immunogenicity. The case studies comprise molecular modeling of the capsular polysaccharides for meningococcal serogroups Y and W and pneumococcal serogroups 6, 19 and 23, as well O-antigens of *Salmonella enterica* and *Shigella flexneri*. Conformational analysis can provide a mechanistic insight into clinical observations on cross-protection and may further indicate the importance of specific structural features, such as substituents, thereby facilitating vaccine design and broadening vaccine coverage.

Introduction

Licensed glycoconjugate vaccines have proven to be highly effective in preventing infectious disease caused by *Haemophilus influenzae* type b, *Neisseria meningitides* and *Streptococcus pneumoniae*. A central challenge for vaccine manufacturers is to achieve the maximum coverage of current and emerging bacterial serotypes with the minimum number of serotypes (valency), as each additional serotype component increases the final cost of the vaccine. The valency of a vaccine can be reduced if there is demonstrated cross-protection between serotypes. Cross-protection refers to the phenomenon whereby a specific vaccine antigen elicits an immune response that protects against infection by closely related (non-vaccine) strains, which typically occurs within the same serogroup. Evidence for cross-protection between vaccine serotypes is necessarily indirect and typically only manifest after widespread vaccine introduction. Historically, cross-protection has been determined indirectly by sera fractionation and pre-clinical studies (*1*, *2*). However, immunological cross-reactivity in humans can only be properly established post-licensure and once sufficient data on efficacy has been collected.

Unfortunately, where available, efficacy data indicate that close structural similarity in carbohydrate antigens does not reliably predict cross-protection (*3–5*). For saccharides, small changes in the molecular structure (in a residue type, linkage position or substituent) can produce significant changes in conformation and flexibility, with consequent alterations in physical properties and antigenicity. A broad assumption is that antigens having both similar primary structures and similar conformations will have immunological cross-reactivity, and thus exhibit cross-protection. Conversely, antigens with similar primary structures but significantly different conformations will provide, at best, limited cross-protection.

Now that many of the capsular polysaccharide serotypes have been precisely chemotyped (*6*), the influence of subtle differences in linkage positions and substituents (e.g. O-acetylation. glycerolphosphate) on conformation and, consequently, potential cross-protection may be evaluated. Such conformational information can potentially provide a mechanistic understanding for clinical observations and may further indicate the importance (or unimportance) of specific substituents or other features. Identification of the conformational effect of specific structural changes in conjugate vaccine antigens may usefully inform the development of conjugate vaccines against either additional bacterial strains or other pathogens. However, the detailed understanding of mechanisms and determinants of polysaccharide recognition by antibodies that is required for a rational selection of vaccine components that provide cross-protection is currently lacking (*7*). Although tools for determining the structure of bacterial polysaccharides are well established, it is difficult to determine the associated molecular conformation and there remains a paucity of information on the conformation of the saccharide antigen component of carbohydrate vaccines. Indeed, the conformational information encoded in carbohydrate sequences in general, the so-called "sugar code", remains largely undeciphered (*8*).

The most common sources of experimental evidence for carbohydrate conformation are nuclear magnetic resonance (NMR) spectroscopy and X-ray crystallography; indirect information on carbohydrate conformation is supplied by other experimental methods, such as fluorescence spectroscopy. While carbohydrates are difficult to crystallize, X-ray crystallography can be a powerful tool if the crystal structure of the carbohydrate antigen-antibody complex is available (*9*). However, the X-ray structure of the bound glycan is often poorly resolved due to carbohydrate flexibility (*10*) and for crystal structures of uncomplexed carbohydrates, there is always uncertainty as to whether these molecules adopt similar conformations in solution as in the solid state. NMR spectroscopy has a key role in determining the primary structure of carbohydrates (*11*), but its application to the elucidation of carbohydrate conformation in solution poses a number of challenges due to the flexibility and dynamic motion of these molecules: solution conformations often represent a dynamic average over all the populated conformational states. For this reason, a variety of different molecular modeling methods have been employed to assist in interpretation of NMR spectra (*12*).

In silico molecular modeling can provide valuable theoretical insights into polysaccharide conformation and hence the differences between structurally similar polysaccharides. Here we focus on atomistic molecular dynamics (MD) simulations, which render detailed information on molecular flexibility and dynamics over nanosecond timescales that is currently inaccessible to any experimental method. MD is a powerful tool for prediction of the conformation of carbohydrate antigens, both as a complement to experiments (*11–18*), and, increasingly, as a stand-alone method (*19–26*). Although recent studies have demonstrated ring flexibility for some monosaccharides over microsecond timescales (*27*), it is generally expected that the conformation of a saccharide is largely determined by the orientation of the constituent glycosidic linkages, while the saccharide rings remain chiefly in low-energy chair conformations. The orientations of the glycosidic linkages are dependent on the nature of the constituent monosaccharides and are affected by the addition of substituents and neighboring branches. Comparative modeling can provide insight into the specific conformational effects of residue substitutions, changes in linkage position and substituents. For example, O-acetylation is a common modification of bacterial polysaccharides, but its impact on conformation is poorly understood and polysaccharide dependent. For the Group B *Salmonella* O-polysaccharide, MD simulations showed that O-acetylation had little effect on the conformation of this polysaccharide, but that the O-acetyl groups were highly solvent exposed and thus potentially important for antigen binding (*28*).

MD simulations can also provide insight into the carbohydrate-antibody binding process that is central to the immune response. For example, different monoclonal antibodies may select different conformers of a sugar from an equilibrium mixture of preexisting conformational epitopes of the antigen in solution, as occurs for some lectins binding the same sugar in different orientations (*8*). This conformational selection mechanism (*29*), whereby an antibody selects a specific conformation of an unbound polysaccharide from its solution populations, has been recently demonstrated in subsequent simulations of *Salmonella* group B O-polysaccharide with a monoclonal antibody (*26*).

Mapping of epitopes recognized by protective antibodies is crucial for understanding the mechanism of action of vaccines and for enabling antigen design. As a method, molecular modeling can provide a rationalization for the existence of cross-protection, or the lack thereof, between specific carbohydrate antigen pairs, and may prove to have predictive capabilities. In this chapter, we describe the general molecular modeling approaches to establish carbohydrate conformations, and then review their application to antigenic polysaccharides of a few key pathogenic bacteria that are the focus of previous or current conjugate vaccine development efforts. These case studies comprise modeling of capsular polysaccharides of *Neisseria meningitides* (serogroups Y and W) and *Streptococcus pneumoniae* (serogroups 6, 19 and 23) as well as lipopolysaccharide O-antigens of *Salmonella* and *Shigella* bacteria.

Molecular Modeling Methods

Molecular modeling methods are under continual development and range from simple approaches analogous to building a physical model of a carbohydrate, to sophisticated quantum mechanical simulations. In general, the more sophisticated the method and the larger the molecular system, the higher the computational cost. As an example, for a 30-unit polysaccharide, currently a static model may be built on a laptop in a few days, including the calculation of the preferred orientations of the constituent glycosidic linkages. In contrast, a molecular dynamics simulation of a microsecond of the oligosaccharide motion in water may take months to complete on a supercomputer; and a quantum mechanical simulation in solution may be infeasibly long (requiring years to run to completion). Here we focus on the methods most suitable for, and most commonly applied to, the simulation of oligo- and polysaccharides.

Structure Visualization

Visualization packages, such as VMD (*30*) and PyMOL (*31*), can be used to rotate and inspect three-dimensional molecular structures and often provide tools to evaluate and analyze simulation trajectories. Interrogation of protein structures is facilitated by abstract visualizations, such as the "cartoon" diagrams, that illustrate secondary structure in proteins. Effective visualizations for carbohydrates should highlight important or relevant aspects of carbohydrate conformation, such as the location and pucker of the sugar rings, and the orientation of the glycosidic linkages. However, far fewer carbohydrate-specific visualization representations have been developed to assist the researcher than is the case for proteins. We designed the *Twister* and *Paperchain* visualizations to highlight the backbone and ring conformations (*32*), respectively, with subsequent implementation into the VMD package (*33*). More recently, Pérez et al. extended this approach to the *SugarRibbons* visualization that is implemented as part of SweetUnityMol: a game-based visualization package for carbohydrates (*34*).

3D representations of common glycan residue symbols have also been developed, such as the 3D–CFG plugin for PyMOL (*35*) and, later, the similar 3D-SNFG representation for VMD (*36*). However, there remains considerable scope for development of novel carbohydrate visualizations analogous to the cartoon and ribbon representations for proteins.

Structure Building

As glycan 3D structure is chiefly determined by the conformational preferences of the constituent glycosidic dihedral angles, rather than by strong inter-residue interactions, approximate low energy conformations of a glycan can be generated by consideration of the preferred orientations of the individual linkages only (*37*). This strategy has been termed the 'pragmatic' approach to generating a three-dimensional model of a carbohydrate: monosaccharides are treated as stable, rigid building blocks and glycosidic linkage dihedral angles are used as the primary variable for building. Software available for building three-dimensional molecular models of carbohydrates currently includes web-based tools (e.g. SWEET-II (*38*) and the GLYCAM-Web carbohydrate builder (*39*)), plugins for visualization packages (e.g. Azhar for PyMOL (*40*)) and standalone software packages (e.g. POLYS (*41*, *42*) and our CarbBuilder tool (*43*)). The software packages differ according to the range of monosaccharides and substituents supported as well as the conformational heuristics used to generate a structure. For example, GLYCAM-Web currently supports 22 basic monosaccharide residues, whereas CarbBuilder allows for 30. Some software packages contain checks to prevent building physically implausible conformations containing self-intersecting atoms: CarbBuilder does a simple atomic collision check, while GLYCAM-Web allows for optimizing a simple model to a local energy minimum: the molecule relaxes until atoms, bonds, angles, and dihedrals are in a locally optimal conformation.

Though of value as a first conformational estimate, these static models do not consider the effect of interactions with the molecular environment (water, ions and other molecules) and thus provide a very limited picture of the solution conformation of flexible carbohydrates. However, building of a suitable initial structure remains an essential first step in performing MD simulations. If an incorrect high-energy conformation is used as a starting point for a simulation, it is unlikely that the molecule will relax to a low energy state and invalid conformational populations will be generated.

Molecular Dynamics: Modeling Flexibility in Solution

From a suitable starting structure for a molecule, molecular dynamics (MD) software uses a potential energy function – a physical model of atoms, their bonded and non-bonded interactions – to calculate the forces and velocities on each atom in the system and then numerically integrates Newton's equations of motion at discrete time steps to produce a time series of the motion and interaction of each atom in the system. The efficacy of molecular dynamics as a method for protein structure prediction has been well established and MD has become the standard method for simulating the conformational dynamics of oligo- and polysaccharides in solution (*44*).

In contrast to proteins, carbohydrates do not typically have single well-defined conformation, but rather a family of conformers. The dynamic behavior of a flexible carbohydrate in water can be probed using MD simulations with explicit solvent (including appropriate counter-ions for ionic antigens). Subsequent analysis of the MD simulation time series for a carbohydrate will reveal the preferred conformational families, the conformational distribution, as well as derived properties, such as viscosity and hydrodynamic size. Such conformational analysis typically uses clustering algorithms to group simulation conformations into families of related structures. Clustering algorithms are sensitive to the clustering metric applied. A common approach is to cluster on a Root Mean Squared Deviation (RMSD) fit to the carbohydrate "backbone" atoms (ignoring side chains, hydrogen atoms and pendant groups such as primary alcohols) with an appropriate cut-off (between 1 and 10 Å, depending on molecular size and motion). Another related approach is to cluster solely on the glycosidic linkages (*45*).

Thus far, MD simulations of polysaccharide antigens have provided support for the general assumption that carbohydrate conformation and dynamics are strongly dependent on the primary structure: a slight change in structure, such as a linkage position, can dramatically alter the molecular conformation (*19–27*). However, it is important to bear in mind that the validity of an MD simulation is strongly dependent on the both the accuracy of the physical model (or force field), as well as the extent of molecular conformational sampling, as discussed below.

Carbohydrate Force Fields: The Physical Model

The core potential energy function, or model, for MD must be parameterized for a specific molecular class in order to accurately represent the biological world. Parameters are derived from both quantum mechanical calculations and empirical data. A parameterized model is usually termed a *force field*.

The development of biomolecular force fields has been especially complex for carbohydrates because of their structural diversity and conformational heterogeneity, as well as the relative scarcity of accurate experimental data suitable for force-field refinement (*46*, *47*). A complication is that carbohydrate force fields are parameterized, and thus dependent on, a specific water model, typically the rigid three-site TIP3P model (*48*). In addition, for studies of carbohydrates in complex heterogeneous systems, the carbohydrate parameters must be properly matched with parameters for other aspects of biological systems, most notably proteins and lipids. Despite these challenges, carbohydrate force fields have matured considerably over the last decade. Currently, the four main carbohydrate-specific force fields are broadly termed CHARMM36 (*47*, *49–51*),

GROMOS (*16*, *52*), GLYCAM06 (*53*) and OPLS (*54*, *55*) – in each case the name refers to the version of these parameter sets optimized specifically for carbohydrates (see a useful review by Xiong et al. (*56*)). A confusion is that these force fields were each developed for a specific MD software package and, in the case of CHARMM36 and GROMOS, often have the same name as the associated software. However, the majority of these force fields may be used with a range of MD simulation programs. Another problem is that there is often ambiguity as to which version of a parameter set is specifically referred to, as the force fields are continually developing and hybrid versions abound. For example, the GROMOS force field has undergone frequent revisions to better model aspects of carbohydrate conformation, moving from the GROMOS 45A4 (*52*) and $45A4_{ASPG}$ (*16*) parameters to successive improvements of the newer GROMOS $56A6_{CARBO}$ parameter set (*46*, *57*). In addition, the CHARMM36 force field is currently being extended to a polarizable Drude force field which provides increased accuracy by supporting electronic polarization in a classical force field, instead of the standard fixed charge approximation. The focus of the early stage of development of this force field was on monosaccharides (*58–60*), with an extension to polysaccharides expected shortly.

Because of the computational expense of long MD simulations, studies typically pick a single, well-established force field for molecular modeling. Carbohydrate force field comparison studies are relatively rare, as they require duplicate simulations and, often, MD packages. However, where it has been undertaken, comparative modeling has sometimes revealed significant differences in conformational ensembles depending on the force field used. For example, for monosaccharides, the GLYCAM06, GROMOS (45a4) and OPLS force fields describe the conformational inversions of the β-D-glucopyranose ring relatively accurately (*61*), but the frequency and stability of ring inversions for other sugar residues is often force field dependent (*57*). For disaccharides and longer chains, force-field dependence of the glycosidic linkage orientations has been noted. For example, bent anti-ψ conformations of the glycosidic linkages are observed to be more favored in the GLYCAM06 model than in either CHARMM36 (*17*, *22*) or OPLS (*24*). In addition, some studies have shown that the GLYCAM06 force field favors more compact oligosaccharide conformations than other carbohydrate force fields, including CHARMM36 (*21*) and OPLS (*24*). Further, comparative simulations of hemicellulose molecules in TIP3P water showed aggregation of the molecules into clusters with GLYCAM06, which did not occur with either CHARMM36 or GROMOS (*62*). These differences have been postulated to be the result of dominating attractive intra/inter molecular interactions in GLYCAM06, likely due to the reduced contribution of electrostatic interactions relative to the other force fields (*21*, *22*, *62*). In a recent comparison of force fields for modeling cyclodextrins, the CHARMM36 force field showed similar performance to the GROMOS parameter sets, albeit with increased conformational flexibility due to tumbling of glucopyranose units (*63*). Comparison of MD simulations of a trisaccharide fragment of a tumor antigen in TIP3P water showed that GLYCAM06 and CHARMM36 predicted similar conformational preferences; both force fields reproduced NMR spectroscopic data well (*17*). Unfortunately, in the absence of unambiguous experimental information and more thorough studies comparing force field performance, it is not currently possible to determine which force field represents carbohydrates most accurately.

An alternative to atom-based models are coarse-grained (CG) models, such as the Martini model for carbohydrates (*64*), which simplify the physical representation of a molecule, combining on average four heavy atoms into a single CG bead. The aim of CG models is to reduce the computational requirements and thus enable either simulation for longer times or bigger molecular systems. For example, Ma et al. extended the Martini force field to model lipopolysaccharide molecules to enable CG simulation of a gram-negative bacterial outer membrane bilayer over 10 μs (*65*). However, while CG approaches are useful for simulating interactions in large systems, the loss of atomic detail makes them often unsuitable for conformational studies. Further, there are indications that the MARTINI force field considerably overestimates the aggregation behavior of even small saccharides (*66*).

For water solution, the TIP3P model is typically employed (*48*). Although more sophisticated water models exist, such as TIP5P (*67*), most of the currently available carbohydrate force fields have been parameterized with the simpler, rigid TIP3P. It should be noted that changing the water model to one that was not used in the carbohydrate force field parameterization may lead to unexpected behavior, such as aggregation. Implicit solvent models may also be employed in oligosaccharide simulations to reduce computational time (*21*), although this practice is questionable for carbohydrates, due to the importance of specific carbohydrate hydroxyl-water interactions.

Conformational Sampling: Simulation Convergence and Free Energy Estimates

Most polysaccharides do not have a single, well-defined structure in solution, but rather a family of conformations. For reliable predictions of conformation, MD simulations must be run for long enough to achieve *convergence*: the simulation must show the correct statistical weighting of all low-energy conformations. As convergence is impossible to prove, statistical convergence is usually determined when the average values for all properties of interest remain roughly constant with increased sampling (*68*). Typically, the variation in over time of a simple conformational measure (such as root-mean-square deviation of a conformation from a reference value) is monitored to estimate convergence. More reliable methods for estimating convergence, such ensuring that the simulation length is considerably longer than the autocorrelation decay for the molecular radius of gyration (*23*), are employed infrequently.

Simulation sampling requirements are more stringent for the calculation of conformational free energies, which require thorough sampling of not only the range of low-energy states, but also the high energy transitional states. In most cases, free energy calculations are aimed at estimating a *potential of mean force* (PMF), which is the change in free energy as a function of either an external parameter or an internal coordinate. A PMF is a detailed description of barrier heights and energy changes along a free energy pathway and allows for precise characterization of a structural change. Free energy calculations for the important degrees of freedom in a carbohydrate allow for identification of the number of low-energy conformers, the flexibility of each conformer, the time spent in each conformation and the rate of conversion between conformers. In solution, a PMF incorporates not only contributions from entropy, but also the averaged solvent–solute interactions. For carbohydrates, 2D PMFs are most commonly calculated to characterize the low energy conformations and flexibility of glycosidic linkages (*69*).

Advances in high-performance computing have enabled MD simulations to be performed for significantly longer periods (up to microseconds), and on larger systems, than was previously feasible. However, in the case of large molecules and high energy barriers along the reaction coordinate, standard MD simulations using Boltzmann sampling are still not sufficient to achieve convergence and biasing methods must be employed to enhance the conformational sampling. Several enhanced sampling methods have been developed: metadynamics and umbrella sampling approaches are typically used to calculate PMFs for mono- and disaccharides (*69*). Replica exchange MD methods (which run multiple copies of a simulation with different temperatures, swapping conformations between them at intervals) are more effective for enhancing sampling of multiple glycosidic linkages in oligosaccharides (*70*). However, standard temperature-based replica exchange requires many replicas and is not practicable for large molecules in explicit solvent. Alternative approaches for enhanced sampling of polysaccharides combine Hamiltonian replica exchange with two-dimensional biasing potentials to improve the sampling efficiency about glycosidic linkages (*71*, *72*). In these methods, rather than the temperature, the underlying potential energy function (i.e. the Hamiltonian) used to calculate the forces is perturbed to drive enhanced sampling. Although highly effective, such approaches are complex to implement and not currently available as a standard option in most MD codes. In the absence of enhanced sampling, long MD simulations (250 ns or more, depending on chain length) are required to expose the conformational behavior of flexible oligosaccharides in solution. This is important as conformational transitions of the glycosidic linkages may only take place on a 10-50 ns time scale.

Molecular Dynamics: Modeling Antibody Binding

Understanding of the structural basis of antibody recognition and immune response against the antigen and specific epitopes is limited by the availability of crystal structures of the relevant monoclonal antibody. For polysaccharide antigens, the structures that have been solved are typically of complexes of an oligosaccharide with a single monoclonal antibody, or fragment of an antibody that can mediate killing. For example, an antibody fragment bound to the Group B streptococcus type III oligosaccharide usefully identified a sialic acid-dependent functional epitope of GBS (*9*).

Where antibody structures are available, modeling carbohydrate-antibody binding becomes a special case of the molecular docking problem, which aims to predict the ligand orientation and conformation, or pose, in the binding site as well as the binding interaction energy. However, carbohydrate-antibody binding is complicated by the extreme flexibility of the carbohydrate ligand/antigen coupled with the, often weak, carbohydrate-antibody binding affinity (*8*, *37*).

Automated docking with programs such as AutoDock (*73*) is an attractive approach because it is computationally cheap, as it uses a simple scoring function for the ligand-protein interaction. However, automated docking routines are primarily parameterized for aligning small relatively rigid ligands with protein binding sites; they are usually less effective for highly flexible carbohydrates, especially as they do not allow for explicit inclusion of water molecules (*37*, *74*). A recent improvement in the carbohydrate scoring functions to incorporate glycosidic torsion angle preferences hopes to address this (*75*), but has not yet been applied to carbohydrate-antibody binding.

MD simulation is a more sophisticated, but more expensive approach for investigating carbohydrate-antibody binding. MD allows for free rotation of the ligand in explicit solvent, as well as induced fit for both antibody and antigen. MD may be used to refine ligand placement, or as a stand-alone method, but requires either long simulation times or enhanced sampling methods to ensure sufficient conformational sampling (*26*). For a thorough treatment, it is also necessary to compare the bound saccharide structure with its unbound conformations in solution. This makes the modeling of carbohydrate-antibody binding very time-consuming.

In general, modeling of carbohydrate antigen-antibody binding is complicated by the fact that an antigen elicits an array of antibodies of differing type, amount and avidity directed against the epitopes presented. A structure of a single complex cannot account for all families of antibodies and alternative epitopes that may be involved in providing vaccine protection. Given the uncertainty in the possibility of the binding of other families of antibodies with alternative epitopes of the carbohydrate, modeling time may be more profitably spent in considering the conformation of the unbound saccharide alone.

Recommended Systematic Modeling Approach

In considering a molecular modeling strategy, there are continual trade-offs between accuracy and computation time. We have identified an effective systematic approach to modeling bacterial polysaccharides, as follows. First, we calculate PMFs for the glycosidic linkages in the representative disaccharide units, in order to determine the preferred linkage conformations and enable building of an initial static model. Then we progress to MD simulations of oligosaccharide chains in solution, to establish the conformations and dynamics of the polymer strand (*22*, *23*). Other studies use a similar approach, focusing first on disaccharide fragments and then on the chain conformations (*21*, *24*).

In a subsequent step, the oligosaccharide may be substituted with O-acetyl groups or other modifications and simulated to determine the effect on conformation. If a crystal structure is available, a final step can then, with knowledge of the unbound conformations, attempt to model carbohydrate binding with an antibody.

The effect of chain length on conformation can also be explored, by comparing simulations of increasing oligosaccharide length. There remains a gap in modeling studies that systematically examine the effect of chain length on conformation, which can be an important question for the manufacture of synthetic carbohydrate vaccines. However, increasing chain length dramatically increases the computational expense, as the number of calculations grows with the square of the number of atoms in the system. This is compounded by the fact that increasing the length of the carbohydrate solute by n atoms increases the number of water molecules required by roughly n^3. Therefore, it is expedient to limit the chain length to the minimum size required to effectively model polysaccharide conformations. For multi-residue repeating unit (RU) antigens, 3RU is often of sufficient length to simulate polysaccharide-like behavior. We have compared simulations of 3RU and 6RU for pneumococcal serotypes 14, Pn19A, Pn19F as well as Group B streptococcus GBS type III and have found no significant differences in the longer chain conformations for these very flexible oligosaccharide strands: in all cases the hydrodynamic behavior of the longer 6RU strands was consistent with the corresponding 3RU simulations (*23*). Conversely, some studies have claimed that much longer strands reveal conformational transitions that do not occur in shorter chains (*24*). It seems likely that the minimum length for representing a polysaccharide conformation will be, to some extent, dependent on the primary structure of the antigen concerned.

Analysis of the simulation trajectories is then also systematic, working from mapping the conformations of individual glycosidic linkages and sugar rings (which may switch to boat or twist-boat conformations for specific residues (*24*, *57*)), to analysis of chain conformations and, most importantly, their clustering into conformational families.

Ideally, molecular simulations will then be corroborated by experimental observations or experimentally derived parameters. NMR NOEs or ^3J couplings are often used for this purpose, but are not unambiguous (*11*). However, performing such comparisons is often not possible. For example, the existence of non-chair conformations of pyranose rings cannot be established with current experimental approaches (*57*). When experimental corroboration is not possible, the speculative nature of predictions gleaned from molecular modeling must always be kept in mind.

Case Study: *Neisseria meningitides*

Infections by *Neisseria meningitidis* (Mn) cause life-threatening illnesses such as meningitis, bacteremia and pneumonia. Humans are the only host and the ease of transmission of bacteria by respiratory droplets leads to outbreaks and epidemics. A combination of non-specific symptoms, rapid onset and increasing antibiotic resistance make vaccination the most cost-effective way to control meningococcal disease.

The capsular polysaccharide (CPS) is the main virulence factor in meningococcus: immunity to infection is conferred by antibodies to the CPS. There are twelve meningococcal serogroups, five of which (A, B, C, Y, and

W) are responsible for the vast majority of disease in children and adults. The polysaccharide vaccines developed in the 1970s have largely been replaced by the more immunogenic corresponding conjugate vaccines: tetravalent meningococcal conjugate vaccines (Menactra®, Menveo™ and Nimenrix®) are available. The high burden of serogroup A disease in the Meningitis Belt of sub-Saharan Africa led to the introduction of MenAfriVac®, which has successfully reduced the cases of group A disease. However, several outbreaks caused by other serogroups have been reported, including those due to serogroup X (*76*, *77*). This has led to the current development of a pentavalent conjugate vaccine (NmCV-5) against serogroups A, C, Y, W and X (*78*).

Structurally, CPSs of the six dominant meningococcal serogroups fall into three pairs: the CPSs for MnA and MnX are phosphodiester-containing homopolymers of amino sugars, MnB and MnC are homopolymers of sialic acid, whereas MnY and MnW are almost identical polymers of hexose-sialic acid, as listed below.

MnA: →6)αDMan*p*NAc(3/4OAc)(1→OPO₃→

MnX: →4)αDGlc*p*NAc(1→OPO₃→

MnB :→8)αDNeu5Ac(2→

MnC :→9)αDNeu5Ac(7/8OAc)(2→

MnW :→6)αDGlc*p*(1→4)αDNeu5Ac(7/9OAc)(2→

MnW :→6)αDGlc*p*(1→4)αDNeu5Ac(7/9OAc)(2→

Serogroups MnY and MnW

The structural similarity of the antigens in meningococcal serogroups Y and W suggests the possibility of cross-protection between MnY and MnW vaccines: the capsular polysaccharides are almost identical polymers of hexose-sialic acid that differ only in the stereochemistry of the C-4 hydroxyl group which is equatorial in group MnY (glucose) and axial in group MnW (galactose). Cross-protection between group MnY and MnW was not evaluated during development of the tetravalent polysaccharide vaccine because the vaccines were licensed together. However, a small scale clinical trial in 1981 tested monovalent and divalent polysaccharide MnY and MnW vaccine formulations (*3*). This study showed that both the group MnY and MnW monovalent polysaccharide vaccines were able to elicit a strong immune response against their respective antigens and the divalent vaccine, as expected, elicited bactericidal antibody against both serogroups. However, interestingly, the monovalent vaccines elicited different levels of cross-protection, as follows. While 71% of volunteers who received the group MnY vaccine had bactericidal antibody against group MnW bacteria after 4 weeks, only 30% of volunteers receiving the group MnW vaccine had bactericidal antibody against group MnY bacteria after the same period. A more recent trial of the HibMenCY-TT conjugate vaccine suggested possible cross-protection by the group MnY vaccine against group MnW (*79*). The vaccinees showed markedly higher seroprotection rates and antibody titer to group MnW compared to the control group, even though neither group had previously received the group MnW vaccine.

Although C-4 stereochemistry is known to affect the orientation and dynamics of the hydroxymethyl group in galactose (axial) and glucose (equatorial) (*80*), it was generally assumed that the similarity in RU for the MnY and MnW CPS produces very similar molecular conformations. This assumption was originally supported by simple models that predicted similar regular helical conformations for the MnY and MnW polysaccharides (with four residues per helical turn) (*81*). However, our recent modeling study (*25*) showed significant differences in the MnY and MnW antigen conformations, as follows.

We performed MD of 3RU oligosaccharide strands of MnY and MnW in explicit water. Although the MnY and MnW polysaccharides are partly O-acetylated at O-7/O-9, we did not consider the effects of O-acetylation on conformation, as low levels of O-acetylation are specified in the WHO guidelines for the polysaccharide vaccine (*82*) and pre-clinical and clinical studies indicate that the de-O-acetylated epitope is the primary target for bactericidal antibodies (*83*).

Long molecular dynamics simulations were required to reveal the conformational differences in the MnY and MnW strands, which arise from distinct conformational populations for the αDNeu5Ac(2→6)αDGlc*p* and αDNeu5Ac(2→6)αDGal*p* linkages in the polysaccharides. The 2D PMFs for the φ and ω dihedrals angles in these linkages (Figure 1) revealed that the αDNeu5Ac(2→6)αDGlc*p* linkage has a single dominant global energy minimum, corresponding to the ω dihedral in a *gg* conformation, whereas in the αDNeu5Ac(2→6)αDGal*p* linkage the *gg*, *gt* and *tg* conformations are all close in energy.

This has consequences for the MnY and MnW CPS: MnY (Glc) has a single dominant strand conformation whereas MnW (Gal) exhibits a variety of conformations, with the principal conformational family of MnW being very similar to the sole MnY conformational family. Representatives of the conformational families for the MnY and MnW CPS are shown in Figure 2. The conformations predicted from our modeling studies were supported by NMR NOESY analysis, which indicated key close contacts for MnW that are not present in MnY.

These conformational differences provide an explanation for the different levels of cross-protection measured for the MnY and MnW monovalent vaccines and the high group MnW responses observed in HibMenCY-TT vaccinees. This can be rationalized by understanding that the MnY CPS would raise antibody against a single conformation, whereas the MnW CPS would raise antibody against a family of conformations, only one of which would correspond to the MnY conformation.

Our MnY and MnW simulations thus revealed a conformational rationale for the observed heterologous immune response. This work also highlighted the importance of performing long MD simulations (250 ns or more) to expose the conformational behavior of flexible oligosaccharides in solution.

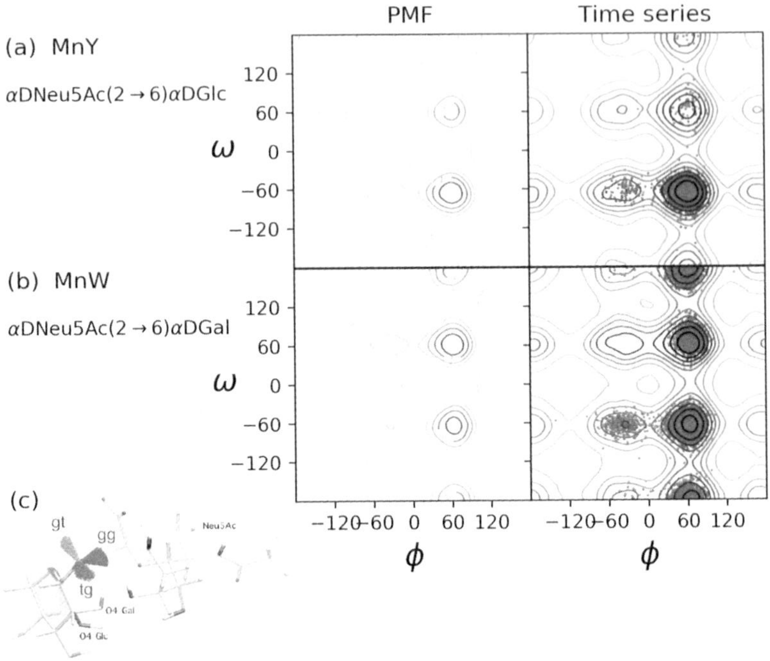

Figure 1. Contoured disaccharide φ, ω PMF surfaces in aqueous solution for (a) αDNeu5Ac(2→6)αDGlcp and (b) αDNeu5Ac(2→6)αDGalp glycosidic linkages. Plots of the simulation time series for both a(2/6) linkages in the 3RU oligosaccharides are superimposed on the corresponding disaccharide solution free energy surfaces in the right column (MnY: top, MnW: bottom). The disaccharide structure shown in (c) illustrates the difference in conformation at the hexose C-4 (which is axial in Gal and equatorial in Glc), as well as the three orientations of the ω dihedral in the glycosidic linkage: gg, gt and tg. (see color insert)

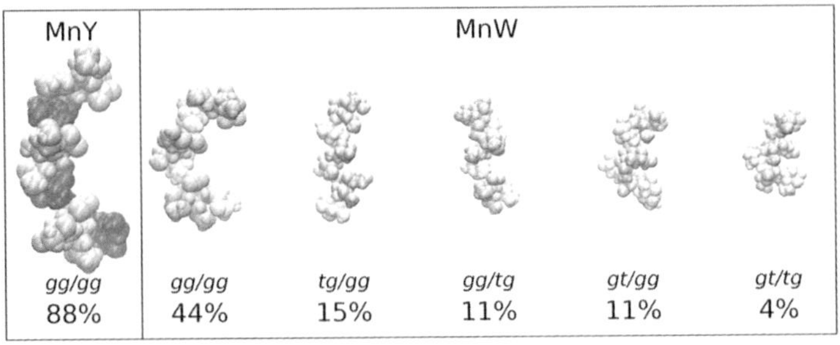

Figure 2. Representative structures from the dominant conformational families for the capsular polysaccharides in MnY (left) and MnW (right). (see color insert)

Case Study: *Streptococcus pneumoniae*

Streptococcus pneumoniae is a globally important encapsulated human pathogen with approximately 100 different serotypes recognized (*84*). The first pneumococcal vaccines (PPVs) were polysaccharide-based; of these, only PPV23 is currently available. PPVs showed poor immunogenicity in infants younger than 2 years, which spurred the development of conjugate vaccines (PCVs). Following the introduction of PCV7, PCV10 and PCV13 into childhood immunization schedules, there has been a significant decrease in invasive pneumococcal disease (IPD) in both targeted populations and unvaccinated adults. The approval of PCV13 for use in adults together with the variable increases in non-vaccine-serotype disease and possible serotype replacement anticipated in the future (*6*, *85*) is expected to drive the development of higher valency PCVs.

Some early modeling studies on pneumococcal polysaccharides combined NMR analysis with simple conformational analysis. For Pn4, a conformational search with a simple interaction model predicted a helical conformation of the polysaccharide backbone, which was little affected by the addition of the 2,3-linked pyruvic acetal on the 4-linked Gal (*86*). However, recent detailed evaluation of synthetic pyruvylated and non-pyruvylated oligosaccharides and derived conjugates showed that the pyruvate is indeed needed for cross-reactivity with the native CPS (*87*, *88*). A similar conformational study was performed on the serogroup 9 polysaccharides – Pn9A, Pn9L, Pn9N and Pn9V – all comprising pentasaccharide RUs (*89*, *90*). Serotypes 9A and 9V differ only in the levels of O-acetylation, Pn9N differs from Pn9L with a 3-linked α-Glc instead of α-Gal. Pn9N and Pn9V cause the most disease and were included in PPV23 (*1*), the licensed conjugate vaccines all contain Pn9V as it was identified as the main serotype associated with IPD in children under five prior to PCV introduction (*91*). NMR and conformational analysis predicted an extended ribbon conformation for all Pn9 polysaccharides, with minor differences localised at the sites of structural variation. The authors concluded that these slight structural variations in sugar type and O-acetylation, rather than conformational differences, determines the antigenic specificity of the serogroup 9 polysaccharides. As there has been considerable development in molecular modeling methods in the last 20 years, there is scope to repeat and extend the early calculations on pneumococcal serotypes 4 and 9. As discussed above for MnY/MnW, the prediction of similar structures based on early models may be overturned by lengthy MD simulations using modern carbohydrate-specific force fields.

A more recent study compared the conformations of the branched capsular polysaccharides in Pn14 and *Streptococcus agalactiae* (usually termed Group B Streptococcus) type III. The similarity of these two polysaccharides has long been of interest: they are identical except for the addition in Group B Streptococcus type III (GBSIII) of a terminating sialic acid in the galactose side chain. Despite their similarity in primary structure, limited cross-protection between GBSIII and Pn14 has been reported. MD simulations of 5RU strands with the GLYCAM_2000 force field suggested that the cause of the antigenic differences between GBSIII and Pn14 capsular polysaccharides is significantly different conformations and dynamic behavior: GBSIII forming a large helical structure due to stabilization by the terminal sialic acid residues and Pn14 having a more flexible and disordered conformation (*14*, *19*). This work suggested that the sialic acid is not directly involved in binding: the anti-GBSIIII and anti-Pn14 antibodies recognize the backbone and not the branch points of the capsular polysaccharide, with a minimal conformational epitope of three RU (*14*). However, more recently Carboni et al. published the structure of a hexasaccharide GBSIII-Fab complex derived from a functional antibody (*9*). This showed that the sialic acid residue participates directly in antigen binding and is essential for the elicitation of protective antibodies, thus calling into question the need for a conjugate vaccine to present the postulated GBSIII conformational epitope. This highlights the potential complexity of antibody-carbohydrate antigen binding and the difficulty in making broad generalizations about the mode of interaction from modeling studies for cases when the backbone conformation may be of little importance for the binding of antibodies to immunodominant side chains of branched polysaccharides.

Below, we discuss in more depth the application of molecular modeling to explore the issue of cross-protection in three key pneumococcal serogroups: Pn23, Pn6 and Pn19.

Pn23

Pneumococcal serogroup Pn23 consists of serotype 23F (present in licensed vaccines) and emerging serotypes Pn23A and Pn23B. Structures for these serotypes have been recently published[18] and are as follows.

Pn23F:
→6)[αLRha(1→2)][Gro2P(O→3)]βDGal(1→4)βLRha(1→4)βDGlc(1→

Pn23A:
→3)[[αLRha(1→2)][Gro2P(O→3)]βDGal(1→4)]βLRha(1→4)βDGlc(1→

Pn23B: →4)[Gro2P(O→3)]βDGal(1→4)βLRha(1→4)βDGlc(1→

We built simple 6RU models of each of these polysaccharides, shown in Figure 3. Static structures can be effective as a first estimate of a structure and are revealing of the clear conformational differences in the Pn23 serogroup, with Pn23F and Pn23B having helical conformations in contrast to the ribbon-like Pn23A. An observation from animal studies is that Pn23F gave small cross-reaction with Pn23A and none with Pn23B (*1*). Our static models reveal that the immunodominant terminal α-Rha in Pn23F (absent in Pn23B) is clearly exposed on the edge of the helix (purple residues, Figure 3 left): Pn23F thus presents a quite different surface for antibody binding than Pn23B.

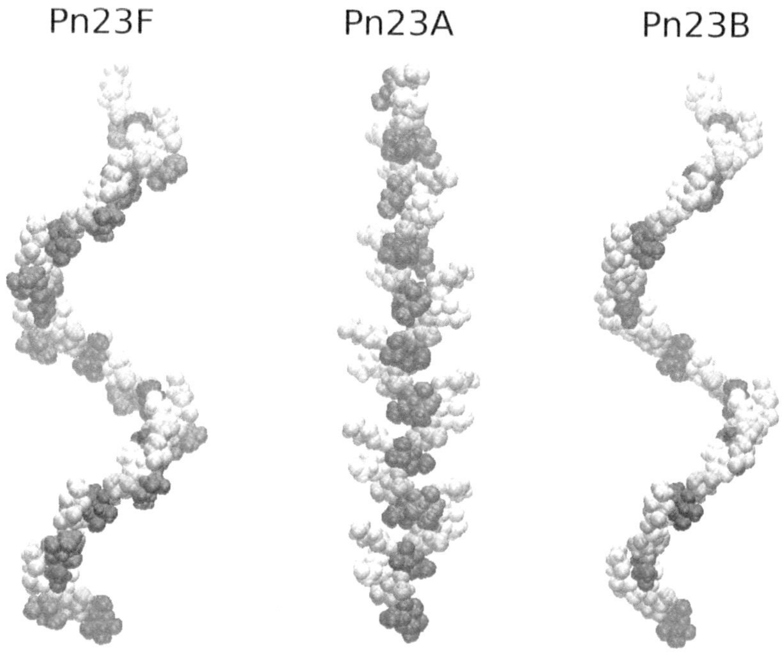

Figure 3. 6RU models of Pn23F, Pn23A and Pn23B. (see color insert)

In contrast to the helical conformations of Pn23F and Pn23B, the model for Pn23A is a slightly twisted flat ribbon, with clear steric crowding at the β-Rha branch point: the β-Glc is in close proximity to β-Gal (< 3 Å). This molecular model thus explains the strong deshielding of H-1 of 2,3-β-Gal observed in the NMR spectrum of polysaccharide Pn23A), as well as the small cross-reaction of Pn23A with Pn23F (*18*). These results suggest little likelihood of cross-protection against 23A or 23B by a 23F vaccine and could explain the emergence of these serotypes in vaccinated populations.

Pn6

A recent review of pneumococcal capsules identified eight serotypes in serogroup 6 (*6*). The four major serotypes — Pn6A, Pn6B, Pn6C and Pn6 — all contain ribitol phosphate in a trisaccharide RU, with chemical structures as follows (differences in bold):

Pn6A: [→2)αDGal*p*(1→3)αDGlc*p*(1→3) αLRha*p*(1→3)DRib-(5P→]

Pn6B: [→2)αDGal*p*(1→3)αDGlc*p*(1→3)αLRha*p***(1→4)**DRib-(5P→]

Pn6C: [→2)αD**Glc***p*(1→3)αDGlc*p*(1→3)αLRha*p*(1→3)DRib-(5P→]

Pn6D: [→2)αD**Glc***p*(1→3)αDGlc*p*(1→3)αLRha*p***(1→4)**DRib-(5P→]

Additional serotypes Pn6F, Pn6G and Pn6H are hybrids and contain mixtures of the 6A/6C, 6B/6D and 6A/6B RUs respectively. In 2016, genomics revealed the worldwide distribution of a new multidrug-resistant serotype designated Pn6E (*92*). However, subsequent NMR and chemical analysis showed that serotype Pn6E produces the serotype Pn6B polysaccharide (*93*), which illustrates the difficulties of assigning new bacterial serotypes based on genetic findings alone.

At the time of formulation of the current PPV23 vaccine, only two cross-reactive serotypes in group 6 were known, Pn6A and Pn6B. The difference between these serotypes is in a single linkage: Pn6B contains Rha(1→4)Rib-5P whereas Pn6A has Rha(1→3)Rib-5P. As the Pn6A antigen is relatively unstable (the free C-4 hydroxyl in Pn6A facilitates cleavage of the phosphodiester linkage at C-5) (*94*), Pn6B was selected for inclusion in the vaccine, on the assumption that it would cross-protect against serotype Pn6A (*1*). The widespread introduction of PCV7 (which includes serotype Pn6B) resulted in an overall reduction of IPD, including IPDs caused by Pn6B and Pn6A. However, there was a gradual subsequent increase in IPD caused by non-vaccine serotypes, including serotype Pn6A. Capsule-specific monoclonal antibodies revealed that the apparent increase in Pn6A disease was due to a new serotype, Pn6C, which has the Pn6A RU, but with the 2-linked α-Gal replaced by α-Glc (*95*, *96*). This led to the postulation, then discovery, of the corresponding 2-linked α-glucose variant of Pn6B, designated Pn6D (*97*, *98*). The inability of the typing sera to differentiate between Pn6A/Pn6C and Pn6B/Pn6D underlines the structural similarity of these pairs of antigens.

Further information is now available on the extent of cross-protection in serogroup 6 for the PCV7 and PCV13 vaccines. Firstly, there is evidence that PCV7 (which contains Pn6B only) does not provide protection against Pn6C disease. American Indian children have a high burden of pneumococcal disease and use of PCV7 in this population resulted in 76% effectiveness against Pn6A IPD, with no impact on 6C disease (*99*). This is consistent with other populations in the USA, which have shown a steady increase in Pn6C prevalence (*4*, *100*). However, evidence from opsonophagocytic killing assays (OPA) on sera from immunized infants indicates that the higher valency PCV13 vaccine (which contains both Pn6B and Pn6A) provides cross-protection against Pn6C (*101*). More recent analysis of sera after vaccination with PCV7 or PCV13 allowed the extent of cross-protection of serotypes by these two vaccines to be evaluated (*5*). Immunogenicity (IgG) was measured by ELISA and functional antibody was assessed by OPA for serotypes Pn6A/B/C. PCV7 vaccination generated functional cross-reacting Pn6A antibodies that correlate with the near elimination of type Pn6A disease and colonization following widespread use of this vaccine. Functional antibodies to Pn6C were also observed, but at significantly lower levels than those seen following PCV13 vaccination. This suggests only partial cross-protection for 6C by Pn6B, which is highest for PCV13 due to the combined cross-protection from both Pn6A and Pn6B. The authors conclude that functional antibody activity against Pn6A/B/C suggests that PCV13 is likely to control the 6C disease that was not prevented by widespread vaccination with PCV7. Further, there is some evidence for cross-protection against serotype 6D by Pn6B (*102*). Surveillance of serogroup 6 in children under five in Israel showed that the incidence of Pn6A, Pn6B and 6D in *otitis media* dropped gradually along with PCV7/13 introduction, whereas Pn6C rates increased in the PCV7 period and then decreased following PCV13 implementation. Thus, the related Pn6C/6D do not act as replacement serotypes for Pn6A/6B following vaccination with PCV13.

To explore the conformations of the main serogroup 6 polysaccharides, we performed 250 ns molecular dynamics simulations in aqueous solution for 3RU of the capsular polysaccharides in Pn6A (in PCV13), Pn6B (in PCV7 and PCV13) and the non-vaccine serotypes Pn6C and Pn6D. Conformations from the 3RU trajectories were clustered into families using VMD's (*30*) internal *measure cluster* command. All simulation conformations were aligned on the middle RU in the 3RU chains. Clustering was performed with a cut-off of 5.5 Å on an RMSD fit to the carbon and oxygen ring- and linkage atoms.

Comparison of the conformational families in these closely related serotypes reveals significant differences in conformation (Figure 4). The effect of replacing galactose with glucose (i.e. Pn6A->Pn6C and Pn6B->Pn6D) is that the equatorial O-4 enables close Glc-Glc interactions between successive RU, resulting in more stable conformations. Substitution of the αLRha*p*(1→3)DRib-5P linkage (Pn6A and Pn6C) with the more constrained α(1→4) linkage (Pn6B and Pn6D) has a more dramatic impact: the compressed hairpin bend conformations of Pn6A and Pn6C become more extended in Pn6B and Pn6D. The combination of these two factors means that the Pn6B CPS has the greatest conformational diversity and Pn6C the least.

The greatest similarity is between the CPS conformations of Pn6A and Pn6C. The primary conformational cluster in Pn6A (59%, Figure 4 top left) – a compact coil with stabilizing inter-residue interactions - corresponds to the sole conformational cluster in Pn6C (88%, Figure 4 bottom left) – suggesting a strong likelihood of cross-protection between these serotypes. This finding is supported by the available clinical evidence for Pn6A-Pn6C cross-protection (*101*). In contrast, Pn6A shows less similarity with Pn6B, with only a minor conformational cluster in Pn6B (9%) corresponding to the secondary conformational cluster in Pn6A (29%). Further, there is no overlap in the conformational clusters of Pn6B and Pn6C. This provides a rationale for the limited Pn6B-Pn6A cross-protection reported (some conformational overlap) and the lack of Pn6B-Pn6C cross-protection (no conformational overlap). The Pn6D CPS shows the same primary conformational cluster as Pn6B, albeit at a higher incidence (65% versus 38%). This suggests that the PCV13 vaccine provides cross-protection against Pn6D from the Pn6B component, as is suggested by recent evidence (*102*). However, there is no conformational overlap between either Pn6A or Pn6C with Pn6D.

In summary, our simulations provide both a rationalization of clinical observations and predictions. The observation that Pn6B protects only partially against Pn6A, but not against Pn6C, whereas Pn6A protects against Pn6C (*101*) can be explained by the close conformational similarity between Pn6A and Pn6C, but not Pn6B. Further, the marked differences in the conformational families obtained from our simulations suggest little likelihood of cross-protection between Pn6D and either Pn6A or Pn6C. Pn6B shows the greatest conformational diversity and overlap with the other serotypes, allowing for at least partial cross-protection against Pn6A and Pn6D (but not Pn6C).

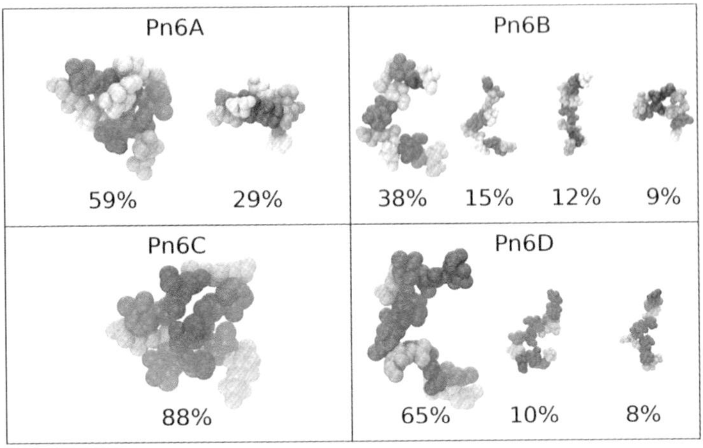

Figure 4. Representative structures from the dominant conformational families for 3RU of the capsular polysaccharides in Pn6A, Pn6B, Pn6C and Pn6D. Residues are colored according to type: glucose blue, galactose yellow, rhamnose magenta and ribitol grey. (see color insert)

Pn19A and Pn19F

Historically, serogroup 19 has been responsible for the bulk of pneumococcal disease, with infection caused chiefly by serotypes Pn19A and Pn19F. The Pn19F and Pn19A polysaccharides comprise very similar trisaccharide repeating units, as follows.

Pn19F: [→4)-αDMan*p*NAc-(1→4)-αDGlc*p*-**(1→2)**-αLRha*p*-(1-P→]

Pn19A: [→4)-αDMan*p*NAc-(1→4)-αDGlc*p*-**(1→3)**-αLRha*p*-(1-P→]

The PCV7 vaccine contained Pn19F, on the assumption that it would provide cross-protection against Pn19A. However, studies showed that PCV7 was only 26% effective against Pn19A IPD (*103*) and provided limited cross-reactive protection against Pn19A disease (*104*). This resulted in serotype Pn19A becoming a leading cause of pneumococcal disease in both vaccinated and unvaccinated individuals.

It was hoped that the new conjugate vaccines – PCV10 (Pn19F) and PCV13 (Pn19A) – would be more cross-protective against Pn19A disease than PCV7 (*104*). Analysis of the sera of American Indian children after vaccination with PCV7 (Pn19F) or PCV13 (Pn19A and Pn19F) after 2010 allowed the extent of cross-protection of serotypes by these two vaccines to be evaluated (*5*). Immunogenicity (IgG levels) of Pn19F and Pn19A was higher in PCV13 than

PCV7. For PCV7, although at least 68% of PCV7 recipients achieved an IgG titre of 0.35 mg/mL for the cross-reactive serotype Pn19A, these antibodies were not functional. In contrast, OPA titers and anti-Pn19A IgG were positively correlated among PCV13 recipients. The study concluded that the functional antibody activity against Pn19A/F suggests that PCV13 was likely to control Pn19A disease.

The impact of PCV10 (19F) and PCV13 (Pn19F and Pn19A) on the worldwide epidemiology of serotype Pn19A has recently been reviewed (*105*). Early effectiveness in vaccinated children using PCV7 or PCV10 against IPD caused by serotype Pn19A shown in case-control studies was not sustained and the vaccines did not conclusively show any reductions of Pn19A carriage, resulting in continued transmission and disease. Despite the ability of PCV7 and PCV10 to raise IgG antibodies to serotype Pn19A, these antibodies appear clinically nonfunctional, and cross-protection cannot be achieved. In contrast, PCV13 serotype Pn19A elicits significantly higher functional immune responses against serotype Pn19A than PCV7 and PCV10. Higher responses are likely to be linked to both direct impact in vaccinated populations and reductions in Pn19A nasopharyngeal carriage in children, thus inducing herd protection and reducing Pn19A IPD in unvaccinated children and adults.

For evaluation of the molecular basis of cross-protection, a key question is whether the switch from the $\alpha(1\rightarrow2)$ linkage in Pn19F to the $\alpha(1\rightarrow3)$ configuration in Pn19A results in a significant change in polysaccharide conformation. Early modeling studies with generic (non-carbohydrate specific) force fields suggested no difference in conformation (*15*, *106*), whereas our subsequent simulations with the CHARMM carbohydrate force field predicted marked differences in conformation and dynamics of the Pn19F and Pn19A polysaccharides. We used a systematic approach for our modelling study, with successive MD simulations of 1RU, 3RU and 6RU strands (*22*, *23*). These calculations revealed that the conformations of the repeating units in the polysaccharides differ. In Pn19F, the rhamnose residue is nearly orthogonal to the other residues, whereas Pn19A has residues in similar orientations (Figure 5). These RU conformations were corroborated by key inter-residue distances calculated from NMR NOESY experiments on the Pn19F and Pn19A polysaccharides, as illustrated in the scatter plot in Figure 5.

In addition to showing conformational differences for the repeating units, the simulations also revealed differences in the chain conformations for the Pn19F and Pn19A 3RU and 6RU saccharides. While both polysaccharides are flexible chains with no single well-defined conformation, Pn19F showed a higher frequency of extended structures and Pn19A more persistent compact structures. These differences are a direct consequence of the repeat unit conformations: the parallel arrangement of residues in Pn19A allows for close stacking of residues in neighboring RU's in tight hairpin bends about the phosphodiester linkage, while close inter-RU contacts are precluded in Pn19F by the more bent RU conformation. Therefore, the conformational differences revealed between Pn19F and Pn19A help to explain the otherwise unexpected limited antibody cross-protection observed between these serotypes.

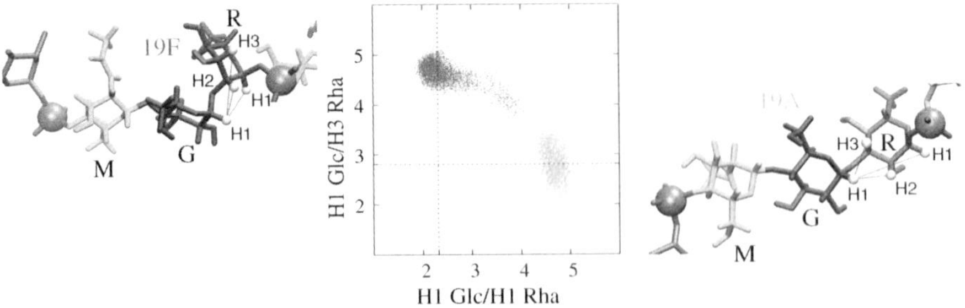

Figure 5. Time series of the H1 Glc/H1 Rha and H1 Glc/H3 Rha distances for the middle repeat unit in the RU3 simulations of Pn19F (red, top and left) and Pn19A (green, bottom and right). NMR NOESY distances are indicated by dashed lines. Representative conformations for the middle RU are shown for Pn19F (left) and Pn19A (right), with key atomic distances indicated. Structures are annotated to indicate residue identity: ManNAc (M, green), Glc (G, blue) and Rha (R, purple). (Reproduced with permission from reference (23). Copyright 2015 Elsevier.) (see color insert)

Case Study: *Salmonella enterica* O-Antigens

Salmonellae are responsible for a huge global disease burden through two forms of invasive illness: enteric fever and invasive non-typhoidal *Salmonella* disease. Enteric fever is principally caused by *Salmonella enterica* serovar *Typhi* (*S. Typhi*) and *S. paratyphi A*. Disease due to both serovars is a major problem in South and South-East Asia, whereas *Salmonella enterica Typhimurium* and *Enteritidis* are the most common serovars responsible for invasive nontyphoidal Salmonella disease in Africa. Two typhoid vaccines against *S. Typhi* are currently recommended: an injectable polysaccharide vaccine based on the purified Vi antigen and a live attenuated oral Ty21a vaccine. However, neither is effective in young children, where the burden of invasive Salmonella disease is highest. This has led to the development of new Vi conjugate vaccines that are expected to have improved immunogenicity and efficacy in young children and infants (*107*).

The protective lipopolysaccharide (LPS) in *Salmonella* is essential for bacterial survival and adaptation within the host. The LPS comprises a fatty acid (Lipid A) buried in the membrane, an oligosaccharide core region and an exposed, antigenic polysaccharide (O-antigen) tail. The O-antigen is the vaccine target for *S. paratyphi A*, *S. Typhimurium* and *S. enteritidis*. The O-antigens share a common backbone –

$$[\rightarrow 2)[\ \alpha\mathrm{DMan}p(1\rightarrow 4)\alpha\mathrm{LRha}p(1\rightarrow 3)\ \alpha\mathrm{DGal}p(1\rightarrow]$$

– but differ in the structure of the side chains at O-3 of mannose (paratose, abequose or tyvelose, respectively) and have variable glucosylation and O-acetylation (*108*). Little is known about the conformation of the LPS O-antigen chains.

Salmonella paratyphi A

As part of vaccine development, the O-polysaccharide structure of *S. paratyphi A* was fully characterized chemically and by 1D and 2D-NMR spectroscopy (*109*). The structure of the pentasaccharide repeating unit is:

[→2)[αDPar*p*(1→3)]αDMan*p*(1→4)αLRha*p*(1→3)[αDGlc*p*(1→6)]αDGal*p*(1→].

The NMR spectra were complex due to the presence of O-acetyl groups on C-2 and C-3 of Rha and incomplete glucosylation on O-6 of Gal. Glucosylation resulted in shielding of the anomeric proton of Man from 5.30 (in the tetrasaccharide) to 5.18 ppm for the pentasaccharide and affected H-4 to H-6 of Rha.

To aid with interpretation of the NMR, in particular to account for the major chemical shift influence observed from glucosylation, we built a 3RU static model with our CarbBuilder software. For this, 2D PMF surfaces were calculated for disaccharides representing all of the backbone glycosidic linkages in the saccharide to identify the low-energy orientations for each linkage. This model, depicted in Figure 6, shows that a structure built with favored orientations of the glycosidic linkages brings the glucose (Glc) residue into close proximity with the backbone Man and Rha residues in the neighbouring repeating unit. This is not apparent from perusal of the primary structure. The close approach of Glc H-1 to Rha H-6 and Man H-1 in the models thus assisted in explaining the chemical shifts of these atoms upon glucosylation.

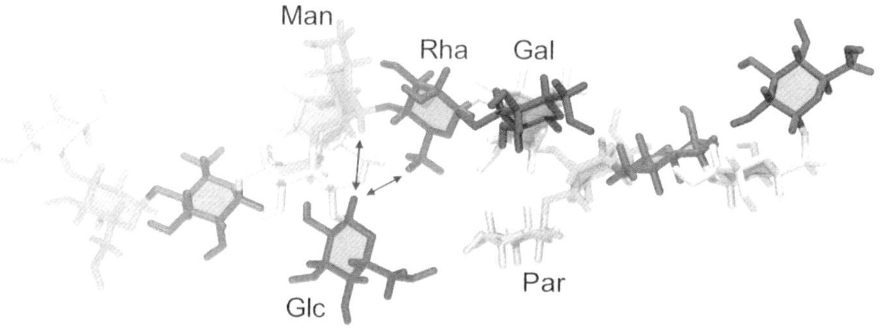

Figure 6. Static structure of S. paratyphi A showing the proximity of the glucose residue (blue) to the backbone rhamnose (purple) and mannose. (see color insert)

Salmonella typhimurium Serogroup B

Salmonella typhi is a serovar of *S. enterica* that is the cause of most cases of enteric fever. Development of a candidate glycoconjugate vaccine against the *Salmonella typhimurium* serogroup B (STB) O-antigen has been recently described (*28*). The STB O-antigen has the following branched tetrasaccharide repeating unit.

STB:

[→2)[αDAbe*p*(1→3)]αDMan*p*(1→4)αLRha*p*(1→3)αDGal*p*(1→]

C-2 O-acetylation can occur on either or both the abequose (Abe) and rhamnose (Rha) residues. Serological studies showed that recognition of STB LPS by monoclonal antibodies is affected by acetylation of Abe, which was postulated to be due to acetylation changing the conformation of the O-antigen (*110*).

Galochkina et al. performed MD simulations of 12 RU of de-O-acetylated STB O-antigen with both the GLYCAM and OPLS force fields at different temperatures (*24*). They found considerable differences in O-antigen behavior depending on the force field used, as follows. At 300 K, the OPLS chain was predominantly in an extended conformation, with occasional formation of hairpin bends, while at 500 K there was reversible formation of a globule conformation. In contrast, the GLYCAM chain collapsed irreversibly to a globule over the course of the 400 ns simulations at both 300 K and 500 K. They concluded that the limited available experimental information did not support the formation of a globule predicted by the GLYCAM force field.

In a separate study performed as part of the development process for an STB O-antigen vaccine, MD simulations with enhanced sampling (via Hamiltonian replica exchange) were run to probe the effect of O-acetylation on the polysaccharide conformations and, by extension, protective immunity (*28*). The simulations, performed with CHARMM36, compared 3RU strands with and without C-2 O-acetylation on both Abe and Rha, as well as with glucosylation on either the central or terminal Gal residues. Conformations of the saccharides were classified using clustering analysis performed on the basis of glycosidic linkages. Interestingly, this thorough simulation predicted that O-acetylation does not affect the conformation of the STB OPS: the 3RU strand formed a single dominant conformation, irrespective of the presence or absence of O-acetylation. This conformation is shown in Figure 7B and C for the de-O-acetylated and O-acetylated oligosaccharides, respectively. However, for the O-acetylated saccharide, the O-acetyl groups were shown to be highly solvent exposed and thus potentially important for antigen binding.

Additional simulations by the same group investigated the effect of O-acetylation on antibody binding by modeling 3RU acetylated and de-O-acetylated strands bound to the monoclonal antibody Se155–4 (*26*). This work showed that abequose is central to the binding, with O-acetylation altering the preferred bound conformation of the saccharide. The antibody binding occurred through a conformational selection mechanism, where the antibody selected a specific conformation of the unbound saccharide from the solution populations. O-acetylation resulted in a minor conformation being bound, which incurred a small entropic penalty (0.5 kcal/mol) in binding energy. Therefore, overall this study did not show that O-acetylation produced a clear difference in binding preference for the monoclonal antibody studied. It is conceivable that the physicochemical properties of O-acetyl groups (e.g., size, partial charge, hydrophobicity) may stabilize the interaction between the polysaccharide and the B-cell receptor and thus account for their pronounced immunogenicity.

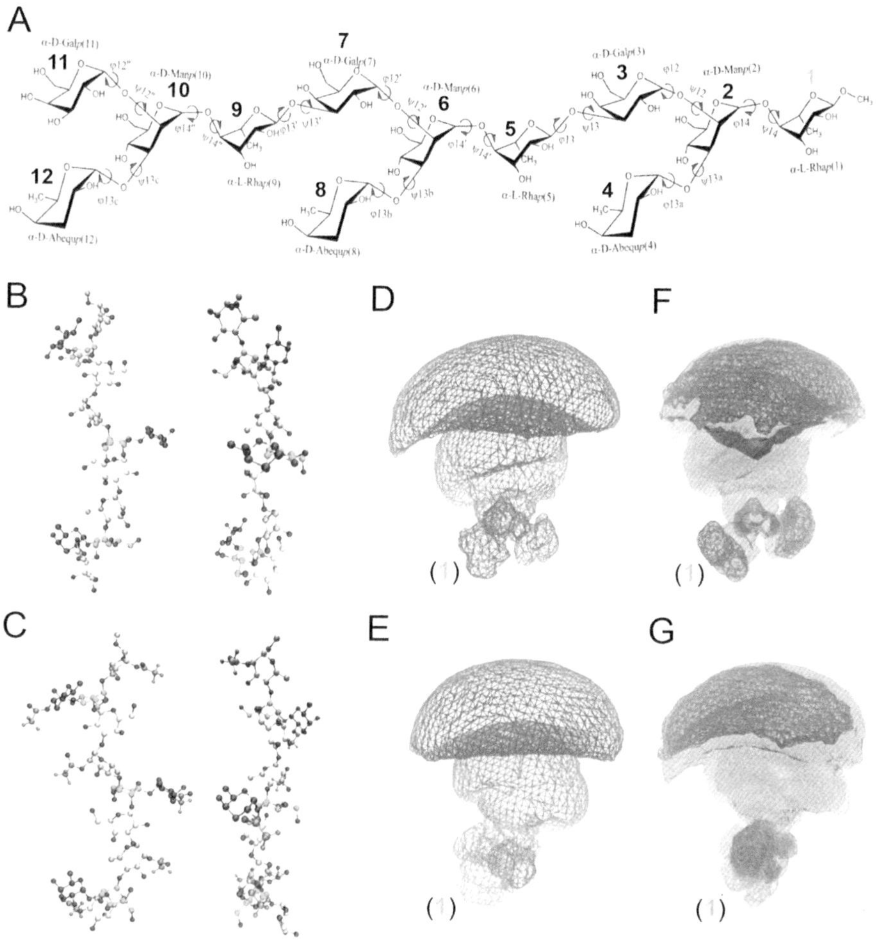

Figure 7. (A) Structure of the base 3-repeat STm polysaccharide unit used for computational analyses. The representative O-acetylated polysaccharide was constructed with the hydroxyl group substituted by an acetyl at the C-2 position of both α-D-Abep and α-L-Rhap (-OH, red). (B, C) Two views of the dominant conformation of the base (B) and O-acetylated (C) polysaccharides. The acetyl group is shown in brown, glucose in blue, mannose in green, galactose in yellow, abequose in purple, and rhamnose in cyan. For clarity, all hydrogen atoms are omitted except those on the acetyl group. (D, E, F, G) Wire frame models of the base (D, E) and O-acetylated (F, G) polysaccharides for all non-hydrogen atoms. Conformational flexibility increases progressively relative to the reducing end anchor point. (Figure reproduced from (28), Copyright 2017 PLOS.) (see color insert)

These detailed simulations of the branched STB O-antigen highlight the potential complexity of antibody-carbohydrate antigen binding and the difficulty in making broad generalizations about the mode of interaction based on antigen primary structure.

Case Study: *Shigella flexneri* O-Antigens

Shigella is one of the five main pathogens causing diarrheal disease, with high morbidity and more than 800 000 fatalities annually, mainly in young children in sub-Saharan Africa and south Asia (*111–113*). A low infectious dose (10 cells) (*114*) allows the disease to be spread effectively and the difficulties of improving sanitation to prevent shigellosis, together with increasing antibiotic resistance of Shigella species, has made vaccine development a high priority for the World Health Organization (*115*). Vaccine strategies include live-attenuated, inactivated whole-cell and, more recently, subcellular and purified subunits, such as the O-antigen conjugate vaccines (*116*). Immunity to *Shigella* appears to be strain specific, thus justifying the O-antigen as a vaccine target, while the success of multivalent conjugate vaccines against meningococcal and pneumococcal disease further validates the O-antigen conjugate vaccine approach. The preparation of conjugate vaccines involves chemical linkage of the saccharide from the bacterial surface carbohydrate (LPS for shigella) to the carrier protein. The saccharide component is either the isolated O-antigen (terminally or randomly activated) or a synthetic oligosaccharide. Alternatively, the intact glycoconjugate can be prepared biosynthetically.

The composition of a protective multivalent *Shigella* vaccine depends on epidemiology. *S. flexneri* is the major cause of shigellosis in endemic countries, accounting for up to 60% cases of shigellosis mainly in developing countries (*115*). The most prevalent *S. flexneri* O serotype is 2a, followed by 3a, 6 and 1b. *S. sonnei* is more common in low- and middle-income countries; and *S. dysenteriae* has not caused epidemics of dysentery since the 1990s.

There is evidence for a large degree of cross-protection between *Shigella* serotypes: the Global Enteric Multicenter Study showed that broad-spectrum vaccine protection against *S. sonnei* and 15 *S. flexneri* serotypes/subserotypes can be achieved with a quadrivalent vaccine comprising O antigens from *S. sonnei*, *S. flexneri* 2a, *S. flexneri* 3a, and *S. flexneri* 6 (*115*). This vaccine can provide broad direct coverage against these most common serotypes and possibly indirect coverage against all but the rare *S. flexneri* 7a subserotype through cross-protection against shared *S. flexneri* group antigens.

Except for serogroup 6, *Shigella flexneri* serotype O antigens share a common tetrasaccharide backbone structure:

$$\rightarrow 2)\text{-}\alpha\text{LRha}p\text{-}(1\rightarrow 2)\text{-}\alpha\text{LRha}p\text{-}(1\rightarrow 3)\text{-}\alpha\text{LRha}p\text{-}(1\rightarrow 3)\text{-}\beta\text{DGlc}p\text{NAc-}(1\rightarrow$$

with antigenic variation provided by site-selective glucosylation(s) and/or O-acetylation. The primary structures of the main *S. flexneri* serotypes considered in this chapter are listed in Table 1.

The final elucidation of the O-acetylation profiles and a survey of the O-antigen structure diversity for *S. flexneri* was published in 2012 by Perepelov et al. (*117*) Since then a recent review has identified a total of 30 O-antigen variants which include glucosylation, additional sites of O-acetylation and phosphorylation with phosphoethanolamine (PEtN) (*118*). Glucosylation of the O-antigen has a significant impact on virulence of *S. flexneri* by changing the O-polysaccharide conformation from a more filamentous to a more compact structure that facilitates bacterial invasion of gut epithelium (*119*), whereas the role of O-acetylation for pathogenesis is yet to be determined. For example, although serotype 2a contains non-stoichiometric O-acetylation, this may not be important for immunogenicity (*120–122*).

Table 1. Primary O-antigen structures of 20 serotypes of *S. flexneri*, *glucosylation branches in italics*, acetylation in bold

Y	→2)αLRha(1→2)αLRha(1→3)αLRha(1→3)βDGlcNAc(1→
Yv	→2)αLRha3PEtN(1→2)αLRha(1→3)αLRha(1→3)βDGlcNAc(1→
X	→2)*[αDGlc(1→3)]*αLRha(1→2)αLRha(1→3)αLRha(1→3)βDGlcNAc(1→
Xv	→2)*[αDGlc(1→3)]*αLRha(1→2)αLRha3PEtN(1→3)αLRha(1→3)βDGlcNAc(1→
1a	→2)αLRha(1→2)αLRha(1→3)αLRha(1→3)*[αDGlc(1→4)]*βDGlcNAc(1→
1b	→2)αLRha(1→2)αLRha(1→3)αLRha2**Ac**(1→3)*[αDGlc(1→4)]*βDGlcNAc(1→
1c	→2)αLRha(1→2)αLRha(1→3)αLRha(1→3)*[αDGlc(1→4)αDGlc(1→4)]*βDGlcNAc(1→
1d	→2)*[αDGlc(1→3)]*αLRha(1→2)αLRha(1→3)αLRha(1→3)*[αDGlc(1→4)]*βDGlcNAc(1→
2a	→2)αLRha(1→2)αLRha(1→3)*[αDGlc(1→4)]*αLRha(1→3)βDGlcNAc(1→
2b	→2)*[αDGlc(1→3)]*αLRha(1→2)αLRha(1→3)*[αDGlc(1→4)]*αLRha(1→3)βDGlcNAc(1→
3a	→2)*[αDGlc(1→3)]*αLRha(1→2)αLRha(1→3)αLRha2**Ac**(1→3)βDGlcNAc(1→
3b	→2)αLRha(1→2)αLRha(1→3)αLRha2**Ac**(1→3)βDGlcNAc(1→
4a	→2)αLRha(1→2)αLRha(1→3)αLRha(1→3)*[αDGlc(1→6)]*βDGlcNAc(1→
4av	→2)αLRha3PEtN(1→2)αLRha(1→3)αLRha(1→3)*[αDGlc(1→6)]*βDGlcNAc(1→
4b	→2)αLRha(1→2)αLRha(1→3)αLRha2**Ac**(1→3)*[αDGlc(1→6)]*βDGlcNAc(1→
5a	→2)αLRha(1→2)*[αDGlc(1→3)]*αLRha(1→3)αLRha(1→3)βDGlcNAc(1→
5b	→2)*[αDGlc(1→3)]*αLRha(1→2)*[αDGlc(1→3)]*αLRha(1→3)αLRha(1→3)βDGlcNAc(1→
6	→2)αLRha3**Ac**(1→2)αLRha(1→4)βDGalA(1→3)βDGalNAc(1→
7a	→2)αLRha(1→2)αLRha(1→3)αLRha(1→3)*[αDGlc(1→2)αDGlc(1→4)]*βDGlcNAc(1→
7b	→2)αLRha(1→2)αLRha(1→3)αLRha2**Ac**(1→3)*[αDGlc(1→2)αDGlc(1→4)]*βDGlcNAc(1→

We built static models of 6RU for the 20 *S. flexneri* serotypes listed in Table 1, visualized in Figure 8. These models were generated using our CarbBuilder software (*43*) supplied with optimal dihedral values obtained from calculations of 2D PMFs for all the backbone glycosidic linkages in the twenty serotypes using the CHARMM36 force field. Our simple models suggest that serogroups 2, 6 and 7 deviate most from the common backbone conformation of an extended helix. This finding supports potential broad coverage against *S. flexneri* in a vaccine comprising serogroups 2a, 3a, and 6 (*115*), although serogroup 7 remains an outlier, but is rare. Serotype 3a is representative of the most common helical conformation. Serogroup 2 is differentiated from the other serogroups by glucosylation forcing the αLRha(1→3)αLRha linkage into the a secondary conformation, which creates a tighter helical backbone. In addition, serotypes 2a and 6 are both helices, but have opposite handedness as a consequence of the O-4 linkage to Gal in serogroup 6 and are otherwise dissimilar: the static models do support serotype cross-protection that has been proposed for these common serotypes and their unique conformations suggest differing antigenicity. Further, these static models predict that the Rha 2-O-acetylation in serotypes 1b and 7b has a marked effect on the backbone conformation relative to their de-O-acetylated counterparts (1a, 3b and 7a). This is not the case for serotypes 3a/3b and 4a/4b, where acetylation does not affect the predicted conformation dramatically. For serotypes 1b and 7b the crowding induced by O-4 glucosylation in combination with acetylation sterically forces the αLRha(1->3) αLRha2Ac linkage into an anti-conformation.

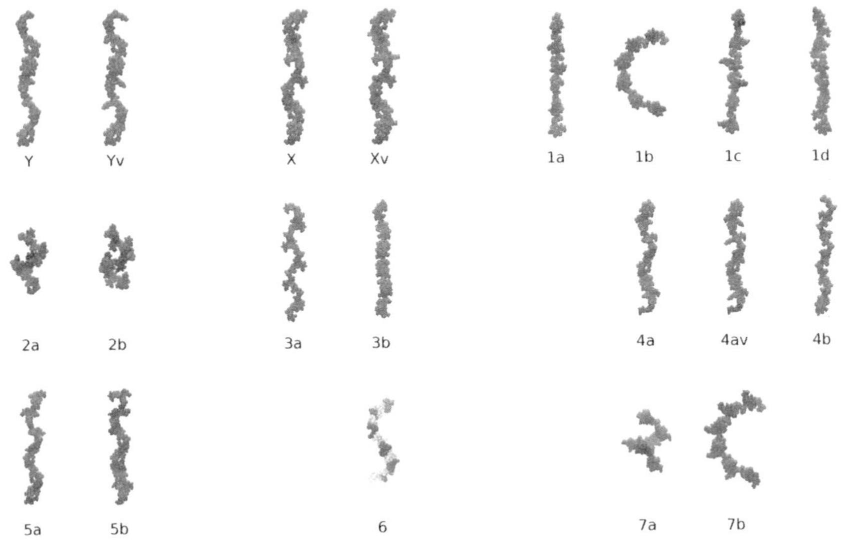

Figure 8. 3D models of six repeating units (6RU) of each of the O-antigens of S. flexneri produced by CarbBuilder from the primary structure. Residues are colored according to type: glucose blue, galactose yellow and rhamnose magenta. (see color insert)

However, static models have limited predictive power, as they consider only steric clashes and are highly sensitive to the specified dihedral values and do not allow for relaxation, with the result that a change in a dihedral conformation can have a dramatic effect on the backbone structure. MD simulations are required to model structural relaxation and flexibility flexibility and thereby obtain the conformational distribution of the *S. flexneri* O-antigens. Previously, Theillet et al. performed a computational investigation of the conformational basis of serotype specificity. The effects of serotype-specific substitutions of the backbone serotypes 1a, 1b, 2a, 2b, 3a, 3b, 4a, 4b, 5a, 5b, X and Y, were explored with 60 ns molecular dynamics simulations of 3RU oligosaccharides using the GLYCAM06 carbohydrate force field (2a and 2b were modeled for 350 ns in order to achieve adequate sampling) (*20*). Conformations of the 1a, 2a, 3a and 5a serotypes were verified by NMR analysis. This work predicted that, in general, branch substitutions have little effect on the backbone conformations: the 3RU strands showed similar backbone conformations in all serotypes, except for serotypes 1a and 1b. Conformational differences in specific backbone linkages were identified for 2a/2b/5a/5b (αLRha(1→2)αLRha), but the consequences for the backbone conformation in a longer strand were not explored. Overall, it is difficult to compare the 3RU conformations from this work with our 6RU static models, as the MD study did not find an overall stable structure for any of the strands and provided no overall analysis of conformational families for the serotypes.

More recently, Kang et al. modeled *S. flexneri* serotype Y O-antigen polysaccharide chains of one to four RU using both atomistic and multiscale modeling approaches (*21*). They identified considerable molecular flexibility in various chains lengths for this serotype, including the formation of hairpin-like bent conformations, which were more dominant in the simulations with the GLYCAM06 force field than in the CHARMM36 carbohydrate force field.

There remain many unanswered questions about the conformational differences and similarities across the *S. flexneri* O-antigens that could be addressed by more detailed MD simulations of *S. flexneri*, using both longer oligosaccharide strands and simulation times.

Conclusions

In this chapter, we have illustrated the valuable role that molecular modeling can play in elucidating conformational contributions to carbohydrate immunogenicity. Firstly, we have shown with examples from Pn23A, *S. paratyphi* A and *S. flexneri* that simple molecular models can be useful tools, both in accounting for NMR chemical shifts that arise from the close proximity of residues and as an initial conformational rationalization of immunological behavior. Secondly, we have demonstrated with examples from meningococcal and pneumococcal polysaccharides and *Samonella* O-antigens that more time-consuming and sophisticated molecular dynamics simulations can reveal how subtle changes in RU structure may result in significant differences in saccharide conformation and dynamics. Such conformational analysis can provide valuable mechanistic insights into clinical observations on cross-protection between carbohydrate antigens. As computational power and the molecular modeling methodology improves, with more structural features incorporated into carbohydrate force fields, we expect molecular modelling to become an increasingly important tool for vaccine design and investigation of immunogenicity.

Acknowledgments

We would like to thank Krishna Prasad of Pfizer for his invitation to contribute this chapter. Our computations were performed using facilities provided by the University of Cape Town's ICTS High Performance Computing team: http://hpc.uct.ac.za.

References

1. Robbins, J. B.; Austrian, R.; Lee, C. J.; Rastogi, S. C.; Schiffman, G.; Henrichsen, J.; Mäkelä, P. H.; Broome, C. V.; Facklam, R. R.; Tiesjema, R. H.; Parke, J. C. *J. Infect. Dis.* **1983**, *148*, 1136–1159.
2. Henrichsen, J. *J. Clin. Microbiol.* **1995**, *33*, 2759–2762.
3. Griffiss, J. M.; Brandt, B. L.; Altieri, P. L.; Pier, G. B.; Berman, S. L. *Infect. Immun.* **1981**, *34*, 725–732.
4. Nahm, M. H.; Lin, J.; Finkelstein, J. A.; Pelton, S. I. *J. Infect. Dis.* **2009**, *199*, 320–325.
5. Grant, L. R.; O'Brien, S. E.; Burbidge, P.; Haston, M.; Zancolli, M.; Cowell, L.; Johnson, M.; Weatherholtz, R. C.; Reid, R.; Santosham, M.; O'Brien, K. L. *PloS One* **2013**, e74906.
6. Geno, K. A.; Gilbert, G. L.; Song, J. Y.; Skovsted, I. C.; Klugman, K. P.; Jones, C.; Konradsen, H. B.; Nahm, M. H. *Clin. Microbiol. Rev.* **2015**, *28*, 871–899.
7. Anish, C.; Schumann, B.; Pereira, C. L.; Seeberger, P. H. *Chem. Biol.* **2014**, *21*, 38–50.
8. Solis, D.; Romero, A.; Menéndez, M.; Jiménez-Barbero, J. In *The Sugar Code*; Gabius, H-. J., Ed.; Wiley-VCH Verlag GmbH & Co. KGaA: Weinheim, 2009; pp 238–241.
9. Carboni, F.; Adamo, R.; Fabbrini, M.; Ricco, R. D.; Cattaneo, V.; Brogioni, B.; Veggi, D.; Pinto, V.; Passalacqua, I.; Oldrini, D.; Rappuoli, R.; Malito, E.; Margarita, I. R.; Berti, F. *Proc. Natl. Acad. Sci. U.S.A.* **2017**, *114*, 5017–5022.
10. Lütteke, T. *Acta Cryst.* **2009**, *D65*, 156–168.
11. Widmalm, G. *Carbohyd. Res.* **2013**, *378*, 123–132.
12. Toukach, F. V.; Ananikov, V. P. *Chem. Soc. Rev.* **2013**, *42*, 8376–8415.
13. Clément, M.-J.; Imberty, A.; Phalipon, A.; Pérez, S.; Simenel, C.; Mulard, L. A.; Delepierre, M. *J. Biol. Chem.* **2003**, *278*, 47928–47936.
14. Kadirvelraj, R.; González-Outeiriõ, J.; Foley, L.; Beckham, M. L.; Jennings, H. J.; Foote, S.; Ford, M. G.; Woods, R. J. *Proc. Natl. Acad. Sci. U.S.A.* **2006**, *10*, 8149–8154.
15. Legnani, L.; Ronch, S.; Fallarini, S.; Lombardi, G.; Campo, F.; Panza, L.; Lay, L.; Poletti, L.; Toma, L.; Ronchetti, F.; Compostella, F. *Synthesis Org. Biomol. Chem.* **2009**, *7*, 4428–4463.
16. Autieri, E.; Sega, M.; Pederiva, F.; Guella, G. *J. Chem. Phys.* **2010**, *133*, 095104.
17. Zaccheus, M.; Pendrill, R.; Jackson, T. A.; Wang, A.; Auzanneau, F.-I.; Widmalm, G. *Eur. J. Org. Chem.* **2012**, *25*, 4705–4715.

18. Ravenscroft, N.; Omar, A.; Hlozek, J.; Edmonds-Smith, C.; Follador, R.; Serventi, F.; Kuttel, M. M.; Cescutti, P.; Faridmoayer, A. *Carbohydr. Res.* **2017**, *450*, 19–29.
19. González-Outeiriõ, J.; Kadirvelraj, R.; Woods, R. J. *Carbohydr. Res.* **2005**, *340*, 1007–1018.
20. Theillet, F. X.; Simenel, C.; Guerreiro, C.; Phalipon, A.; Mulard, L. A.; Delepierre, M. *Glycobiology* **2011**, *21*, 109–121.
21. Kan, K.; Barbirz, S.; Lipowsky, R.; Santer, M. *J. Phys. Chem. B.* **2014**, *118*, 2523–2534.
22. Kuttel, M. M.; Gordon, M.; Ravenscroft, N. *Carbohydr. Res.* **2014**, *390*, 20–27.
23. Kuttel, M. M.; Jackson, G. E.; Mafata, M.; Ravenscroft, N. *Carbohydr. Res.* **2015**, *406*, 27–33.
24. Galochkina, T.; Zlenko, D.; Nesterenko, A.; Kovalenko, I.; M. Strakhovskaya, M.; Averyanov, A.; Rubin, A. *ChemPhysChem* **2016**, *17*, 2839–2853.
25. Kuttel, M. M.; Timol, Z.; Ravenscroft, N. *Carbohyd. Res.* **2017**, *446-447*, 40–47.
26. Yang, M.; Simon, R.; MacKerell, A. D., Jr. *J. Phys. Chem. B* **2017**, *121*, 3412–3423.
27. Sattelle, B. M.; Almond, A. *Carbohydr. Res.* **2014**, *383*, 34–42.
28. Baliban, S. M.; Yang, M.; Ramachandran, G.; Curtis, B.; Shridhar, S.; Laufer, R. S.; Wang, J. Y.; Van Druff, J.; Higginson, E. E.; Hegerle, N.; Varney, K. M.; Galen, J. E.; Tennant, S. M; Lees, A.; MacKerell, A. D., Jr.; Levine, M. M.; Simon, R. *PLoS Neglected Trop. Dis.* **2017**, *11*, e0005493.
29. Boehr, D. D.; Nussinov, R.; Wright, P. E. *Nat. Chem. Biol.* **2009**, *5*, 789–796.
30. Humphrey, W.; Dalke, A.; Schulten, K. *J. Mol. Graphics* **1996**, *14*, 33–38.
31. *The PyMOL Molecular Graphics System*, Version 2.0, Schrödinger, LLC.
32. Kuttel, M.; Gain, J.; Burger, A.; Eborn, I. *J. Mol. Graphics Modell.* **2006**, *25*, 380–388.
33. Cross, S.; Kuttel, M.; Stone, J.; Gain, J. *J. Mol. Graphics Modell.* **2009**, *28*, 131–139.
34. Pérez, S.; Tubiana, T.; Imberty, A.; Baaden, M. *Glycobiology* **2015**, *25*, 483–49.
35. Pendrill, R.; Jonsson, K. H. M.; Widmalm, G. *Pure Appl. Chem.* **2013**, *85*, 1759–1770.
36. Thieker, D. F.; Hadden, J. A.; Schulten, K.; Woods, R. J. *Glycobiology* **2016**, *26*, 786–787.
37. Grant, O. C.; Woods, R. J. *Curr. Opin. Struct. Biol.* **2014**, *28*, 47–55.
38. Bohne, A.; Lang, E.; von der Lieth, C. W. *Bioinformatics* **1999**, *15*, 767–768.
39. Woods Group (2005–2017) GLYCAM Web. Complex Carbohydrate Research Center, University of Georgia, Athens, GA. http://glycam.org/tools/molecular-dynamics/oligosaccharide-builder/build-glycan?id=1 (accessed April 6, 2018).
40. Arroyuelo, A.; Vila, J. A.; Martin, O. A. *J. Comput.-Aided Mol. Des.* **2016**, *30*, 619–624.
41. Englesen, S. B.; Hansen, P. I.; Pérez, S. *Biopolymers* **2014**, *101*, 733–743.

42. Englesen, S. B.; Cros, S.; Mackie, W.; Pérez, S. *Biopolymers* **1996**, *39*, 417–433.
43. Kuttel, M. M.; Ståhle, J.; Widmalm, G. *J. Comput. Chem.* **2016**, *37*, 2098–2105.
44. Frank, M. In *Glycoinformatics*; Lütteke, T., Frank, M., Ed.; *Methods in Molecular Biology*; Springer: New York, 2015; Vol. 1273, pp 359–377.
45. Yang, M.; Angles d'Ortoli, T.; Sawen, E.; Jana, M.; Widmalm, G.; MacKerell, A. D. *Phys. Chem. Chem. Phys.* **2016**, *18*, 18776–18794.
46. Hansen, H. S.; Hünenberger, P. H. *J. Comput. Chem.* **2011**, *32*, 998–1032.
47. Guvench, O.; Mallajosyula, S. S.; Raman, E. P.; Hatcher, E.; Vanommeslaeghe, K.; Foster, T. J.; Jamison, F. W., II; MacKerell, A. D., Jr. *J. Chem. Theory Comput.* **2011**, *7*, 3162–3180.
48. Jorgensen, W. L.; Chandrasekhar, J.; Madura, J. D.; Impey, R. W.; Klein, M. L. *J. Chem. Phys.* **1983**, *79*, 926–935.
49. Guvench, O.; Hatcher, E.; Venable, R. M.; Pastor, R. W.; MacKerell, A. D. *J. Chem. Theory. Comput.* **2009**, *5*, 2353–2370.
50. Hatcher, E. R.; Guvench, O.; MacKerell, A. D. *J. Phys. Chem. B* **2009**, *113*, 12466–12476.
51. Hatcher, E. R.; Guvench, O.; MacKerell, A. D. *J. Chem. Theory. Comput.* **2009**, *5*, 1315–1327.
52. Lins, R. D.; Hünenberger, P. H. *J. Comput. Chem.* **2005**, *26*, 1400–1412.
53. Kirschner, K. N.; Yongye, A. B.; Tschampel, S. M.; J González-Outeiriõ, J.; Daniels, C. R.; Foley, B. L.; Woods, R. J. *J. Comput. Chem.* **2008**, *29*, 622–655.
54. Damm, W.; Frontera, A.; Rivado-Rives, J. *J. Comput. Chem.* **1997**, *18*, 1955–1970.
55. Kony, D.; Damm, W.; Stoll, S.; Van Gunsteren, W. F. *J. Comput. Chem.* **2002**, *23*, 1416–1429.
56. Xiong, X.; Chen, Z.; Cossins, B. P.; Xu, Z.; Shao, Q.; Ding, K.; Zhu, W.; Shi, J. *Carbohydr. Res.* **2015**, *401*, 73–81.
57. Plazinski, W.; Lonardi, A.; Hünenberger, P. H. *J. Comput. Chem.* **2016**, *37*, 354–365.
58. Patel, D. S.; He, X.; MacKerell, A. D., Jr. *J. Phys. Chem. B* **2015**, *119*, 637–652.
59. Jana, M.; MacKerell, A. D., Jr. *J. Phys. Chem. B* **2015**, *119*, 7846–7859.
60. He, X.; Lopes, P. E.; Mackerell, A. D., Jr *Biopolymers* **2013**, *99*, 724–738.
61. Spiwok, V.; Kralova, B.; Tvaroska, I. *Carbohydr. Res.* **2010**, *345*, 530–537.
62. Sauter, J.; Grafmüller, A. *J. Chem. Theory Comput.* **2015**, *11*, 1765–1774.
63. Gebhardt, J.; Kleist, C.; Jakobtorweihen, S.; Hansen, N. *J. Phys. Chem. B.* **2018**, *122* (5), 1608–1626.
64. Lopez, C. A.; Rzepiela, A. J.; de Vries, A. H.; Dijkhuizen, L.; Huenenberger, P. H.; Marrink, S. J. *J. Chem. Theory Comput.* **2009**, *5*, 3195–3210.
65. Ma, H.; Irudayanathan, F. J.; Jiang, W.; Nangia, S. J. *J. Phys. Chem. B* **2015**, *119*, 14668–14682.
66. Schmalhorst, P. S.; Deluweit, F.; Scherrers, R.; Heisenberg, C. -P.; Sikora, M. *J. Chem. Theory Comput.* **2017**, *13*, 5039–5053.

67. Mahoney, M. W.; Jorgensen, W. L. *J. Chem. Phys* **2000**, *112*, 8910–8922.
68. Cramer, C. J.; *Essentials of Computational Chemistry: Theories and Models*; Wiley: Chichester, England. 2002; pp 93–96.
69. Kuttel, M. *Mini Rev. Org. Chem.* **2011**, *8*, 256–262.
70. Nishima, W.; Miyashita, N.; Yamaguchi, Y.; Sugita, Y.; Re, S. *J. Phys. Chem. B* **2012**, *116*, 8504–8512.
71. Mishra, S. K.; Kara, M.; Zacharias, M.; Koca, J. *Glycobiology* **2014**, *24*, 70–84.
72. Yang, M.; MacKerell, A. D., Jr. *J. Chem. Theory Comput.* **2015**, *11*, 788–799.
73. Huey, R.; Morris, G. M.; Olson, A. J.; Goodsell, D. S. *J. Comput. Chem.* **2009**, *30*, 2785–2791.
74. Sapay, N.; Nurisso, A.; Imberty, A. *Methods Mol. Biol.* **2013**, *924*, 469–483.
75. Nivedha, A. K.; Thieker, D. F.; Makeneni, S.; Hu, H.; Woods, R. J. *J. Chem. Theory Comput.* **2016**, *12*, 892–901.
76. Borrow, R.; Alarcón, P.; Carlos, J.; Caugant, D. A.; Christensen, H.; Debbag, R.; De Wals, P.; Echániz-Aviles, G.; Findlow, J.; Head, C.; Holt, D. *Expert Rev. Vaccines* **2016**, *22*, 1–16.
77. Xie, O.; Pollard, A. J.; Mueller, J. E.; Norheim, G. *Vaccine* **2013**, *31*, 2852–2861.
78. La Force, M. *Progress on MenACWYX Vaccine: Phase 1 Trial Results*. Presented at MRF Conference, London, November 14–15, 2017.
79. Leonardi, M.; Latiolais, T.; Sarpong, K.; Simon, M.; Twiggs, J.; Lei, P.; Rinderknecht, S.; Blatter, M.; Bianco, V.; Baine, Y.; Friedland, L. R. *Vaccine* **2015**, *33*, 933–941.
80. Kuttel, M.; Brady, J. W.; Naidoo, K. J. *J. Comput. Chem.* **2002**, *23*, 1236–1243.
81. Moore, S. L.; Uitz, C.; Ling, C. C.; Bundle, D. R.; Fusco, P. C.; Michon, F. *Clin. Vaccine Immunol.* **2007**, *14*, 1311–1317.
82. Jones, C.; Lemercinier, X. *J. Pharm. Biomed. Anal.* **2002**, *30*, 1233–1247.
83. Fusco, P.; Farley, E. K.; Huang, C. H.; Moore, S.; Michon, F. *Clin. Vaccine Immunol.* **2007**, *14*, 577–584.
84. van Tonder, A. J.; Bray, J. E.; Quirk, S. J.; Haraldsson, G.; Jolley, K. A.; Maiden, M. C.; Hoffmann, S.; Bentley, S. D.; Haraldsson, Á.; Erlendsdóttir, H.; Kristinsson, K. G. *Microb. Genom.* **2016**, *2*, 1–17.
85. Feikin, D. R.; Kagucia, E. W.; Loo, J. D.; Link-Gelles, R.; Puhan, M. A.; Cherian, T.; Levine, O. S.; Whitney, C. G.; O'Brien, K. L.; Moore, M. R. *PLoS Med.* **2013**, *10*, e1001517.
86. Jones, C.; Currie, F.; Forster, M. J. *Carbohydr. Res.* **1991**, *221*, 95–121.
87. Pereira, C. L.; Geissner, A.; Anish, C.; Seeberger, P. H. *Angew. Chem. Int., Ed. Engl.* **2015**, *54*, 10016–10019.
88. Geissner, A.; Pereira, C.;L.; Leddermann, M.; Anish, C.; Seeberger, P. H. *ACS Chem. Biol.* **2016**, *11*, 335–344.
89. Rutherford, T. J.; Jones, C.; Davies, D. B.; Elliott, A. C. *Carbohydr. Res.* **1994**, *265*, 97–111.
90. Rutherford, T. J.; Jones, C.; Davies, D. B; Elliott, A. C. *Carbohydr. Res.* **1994**, *265*, 79–96.

91. Pilishvili, T.; Lexau, C.; Farley, M. M.; Hadler, J.; Harrison, L. H.; Bennett, N. M.; Reingold, A.; Thomas, A.; Schaffner, W.; Craig, A. S.; Smith, P. J. *J. Infect. Dis.* **2010**, *201*, 32–41.
92. van Tonder, A. J.; Bray, J. E.; Roalfe, L.; White, R.; Zancolli, M.; Quirk, S. J.; Haraldsson, G.; Jolley, K. A.; Maiden, M. C.; Bentley, S. D.; Haraldsson, Á.; Erlendsdóttir, H.; Kristinsson, K. G.; Goldblatt, D.; Brueggemann, A. B. *J. Clin. Microbiol.* **2015**, *53*, 2271–2285.
93. Burton, R. L.; Geno, K. A.; Saad, J. S.; Nahm, M. H. *J. Clin. Microbiol.* **2016**, *54*, 967–971.
94. Zon, G.; Szu, S. C.; Egan, W.; Robbins, J. D.; Robbins, J. B. *Infect. Immun.* **1982**, *37*, 89–103.
95. Park, I. H.; Pritchard, D. G.; Cartee, R.; Brandao, A.; Brandileone, M. C. C.; Nahm, M. H. *J. Clin. Microbiol.* **2007**, *45*, 1225–1233.
96. Park, I. H.; Moore, M. R.; Treanor, J. J.; Pelton, S. I.; Pilishvili, T.; Beall, B.; Shelly, M. A.; Mahon, B. E.; Nahm, M. H. *J. Infect. Dis.* **2008**, *198*, 1818–1822.
97. Bratcher, P. E.; Kim, K. H.; Kang, J. H.; Hong, J. Y.; Nahm, M. H. *Microbiology* **2010**, *156*, 555–560.
98. McEllistrem, M. C. I.; Nahm, M. H. *Clin. Infect. Dis.* **2012**, *55*, 1379–1386.
99. Millar, E. V.; Pimenta, F. C.; Roundtree, A.; Jackson, D.; da Gloria Carvalho, M.; Perilla, M. J.; Reid, R.; Santosham, M.; Whitney, C. G.; Beall, B. W.; O'Brien, K. L. *Clin. Infect. Dis.* **2010**, *51*, 258–1265.
100. *MMWR Morb. Mortal. Wkly. Rep.* **2008**, *57*, 144–148.
101. Cooper, D.; Yua, X.; Sidhua, M.; Nahm, M. H.; Fernsten, P.; Jansen, K. U. *Vaccine* **2011**, *29*, 7207–7211.
102. Porat, N.; Benisty, R.; Givon-Lavi, N.; Trefler, R.; Dagan, R. *Vaccine* **2016**, *34*, 2806–2812.
103. Whitney, C. G.; Pilishvili, T.; Farley, M. M.; Schaffner, W.; Craig, A. S.; Lynfield, R.; Nyquist, A. C.; Gershman, K. A.; Vazquez, M.; Bennett, N. M.; Reingold, A.; Thomas, A.; Glode, M. P.; Zell, E. R.; Jorgensen, J. H.; Beall, B.; Schuchat, A. *Lancet* **2006**, *368*, 1495–1502.
104. Hausdorff, W. P.; Hoet, B.; Schuerman, L. *BMC Pediatr.* **2010**, *10*, 4.
105. Isturiz, R.; Sings, H. L.; Hilton, B.; Arguedas, A.; Reinert, R. R.; Jodar, L. *Expert Rev. Vaccines.* **2017**, *16*, 1007–1027.
106. Ciuffreda, P.; Colombo, D.; Ronchetti, F.; Toma, L. *Carbohydr. Res.* **1992**, *232*, 327–339.
107. MacLennan, C. A.; Martin, L. B; Micoli, F. *Hum. Vaccin. Immunother.* **2014**, *10*, 1478–1493.
108. Liu, B.; Knirel, Y. A.; Feng, L.; Perepelov, A. V.; Senchenkova, S. Y. N.; Reeves, P. R.; Wang, L. *FEMS Microbiol. Rev.* **2014**, *38*, 56–89.
109. Ravenscroft, N.; Cescutti, P.; Gavini, M.; Stefanetti, G.; MacLennan, C. A.; Martin, L. B.; Micoli, F. *Carbohydr. Res.* **2015**, *404*, 108–116.
110. Slauch, J. M.; Mahan, M. J.; Michetti, P.; Neutra, M. R.; Mekalanos, J. J. *Infect. Immun.* **1995**, *63*, 437–441.
111. Kotloff, K. L.; Nataro, J. P.; Blackwelder, W. C.; Nasrin, D.; Farag, T. H.; Panchalingam, S.; Wu, Y.; Sow, S. O.; Sur, D.; Breiman, R. F.; Faruque, A. S. *Lancet* **2013**, *382*, 209–222.

112. Platts-Mills, J. A.; Babji, S.; Bodhidatta, L.; Gratz, J.; Haque, R.; Havt, A.; McCormick, B. J.; McGrath, M.; Olortegui, M. P.; Samie, A.; Shakoor, S. *Lancet Global Health* **2015**, *3*, e564–575.
113. Liu, L.; Johnson, H. L.; Cousens, S.; Perin, J.; Scott, S.; Lawn, J. E.; Rudan, I.; Campbell, H.; Cibulskis, R.; Li, M.; Mathers, C. *Lancet* **2012**, *379*, 2151–2161.
114. Dupont, H. L.; Levine, M. M.; Hornick, R. B.; Formal, S. B. *J. Infect. Dis.* **1989**, *159*, 1126–1128.
115. Livio, S.; Strockbine, N. A.; Panchalingam, S.; Tennant, S. M.; Barry, E. M.; Marohn, M. E.; Antonio, M.; Hossain, A.; Mandomando, I.; Ochieng, J. B.; Oundo, J. O. *Clin. Infect. Dis.* **2014**, *59*, 933–941.
116. Walker, R. I. *Vaccine* **2015**, *33*, 954–965.
117. Perepelov, A. V.; Shekht, M. E.; Liu, B.; Shevelev, S. D.; Ledov, V. A.; Senchenkova, S. Y. N.; L'vov, V. L.; Shashkov, A. S.; Feng, L.; Aparin, P. G.; Wang, L. *FEMS Immunol. Med. Microbiol.* **2012**, *66*, 201–210.
118. Knirel, Y. A.; Sun, Q.; Senchenkova, S. N.; Perepelov, A. V.; Shashkov, A. S.; Xu, J. *Biochemistry (Moscow)* **2015**, *80*, 901–914.
119. West, N. P.; Sansonetti, P.; Mounier, J.; Exley, R. M.; Parsot, C.; Guadagnini, S.; Prévost, M. C.; Prochnicka-Chalufour, A.; Delepierre, M.; Tanguy, M.; Tang, C. M. *Science* **2005**, *307*, 1313–1317.
120. Gauthier, C.; Chassagne, P.; Theillet, F. X.; Guerreiro, C.; Thouron, F.; Nato, F.; Delepierre, M.; Sansonetti, P. J.; Phalipon, A.; Mulard, L. A. *Org. Biomol. Chem.* **2014**, *12*, 4218–4232.
121. Van Der Put, R. M.; Kim, T. H.; Guerreiro, C.; Thouron, F.; Hoogerhout, P.; Sansonetti, P. J.; Westdijk, J.; Stork, M.; Phalipon, A.; Mulard, L. A. *Bioconjug. Chem.* **2016**, *27*, 883–892.
122. Riddle, M. S.; Kaminski, R. W.; Di Paolo, C.; Porter, C. K.; Gutierrez, R. L.; Clarkson, K. A.; Weerts, H. E.; Duplessis, C.; Castellano, A.; Alaimo, C.; Paolino, K. *Clin. Vaccine Immunol.* **2016**, *23*, 908–917.

Chapter 8

Advances in Synthetic Approaches towards Glycoantigens

A. R. Vartak and S. J. Sucheck*

Department of Chemistry and Biochemistry, University of Toledo, 2801 W. Bancroft, Toledo, Ohio 43606, United States
***E-mail: steve.sucheck@utoledo.edu.**

The interest and progress in the field of synthetically derived saccharide antigens has gained considerable momentum in the past decade, in particular. Efficient oligosaccharide synthesis with desired stereo- and regio-selectivity is necessary to access high value targets for use in drug discovery and biology. However, creating a synthesis platform that accommodates the large variation in oligosaccharide conformation and connectivity is a daunting task. Nonetheless, over the last few decades researchers have made great progress towards achieving this goal. In this book chapter, we have surveyed recent advances in synthetic methods for oligosaccharide synthesis including the glycal assembly method, one-pot glycosylation, and solid-supported synthesis and their applications toward the synthesis of glycoantigens.

Introduction

Oligosaccharides are critical for a number of biological processes such as cellular recognition and signalling (*1*, *2*). In addition, these biopolymers have been increasingly exploited as antibiotics, anti-cancer agents and as components of immunotherapeutics. For many years, the field of glyco-medicine was hampered by the difficulties in isolation and characterization of pure, homogenous oligosaccharides. Breakthroughs in synthetic and enzymatic approaches for synthesizing oligosaccharides have lead to a greater access of homogeneous complex carbohydrates (*3–6*). Synthetically derived saccharides have garnered significant interest due to their potential use as antigens to generate prophylactic (*Neisseria meningitidis* serogroup A, *Shigella flexneri*, etc) as well as therapeutic (tumor associate carbohydrate antigens, TACA) glycoconjugate vaccines. Recent developments in selected methodologies for oligosaccharide synthesis with related application examples are reviewed in this chapter.

Glycal Assembly Method

The use of glycals for oligosaccharide synthesis was pioneered by Danishefsky and co-workers in early 90's (*7–9*). The strategy involved use of glycals as both glycosyl donors and acceptors under wide range of reaction conditions. The use of the glycal moiety at the reducing end can serve as a protecting group and a potential conjugation handle. A generalized scheme for glycal assembly is shown in Figure 1. The glycal is first converted to an *in situ* donor or isolable donor using electrophilic activation. The donor then reacts with second glycal containing free hydroxyl group. The resulting disaccharide can be activated in an iterative fashion to synthesize oligosaccharides. The presence of a glycal moiety in the acceptor can be problematic as it can potentially get activated to form self-coupled products. To distinguish between glycal donors and acceptors selective protecting groups have been utilized (*10*). The presence of electron withdrawing protecting groups such as acyl (-Ac) and benzoyl (-Bz) were observed to lower the nucleophlicity of the glycal towards the activating electrophile. As a result, a desired glycal can be activated selectively.

Many biologically relevant oligosaccharide based glycoantigens consist of aminosugar derivatives (Figure 2). Access to the aminosugar intermediates from glycals has improved the scope of the glycal assembly method. One of the common methods to introduce nitrogen at the C-2 position is azidonitration of glycal. Lemieux *et al.* reported the azidonitration reaction using cerric ammonium nitrate (CAN) and sodium azide in acetonitrile to produce 2-azido-2-deoxysugars from glycals (*11*). The nitrate group at the reducing end can be converted into the desired donor functionality such as halide, acetate or thioglycoside (*11–13*).

Danisheflsky and co-workers developed a sulfonamidation protocol to afford *trans*-diaxial iodosulfonamides using iodonium di-*sym*-collidine perchlorate and benzenesulfonamide (*14*, *15*). The stereoselective migration of sulfonamide (C-1 to C-2) via an aziridine intermediate with subsequent glycosylation afforded β-amidoglycoside with several *O*-, *S*- and *N*-nucleophiles (Figure 3). The methodology is an improvised version of a phosphoramidation strategy developed by Lafont *et al.* (*16*, *17*) Several approaches to access 2- or 3-aminosugars from a glycal have been reported since and have been recently reviewed by Goti *et al.* (*18*)

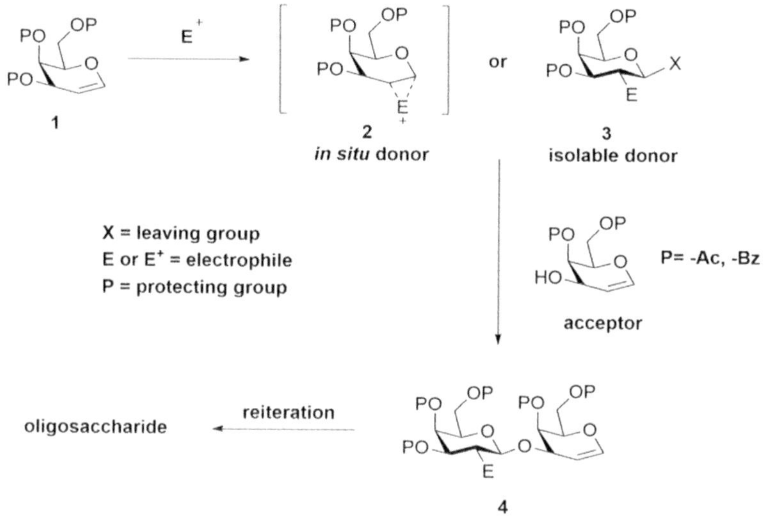

Figure 1. The glycal assembly method.

Case Studies

Danishefsky and co-workers reported the synthesis of several tumor markers and blood group determinants with the use of sulfonamidation and glycal epoxides (*4*). The group capitalized on the oxirane chemistry of glycals developed by Halcomb *et al.* and Murray *et al.* to synthesize α or β-glycosides with or without the C-2 hydroxyl group (*19*, *20*).

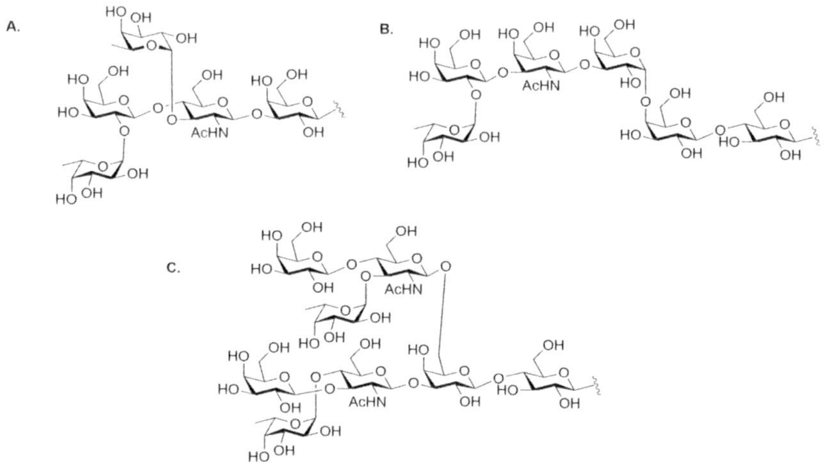

*Figure 2. Structure of **A**. Lewis^{y} (Le^{y}) antigen. **B**. Human milk derived N_3 antigen **C**. Globo-H antigen.*

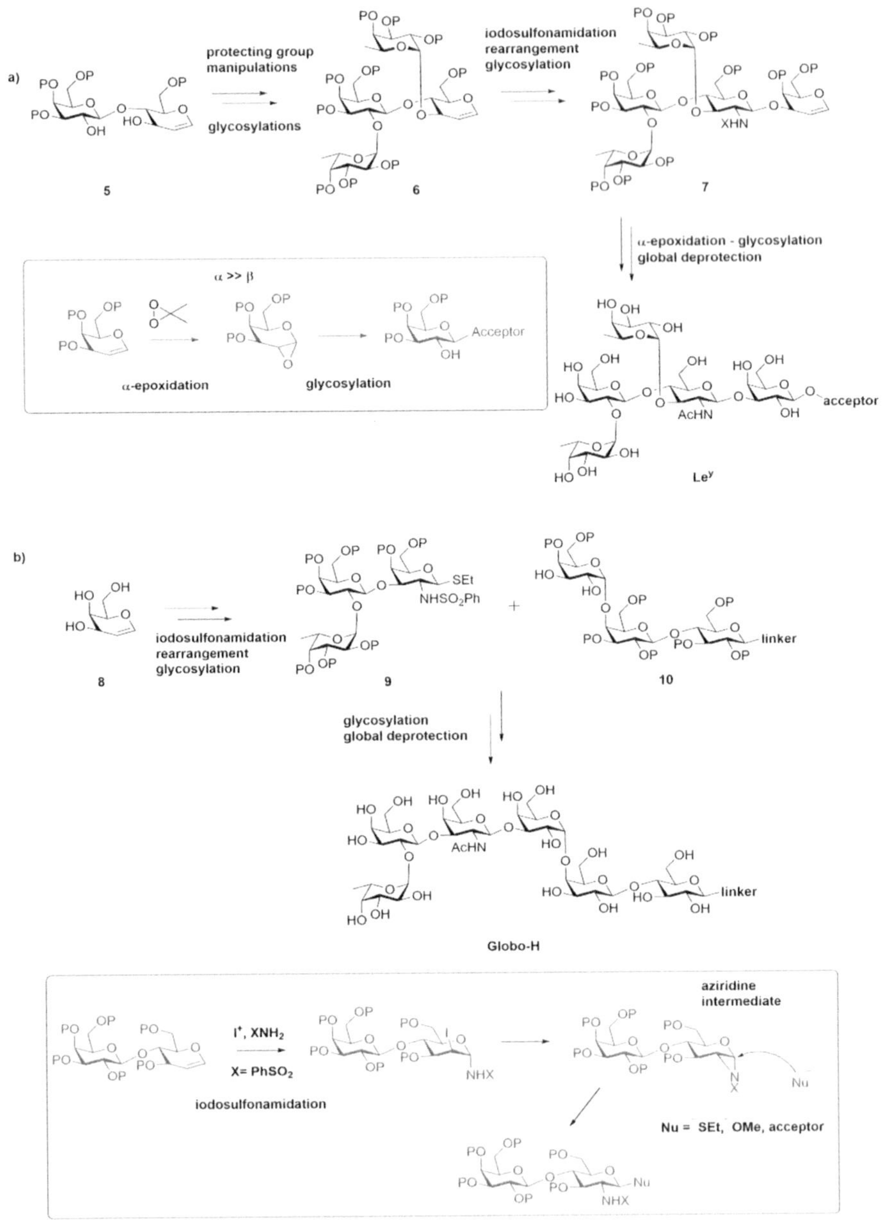

Figure 3. a) Linear synthesis of Lewis^y antigen; b) Convergent synthesis of Globo-H antigen using glycal-assembly method.

One example is synthesis of Lewis determinant, Lewis^y (*21–23*). The synthetic route begins with a lactal **5** (Figure 3). The lactal **5** is glycosylated using a donor to afford tetrasaccharide **6** after selective protecting group transformation. The glycal in the resulting tetrasaccharide **6** is subjected to iodosulfonamidation followed by glycosylation with glycal acceptor to synthesize the Lewis^y antigen.

The terminal glycal of the antigen is treated with 2,2-dimethyldioxirane (DMDO) to obtain the 1,2-epoxide ($\alpha >> \beta$) which is opened using a nucleophilic aglycon to afford protein conjugated version of Lewis^y. A similar strategy was applied to synthesize the Globo-H antigen (*24–27*).

One-Pot Glycosylation Strategies

The first sequential one-pot glycosylation (OPG) was reported by Kanhe and Raghavan in 1993. OPG has several advantages over the conventional method of oligosaccharide assembly (*28*). These advantages include bypassing cumbersome separation and purification procedures. OPG strategies are based on three major concepts: (1) Chemoselectivity (armed-disarmed thioglycosides), (2) Orthogonal donors and activating conditions, (3) Pre-activation of donors.

Chemoselective OPG Strategies

Chemoselective OPG is based on armed/disarmed glycoside donors, a concept introduced by Fraiser-Reid *et al.* (*29*, *30*) The effect of protecting groups on the activity of the glycosyl donor is key to the chemoselective glycosylations. The presence of electron donating groups activate the donor while an electron withdrawing group decativates. The activated donor can react preferentially with the promoter species followed by glycosylation of the lower activity donor-acceptor (Figure 4). Protecting groups at the C-2 position of the saccharide have the most significant effect on reactivity (*31*).

Thioglycosides are one of the major donor building blocks used in chemoselective OPG. The longer shelf-life, ease of preparation, stability towards most oligosaccharide transformations including protection and de-protection chemistry and number of thioglycoside promoter systems make them easy to use in this multi-variable system. To obtain the desired product selectively with minimum side-products, the relative reactivities of donors with different protecting group should be known. Lay and co-workers provided the data-tables for reactivities with different ethyl thioglycosides with help of NMR monitoring (*32*). Later, Wong and group reported relative reactivity values (RRV's) for more than 600 *S*-tolyl donors (*31*, *33–35*). The study involved comparison of di- and trisaccharide donors as well with thiotoluoyl glycosides as HPLC monitoring handle. The least reactive donor has a normalized RRV of 1.0. Based on these relative values, some interesting trends in reactivities are observed:

- Reactivity order of different thiopyranosides with the same protecting groups; fucose > galactose > glucose > mannos > silaic acid.
- De-activating effect of C-2 protecting groups on thiogalactopyranoside; OClAc > OBz > OAc > NHTroc > OBn > OH > OSilyl > H.
- Reactivity of aminosugars based on protecting groups used; NHCbz > NHTroc > NHPhth > N_3 > NHAc.
- The degree of deactiavation influenced by position of -Bz group on carbon number C4 > C3 > C2 > C6 for a thiogalactoside.

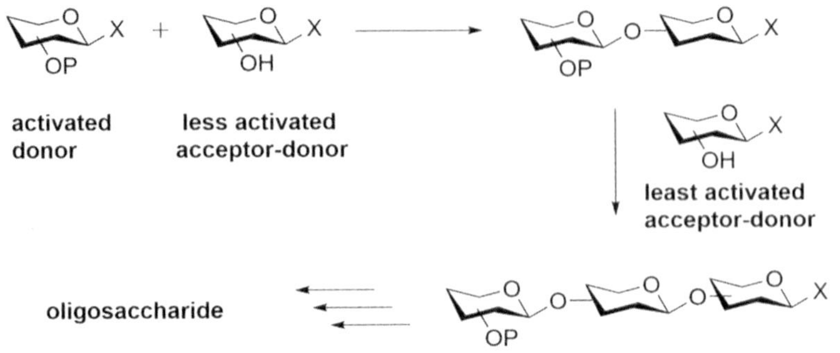

Figure 4. The chemoselective OPG method.

Researchers are studying effects of sterically hindered protecting groups, (*36*, *37*), confirmations of pyranosides (*38*), solvent, and temperature to refine the currently available tool-box.

Orthogonal Strategies

Orthogonal strategies for OPG use different donor leaving groups that can be selectively activated for the reaction. The major advantage of the orthogonal strategies over the chemoselective methods is that the glycosylation sequence is independent of the reactivity of the donors. Using the orthogonal methodology, a disarmed donor can be activated while armed donor acts as glycosyl acceptor. A variety of orthogonal donor combinations have been used to access complex oligosaccharides in a one-pot synthesis (Figure 5). The use of thioglycosides and glycosyl fluorides is one combination described by Ogawa *et al.* (*39*) The thioglycosides can be activated using NIS-AgOTf keeping the fluoride donor intact, which can be activated later using $Cp_2Hf_2Cl_2$-AgOTf. The orthogonal activation using anomeric carbonates and 6-nitro-2-benzothiazoates was developed by Mukaiyama *et al.* The group used carbonate and ethylthioglycoside orthogonally to synthesize a mucin-related F1α antigen in a one-pot assembly (*40*). Recently, Misra *et al.* synthesized a portion of the O-antigen of *Escherichia coli* O59 using orthogonal activation of trichloroacetimidates and ethyl thioglycosides (*41*). Bernardes *et al.* demonstrated use of the *O*-mesitylenesulfonylhydrxylamine (MSH) as a reagent to selectively activate *S*-ethyl in presence of *S*-phenyl donors (*42*). Demchenko *et al.* added another variation using *S*-benzimidazolyl (*S*Biz) donors which can be activated using MeI (*43*). A number of other combinations such as propargyl and *n*-pentenyl donors, *S*-benzoxzolyl (*S*Box) and *S*-thiazolinyl (*S*Taz) donors have been successfully used (*44–47*). Demchenko *et al.* has published a recent review on oligosaccharide synthesis using selective orthogonal methodologies (*48*). There is great degree of overlap between the chemoselective and orthogonal methodologies. All orthogonal strategies are chemoselective; however, the inverse is not always true.

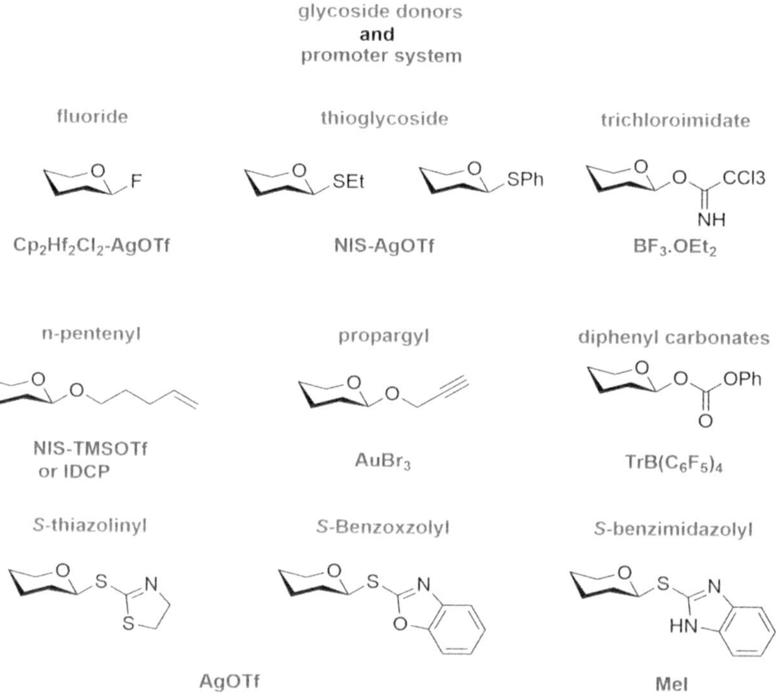

Figure 5. Common glycoside donors and promoter systems used in the orthogonal OPG.

Preactivation Strategies

The pre-activation methodology was developed to circumvent the specific requirements needed for chemoselective and orthogonal methods. Time consuming synthesis of armed/disarmed donors (chemoselective) or the use of different promoters for glycosylations (orthogonal) is avoided with the pre-activation strategy. The pre-activation methodology involves activation of a donor glycoside resulting in a reactive glycosylating moiety which then reacts with a sequentially added acceptor molecule. The acceptor is chosen in such a way that the product saccharide can be activated using addition of second equivalent of promoter system in a repetitive fashion.

Huang and co-workers used the term 'iterative one-pot glycosylation' based on pre-activation of thioglycoside donors using a *p*-toluenesulfenyl chloride (*p*-TolSCl) - silver triflate (AgOTf) promoter system (*49–51*). The Van der Marel group utilized benzenesulfinyl piperidine (BSP) - triflic anhydride (Tf_2O) as well as diphenyl sulfoxide (Ph_2SO) - triflic anhydride (Tf_2O) as promoter systems for iterative one-pot oligosaccharide assembly (*52, 53*). The latter one was found to be effective when disarmed donors were involved (*54*). Several other promoter systems have been successfully evaluated for iterative one-pot glycosylation strategies (Figure 6) (*50, 55, 56*).

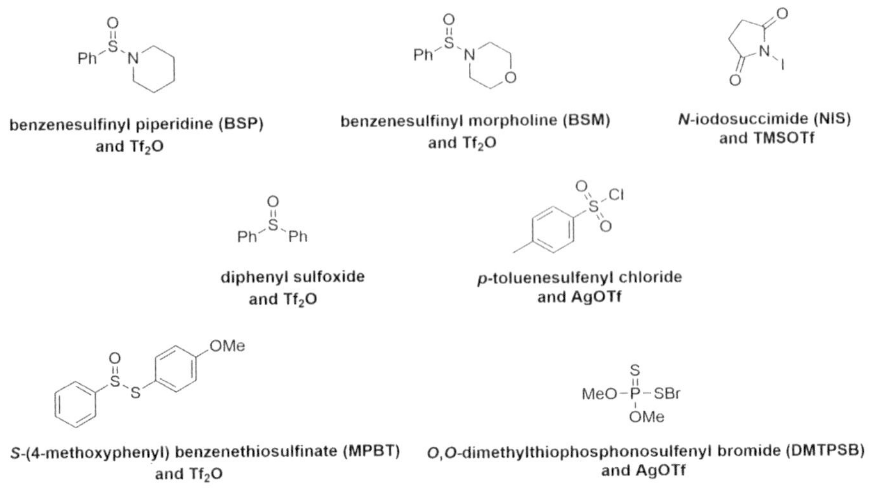

Figure 6. Common promoter systems used in pre-activation strategies.

However, one limitation of these methods is aglycon transfer which can lead to undesired side-products (*57–59*). Furthermore, the promoter system used for the glycosylation has to be used in stoichiometric proportion with the donor. Traces of excess activator create undesired products. Despite these hurdles, pre-activation strategies are efficient and high-yielding methodologies for oligosaccharide synthesis.

Case Studies

In 2017, Ghosh and co-workers synthesized pentasaccharide **11** related to the O-antigen of *Escherichia coli* O120 using sequential OPG methods (Figure 7) (*60*). The first strategy involved a three component [1 + 2 + 2] sequence using orthogonal strategies involving trichloroacetimidate **14** and thioglycoside **13** donors. Trichloroacetimidate donor **14** is activated using $FeCl_3$ at -60 °C followed by addition of disaccharide acceptor-donor **13**. The resulting trisaccharide thioglycoside donor is then activated using NIS - $FeCl_3$ and reacted with acceptor **12** at 0 °C to afford pentasaccharide **11**.

The alternative four component [1 + 2 + 1 + 1] strategy is more economic and uses a pre-activation OPG methodology. Trisaccharide thioglycoside donor is synthesized from **13** and **14** using a [1+ 2] glycosylation. The resulting thioglycoside donor is pre-activated using Ph_2SO, 2,4,6-tri-*tert*-butylpyrimidine (TTBP) and triflic anhydride (Tf_2O) followed by addition of acceptor **16** at -40 °C. The final glycosylation between tetrasaccharide donor and monosaccharide acceptor **15** is carried out in presence of NIS - $FeCl_3$ at 0 °C. Regio-selective oxidation of the primary alcohol followed by debenzylation of pentasaccharide afforded final target O-antigen as its *p*-methoxyphenyl (PMP) glycoside.

Zhongwu *et al.* developed a one-pot pre-activation glycosylation [2 + 1 + 4] strategy to synthesize the heptasaccharide repeating unit of type V group B *Streptococcus* capsular polysaccharide (CPS) (Figure 8) (*61*). Tetrasaccharide unit **20** is synthesized using traditional sequential glycosylation with orthogonal protecting groups. Dibutylphosphate sialoside donor is activated using TMSOTf and reacted with galactoside acceptor to obtain disaccharide donor **18**. With all three components in hand, the authors used *p*-toluenesulfenyl triflate for pre-activation and obtained protected heptasaccharide **17** in one-pot with high yield. The authors outlined a five step protocol involving deprotection of carboxylate, carbamate, amido, benzylidene, benzyl, azido, and acyl groups to afford the final target.

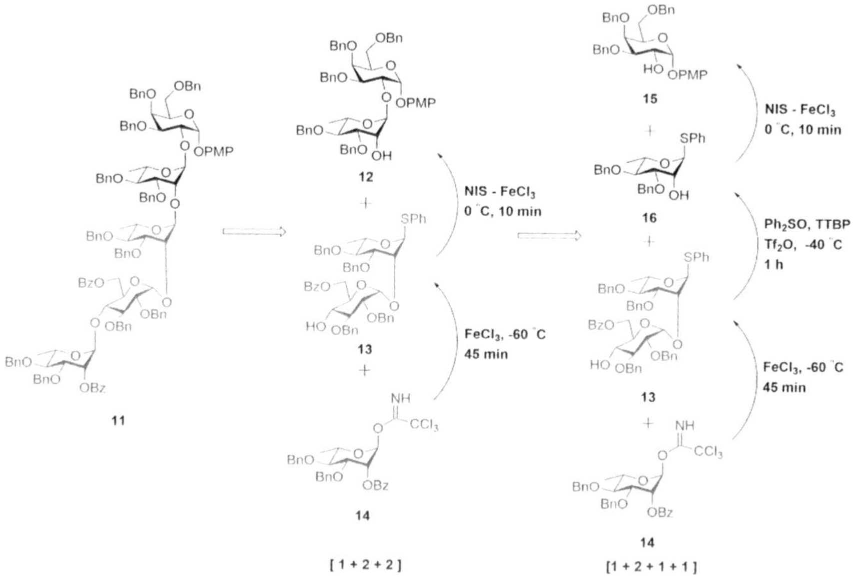

Figure 7. Retro-synthesis of a pentasaccharide related to the O-antigen of E. coli 120 using orthogonal OPG strategy.

Over last two decades, many researchers have implemented OPG strategies to afford high value, complex oligosaccharides owing to their ease of synthesis, efficiency and lack of multiple purifications (*3*, *10*, *62–66*). Takahashi and co-workers reported the synthesis of the Forssman pentasaccharide antigen using OPG orthogonal and pre-activation strategies (*67*). On the other hand, Huang *et al.* synthesized tumor - associated carbohydrate antigen Globo-H using a OPG pre-activation strategy based on thioglycoside donors (*64*). Wong and co-workers have reported an OPG-based synthesis of fucose GM_1 oligosaccharide, a sialylated epitope of small-cell lung cancer (*68*). The group used RRVs of thioglycosides and observed improved results with the BSP-Tf_2O promoter system compared to NIS and dimethyl(methylthio)sulfonium triflate (DMTST) activators. The use of the pre-activation strategy to synthesize a mycobacterial arabinogalactan containing 92 monomer units is the recent highlight of the methodology (*69*).

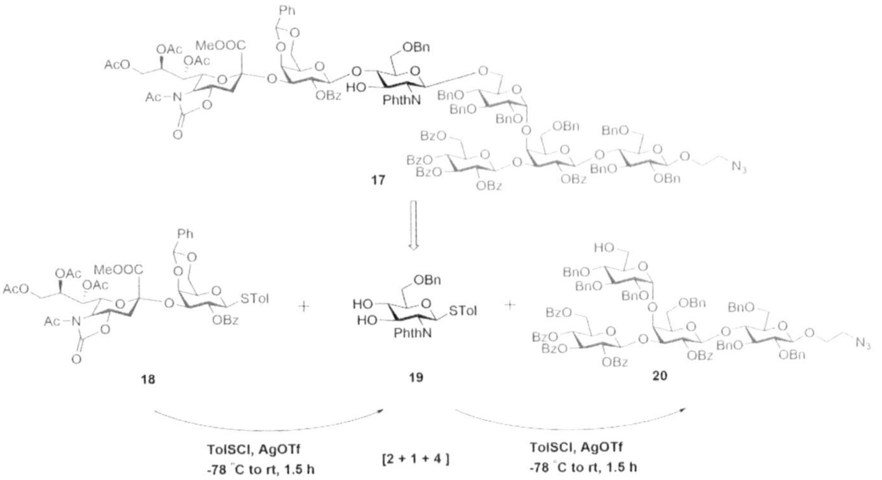

Figure 8. Synthesis of a heptasaccharide repeating unit of type V group B Streptococcus capsular polysaccharide by preactivation OPG.

Solid Supported Oligosaccharide Synthesis and Automation

Access to the other two classes of biopolymers, oligonucleotides and peptides, through automated synthesis were major breakthroughs in genomics and proteomics, respectively. On the other hand, automated oligosaccharide synthesis is in the early phase of development and needs further improvements in terms of available building blocks, efficient glycosylation strategies and better stereo-control in some specific cases. Formation of a new stereogenic centre after each glycosylation, number of possible branching points, and variety of available monosaccharide building blocks makes oligosaccharide synthesis on solid phase challenging compared to their biopolymer counterparts.

Any solid phase oligosaccharide assembly contains three major components: (1) Solid support, (2) Linker chemistry, (3) Building block's protecting group selection.

Solid Support

Merrifield's resin (polystyrene) was the obvious first choice for solid phase oligosaccharide synthesis as it is widely used in peptide synthesis (*70, 71*). The insoluble matrix is compatible with most oligosaccharide transformations and is commercially available. The loading capacity of resin depends on the swelling factor in different solvents (*72*). The limited number of solvents with a high swelling factor for Merrifield's resin restricts its universal use and led to further modification of the support. Polyethylene glycol (PEG) grafted onto a polystyrene backbone, also known as TentaGel, has improved physiological properties. The copolymer has consistent swelling irrespective of solvent and the kinetic behaviour of end-groups is more solution like giving more consistent results in some cases (*73, 74*). Other resins such as HypoGel, JandaJel, and many variations have shown satisfactory results in small molecule and oligosaccharide synthesis but their incorporation into the automated synthesis still needs more optimization (*75*, *76*).

Substantial need for reaction optimization, use of excessive reagents and slow reaction kinetics are some of the limitations of insoluble solid phase resins. Among several other platforms developed, (*77–80*), soluble polymer-bound synthesis has gained significant attention. The solubility of these polymers in organic solvents helps to reproduce the solution - based chemistry and kinetics more reliably. Also, the macromolecular hydrophobic nature of many soluble polymers facilitates purification using anti-solvents such as methanol or ether. Poly (ethylene glycol) methyl ether (mPEG) is the most widely used soluble polymer for oligosaccharide synthesis (*58*, *81–84*). In recent years, other polymers such as polyacrylamide, polyvinyl alcohol (PVA), hyper-branched PEG have been developed for oligosaccharide synthesis (*75*, *85*, *86*). To overcome inconsistent precipitation from anti-solvents (solvents with less polymer solubility), integrating recent advances in ultrafiltration membrane technology could be very useful improving recovery and efficiency.

Linker Chemistry

The selection of the linker is extremely crucial irrespective of the type of solid support used. The chemical nature of the linker dictates the strategic assembly and tolerable reaction conditions throughout the process. A linker connecting the support to a glycosyl acceptor must be stable throughout the repeated deprotection and glycosylation transformations and be orthogonally cleavable to other protecting groups. The purification of cleaved protected glycan followed by global deprotection affords desired oligosaccharide.

The desired monosaccharide loading on the solid support through linker can be achieved using two methods. The support can be loaded with linker first, followed by attachment of monosaccharide or the loading of monosaccharide-linker conjugate on to the solid support directly. The position of attachment of the linker on the sugar can be varied as well, but attachment at the reducing end is very common with few exceptions.

One of the example of non-reducing end attached linking chemistry is use of silyl based linkers developed by Danishefsky *et al* (**entry 1**, Table 1). In 1999, Danishefsky and co-workers used diisopropylsilyl linker, an improvised version of their previous diphenylsilyl linker, to synthesize N-linked disaccharide glycopeptides on benzyl hydroxy resin (*87*, *88*). The galactal was attached to the resin through the C-6 position, previously protected as a chlorodiisopropylsilyl ether. The glycal assembly method described earlier was used to attach another glucal unit. The linker can be cleaved using a fluoride source such as TBAF or HF in pyridine. Recently, Nieto *et al.* used this siloxane linker at reducing end to synthesize trisaccharide repeating unit of the capsular polysaccharide of *Neisseria meningitis* using commercially available soluble polymer PEG, as a polymer support.[89]

Boons and co-workers utilized polystyrene boronic acid as a polymer support and attached it to 4,6-*O* positions on the sugar using pyridine (*90*). The ease of removal using acetone-water at 60 °C was capitalized by the group (**entry 2**, Table 1). In case of anomeric mixtures, the product can be cleaved off, purified and re-loaded on to the same support. Jensen *et al.* developed the tris(alkoxy)benzylamine linker (BAL) connecting amino sugars through the C-2 amine using reductive amination (*91*). The stability of the linker under acidic conditions allows the use of excess lewis acid during glycosylations (**entry 3**, Table 1). Fukase *et al.* synthesized an alkyne type linker which can be attached to the resin using a Pd(0) catalyzed Sonogashira coupling (*92*). The alkyne linker (**entry 4**, Table 1) is stable against acids but is readily cleavable using TFA after complex formation with $Co_2(CO)_8$. Bennett and co-workers have developed a thioether based linker (**entry 5**, Table 1) attached to the reducing end (*93*). The activation of thio-linker using BSP and Tf_2O, followed by addition of small molecule acceptor acceptor leading to direct transfer of oligosaccharide to the desired aglycon.

One of the most widely used linkers in solid phase oligosaccharide synthesis is the 4-octenediol linker (**entry 6**, Table 1) developed by Seeberger *et al.* (*94*) The linker is stable to all standard protecting group transformations in solid phase oligosaccharide synthesis. The only limitation is its susceptibility to addition reactions under electrophilic conditions required for thioglycoside activation. The orthogonal cleavage of the linker is achieved by alkene metathesis using Grubb's catalyst in presence of ethylene. The deactivation of catalyst by the resin was addressed by the next generation self-cleavable linkers (**entry 9** and **10**, Table 1) developed by Schimdt *et al.* (*95*) and Van der marel *et al.* (*96*)

Ester-type linkers are another common type found in the literature. Base labile bi-functional linker (**entry 7**, Table 1) developed by Seeberger *et al.* (*97*) is one important example. The alkyl amine functionality at the reducing end of the final oligosaccharide product is utilized for further conjugation with proteins and carriers for vaccine purposes (*95*, *98*). A photo-cleavable linker (**entry 8**, Table 1) introduced by Seeberger *et al.* has also been an important addition to the linking chemistry toolbox (*99*, *100*). The linker is used widely for the synthesis of aminoglycans and sialic acid-containing oligosaccharides as the linker provides better orthogonality. However, correct selection of light and exposure time requires optimization which can affect cleavage efficiency significantly. The 4-octenediol and photo-cleavable linkers have been favoured for automation compared to other linkers.

Table 1. Selective linkers and their clevage conditions

Entry	Linker	Connectivity	Deprotection condition	Ref.
1	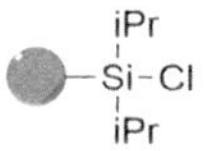	Reducing end, C-6 position	TBAF (fluoride ion source)	45, 46
2		C-4, 6 positions	Acetone/ H_2O 60 °C	47
3	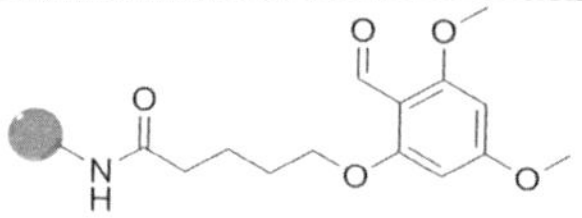	C-2 amine	TFA: H_2O (19:1)	48
4	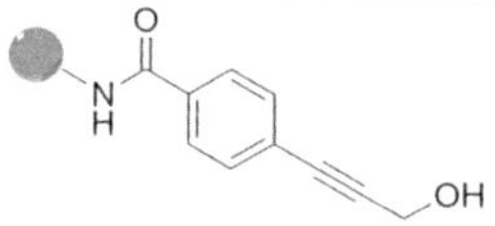		$Co_2(CO)_8$, TFA	49
5	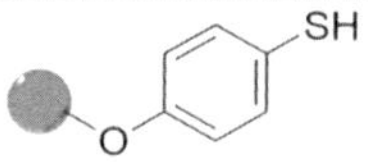		BSP, Tf_2O, Aglycon acceptor	50
6	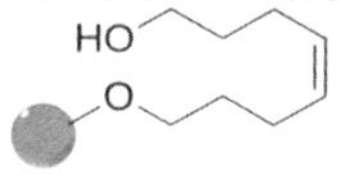		Olefin metathesis (Grubbs cat.)	
7	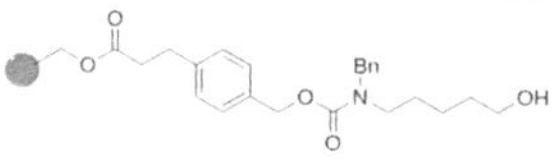		NaOMe/MeOH, H_2, Pd/C	51, 54, 56
8	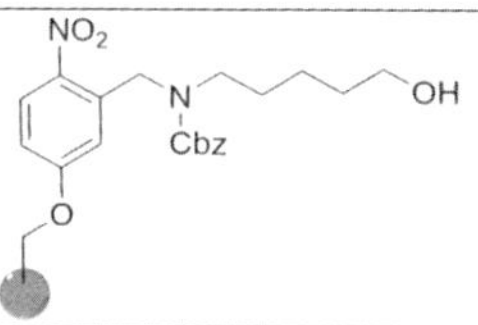	Reducing end	Hv	
9	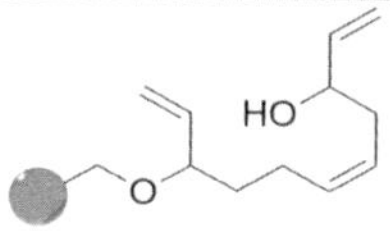		Olefin metathesis (Grubbs cat.)	53
10	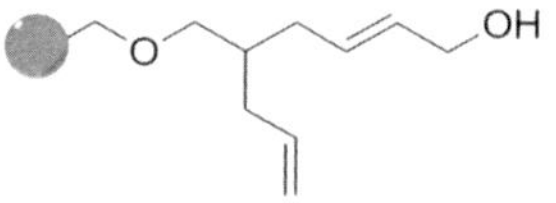		Olefin metathesis (Grubbs cat.)	52

Building Block Protecting Group Selection

The choice of donor building block depends on the solid support and linker chemistry chosen for oligosaccharide synthesis. Thioglycosides, glycosyl trichlroacetimidates, phosphates and thioimidates are some of the common glycosyl donors used in solid phase synthesis. Selection of protecting groups plays a crucial role in designing the experimental sequence as permanent and temporary groups should be orthogonal to each other. In addition, the presence of participating versus non-participating groups affects the stereochemical outcome of the reaction. Benzyl ethers and benzoyl esters are commonly used permanent protecting groups, with the latter being participating. For temporary protection, 9-fluorenylmethoxycarbonate (Fmoc), levulinate (Lev) and 2-methyl naphthyl (NAP) are used frequently in solid phase synthesis. These temporary protecting groups have been used for solid phase reaction monitoring as well. The deprotected by-product of Fmoc group, dibenzofulvene is UV active and can be used colorimetrically to quantify the extent of reaction on solid phase (*101*). The lack of stability of the group under basic conditions is the limitation of the method. Pohl *et al.* described the use of nitrophthalimidobutyric ester (NPB) for reaction monitoring (*102*). The protecting group has better stability and can be easily removed using hydrazine acetate producing orange-coloured nitrophthalhydrazide. The use of disperse red dye and cyanuric chloride for the detection of free hydroxyl and amine groups was described by Ito *et al.* (*103*) Similarly, an on-resin color test developed by the Ito group involved use of a chloroacetyl group as a reaction handle with nitro-benzylpyridine to observe a red-coloured resin which disappears on deprotection. Recently, a non-destructive colorimetric modification of the disperse red reagent was developed by Shin *et al.* (*104*) The dispersive red dye conjugate is removable using TBAF and the resin can be carried forward for the next reaction. The method is limited to detection of primary and secondary amines as well as thiols. Apart from these spectrophotometric methods, NMR techniques such as solid state, ^{13}C and ^{19}F NMR are also useful for analyzing solid-supported transformations (*105*, *106*).

Case Studies

Seeberger *et al.* is one of the major contributors in developing automated glycan assembly (AGA) on a solid support. Over the years, the group has optimized a variety of linkers, building blocks and solid support resins for automation. Recently, the group synthesized complex oligosaccharides related to blood group determinants (Figure 9) (*100*). Merrifield resin modified with a photolabile linker **21** was used to assemble the desired combination of monosaccharide building blocks **22-28**. The synthesis protocol established the stability of photo-labile linker under common acidic promoters (e.g., TMSOTf and TfOH) used in glycosylations as well as identified building blocks suitable for AGA.

The importance of suitable building blocks in AGA is seen clearly in the synthesis of tumor associated antigens Gb-3 and Globo-H developed by the group (*107*). High stereo-selectivity is required in solid phase synthesis since purification after each step is not possible. Formation of *cis*-glycosidic linkages with non-participating groups at the C-2 is proven difficult and a bottle neck for high yields. The group observed different selectivity with different glycosyl donors as well for their corresponding α, β anomers (Figure 10). The β anomer of glycosyl phosphate donor **30** gave significantly better selectivity (α:β 14:1) compared to α-anomer of the donor **31** (α:β 4:1). The 4-octenediol linker **29** was cleaved off in the end using a Grubb's catalyst to obatin protected Gb3 antigen with 46% overall yield.

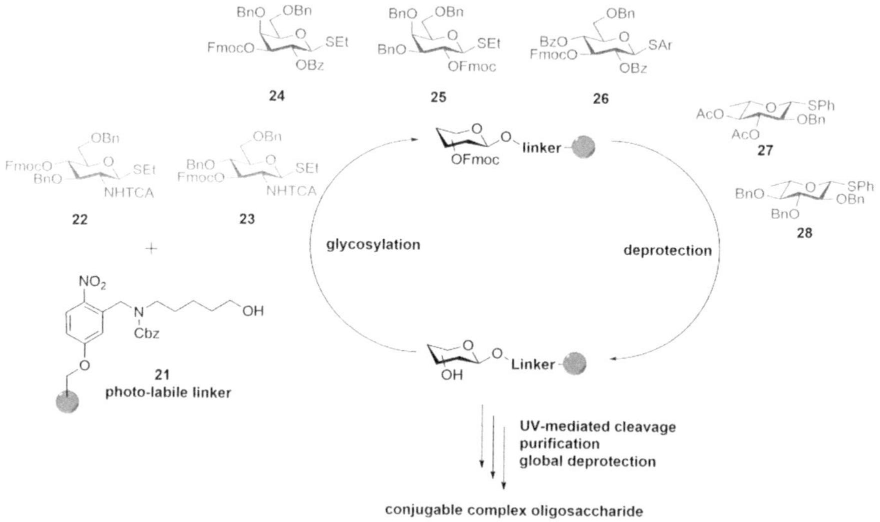

Figure 9. Automated glycan assembly related to blood group determinants Lewis[a], Lewis[b], Lewis[x], Lewis[y], H-type-I and II.

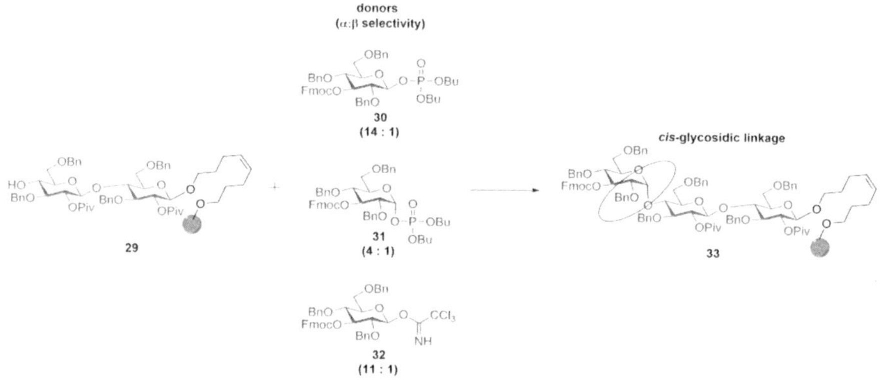

Figure 10. Solid supported synthesis of protected Gb3 antigen using 4-octenediol linker.

Chemo-Enzymatic Synthesis

Synthesis of a specific oligosaccharide with desired chemo- and regio-selectivity using the chemical methodologies described so far is a challenging task. Chemical methods require careful consideration of all the factors such as protection and de-protection chemistry, reaction optimization for stereo-selectivity and selection of promoter system. Even with automation, tuning the desired building block is inevitable and certainly time consuming. Chemo-enzymatic synthesis is an attractive alternative to access oligosaccharides in a potentially less laborious way.

Substrate specific glycoside hydrolase (GH) enzymes can be engineered and used for trans-glycosylation instead of their regular function (*108*). These hydrolases are called glycosynthases as they can form site specific glycosidic linkages using the activated donor with opposite anomeric configuration to that of substrate and the acceptor. Glycosyl transferase enzymes are the ideal candidates for these transformations; nonetheless, difficulties in expressing these enzymes and expensive scale-up have lead to other alternatives such as glycosynthases and glycoside phosphorylases. The detailed description of these enzymes and their substrate selectivity is beyond the scope of this chapter and has been recently reviewed in multiple reports (*109–111*).

Recent Developments

Continuous flow reactors have advantages over conventional batch reactors. Heat and mass transfer in flow reactors is very efficient compared to round bottom flasks. As a result, greater control over temperature, reaction time, and concentration is achieved improving overall yield and diastereoselectivity. Flow reactions can be scaled up using multiple parallel micro-reactors in a small space. Seeberger and group used a silicon-glass micro-reactor along with fluorous-based purification. The authors synthesized a tetrasaccharide composed of β-(1-6) linked D-glucopyranoside using phosphate donors in excellent yields (*112*). Recently, the same group described gold-catalyzed glycosylations under continuous flow systems (*113*). Beau *et al.* reported disaccharide synthesis using *N*-acetyl glucosamine with various acceptors using catalytic iron (III) triflate in a continuous flow process (*114*). Nonetheless, examples of syntheses of complex oligosaccharides using flow chemistry is still scarce (*115*, *116*).

Recently, our group synthesized trisaccharide and tetrasaccharide fragments from the outer core domain of *Pseudomonas aeruginosa* lipopolysaccharide (LPS) common to glycoform I and II (*117*). We envisioned a traditional linear synthesis with a novel reducing end capping group, TBDPS-protected hydroquinone (TPH). The TPH group was used as a purification handle and was stable throughout the chemical transformations. The group can be easily removed via CAN oxidation to obtain free reducing end oligosaccharide. The hydroquinone functional group at the reducing end would be crucial for mild conjugation strategies. The TPH group may easily be intergrated into solid, solution or fluorous-based approaches.

Summary and Outlook

Significant breakthroughs have been achieved over last few decades in the field of oligosaccharide assembly. Recent synthesis of mycobacterial arabinogalactan containing 92 monomer units along with many other examples are indicative of the tremendous progress that has been made in chemical as well as enzymatic methodologies. The current progress in automation, flow chemistry, novel glycosylation reagents, innovative separation and purification techniques will enable new biological and medical research applications for these biopolymers in the near future.

References

1. Cipolla, L.; Araújo, A. C.; Bini, D.; Gabrielli, L.; Russo, L.; Shaikh, N. *Expert Opin. Drug Discovery* **2010**, *5*, 721–737.
2. Hyun, J. Y.; Pai, J.; Shin, I. *Acc. Chem. Res.* **2017**, *50*, 1069–1078.
3. Codée, J. D.; Litjens, R. E.; van den Bos, L. J.; Overkleeft, H. S.; van der Marel, G. A. *Chem. Soc. Rev.* **2005**, *34*, 769–782.
4. Danishefsky, S. J.; Bilodeau, M. T. *Angew. Chem., Int. Ed.* **1996**, *35*, 1380–1419.
5. Li, X.; Zhu, J. *J. Carbohydr. Chem.* **2012**, *31*, 284–324.
6. McKay, M. J.; Nguyen, H. M. *ACS Catal.* **2012**, *2*, 1563–1595.
7. Kim, H. M.; Kim, I. J.; Danishefsky, S. J. *J. Am. Chem. Soc.* **2001**, *123*, 35–48.
8. Roberge, J.; Beebe, X.; Danishefsky, S. *J. Am. Chem. Soc.* **1998**, *120*, 3915–3927.
9. Wang, Z.-G.; Warren, J. D.; Dudkin, V. Y.; Zhang, X.; Iserloh, U.; Visser, M.; Eckhardt, M.; Seeberger, P. H.; Danishefsky, S. J. *Tetrahedron* **2006**, *62*, 4954–4978.
10. Fraser-Reid, B.; López, J. C. In *Reactivity Tuning in Oligosaccharide Assembly*; Fraser-Reid, B.; López, J. C; Springer: New York, 2010; p 1–29.
11. Lemieux, R.; Ratcliffe, R. *Can. J. Chem.* **1979**, *57*, 1244–1251.
12. Marra, A.; Gauffeny, F.; Sinaÿ, P. *Tetrahedron* **1991**, *47*, 5149–5160.
13. Paulsen, H.; Lorentzen, J. P.; Kutschker, W. *Carbohydr. Res.* **1985**, *136*, 153–176.
14. Griffith, D. A.; Danishefsky, S. J. *J. Am. Chem. Soc.* **1990**, *112*, 5811–5819.
15. McDonald, F. E.; Danishefsky, S. J. *J. Org. Chem.* **1992**, *57*, 7001–7002.
16. Lafont, D.; Descotes, G. *Carbohydr. Res.* **1987**, *166*, 195–209.
17. Lafont, D.; Descotes, G. *Carbohydr. Res.* **1988**, *175*, 35–48.
18. Mirabella, S.; Cardona, F.; Goti, A. *Org. Biomol. Chem.* **2016**, *14*, 5186–5204.
19. Halcomb, R. L.; Danishefsky, S. J. *J. Am. Chem. Soc.* **1989**, *111*, 6661–6666.
20. Murray, R. W.; Jeyaraman, R. *J. Org. Chem.* **1985**, *50*, 2847–2853.
21. Behar, V.; Danishefsky, S. J. *Angew. Chem., Int. Ed.* **1994**, *33*, 1468–1470.
22. Danishefsky, S. J.; Behar, V.; Randolph, J. T.; Lloyd, K. O. *J. Am. Chem. Soc.* **1995**, *117*, 5701–5711.

23. Randolph, J. T.; Danishefsky, S. J. *Angew. Chem., Int. Ed.* **1994**, *33*, 1470–1473.
24. Allen, J. R.; Allen, J. G.; Zhang, X. F.; Williams, L. J.; Zatorski, A.; Ragupathi, G.; Livingston, P. O.; Danishefsky, S. J. *Chem. Eur. J.* **2000**, *6*, 1366–1375.
25. Jeon, I.; Iyer, K.; Danishefsky, S. J. *J. Org. Chem.* **2009**, *74*, 8452–8455.
26. Ragupathi, G.; Park, T. K.; Zhang, S.; Kim, I. J.; Graber, L.; Adluri, S.; Lloyd, K. O.; Danishefsky, S. J.; Livingston, P. O. *Angew. Chem., Int. Ed.* **1997**, *36*, 125–128.
27. Slovin, S.; Ragupathi, G.; Adluri, S.; Ungers, G.; Terry, K.; Kim, S.; Spassova, M.; Bornmann, W.; Fazzari, M.; Dantis, L. *Proc. Natl. Acad. Sci.* **1999**, *96*, 5710–5715.
28. Raghavan, S.; Kahne, D. *J. Am. Chem. Soc.* **1993**, *115*, 1580–1581.
29. Fraser-Reid, B.; Wu, Z.; Udodong, U. E.; Ottosson, H. *J. Org. Chem.* **1990**, *55*, 6068–6070.
30. Mootoo, D. R.; Konradsson, P.; Udodong, U.; Fraser-Reid, B. *J. Am. Chem. Soc.* **1988**, *110*, 5583–5584.
31. Zhang, Z.; Ollmann, I. R.; Ye, X.-S.; Wischnat, R.; Baasov, T.; Wong, C.-H. *J. Am. Chem. Soc.* **1999**, *121*, 734–753.
32. Douglas, N. L.; Ley, S. V.; Lücking, U.; Warriner, S. L. *J. Am. Chem. Soc., Perkin Trans. 1* **1998**, 51–66.
33. Koeller, K. M.; Wong, C.-H. *Chem. Rev.* **2000**, *100*, 4465–4494.
34. Ritter, T. K.; Mong, K. K. T.; Liu, H.; Nakatani, T.; Wong, C. H. *Angew. Chem., Int. Ed.* **2003**, *42*, 4657–4660.
35. Yu, C. S.; Niikura, K.; Lin, C. C.; Wong, C. H. *Angew. Chem., Int. Ed.* **2001**, *40*, 2900–2903.
36. Crich, D.; Smith, M. *J. Am. Chem. Soc.* **2002**, *124*, 8867–8869.
37. Crich, D.; Sun, S. *J. Am. Chem. Soc.* **1998**, *120*, 435–436.
38. Angles d'Ortoli, T.; Hamark, C.; Widmalm, G. *J. Org. Chem.* **2017**, *82*, 8123–8140.
39. Kanie, O.; Ito, Y.; Ogawa, T. *J. Am. Chem. Soc.* **1994**, *116*, 12073–12074.
40. Mukaiyama, T.; Ikegai, K.; Jona, H.; Hashihayata, T.; Takeuchi, K. *Chem. Lett.* **2001**, *30*, 840–841.
41. Si, A.; Misra, A. K. *Tetrahedron* **2016**, *72*, 4435–4441.
42. Kitowski, A.; Jiménez-Moreno, E.; Salvadó, M.; Mestre, J.; Castillón, S.; Jiménez-Osés, G.; Boutureira, O.; Bernardes, G. A. J. *Org. Lett.* **2017**, *19*, 5490–5493.
43. Hasty, S. J.; Bandara, M. D.; Rath, N. P.; Demchenko, A. V. *J. Org. Chem.* **2017**, *82*, 1904–1911.
44. Kaeothip, S.; Pornsuriyasak, P.; Demchenko, A. V. *Tetrahedron Lett.* **2008**, *49*, 1542–1545.
45. Kaeothip, S.; Pornsuriyasak, P.; Rath, N. P.; Demchenko, A. V. *Org. Lett.* **2009**, *11*, 799–802.
46. Pornsuriyasak, P.; Demchenko, A. V. *Chem. Eur. J.* **2006**, *12*, 6630–6646.
47. Vidadala, S. R.; Thadke, S. A.; Hotha, S. *J. Org. Chem.* **2009**, *74*, 9233–9236.
48. Kaeothip, S.; Demchenko, A. V. *Carbohydr. Res.* **2011**, *346*, 1371–1388.

49. Huang, L.; Huang, X. *Chem. Eur. J.* **2007**, *13*, 529–540.
50. Huang, L.; Wang, Z.; Li, X.; Ye, X.-s.; Huang, X. *Carbohydr. Res.* **2006**, *341*, 1669–1679.
51. Huang, X.; Huang, L.; Wang, H.; Ye, X. S. *Angew. Chem.* **2004**, *116*, 5333–5336.
52. Codée, J. D.; van den Bos, L. J.; Litjens, R. E.; Overkleeft, H. S.; van Boeckel, C. A.; van Boom, J. H.; van der Marel, G. A. *Tetrahedron* **2004**, *60*, 1057–1064.
53. Codée, J. D.; van den Bos, L. J.; Litjens, R. E.; Overkleeft, H. S.; van Boom, J. H.; van der Marel, G. A. *Org. Lett.* **2003**, *5*, 1947–1950.
54. Codée, J. D.; Litjens, R. E.; den Heeten, R.; Overkleeft, H. S.; van Boom, J. H.; van der Marel, G. A. *Org. Lett.* **2003**, *5*, 1519–1522.
55. Crich, D.; Smith, M. *Org. Lett.* **2000**, *2*, 4067–4069.
56. Peng, P.; Ye, X.-S. *Org. Biomol. Chem.* **2011**, *9*, 616–622.
57. Codée, J. D.; Stubba, B.; Schiattarella, M.; Overkleeft, H. S.; van Boeckel, C. A.; van Boom, J. H.; van der Marel, G. A. *J. Am. Chem. Soc.* **2005**, *127*, 3767–3773.
58. Geurtsen, R.; Boons, G.-J. *Tetrahedron Lett.* **2002**, *43*, 9429–9431.
59. Li, Z.; Gildersleeve, J. C. *J. Am. Chem. Soc.* **2006**, *128*, 11612–11619.
60. Mukherjee, M. M.; Ghosh, R. *J. Org. Chem.* **2017**, *82*, 5751–5760.
61. Gao, J.; Guo, Z. *Org. Lett.* **2016**, *18*, 5552–5555.
62. Podilapu, A. R.; Kulkarni, S. S. *Org. Lett.* **2017**, *19*, 5466–5469.
63. Wang, Y.; Huang, X.; Zhang, L.-H.; Ye, X.-S. *Org. Lett.* **2004**, *6*, 4415–4417.
64. Wang, Z.; Zhou, L.; El-Boubbou, K.; Ye, X.-s.; Huang, X. *J. Org. Chem.* **2007**, *72*, 6409–6420.
65. Yang, W.; Yang, B.; Ramadan, S.; Huang, X. *Beilstein J. Org. Chem.* **2017**, *13*, 2094–2099.
66. Zulueta, M. M. L.; Janreddy, D.; Hung, S. C. *Isr. J. Chem.* **2015**, *55*, 347–359.
67. Tanaka, H.; Takeuchi, R.; Jimbo, M.; Kuniya, N.; Takahashi, T. *Chem. Eur. J.* **2013**, *19*, 3177–3187.
68. Mong, T. K.-K.; Lee, H.-K.; Durón, S. G.; Wong, C.-H. *Proc. Natl. Acad. Sci.* **2003**, *100*, 797–802.
69. Wu, Y.; Xiong, D.-C.; Chen, S.-C.; Wang, Y.-S.; Ye, X.-S. *Nat. Commun.* **2017**, *8*, 14851–14858.
70. Gisin, B.; Merrifield, R.; Tosteson, D. *J. Am. Chem. Soc.* **1969**, *91*, 2691–2695.
71. Merrifield, R. B. *J. Am. Chem. Soc.* **1963**, *85*, 2149–2154.
72. Collot, M.; Eller, S.; Weishaupt, M.; Seeberger, P. H. *Beilstein J. Org. Chem.* **2013**, *9*, 97–105.
73. Ako, T.; Daikoku, S.; Ohtsuka, I.; Kato, R.; Kanie, O. *Chem. Asian J.* **2006**, *1*, 798–813.
74. Bayer, E. *Angew. Chem., Int. Ed.* **1991**, *30*, 113–129.
75. Lu, J.; Toy, P. H. *Chem. Rev.* **2009**, *109*, 815–838.
76. Seeberger, P. H.; Haase, W.-C. *Chem. Rev.* **2000**, *100*, 4349–4394.
77. Galan, M. C.; Jones, R. A.; Tran, A.-T. *Carbohydr. Res.* **2013**, *375*, 35–46.

78. Ganesh, N. V.; Fujikawa, K.; Tan, Y. H.; Stine, K. J.; Demchenko, A. V. *Org. Lett.* **2012**, *14*, 3036–3039.
79. Ko, K.-S.; Jaipuri, F. A.; Pohl, N. L. *J. Am. Chem. Soc.* **2005**, *127*, 13162–13163.
80. Pornsuriyasak, P.; Ranade, S. C.; Li, A.; Parlato, M. C.; Sims, C. R.; Shulga, O. V.; Stine, K. J.; Demchenko, A. V. *Chem. Commun.* **2009**, 1834–1836.
81. Ito, Y.; Kanie, O.; Ogawa, T. *Angew. Chem., Int. Ed.* **1996**, *35*, 2510–2512.
82. Osborn, H. M.; Khan, T. H. *Tetrahedron* **1999**, *55*, 1807–1850.
83. Wang, Z.-G.; Douglas, S. P.; Krepinsky, J. J. *Tetrahedron Lett.* **1996**, *37*, 6985–6988.
84. Zhu, T.; Boons, G.-J. *J. Am. Chem. Soc.* **2000**, *122*, 10222–10223.
85. Majumdar, D.; Zhu, T.; Boons, G.-J. *Org. Lett.* **2003**, *5*, 3591–3594.
86. Rodebaugh, R.; Joshi, S.; Fraser-Reid, B.; Geysen, H. M. *J. Org. Chem.* **1997**, *62*, 5660–5661.
87. Danishefsky, S. J.; McClure, K. F.; Randolph, J. T.; Ruggeri, R. B. *Science* **1993**, *260*, 1307–1310.
88. Savin, K. A.; Woo, J. C.; Danishefsky, S. J. *J. Org. Chem.* **1999**, *64*, 4183–4186.
89. Mar Kayser, M.; de Paz, J. L.; Nieto, P. M. *Eur. J. Org. Chem.* **2010**, *2010*, 2138–2147.
90. Belogi, G.; Zhu, T.; Boons, G.-J. *Tetrahedron Lett.* **2000**, *41*, 6969–6972.
91. Tolborg, J. F.; Jensen, K. J. *Chem. Commun.* **2000**, 147–148.
92. Izumi, M.; Fukase, K.; Kusumoto, S. *Synlett* **2002**, *9*, 1409–1416.
93. Nguyen, S. H.; Trotta, A. H.; Cao, J.; Straub, T. J.; Bennett, C. S. *Org. Biomol. Chem.* **2012**, *10*, 2373–2376.
94. Routenberg Love, K.; Seeberger, P. H. *Angew. Chem.* **2004**, *116*, 612–615.
95. Knerr, L.; Schmidt, R. R. *Eur. J. Org. Chem.* **2000**, 2803–2808.
96. Timmer, M. S.; Codée, J. D.; Overkleeft, H. S.; van Boom, J. H.; van der Marel, G. A. *Synlett* **2004**, *12*, 2155–2158.
97. Kröck, L.; Esposito, D.; Castagner, B.; Wang, C.-C.; Bindschädler, P.; Seeberger, P. H. *Chem. Sci.* **2012**, *3*, 1617–1622.
98. Paulson, J. C.; Blixt, O.; Collins, B. E. *Nat. Chem. Biol.* **2006**, *2*, 238–248.
99. Calin, O.; Eller, S.; Seeberger, P. H. *Angew. Chem., Int. Ed.* **2013**, *52*, 5862–5865.
100. Hahm, H. S.; Liang, C.-F.; Lai, C.-H.; Fair, R. J.; Schuhmacher, F.; Seeberger, P. H. *J. Org. Chem.* **2016**, *81*, 5866–5877.
101. Newcomb, W. S.; Deegan, T. L.; Miller, W.; Porco, J. A. *Biotechnol. Bioeng.* **1998**, *61*, 55–60.
102. Ko, K.-S.; Park, G.; Yu, Y.; Pohl, N. L. *Org. Lett.* **2008**, *10*, 5381–5384.
103. Manabe, S.; Ito, Y. *J. Am. Chem. Soc.* **2002**, *124*, 12638–12639.
104. Yang, S.-J.; Tian, X. Z.; Shin, I. *Org. Lett.* **2009**, *11*, 3438–3441.
105. Kanemitsu, T.; Wong, C.-H.; Kanie, O. *J. Am. Chem. Soc.* **2002**, *124*, 3591–3599.
106. Seeberger, P. H.; Beebe, X.; Sukenick, G. D.; Pochapsky, S.; Danishefsky, S. J. *Angew. Chem., Int. Ed.* **1997**, *36*, 491–493.

107. Werz, D. B.; Castagner, B.; Seeberger, P. H. *J. Am. Chem. Soc.* **2007**, *129*, 2770–2771.
108. Mackenzie, L. F.; Wang, Q.; Warren, R. A. J.; Withers, S. G. *J. Am. Chem. Soc.* **1998**, *120*, 5583–5584.
109. Danby, P. M.; Withers, S. G. *ACS Chem. Biol.* **2016**, *11*, 1784–1794.
110. Hayes, M. R.; Pietruszka, J. *Molecules* **2017**, *22*, 1434–1454.
111. Jahn, M.; Withers, S. G. *Biocatal. Biotransform.* **2003**, *21*, 159–166.
112. Carrel, F. R.; Geyer, K.; Codée, J. D.; Seeberger, P. H. *Org. Lett.* **2007**, *9*, 2285–2288.
113. Matthies, S.; McQuade, D. T.; Seeberger, P. H. *Org. Lett.* **2015**, *17*, 3670–3673.
114. Xolin, A.; Stévenin, A.; Pucheault, M.; Norsikian, S.; Boyer, F.-D.; Beau, J.-M. *Org. Chem. Front.* **2014**, *1*, 992–1000.
115. Cancogni, D.; Lay, L. *Synlett* **2014**, *25*, 2873–2878.
116. Geyer, K.; Gustafsson, T.; Seeberger, P. H. *Synlett* **2009**, 2382–2391.
117. Vartak, A.; Hefny, F. M.; Sucheck, S. J. *Org. Lett.* **2018**, *20*, 353–356.

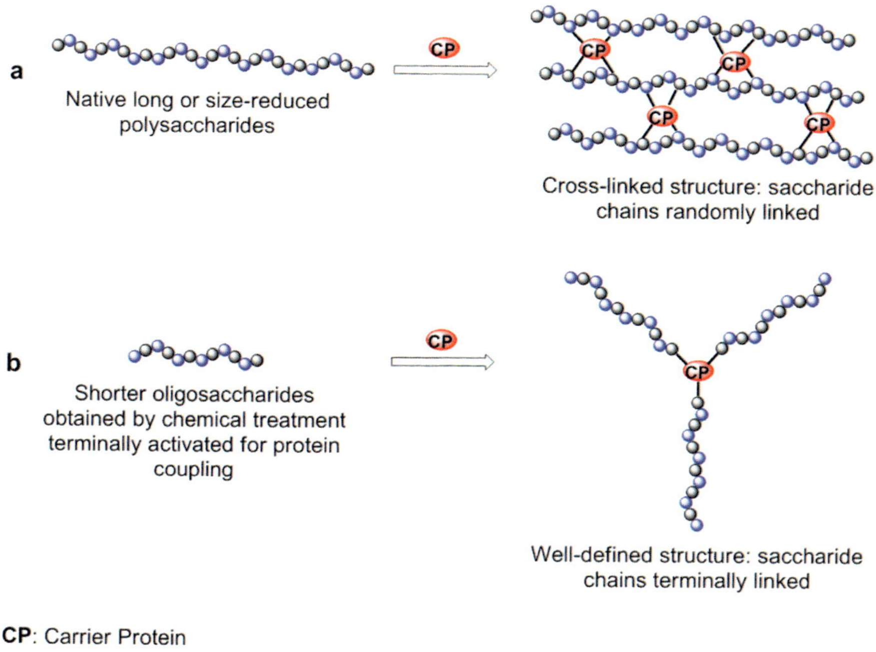

Figure 1. Schematic representation of NmACWY conjugate vaccines prepared by using different conjugation strategies. (Chapter 6)

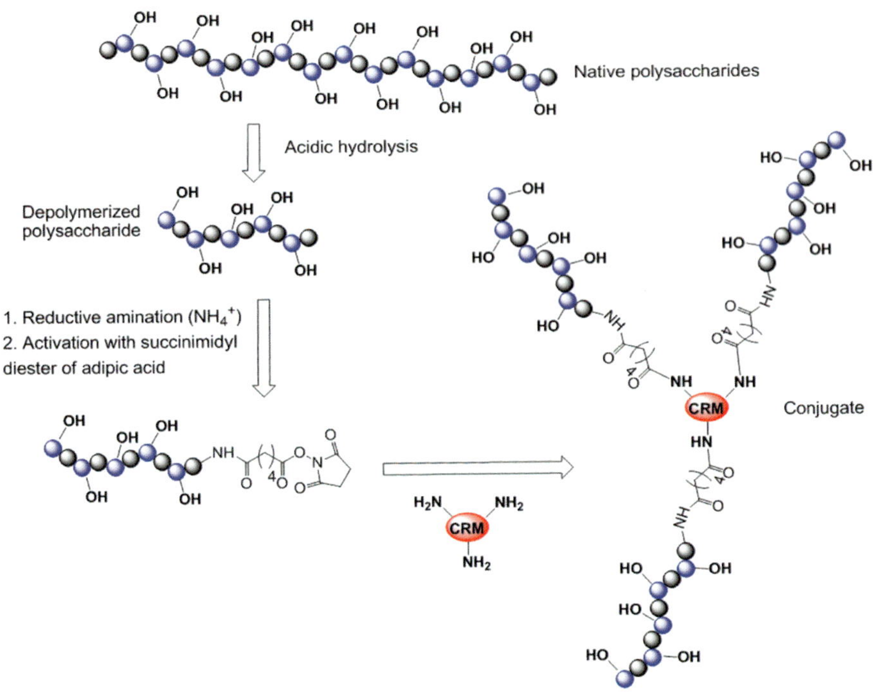

Figure 2. Schematic representation of NmACWY conjugation chemistry used for Menveo® antigens. (Chapter 6)

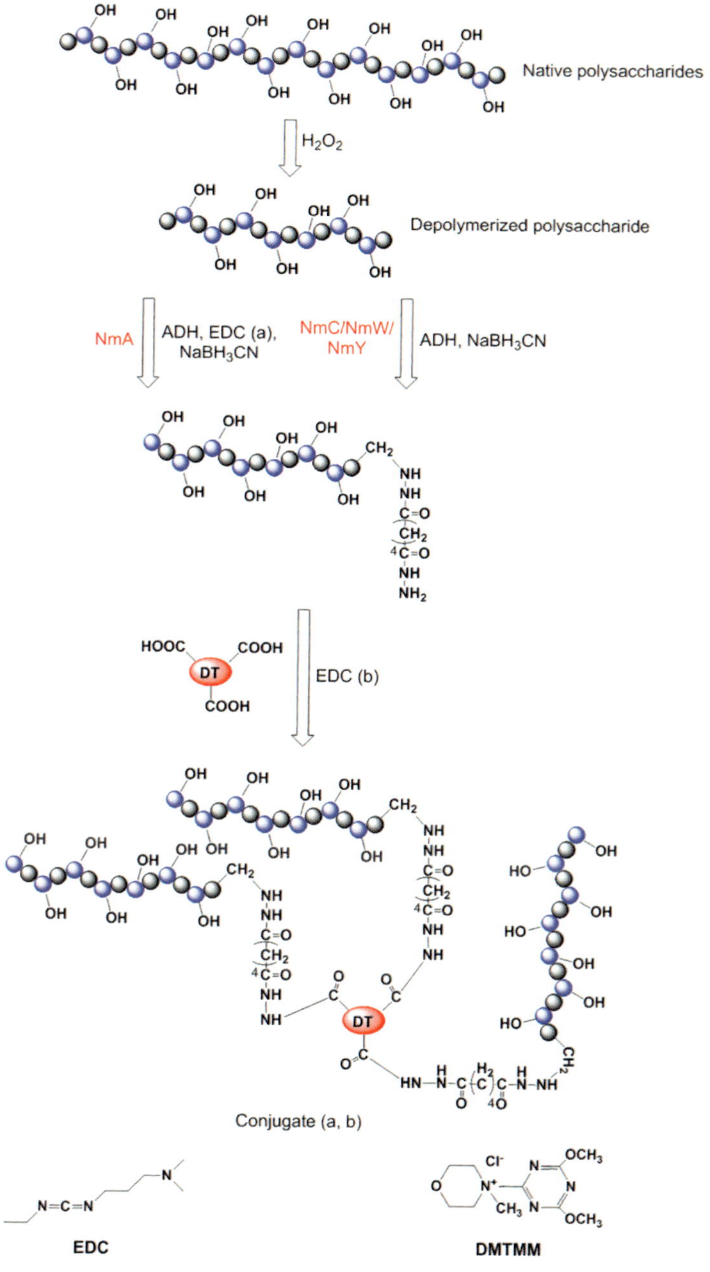

Figure 3. Schematic representation of NmACWY-DT conjugation chemistry using meningococcal polysaccharides firstly depolymerized by hydrogen peroxide treatment. The chemical structure of EDC and DMTMM zero-length crosslinker reagents are reported. (a) The use of EDC allows the NmA phosphodiester group coupling to the amine groups of ADH. (b) The use of EDC allows the NmA phosphodiester group and carboxyl groups of NmCWY coupling to amine groups of DT resulting in a cross-linked conjugate. (Chapter 6)

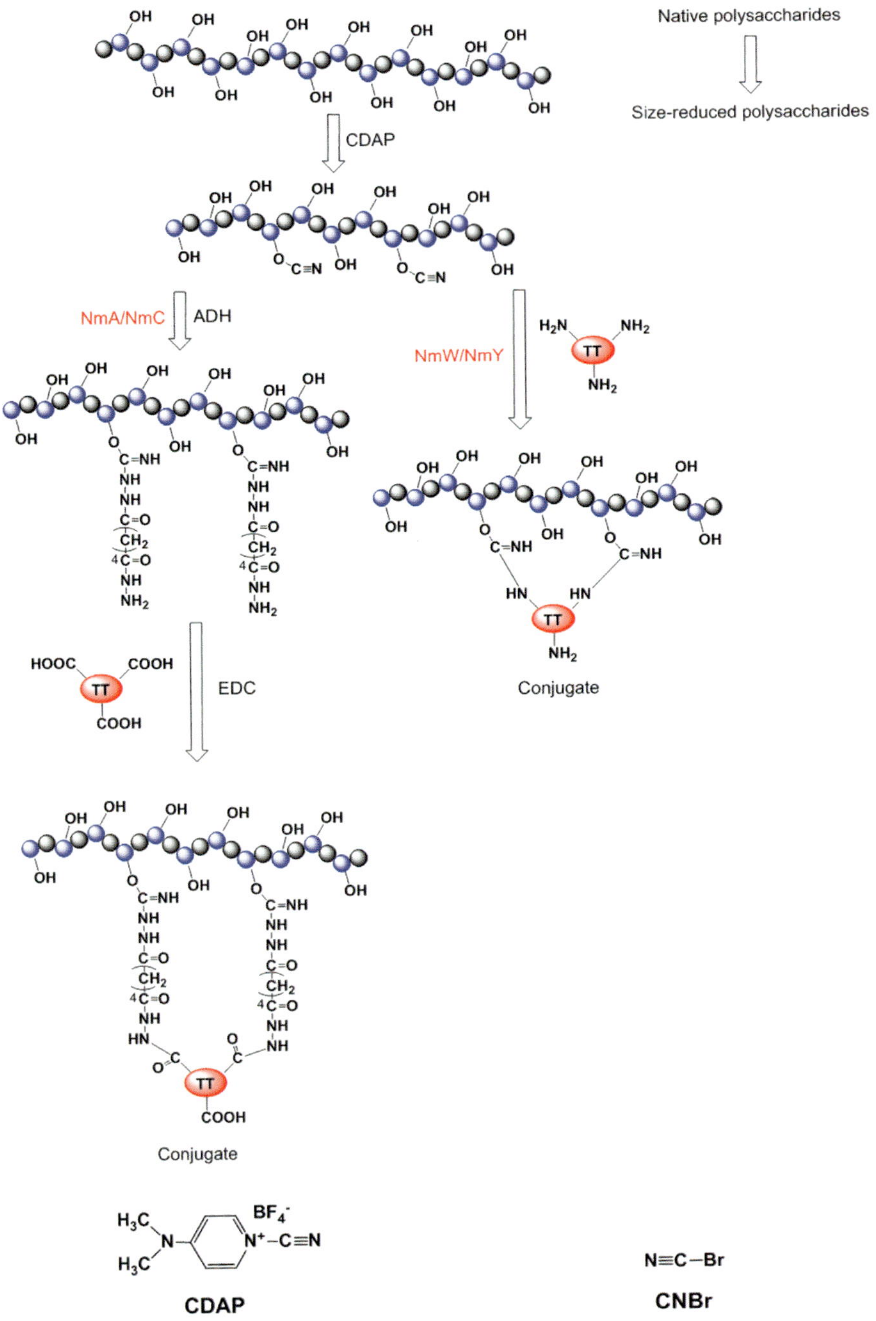

Figure 4. Schematic representation of NmACWY-TT conjugation chemistry using native/size-reduced polysaccharides. (Chapter 6)

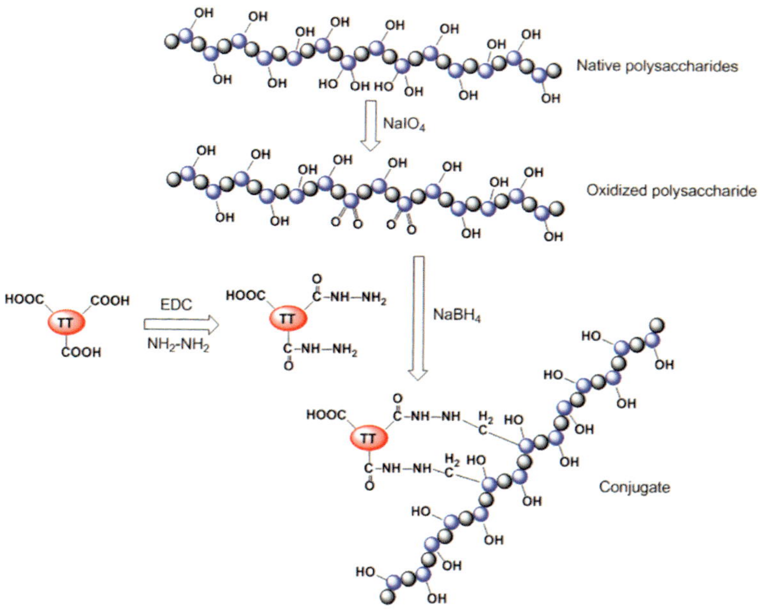

Figure 5. Schematic representation of NmA conjugation chemistry used for preparing NmA-TT conjugate of MenAfriVac®. (Chapter 6)

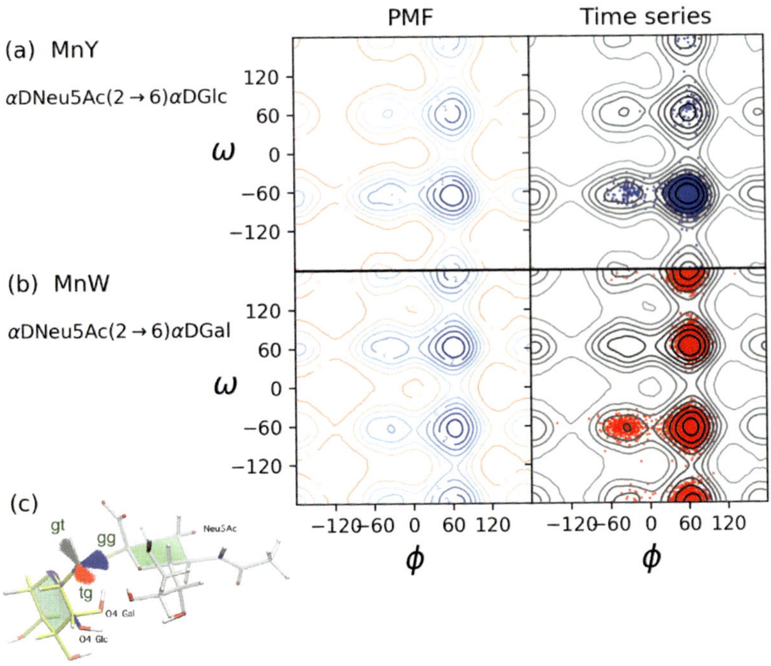

Figure 1. Contoured disaccharide φ, ω PMF surfaces in aqueous solution for (a) αDNeu5Ac(2→6)αDGlcp and (b) αDNeu5Ac(2→6)αDGalp glycosidic linkages. Plots of the simulation time series for both a(2/6) linkages in the 3RU oligosaccharides are superimposed on the corresponding disaccharide solution free energy surfaces in the right column (MnY: top, MnW: bottom). The disaccharide structure shown in (c) illustrates the difference in conformation at the hexose C-4 (which is axial in Gal and equatorial in Glc), as well as the three orientations of the ω dihedral in the glycosidic linkage: gg, gt and tg. (Chapter 7)

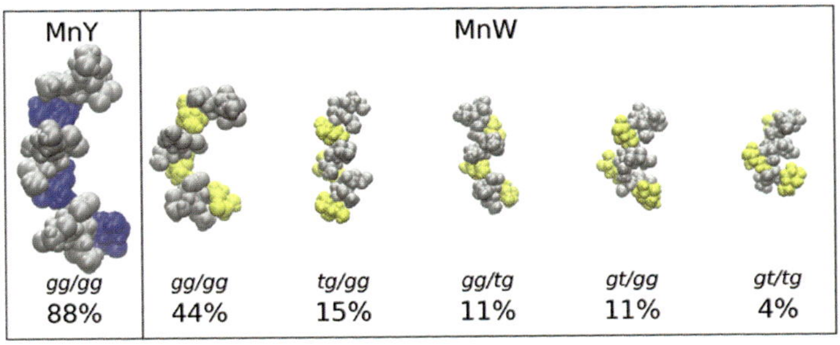

Figure 2. Representative structures from the dominant conformational families for the capsular polysaccharides in MnY (left) and MnW (right). (Chapter 7)

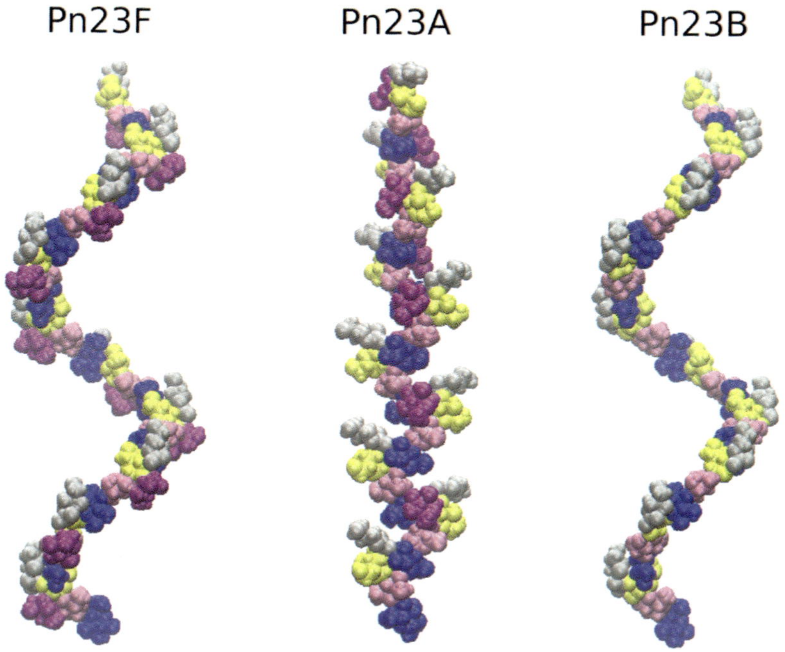

Figure 3. 6RU models of Pn23F, Pn23A and Pn23B. (Chapter 7)

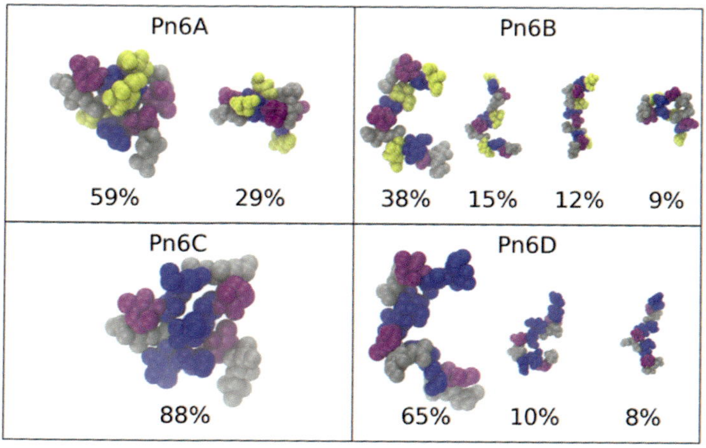

Figure 4. Representative structures from the dominant conformational families for 3RU of the capsular polysaccharides in Pn6A, Pn6B, Pn6C and Pn6D. Residues are colored according to type: glucose blue, galactose yellow, rhamnose magenta and ribitol grey. (Chapter 7)

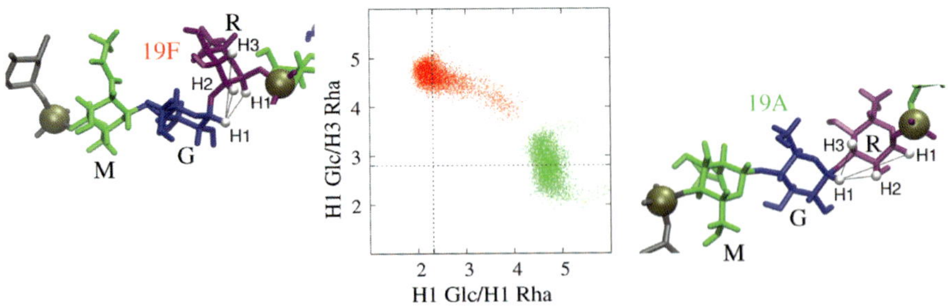

Figure 5. Time series of the H1 Glc/H1 Rha and H1 Glc/H3 Rha distances for the middle repeat unit in the RU3 simulations of Pn19F (red, top and left) and Pn19A (green, bottom and right). NMR NOESY distances are indicated by dashed lines. Representative conformations for the middle RU are shown for Pn19F (left) and Pn19A (right), with key atomic distances indicated. Structures are annotated to indicate residue identity: ManNAc (M, green), Glc (G, blue) and Rha (R, purple). (Reproduced with permission from reference 23. Copyright 2015 Elsevier.) (Chapter 7)

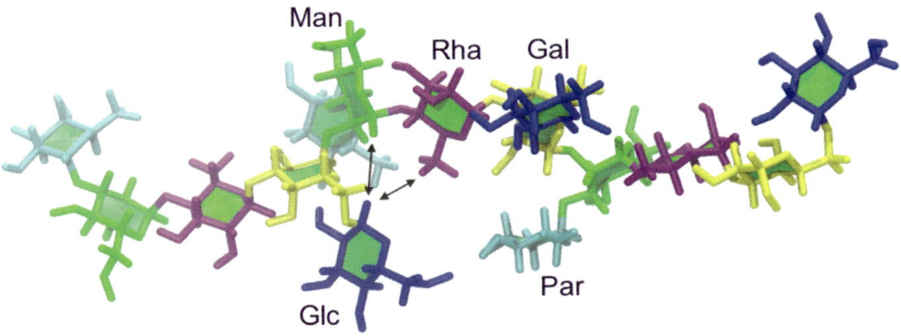

Figure 6. Static structure of S. paratyphi A showing the proximity of the glucose residue (blue) to the backbone rhamnose (purple) and mannose. (Chapter 7)

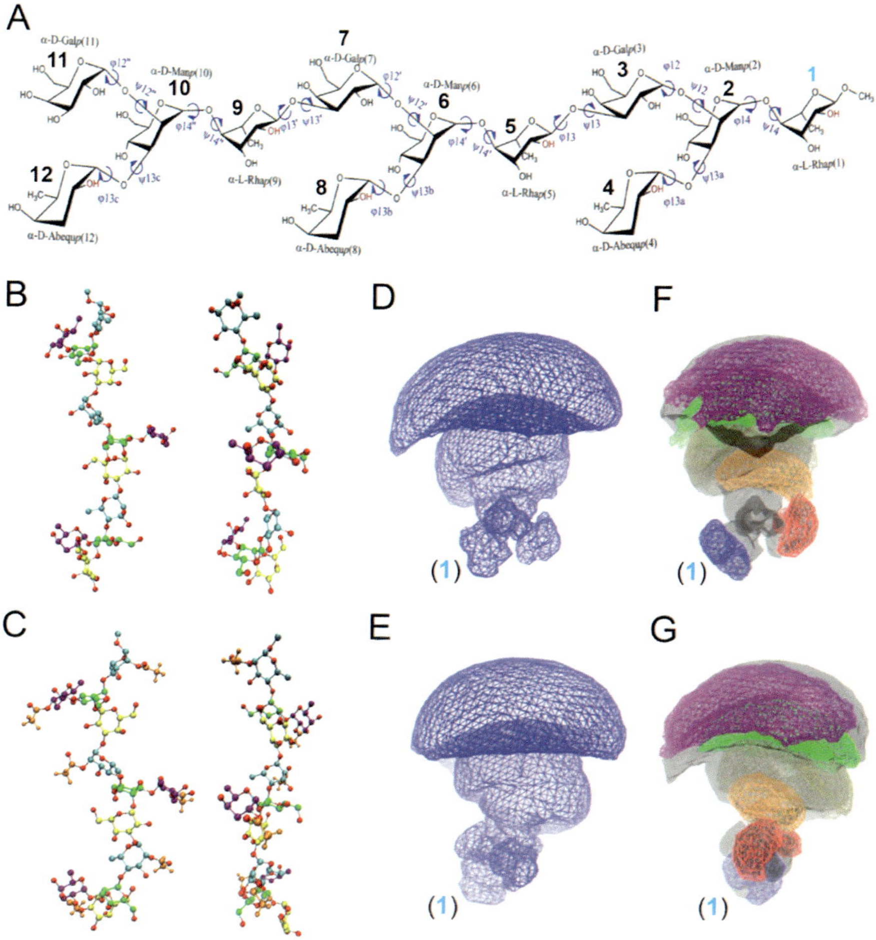

Figure 7. (A) Structure of the base 3-repeat STm polysaccharide unit used for computational analyses. The representative O-acetylated polysaccharide was constructed with the hydroxyl group substituted by an acetyl at the C-2 position of both α-D-Abep and α-L-Rhap (-OH, red). (B, C) Two views of the dominant conformation of the base (B) and O-acetylated (C) polysaccharides. The acetyl group is shown in brown, glucose in blue, mannose in green, galactose in yellow, abequose in purple, and rhamnose in cyan. For clarity, all hydrogen atoms are omitted except those on the acetyl group. (D, E, F, G) Wire frame models of the base (D, E) and O-acetylated (F, G) polysaccharides for all non-hydrogen atoms. Conformational flexibility increases progressively relative to the reducing end anchor point. (Figure reproduced from reference 28, Copyright 2017 PLOS.) (Chapter 7)

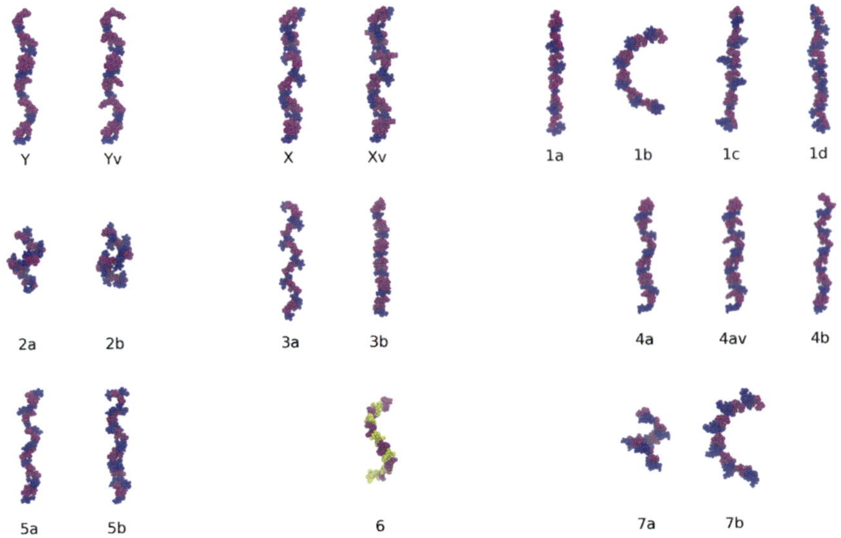

Figure 8. 3D models of six repeating units (6RU) of each of the O-antigens of S. flexneri produced by CarbBuilder from the primary structure. Residues are colored according to type: glucose blue, galactose yellow and rhamnose magenta. (Chapter 7)

Chapter 9

Formulation Development of Glycoconjugate Vaccines for Low- and Middle-Income Countries

Lakshmi Khandke,* Jo Anne Welsch, and Mark R. Alderson

Center for Vaccine Innovation and Access, PATH, Seattle, Washington 98121, United States
***E-mail: lkhandke@path.org.**

Developing vaccines that are safe, efficacious, and affordable for low-resource countries is complex and costly. There are challenges with glycoconjugate vaccines, particularly multivalent vaccines, due to their manufacturing complexity. Formulation development is one of the critical steps in developing a stable glycoconjugate vaccine and in order to be successful, a rational, designed approach is imperative. The approach must leverage current understanding around the critical quality attributes and cost drivers that are important for the product as well as the immunogenicity of the vaccine in clinical studies. With an emphasis on developing country vaccine manufacturers, this review describes formulation strategies and approaches to characterizing glycoconjugate vaccines using a broad analytical toolkit for development and transferring robust processes for production to developing country vaccine manufacturers.

Introduction

Without a doubt, vaccines have made huge contributions toward improving the quality of human life—with their greatest impact on public health in the past century. Immunization has prevented more than a million child deaths each year and protected millions more from illness and/or disability. Countries across income strata have benefitted. Yet every year, three million people still die from diseases that can be prevented by vaccines. More than half of these deaths occur in children younger than five years of age (*1*). Children living in the poorest countries and countries in conflict are at greatest risk but are also the least likely to be immunized.

Although it is well recognized and acknowledged that vaccines lead to much needed economic development in low-income countries due to their public health impact, there are major challenges to their effective implementation for global access and the realization of the enormous potential benefits (*2–7*). Licensed multivalent glycoconjugate vaccines are very complex from a manufacturing perspective and as such, price and supply issues have limited access in underdeveloped countries and have failed to meet the current needs.

a. *Effectiveness of the vaccine and suitability to induce an immune response in the target population that provides broad protective coverage against the prevalent strains of the pathogen targeted by the vaccine and new strains that might emerge following the introduction of the vaccine*: This may be for several reasons such as chronic environmental enteropathy, malnutrition, insufficient maternal antibodies, and host genetic factors (*7, 8*).
b. *Affordability of the vaccine in low resource countries*: The price of manufacturing is a major cost driver from the perspective of the formulation of the fill-finish process and the stability of the vaccine.
c. Most of the outputs (patents, intellectual property rights) from vaccine research and development technologies are under the control of the multinational vaccine manufacturers, including manufacturers of vaccines that are of enormous importance to the developing world and of commercial value to the developed markets.
d. Vaccines are typically initially developed by multinational pharmaceutical companies with a focus on industrialized country markets. In some cases, however, the vaccine formulation needs to be targeted to specific regional or country needs. An example of such a vaccine is the serogroup A meningococcal conjugate vaccine MenAfriVac®, specifically designed for use in the African Meningitis Belt where this serogroup has been responsible for the majority of meningitis epidemics (see MenAfriVac case study as follows).
e. Clinical trials can be complex with multivalent vaccines. In addition, the serotype coverage and efficacy for the vaccine is important and might have regional differences that are important (see Prevnar case study as follows).
f. *Vaccine stability to ensure it can be transported and provided in remote locations*: The "controlled temperature chain" (CTC) is an innovative approach to vaccine management allowing vaccines to be kept at temperatures outside of the traditional cold chain of +2°C to +8°C for a limited period of time under monitored and controlled conditions (see more detailed discussion of CTC as follows).
g. The regulatory authorities in some Low Middle Income Countries (LMICs) do not always have the know-how and experience to provide the guidance required to develop and license a vaccine within their country.

Advisory boards and World Health Organization (WHO) National Regulatory Agencies (NRA) strengthening initiatives could consist of representatives of the ministry of health, ministry of higher education, public health experts, public–private partnerships with experiences in vaccination, clinical researchers, and safety pharmacovigilance experts.

h. *Lack of infrastructure for low-cost manufacturing and/or facilities in many under-developed countries*: Investment is needed to develop technologies that are simple, cost- effective, and ensuring that the production facilities meet the quality systems required for manufacturing (*9–12*).
i. The price of manufacturing is a major cost driver from the perspective of the formulation of the fill-finish process and the stability of the vaccine.
j. Complex vaccines, such as multivalent glycoconjugate vaccines, involve production of multiple components, an example being Prevenar 13, where polysaccharides from 13 different serotypes must be produced, conjugated, and formulated.

Glycoconjugate Vaccines Licensed and Available in LMIC

Globally, several pathogens cause serious infections that are preventable or potentially preventable with glycoconjugate vaccines, and these include *Haemophilus influenzae* type b, *Streptococcus pneumoniae*, Group B *Streptococcus* (*Streptococcus agalactiae)*, and *Neisseria meningitidis*. All have polysaccharide capsules (with different unique complex structures and sugar composition), which are key virulence determinants and targets for protective antibodies. Although polysaccharides can be immunogenic on their own, conjugation of polysaccharides to protein carriers has been used to improve immunogenicity, particularly in young children. The carrier protein can be either a related protein antigen from the target pathogen, boosting the specific immune response to that pathogen, or a generally immunogenic protein that serves simply as a carrier. Chemically conjugating individual capsular polysaccharides to a carrier protein renders the immune response T-cell dependent, thereby making a conjugate vaccine capable of stimulating antibody responses in infants and priming for a memory response upon vaccine boosting or challenge with the organism (*13–17*). The current list of glycoconjugate vaccines that are licensed and available in the LMICs are summarized in Table 1. This chapter focuses on the challenges of formulation development of a glycoconjugate vaccine drug product as it applies to pharmaceutical manufacturing.

Table 1. List of Glycoconjugate Vaccines Prequalified by WHO (as of January 2018)

Commercial name	*Manufacturer*	*Dosage form*	***Formulation*** *(Each 0.5 mL dose)*	*Total Poly µg/dose*	*Protein µg/dose*	*Aluminum (mg/dose)*	*Preservative (mg/dose)*
Pneumococcal conjugate vaccines							
Prevnar 13®	Pfizer	Liquid/ Single/Multi dose vials	2.2 µg of polysaccharide from serotypes 1, 3,4, 5, 6A, 7F, 9V, 14, 18C, 19A, 19F, and 23F and 4,4µg from serotype 6B conjugated to $CRM_{197}$with 0.125 mg Al formulated in succinate buffer and polysorbate 80	31	31	$AlPO_4$ (0.125)	2-PE (4.0)
Synflorix®	GSK	Liquid/ Single/Multi dose vials	1 µg of polysaccharide from serotypes 1, 5, 6B, 7F, 9V, 14, 18C, 23F, and 3 µg from serotypes 4, 18C, and 19F conjugated to protein D (from non-typeable *Haemophilus influenzae*), tetanus toxoid (19F) and diphtheria toxoid with 0.5 mg Al formulated in saline	10	26 (13, 8, and 5)	$AlPO_4$ (0.5mg Al)	2-PE (5mg)

Commercial name	***Manufacturer***	***Dosage form***	***Formulation*** *(Each 0.5 mL dose)*	***Total Poly µg/dose***	***Protein µg/dose***	***Aluminum (mg/dose)***	***Preservative (mg/dose)***
Nimenrix®	Pfizer	Lyo	5 µg each of meningococcal A, C, Y and W polysaccharide conjugated to tetanus toxoid and lyophilized in sucrose and trometamol buffer and reconstituted with isotonic sodium chloride	20	44	NA	NA
Menveo®	Novartis	Lyo/Liquid Single dose vials	Meningococcal A oligosaccharide10 µg, conjugated to CRM_{197} with excipients: mannitol, sucrose and Tris (hydroxymethyl) aminomethane. Men CYW liquid conjugate vaccine is formulated as a liquid at 5 µg/serotype	25	32.7 to 64.1	NA	NA

Continued on next page.

Table 1. (Continued). List of Glycoconjugate Vaccines Prequalified by WHO (as of January 2018)

Commercial name	*Manufacturer*	*Dosage form*	***Formulation*** *(Each 0.5 mL dose)*	*Total Poly µg/dose*	*Protein µg/dose*	*Aluminum (mg/dose)*	*Preservative (mg/dose)*	
Menactra®	Sanofi Pasteur Inc.	Liquid	4 µg each of meningococcal A, C, Y, and W polysaccharides conjugated to diphtheria toxoid and formulated in sodium phosphate buffered isotonic sodium chloride	16	48	NA	NA	
MenAfriVac®	Serum Institute of India Pvt Ltd	Lyo multidose vials1	Meningococcal A polysaccharide 10 µg conjugated to tetanus toxoid lyophilized in mannitol, sucrose and Tris (hydroxymethyl) aminomethane	10 and 5 (adult and pediatric formula-tions)	5 to 16.5	$AlPO_4$ 0.5	Thimerosal	
Commercial name	**Manufacturer**	**Dosage form**	**Formulation (Each 0.5 mL dose)**	**Total Poly µg/dose**	**Protein µg/dose**	**Aluminum (mg/dose)**	**Preservative**	**Dose**
PedvaxHIB®	Merck	Liquid Single dose	7.5 µg of *Haemophilus influenzae* b (Hib) PS, conjugated to *Neisseria meningitidis* OMPC and Al in 0.9% sodium	7.5	125	AAHSA@0.225	NA	NA

Commercial name	***Manufacturer***	***Dosage form***	***Formulation*** *(Each 0.5 mL dose)*	***Total Poly µg/dose***	***Protein µg/dose***	***Aluminum (mg/dose)***	***Preservative (mg/dose)***	
Sii HibPRO®	Serum Institute of India Pvt ltd	Lyophilized Single dose	10 µg of Hib PS conjugated to tetanus Diluent: Reconstitute with Diluent for *Haemophilus influenzae* type b Conjugate Vaccine	10	19 to 33	NA	NA	NA
Vaxem Hib®	Novartis	Liquid/ Single/Multi dose	10 µg of Hib oligosaccharide conjugated to CRM_{197} formulated in sodium chloride, monobasic sodium phosphate, disodium phosphate dihydrate, Polysorbate 80, water for injection.	10	25	$AlPO_4$1.36 as $AlPO_4$	Thimerosal	0.05%

Continued on next page.

Table 1. (Continued). List of Glycoconjugate Vaccines Prequalified by WHO (as of January 2018)

Commercial name	***Manufacturer***	***Dosage form***	***Formulation** (Each 0.5 mL dose)*	***Total Poly µg/dose***	***Protein µg/dose***	***Aluminum (mg/dose)***	***Preservative (mg/dose)***	
Act-HIB®	Sanofi-Pasteur	Lyophilized	Hib PS conjugated to tetanus toxoid. ActHIB vaccine is reconstituted with saline diluent, each single dose of 0.5 mL is formulated to contain -10 µg of Hib PS and 8.5% of sucrose.	10	24	NA	NA	NA
Typbar-TVC®	Bharat Biotech International Limited	Liquid/ Single/Multi dose	25 µg Salmonella typhi Ty2 PS conjugated to tetanus toxoid formulated in sodium phosphate buffered saline	25	No information	NA	2-PE	5 mg

Considerations and Challenges in Formulation Process Development of Glycoconjugate Vaccines

Creating a formulation for a glycoconjugate-based vaccine is an integral part of the overall product development and includes activities from bulk formulation to fill finish, storage, handling, and shipment to the intended target populations. The glycoconjugates must have a defined chemical composition and structure, and the vaccine must be safe and immunogenic in the target population by inducing high avidity bactericidal or opsonic antibodies (*18*). The manufacturing process is highly complex and costly and comes with the risks associated with the development process. The formulation development of multivalent glycoconjugate vaccines follows the general principles of other vaccines yet differs from the formulation of a single active component seen in many vaccines. The major challenges to developing a multivalent glycoconjugate vaccine are:

a. Glycoconjugates elicit an immune response but do not have an inherent functional activity which can be monitored *in vitro,* nor are there efficient *in vitro* or preclinical animal models that can distinguish between a good and a sub potent lot.
b. Unlike recombinantly expressed proteins, glycoconjugates are not classified as well-characterized biologicals, and as such, the process of consistent manufacturing defines the product.
c. The regulatory oversight is high, as glycoconjugate vaccines are not well characterized.
d. The pathway to licensure of vaccines may require extensive and expensive clinical trials to demonstrate vaccine efficacy, unless there is a clearly defined correlate of protection.

The dose of the polysaccharides in the vaccine is typically determined based on guidance from prior knowledge of previous clinical and preclinical studies.

Generally, the dose levels of individual antigens of multivalent vaccines are less than 10 μg per dose (Table 1). Once the dose of the vaccine and the serotypes are selected, each of the individual capsular polysaccharides are purified from the individual serotypes followed by the conjugation and formulation of the vaccine at the intended dose by combining components into a drug product. After selecting the vaccine antigen(s) and adjuvant (if needed to enhance the immune response), the formulation must be developed, which should take into consideration the need for optimal excipients, the level of adjuvant required, the dosage form for stability, preservatives, and container closures. The nature of the interaction (or non-interaction) of the vaccine antigen(s) with or without an adjuvant, needs to be well characterized and maintained at its optimal state to ensure that the immune responses elicited for each of the antigens are consistent through the end of the shelf-life of the vaccine. The individual components of the vaccine must maintain a predictable level of potency (strength) and purity and must be packaged appropriately for long-term storage and distribution for administration to the target populations (*19*).

Formulation Composition Considerations

Serotype Considerations

Glycoconjugates are often multivalent to provide broad coverage against the dominant disease-causing serotypes. For example, the current licensed pneumococcal conjugate vaccines are 10-13 valent and are designed to cover the dominant invasive disease-causing serotypes out of the more than 90 serotypes that have been identified. The composition of the vaccine should be, ideally, based on the epidemiology and disease burden in the countries or regions of the world in which the vaccine is intended for use. However, producing a glycoconjugate vaccine with multiple serotypes tailored for use in different individual countries can become very costly for a manufacturer, as each of the different formulations would have to be licensed for use with multiple production processes. As such, most glycoconjugates are designed with global epidemiology in mind.

The challenge with multivalent glycoconjugate vaccines is their complexity and cost, and thus there are instances where a vaccine targeted to regional needs can have a major impact at an affordable cost. An example is with the meningitis belt in Africa where until recently, the dominant cause of meningitis epidemics was serogroup A meningococcus (*20*). As the multivalent meningococcal vaccines produced by the multinational vaccine manufacturers were not affordable for use in meningitis belt countries, a low-cost monovalent MenA conjugate vaccine (MenAfriVac®) was developed specifically for Africa (*21*, *22*).

Drug Substance—Glycoconjugates

The criteria that define the characteristics of glycoconjugate vaccines are typically well documented and are dependent on the conjugation chemistry and process used in the manufacturing. The critical quality attributes of a glycoconjugate vaccine (*23–25*) that indicate stability and that can be measured in the drug substance are as follows:

a. molecular size of the polysaccharide or oligosaccharide;
b. chemistry for activation of the polysaccharide;
c. choice of carrier protein;
d. saccharide—protein conjugation chemistry;
e. saccharide to carrier protein ratio;
f. levels of free saccharide; and
g. levels of other contaminants such as host protein and nucleic acids, endotoxin and conjugation chemistry by-products.

These critical quality attributes apply to each of the individual conjugates in the vaccine and play a role in the design of the formulation. A combination of multiple conjugates from different serotypes (each with its unique structural characteristics) makes the formulation more complex and difficult to characterize.

Formulation Process

The general guidelines for developing an optimized process for a multivalent glycoconjugate formulation need to take all aspects of components into consideration. These are:

a. understanding the physical and chemical nature of each of the individual glycoconjugates as drug substances intended to be in the vaccine;
b. selection of excipients/container closures /storage; and
c. stability (real time and accelerated temperature for process, freeze thaw if needed, impact of process such variables as agitation).

Excipient Selection

At the pre-formulation stage one must optimize the buffer (concentration and pH) to ensure the conjugates do not aggregate. Stabilizers are added as needed based on data generated on the stability of the individual conjugates at the drug substance level. The role of the excipients is summarized in Table 2. The studies conducted to identify the excipients are generally performed under accelerated stability conditions using a design-based approach using the key stability indicating parameters to obtain a pH range for process operations and to provide guidance to inform the boundaries within which the optimal drug product will be formulated and can be held longer term. The most commonly utilized excipients are those that are considered "generally regarded as safe" (GRAS) excipients, as they have been classified as safe for food consumption (*26–28*).

Given the complexity of the formulation with multiple glycoconjugates from different serotypes, one may need to strategize and drive the formulation studies with informed decisions based on the stability of each of the individual conjugates at higher concentrations.

Table 2. Role of Excipients in Formulations

Excipient	*Function*
Buffer	pH control for solubility and stability
Ionic strength modifier	Isotonicity and stability of the drug substance
Surfactant	Prevents aggregation and increase solubility during process as well as when in contact with multiple surface areas
Bulking agents	Stability and improved appearance of lyophilized product if a liquid vaccine is generally not stable
Stabilizers	Antioxidants, cryoprotectants, etc. to improve stability in solution and/or lyophilized product

The use of single-use (i.e., disposable) technology is becoming increasing popular as there are many advantages for storage of both the polysaccharides and the glycoconjugate bulks. The bulk conjugate drug substances need to be appropriately handled until they are ready to be formulated, sterile filtered, followed by the addition of aluminum (as necessary) and filling of the vaccine. During the process of bulk formulation at the manufacturing scale, the product can go through stress due to exposure to liquid–air, liquid– solid, and liquid–liquid interfaces through mechanical stresses such as stirring and pumping. The rate of aggregation can be impacted by the solution conditions. Physico-chemical analysis of pneumococcal and meningococcal glycoconjugates has demonstrated that the stability of the conjugate is dependent on the conjugation chemistry and the carrier protein (*29*, *30*). Berti et al. (*31*) have shown with meningococcal CRM_{197} conjugates that the insertion of polysaccharides chains alters dramatically the hydrodynamic properties of the protein that can lead to reduced protein hydration with respect to the carrier protein alone, which is much larger than flexibility of the conjugates with respect to a compact macromolecule of the same molecular weight and a strong tendency to aggregate.

The bulk monovalent glycoconjugates are either held frozen or at 2-8°C, depending on the stability of the conjugates and the anticipated time period before drug product formulation. For freezing, a freeze–thaw process needs to be established to ensure the conjugates do not aggregate upon thawing. Studies need to be conducted early in the process development stage to determine the freeze–thaw aspects to ensure they are considered while planning the process. Aggregation of the conjugates can make sterile filtration challenging and lead to loss of product during processing. The data from accelerated stability studies becomes valuable to support the development of process parameters such as mixing speed and time, bulk hold time, and time taken for each unit operation.

Drug Product Formulation Process

The development of the drug product formulation with the multivalent glycoconjugate needs to consider both the formulation components as well as the process design. Some of the key aspects that are critical to define the process are:

- selection of the optimal adjuvant, if needed;
- formulation process design and process flow for formulated bulk and fill-finish;
- definition of process parameters for mixing, order of addition of components and filtration;
- selection of filters based on product recovery;
- over all process recovery—minimum line loses;
- in process assays used during the formulation—fill finish process;
- preservatives for a multi-dose vaccine, if applicable;
- identification of critical quality attributes for the vaccine;
- identification of the delivery system and selection of the optimal route of administration;

- container closures for the vaccine suited for a liquid; and lyophilized, multi-dose vaccine and for a diluent if needed;
- storage / distribution / CTC.

The process flow for the formulation and fill finish operations requires a well-thought process design and well designed, with built in modular systems to accommodate addition of the glycoconjugates with varied chemistries and unique characteristics and the adjuvant if needed. A facility limitation with a smaller manufacturing space or footprint may have to be considered to effectively manage the process design, prefiltered formulation and final formulated bulk vessels and ensure that the orderly addition and uniform mixing of multiple components and filtration are appropriate.

Pre-Formulation

The rational approach to design a formulation from the pre-formulation stage to licensure is depicted pictorially in Figure 1.

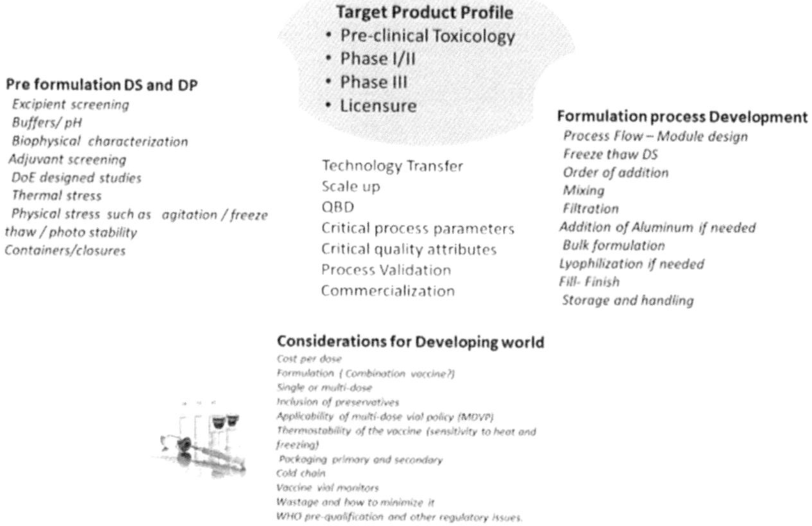

Figure 1. Rational approach to vaccine formulation development.

Pre-formulation or early drug product development includes bulk drug substance formulation, final dosage form development, and process and fill-finish. This is done early in development in parallel with the conjugation process development to help guide the next step in manufacturing the product. If the number of serotypes in the vaccine increases, the complexity of the formulation also increases because the product development and stability is dependent on an understanding of the individual conjugates in the vaccine. The least stable conjugate becomes the driver for the final presentation of the drug product as a liquid or lyophilized vaccine. Optimizing the formulation should be based on a designed approach using first principles that focus on the least stable components to prevent losing time in the development process until the critical quality attributes can be established with clinical studies. The role of biophysical tools in developing a vaccine formulation is illustrated in Figure 2.

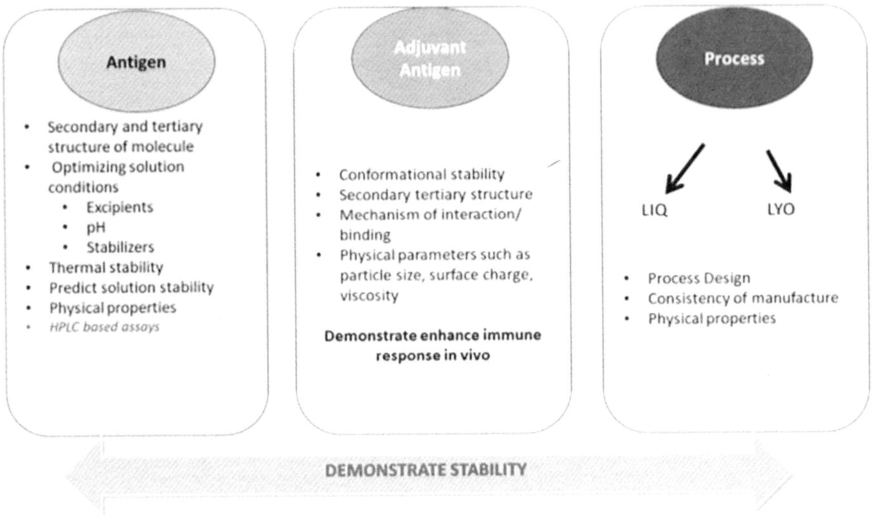

Figure 2. Pre-formulation characterization of vaccines.

Liquid versus Lyophilization

In general, the preference is to formulate glycoconjugate vaccines as a liquid rather than a lyophilized product unless the stability of one or more of the conjugate vaccine components is in question or there is a need for controlled temperature shipments. Freeze-drying, or lyophilization, is widely used for biopharmaceuticals to improve the long-term storage stability of temperature labile molecules. Lyophilization enhances the stability of glycoconjugates and in addition may offer an opportunity to store the vaccine under ambient conditions, which can be helpful in developing countries where maintenance of the cold chain for storage and transportation to remote settings can be difficult (see MenAfriVac case study). The downside to a lyophilized vaccine formulation is that it requires a separate diluent, which increases the burden on packaging, shipment, storage, and mixing with diluent prior to administration. Stability data at the earlier stages of formulation development inform whether a lyophilized formulation needs to be considered. The formulation then requires the screening of excipients suitable for lyophilization as well as stability of the conjugates. Considerations for a lyophilized formulation are shown in Table 3.

Table 3. Considerations for Developing a Lyophilized Glycoconjugate Formulation

	Parameters	*Considerations*
1.	Dosage	The concentrations of a conjugate vaccine are very low, typically in the 1-10 μg range per serotype.
2	Buffer selection	Buffers that do not show a pH drift on freezing are generally preferred.
3	Bulking agent /Stabilizers	Sugars used as bulking agents and/or stabilizers need to be at concentrations that impart stability and good cake cosmetics together with minimal reconstitution time .
4	Lyophilization matrix	The thermal properties of the formulation impact the development of the lyophilization cycle as the excipients require different temperatures, sublimation rates, and processing steps and duration. The cycle parameters are determined by the thermal properties, which can be investigated by use of differential scanning calorimetry (DSC) or freeze microscopy.
5	Lyophilization cycle parameters	Optimization of the lyophilization cycle parameters such as freezing, annealing, primary drying, and secondary drying process parameters that balances the product stability, cake cosmetics, and moisture levels. The ideal range for moisture content must be determined and maintained during storage to prevent degradation, particularly for formulations without stabilizers.
6	Container Closures	The optimal container closures depend on whether the vaccine will be for delivering a single dose or multi-dose with minimal loses.
7	Analytics	Quantitation or strength assays for the individual glycoconjugates in the vaccine may require immunoassays as the presence of multiple polysaccharides makes it difficult to use sugar-based chemical assays.
8	Diluent for reconstitution	Selection of the diluent for reconstitution needs to be considered and it may be an aluminum adjuvant, WFI, or saline. It is important to consider the final osmolality of the formulation when selecting a diluent in order to maintain isotonicity.
9	Post-reconstituted vaccine stability	The stability of the reconstituted vaccine must be assessed to determine how long the vaccine can be held after reconstitution. There are guidelines provided by the CDC and WHO on how long the vaccine can be held post reconstitution.

The drug product formulation is optimized in a series of experiments, typically using the Design of Experiment (DOE) or a Quality by Design (QbD) approach, although discrete studies may also be conducted as appropriate, and the resulting samples are assessed under accelerated stress conditions to select the most stable and soluble formulations. Production of a vaccine formulation containing multiple glycoconjugates is challenging because of the inherent unknown stability of these complex molecules as they go through the manufacturing process, especially at the low concentrations that are present in the final product.

Each of the intermediates, such as the polysaccharides and the drug substances (individual conjugates), must be released prior to use. This translates into multiple GMP process steps for each of the serotypes. After manufacturing of all the bulk glycoconjugates, the vaccine is then formulated with or without an adjuvant based on the defined stability. The drug product must then be tested and released based on criteria defined for the vaccine.

Adjuvants

Not all glycoconjugate vaccines require an adjuvant for eliciting optimal immune responses. Indeed, none of the three licensed quadrivalent meningococcal conjugate vaccines contain an adjuvant. The need for an adjuvant in a formulation typically must be demonstrated in clinical studies for immunogenicity or as an excipient required for the stability of the formulation. Adjuvants have been used in vaccines for more than 90 years and in the early days their addition was based on an empirical approach. Today adjuvants in vaccine formulations are tailored to obtain the desired clinical immune response outcome (*32–35*). Historically, adjuvants were often added based upon preclinical data, but more recently, regulators have insisted that the benefit be demonstrated in clinical studies, and at least with glycoconjugates, the immunological benefit of the adjuvant in preclinical studies is often not replicated in clinical studies.

There are only a handful of adjuvants that have been licensed for human use and for glycoconjugates it is currently limited to aluminum salts (aluminum phosphate or aluminum hydroxide).

Indeed, aluminum salts are the most commonly used and well accepted of all the adjuvants in human vaccines. This is primarily due to their excellent safety record over more than ~70 years of use in a wide variety of childhood vaccines. The adjuvant most often used in the currently licensed glycoconjugate vaccines is aluminum phosphate.

Despite intense activity in the field of adjuvant research and development, there remain a relatively small number of adjuvants used clinically. The main challenge for new formulations to enter clinical practice is that the safety requirements for prophylactic vaccines are extraordinarily high, particularly for vaccines aimed for use in healthy infants. However, as more and safer adjuvants are being development and used in commercial vaccines, the general acceptance of vaccine adjuvants appears to be increasing. The documented human clinical studies conducted with different adjuvants other than aluminum for glycoconjugates have been in early clinical trials using QS 21, MF59,

Monophosphoryl lipid A (MPL) and synthetic oligonucleotides CpG but have not been successful enough to be pursued to licensure (*36–39*).

Aluminum in Glycoconjugate Vaccines

The physical properties of aluminum salts have been well studied. The process for producing aluminum phosphate can play a role in determining the properties of the bulk salt as well as its behavior when added to the glycoconjugates. The differences in physical properties greatly influence the interaction of the vaccine antigens with different aluminum salts. Binding of the conjugates to aluminum occurs through the protein moiety. Aluminum adjuvants such as aluminum hydroxide may bind to the conjugates so tightly that it may be difficult for the right epitopes or antigens to be available to the immune cells, and that may impact the stability for certain glycoconjugates. The formulation process, excipients, amount of aluminum, presence of multiple conjugates with varied levels of saccharide to protein ratios, surface charge of the antigen, inherent large molecular size contributing to steric factors, and tertiary structures all play a role in the level of binding of each of the conjugates. For vaccines containing aluminum, the level of binding of each of the conjugates to the aluminum matrix must be measured to ensure lot-to-lot consistency.

Depending on the formulation conditions and the associated electrostatic interactions of the adjuvant particles, the vaccine suspension may transition between flocculated and deflocculated states. The impact of practical formulation parameters, including pH, ionic strength, and the presence of model antigens, has been correlated to the sedimentation behavior of aluminum phosphate suspensions. A novel approach for the characterization of suspension properties of Alum has been developed to predict the flocculated state of the system using a sedimentation analysis-based tool. The manufacturing can be based on:

a. Aseptic processing: The process involves formulating batches of 4-5 glycoconjugates from different serotypes with excipients and aluminum phosphate to dose and then aseptically mixing the different batches together to formulate the final multivalent vaccine with all the intended serotypes, blending and filling the formulated vaccine. This process may or not lend itself to a completely closed manufacturing process.
b. Closed process: The other option is to develop a completely closed process where the buffer /excipients and conjugates are added to the formulation vessel, filtered into another vessel, and this is followed by the addition of the aluminum phosphate mixing and filling of the formulated vaccine.

Aluminum-containing vaccines are particulate and hence require that the product be mixed well during the filling process to ensure content uniformity. If interruptions occur during the filling process and the filling must be stopped, then the aluminum can settle in the lines or filling needles. The process of mixing in large formulation vessels can utilize top or bottom mounted impellers; in addition, the product can be recirculated while it is being filled into vials or prefilled syringes. As larger formulation volumes require a longer filling time, the product is exposed to additional shear due to the range of solid–liquid interfacial shear forces, which in turn is impacted by the shear sensitivity of the product. Excessive mixing can lead to aggregation of the conjugates on the aluminum. The particles of aluminum with the active components can be reduced in size, which can result in a change in the electrostatic interactions or flocculation behavior, morphology, and the surface properties (charge, viscosity, surface tension, etc.). This can result in change in the visual characteristics of the product and its resuspension behavior (*40–46*). The measurement of particle size and other physical parameters during mixing can help determine the optimal mixing conditions and the impact of shear on the aluminum particles. During the filling process, the uniformity of the fill must be measured as a surrogate for content uniformity. The turbidity associated with the aluminum can be used as a measurement of uniformity, as it is dependent on the concentration of the aluminum. Vials filled with the vaccine can be taken at defined intervals and monitored and the turbidity levels in the bulk vaccine versus the filled vials. While developing a formulation process, one can utilize multiple biophysical tools to understand the physical characteristics of the aluminum-containing vaccines to develop a robust tool to monitor the process. Examples of the tools and their applications during process development are shown in Figure 3. Each container in each lot should be inspected visually (manually or with automatic inspection systems), and those showing abnormalities (such as improper sealing, lack of integrity, and, if applicable, clumping or the presence of particles) should be discarded. The visual observation of an aluminum-containing vaccine is extremely important prior to delivery for immunization. When the vaccine is in storage the aluminum in the vaccine will settle in the container closure (i.e., vial) or a prefilled syringe. The general instruction provided in the product inserts or by Center for Disease Control (CDC) is that prior to immunization, one must shake the vial or the prefilled syringe to uniformly resuspend the vaccine to ensure homogeneity and ensure there are no particulates prior to immunization (*47*).

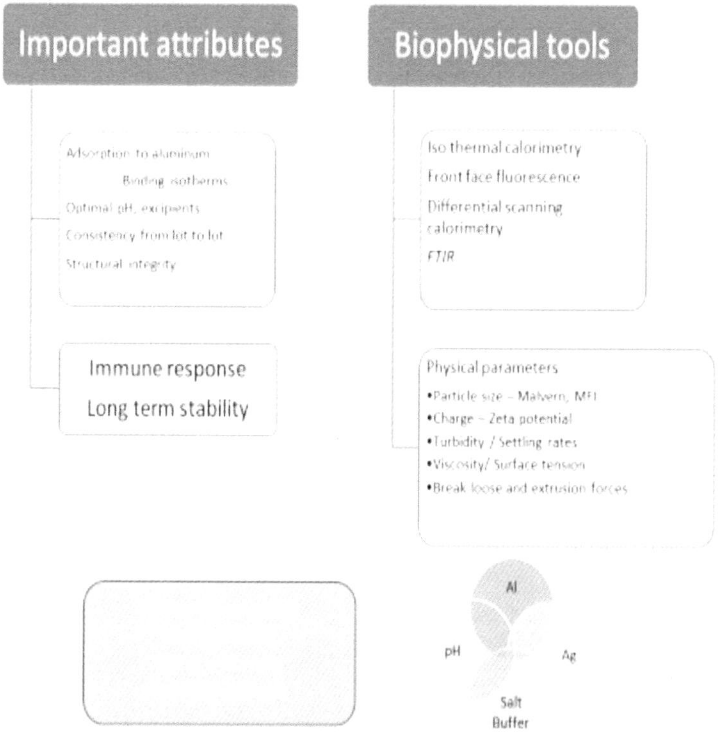

Figure 3. Biophysical tools to develop an aluminum-based formulation for conjugate vaccines.

Preservatives

The general choice of preservatives in WHO-qualified vaccines has been Thimerosal. MenAfriVac is a lyophilized multi-dose (10 doses per vial) vaccine. The preservative Thimerosal is added at 0.01% (vol/vol) to the diluent used to reconstitute the vaccine (*43*). More recent work conducted for the development of Prevenar 13® as a multi-dose for supply to the Global Alliance for Vaccines and Immunization (GAVI) countries has shown 2-phenoxyethanol (2-PE) as an optimal preservative for pneumococcal conjugate vaccines. The studies showed that the dose of Thimerosal used as a preservative did not meet European Pharmacopoeia antimicrobial effectiveness acceptance criteria. The rate of growth inhibition of Thimerosal compared to 2-PE on *Staphylococcus aureus*, a resilient organism in these tests, was significantly slower in single and multi-challenge studies. These results indicate that 2-PE provides a superior antimicrobial effectiveness over Thimerosal for this vaccine formulation (*48–52*). Synflorix® was initially provided as a two-dose vial without preservatives (*53*), but recently a four-dose vial containing 2-PE was WHO prequalified. The typhoid conjugate vaccine Typbar-TVC also contains 2-PE as a preservative (*54*).

The manufacturer has a choice of possible preservatives and consideration should be given to the stability of the chosen preservative and possible interactions between the vaccine components and the preservative. The concentration of the preservative must be approved by the local National Regulatory authority (NRA) and should be assessed in regulatory toxicology studies prior to human clinical studies.

Quality Control of Conjugate Vaccines (Drug Product) and Stability

During the development of a glycoconjugate vaccine an analytical control strategy needs to be considered and methods need to be developed to monitor stability or changes that could potentially impact the functionality of the vaccine. Glycoconjugate vaccines must follow guidelines that are defined and meet the expectations for the control of product by various agencies such as the FDA, WHO, and International Conference on Harmonization (ICH) (*55–59*). These authorities provide specific guidance on the selection of batches, the stability-indicating assays, storage conditions, testing frequency, labeling, test procedures and criteria, specifications, long-term and real-time stability assessment, stress and accelerated testing, etc.

The polysaccharide component of conjugate vaccines might be subject to gradual hydrolysis at a rate which may vary depending upon the type of conjugate, the type of formulation or adjuvant, the type of excipients, and conditions of storage. The hydrolysis may result in reduced molecular size of the polysaccharide component, an increase in free polysaccharide, or change in the molecular size of the conjugate. Stability studies need to be conducted as a part of the vaccine development to arrive at the projected shelf life for the product. Ideally, stability of the drug product should be monitored at its real time storage condition from at least three lots formulated from different independent bulk conjugates.

During the development of the product, a control strategy needs to be in place to define and ensure a robust manufacturing process. The critical quality attributes of the vaccine must be defined, and the manufacturing process can be controlled with the appropriate analytical control strategy. The critical quality attributes are linked to clinical performance through the different stages of clinical trials. The level of free saccharide in a glycoconjugate vaccine is an important quality attribute. Hence controlling the level of free saccharide in a formulation through the shelf life of the product, together with clinical performance, are important aspects of formulation development. The range of free saccharide in the different drug substance lots used to formulate multiple drug product lots through the development process from preclinical toxicology stage through Phase 1 to Phase 3 clinical studies will provide guidance on the range of acceptable levels of free saccharide in the individual serotypes. A similar approach can be used to define the molecular size of the individual conjugates in the vaccine used for formulating the clinical lots. The final specifications for a product at licensure are derived from analyses of multiple lots produced during the development process. A formulation of a glycoconjugate vaccine with multiple components may be difficult to place in the class of well-characterized product. The use of chemical or high pressure liquid chromatography (HPLC)-based assays is not feasible because of the low dose levels with multivalent glycoconjugate vaccines and immunoassays are therefore typically used for identity, quantitation, and stability. The immune assays that are developed should be based on using reagents, such as monoclonal antibodies or polyclonal antisera, that are highly serotype specific and can be consistently made in sufficient quantities to be used through the shelf life of the product.

Preclinical Models

Preclinical animal models have an important but limited role in the development of glycoconjugate vaccines as follows:

- Limited or lack of suitable animal models may or may not have the sensitivity to detect product changes based on an immune response.
- Often the dose levels that need to be used for immunization are very low, so that can differentiate a response while comparing various formulations, and these may or may not be relevant to humans.
- Early preclinical studies can, however, provide guidance in terms of the conjugation chemistry and/or formulation.

The analytical control strategy for a glycoconjugate vaccine drug product is shown in Table 4.

Developing country manufacturers often face challenges in their ability to conduct some of the more complex analytical methods such as size exclusion MALS (multi angle light scattering), measuring particle size, and/or obtaining reagents such as monoclonal or polyclonal antisera for assays.

Risk assessment of the quality and process attributes can be performed at different stages of the clinical trials where the formulation/ manufacturing/ analytical teams can work together to leverage their understanding to identify gaps and address them as needed to optimize the process and identify critical attributes. If the critical attributes for both the product and the process can be identified early in development, it will help to define the control strategy, which in turn will help to define the formulation taking the manufacturing into consideration. When possible, animal studies can be used to study the impact of the process changes to support the development of a design space for the formulation. The process consistency with controlled attributes, in conjunction with clinical performance, becomes important in defining the final vaccine product.

Table 4. Control of Product/Tests for Vaccine Drug Product Containing Glycoconjugates

Quality Attributes	***Methods***
Description Physicochemical characteristics	
pH	If the vaccine is a liquid preparation, the pH of each final lot should be tested and shown to be within the range of values found for vaccine lots shown to be safe and effective in the clinical trials and in stability studies. For a lyophilized preparation, the pH should be measured after reconstitution with the appropriate diluent.
Appearance	Visual comparison to reference for clarity or turbidity levels.
Osmolality	The osmolality should be consistent from lot to lot.
Identity An identity test should be performed that demonstrates that all of the intended polysaccharide serotypes and carrier protein(s) are present in the final product.	Immunoassays specific to each of the components Polysaccharide (s) (serotype specific) Carrier protein(s).
Strength (content) of individual polysaccharides The amount of each pneumococcal polysaccharide in the final containers should be determined and shown to be within the specifications agreed to by the NRA.	Product specific and might include chromatographic or serological methods. Immunological assays such as rate nephelometry or ELISA using serotype specific polyclonal or monoclonal antibodies.
Free sugar	Based on immunoassays or HPLC Methods and requirements based on NRA.
Molecular size	Size exclusion HPLC /Multi angle light scattering or in combination with immune assays.
Concentration of total carrier protein	Chemical assay such as Folin Lowry or bicinchoninic acid assay (BCA) If there is more than one carrier protein, then one may need an immune assay to quantitate each of the proteins.
Purity	(may need to leverage data obtained from the drug substance) Methods to monitor levels of free saccharide may need to be developed.
Moisture (lyophilized product)	2.5% and no vial should be found to have a residual moisture content of 3% or greater.

Continued on next page.

Table 4. (Continued). Control of Product/Tests for Vaccine Drug Product Containing Glycoconjugates

Quality Attributes	*Methods*
Percent antigen or serotype bound to aluminum as applicable	Method agreed by NRA.
Adjuvant Content	If aluminum compounds are used as adjuvants, the amount of aluminum should not exceed 1.25 mg per single human dose method agreed by NRA.
Endotoxin	Compendial test methods Endotoxin content or pyrogenic activity should be consistent with levels found to be acceptable in vaccine lots used in clinical trials and approved by the national control authority.
General safety/ abnormal toxicity	Compendial methods as per local NRA requirements
Dose or content uniformity	Compendial
Deliverable dose volume	Compendial
Preservative content	Method agreed by NRA
Preservative effectiveness agreed by NRA	Compendial
Animal immunogenicity / potency assays	As applicable to the vaccine
Guideline: WHO Recommendations to assure the quality, safety and efficacy of pneumococcal conjugate vaccines-proposed replacement of TRS 927, Annex 2, ECBS, 19 to 23 October 2009 http://www.who.int/biologicals/publications/trs/areas/vaccines/pneumo/en/ March 23, 2018	

Container Closures/Vaccine Presentation

The preferred characteristics for a vaccine that is targeted for the LMICs are shown in Table 5.

Table 5. Preferred Product Characteristics for a Glycoconjugate Vaccine

Container closures	• Unit dose as prefilled syringes or single dose vials or disposable units. Prefilled syringes are expensive and increase the cost of secondary packaging and shipping the vaccine. Prefilled syringes should also have auto disable capability. • Alternate choices such as the BD Uniject™ auto-disable pre-fillable injection system is an all-in-one, auto-disable (nonsyringe) drug delivery system for intramuscular (IM) or subcutaneous injections. • Materials for delivery devices, primary containers and secondary and tertiary packaging that minimize environmental impact of waste disposal is preferred.
Lyophilized products	• Lyophilized products will require a diluent (water for injection, saline based or with adjuvant) for reconstitution prior to immunization.
Dose volume	• The dose of volume of less than 1 mL is preferred for infants and children younger than 1 year of age.
Route of immunization	• Intramuscular, intradermal and sub cutaneous

Multi-Dose Vials

To deliver vaccines to the developing world in a cost-effective efficient manner with minimum waste, it is important to consider a multi-dose vial format. Single-dose vaccine formats (vials and pre-filled syringes) can prevent clinic-level vaccine waste but may incur higher production, medical waste disposal, and storage costs than multi-dose formats. The vaccine presented for WHO prequalification should be adequately preserved (WHO/EPI). The preservative efficacy should be tested using the methodology described in the European Pharmacopoeia (Ph Eur) (a challenge test lasting 28 days with specified microbes) and should demonstrate compliance with the "B" criteria of acceptance, or if justified, the criteria stated in the Ph. Eur monograph "Vaccines for Human Use." The criteria for the use of the vial is outlined by WHO (*47*). All opened WHO-prequalified multi-dose vials of vaccines should be discarded at the end of the immunization session, or within six hours of opening, whichever comes first, unless the vaccine meets all six of the criteria listed as follows. If the vaccine meets these four criteria, the opened vial can be kept and used for up to 28 days after opening:

a. The vaccine is currently prequalified by WHO.
b. The vaccine is approved for use for up to 28 days after opening the vial, as determined by WHO.
c. The expiration date of the vaccine has not been passed.
d. The vaccine vial has been, and will continue to be, stored at WHO- or manufacturer- recommended temperatures.
e. The vaccine vial monitor, if one is attached, is visible on the vaccine label.
f. The vaccine has not been damaged by freezing.

Multi-dose vials should preferably contain a preservative that meets the requirement of the pharmacopoeia and can be shown to be stable and effective through the shelf life of the product. In addition, it is important that the preservative does not negatively impact the stability of the vaccine.

Shelf Life of Current Marketed Glycoconjugate Vaccines

The shelf life of the product is defined based on the stability data obtained from multiple lots of the same vaccine manufactured consistently and well controlled for its quality attributes and demonstrate comparability. The stability of the vaccine is the most important concern throughout the development process, from the early research phase through to the licensure of the vaccine. During early formulation development, accelerated stability studies are conducted to define the stability profiles and direct the optimization of a formulation. Although the most acceptable storage temperature is 2-8°C, stability at 25°C and 40°C for short periods of defined time help can ensure the vaccine is stable enough to support its delivery during immunization as well as temperature excursions that may occur during shipments. During the commercial manufacturing process (inspection, labeling, packaging, and shipment), the general preferred temperature for processing is room temperature. Aluminum-containing vaccines cannot be subjected to freezing temperatures, as this can lead to aggregation of the aluminum and affect its characteristics of binding to protein antigens (*24*, *55*, *60–62*). The shelf life and temperature stability and freeze sensitivity of some conjugate vaccines (*55*) are shown in Table 6.

Table 6. Stability Profile of Glycoconjugate Vaccines

Vaccine	*Commercial Name*	*Storage temperature °C*	*Shelf life (years)*	*Freeze sensitivity*	*Accelerated stability*
Meningococcal ACYW	Menactra® Liquid	2-8	1	Yes	
	Menveo® Lyo	2-8	2	Yes	
MenA	MenAfriVac®	2-8	3	No	40°C 4 weeks 25°C 6 months
Haemophilus influenza type B	Pedvax®	2-8	3	Yes	
	Vaxem-Hib®	2-8			
Pneumococcal	Synflorix®	2-8	3	Yes	37°C – 1 day 25°C – 3 days
	Prevenar 13®	2-8	3	Yes	40°C – 4 days 25°C – 3 days
Reference (*63*)					

Case Studies for Vaccines Targeted for LMICs

Case Study 1: MenAfriVac

The Meningococcal Serogroup A conjugate vaccine MenAfriVac® was the first vaccine developed specifically for African populations. MenAfriVac was developed through the WHO-PATH Meningitis Vaccine Project, manufactured by the Serum Institute of India Ltd and licensed in 2010. With a target price of no more than USD $0.50 per dose and a target population within the African meningitis belt, a number of parameters were considered when designing and manufacturing the vaccine in order to meet these population and price targets. These included the yield of polysaccharide and efficiency of conjugation, formulation, along with need for an aluminum adjuvant, whether to use liquid versus lyophilized, or multi-dose and therefore preservative. Stabilizers and bulking agents incur cost but are necessary for lyophilization, but certain agents are more cost-effective than others (sucrose vs. mannitol). Lot release testing of the different components of the meningococcal group A conjugate was performed following the WHO guidelines (2006 WHO recommendations for production of group A meningococcal conjugate vaccines).

The "controlled temperature chain" (CTC) has been an innovative approach to vaccine management allowing vaccines to be kept at temperatures outside of the traditional cold chain of +2°C to +8°C for a limited period under monitored and controlled conditions, as appropriate to the stability of the antigen. A CTC typically involves a single excursion of the vaccine into ambient temperatures not exceeding +40°C and for the duration of a specific number of days, just prior to administration. The World Health Organization has established programmatic criteria for a vaccine to be labeled for and used in a CTC (*64, 65*).

MenAfriVac can be used in a controlled temperature chain for up to four days at ambient temperatures not exceeding 40°C, which is an important consideration for some of the remote target populations in Africa. A CTC is initiated immediately prior to administration, provided that the vaccine has not reached its expiration date and the vaccine vial monitor is still valid. MenAfriVac is the first vaccine to be approved for use in a controlled-temperature chain (CTC), allowing the vaccine to be kept at a broader range of temperatures than the traditional cold chain for a limited period of time under monitored and controlled conditions (*66–69*).

Case Study 2: Prevnar

Prevnar (original 7 valent PCV) was originally targeted to those pneumococcal serotypes causing the most invasive disease in North America and Europe but had more limited coverage for other parts of the world, like Africa. The advance market commitment (AMC), designed to encourage manufacturers to develop and provide low cost PCVs for GAVI-supported countries, developed a target product profile (TPP) that specified serotypes 1, 5, and 14 and 60% regional coverage, and hence Prevnar didn't fulfill the TPP requirements because it lacked serotypes 1 and 5. Prevenar 13®, which included serotypes 1 and 5, was licensed in 2010 in single-dose prefilled syringes by Wyeth Pharmaceuticals (*70, 71*) and subsequently it became WHO prequalified. However, since 2010, more than 40 low- and lower middle-income countries have launched pneumococcal immunization programs with Prevenar 13®, via the AMC. In mid-2013, the price per dose was reduced from $3.50 to $3.40 and again in 2014, from $3.40 to $3.30. A four-dose, preservative (2-PE) containing vial presentation of Prevenar 13® was WHO prequalified in 2016, which allowed the global use of Prevenar 13® multidose vial (MDV) by United Nations agencies and countries worldwide that require WHO prequalification. This also further reduced the price to $3.05 per dose. Filling four doses of the vaccine in a single vial configuration quadrupled the number of doses that could be delivered using the same packaging as the single dose vials. This helped to minimize waste of the vaccine. The multi-dose presentation of Prevenar 13® offers significant benefits to developing countries, including a 75 percent reduction in the need for temperature-controlled supply chain requirements, United Nations Children's Fund (UNICEF) shipping costs, and storage requirements at the national, regional, district, and community levels (*72–75*).

Summary

Among the current licensed vaccines, the production of multivalent glycoconjugate drug products are the most complex. The challenges are not limited to manufacturing but also to facilities to handle the production of multiple components such as the polysaccharides, conjugate drug substances, and adjuvants in a timely manner for formulation of the multicomponent product in the presence of an adjuvant. To date only limited glycoconjugate vaccines have been licensed by developing country manufacturers, and these are all monovalent conjugates that include Hib, meningococcal, and typhoid vaccines. Indeed, no multivalent vaccines have yet been licensed by developing country manufacturers, although many are in development in China, India, Brazil etc. To enhance the production of vaccines in LMIC, it is important that the infrastructure of manufacturing facilities be carefully monitored, and this would include the facility design and capacity for manufacturing (drug substance, drug product, and adjuvant if needed) for the required number of doses required followed by appropriate temperature controlled storage areas, packaging, shipping, handling, and distribution.

It is also important to design new glycoconjugate vaccine formulations with attributes and production processes that will help to mitigate challenges that immunization programs face as they introduce new vaccines. To make true inroads, contribute to global health, and ensure an adequate supply of vaccines to the countries that need it most, there are requirements not only concerning monetary commitment, but also about providing support toward developing manufacturing capabilities and infrastructure, along with building the quality systems required to maintain consistent and efficient production processes.

References

1. World Health Organization. *The Global Burden of Disease: 2004 Update*. http://www.who.int/healthinfo/global_burden_disease/2004_report_update/en/ (accessed April 20, 2018).
2. Vella, M.; Pace, D. *Expert Opin. Biol. Ther.* **2015**, *15*, 529–46.
3. Pang, T. Vaccination in Developing Countries: Problems, Challenges and Opportunities, Vol II; *Global Perspectives in Health*, 2011.
4. Bloom, D. E.; Canning, D.; Weston, M. *World Economics* **2005**, *6*, 15–39.
5. United Nations. *Millennium Development Goals Report 2012*. www.un.org/millenniumgoals/pdf/MDG%20Report%202012.pdf (accessed April 20, 2018).
6. Gavi, The Vaccine Alliance. *Progress Report 2012*. https://www.gavi.org/library/publications/gavi-progress-reports/gavi-alliance-progress-report-2012/ (accessed February 16, 2018).
7. Barnighausen, T.; Bloom, D. E.; Canning, D.; Friedman, A.; Levine, O. S.; O'Brien, J.; Privor-Drumm, L.; Walker, D. *Vaccine* **2011**, 2371–2380.
8. *Encyclopedia Britannica*. Gates Foundation. https://www.britannica.com/topic/Gates-Foundation (accessed April 16, 2018).
9. MacLennan, C. A. *Semin. Immunol.* **2013**, *25*, 114–123.

10. Holmgren, J.; Svennerholm, A. M. *Curr. Opin. Immunol.* **2012**, 343–353.
11. Sow, S. O.; Okoko, B. J.; Diallo, A.; Viviani, S.; Borrow, R.; Carlone, G.; Tapia, M.; Akinsola, A. K.; Arduin, P.; Findlow, H.; Elie, C.; Cheick Haidara, F.; Adegbola, R. A.; Diop, D.; Parulekar, V.; Chaumont, J.; Martellet, L.; Diallo, F.; Idoko, O. T.; Tang, Y.; Plikaytis, B. D.; Kulkarni, P. S.; Marchetti, E.; LaForce, M.; Preziosi, M. P. *New Eng. J. Med.* **2011**, 2293–2304.
12. Meningitis Vaccine Project. *Announcement: Dr. Marie-Pierre Preziosi to be new director of Meningitis Vaccine Project*; Seattle/Ferney-Voltaire/Geneva; March 1, 2012. https://www.meningvax.org/files/PreziositoleadMVP_1March2012_En.pdf (accessed June 20, 2018).
13. Novak, R. T.; Kambou, J. L.; Diomande, F. V.; Tarbangdo, T. F.; Ouedraogo-Traore, R.; Sangare, L.; Lingani, C.; Martin, S. W.; Hatcher, C.; Mayer, L. W.; LaForce, F.; Avokey, F.; Djingarey, M. H.; Messonnier, N. E.; Tiendrebeogo, S. R.; Clark, T. A. *Lancet Infectious Diseases* **2012**DOI:10.1016/S1473-3099(12)70168-8.
14. Kristiansen, P. A.; Diomande, F.; Ba, A. K.; Sanou, I.; Ouedraogo, A. S.; Ouedraogo, R.; Sangare, L.; Kandolo, D.; Ake, F.; Saga, I. M.; Clark, T. A.; Misengades, L.; Martin, S. W.; Thomas, J. D.; Tiendrebeogo, S. R.; Hassan-King, M.; Djingarey, M. H.; Messonnier, N. E.; Preziosi, M. P.; LaForce, F. M.; Caugant, D. A. *Clin. Infect. Dis.* **2012**DOI:10.1093/cid/cis892.
15. Goldblatt, D. *Clin. Exp. Immunol.* **2000**, *119*, 1–3.
16. Eby, R. In *Vaccine Design-The Subunit and Adjuvant Approach*; Powell, M. F.; Newman, M. J., Eds; Plenum Press: New York, 1995; Vol. 6, pp 695–718.
17. Edwards, K. M.; Griffin, M. R. *N. Engl. J. Med.* **2003**. *Promises and Challenges of Pneumococcal Conjugate Vaccines for the Developing World.* https://scholar.google.com/scholar?q=N+Engl+J+Med,+2003,+Promises+and+Challenges+of+Pneumococcal+Conjugate+Vaccines+for+the+Developing&hl=en&as_sdt=0&as_vis=1&oi=scholart (accessed May 24, 2018).
18. Lipsitch, M. *Emerg. Infect. Dis.* **1999**, *5*, 336–345.
19. Harrison, L. H.; Trotter, C. L.; Ramsay, M. E. *Vaccine* **2009**, *27*, B51–B631.
20. Frasch, C. E.; Preziosi, M. P.; LaForce, F. M. *Hum. Vaccin. Immunother.* **2012**, *8*, 715–724.
21. Greenwood, B. *Epidemiol. Infect.* **2007**, *135*, 703–705.
22. Maiden, M. C.; Ibarz-Pavon, A. B.; Urwin, R.; Gray, S. J.; Andrews, N. J.; Clarke, S. C.; Walker, A. M.; Evans, M. R.; Kroll, J. S.; Neal, K. R.; Ala'aldeen, D. A.; Crook, D. W.; Cann, K.; Harrison, S.; Cunningham, R.; Baxter, D.; Kaczmarski, E.; Maclennan, J.; Cameron, J. C.; Stuart, J. M. *J. Infect. Dis.* **2008**, *197*, 737–743.
23. Frash, C. *Analytical and Manufacturing Challenges: Preparation of Bacterial Glycoconjugates, Presentation at Vaccine Technology II*, Albufeira, Portugal, June 1–6, 2008.
24. Frasch, C. E. *Vaccine* **2009**, *27*, 6468.
25. Ozan, S.; Kumru, A.; Sangeeta, B.; Joshi, A.; Dawn, E.; Smith, B.; Russell, C.; Middaugh, A.; Prusik, T.; David, B.; Volkin, B. *Biologicals* **2014**, *42*, 237–259.

26. U.S. Department of Health and Human Services. FDA/Center for Drug Evaluation and Research, Office of Policy for Pharmaceutical Quality; *Inactive Ingredient Search for Approved Drug Products*, Data through March 5, 2018, Database Last Updated: April 6, 2018. http://www.accessdata.fda.gov/scripts/cder/iig/index.cfm (accessed Jan. 10, 2018).
27. Kamerzell, T. J.; Esfandiary, R.; Joshi, S. B.; Middaugh, C. R.; Volkin, D. B. *Adv. Drug Delivery Rev.* **2011**, *63*, 1118–1159.
28. U.S. Food and Drug Administration. *Inactive Ingredient Search for Approved Drug Products*. https://www.accessdata.fda.gov/scripts/cder/iig/index.cfm (accessed Jan. 11, 2018).
29. Gao, F.; Lockyer, K.; Burkin, K.; Crane, D. T.; Bolgiano, B. *Hum. Vaccin. Immunother.* **2014**, *10*, 2744–2753.
30. Ho, M. M.; Lemercinier, X.; Bolgiano, B.; Crane, D.; Corbel, M. J. *Biotechnol. Appl. Biochem.* **2001**, *33*, 91–98.
31. Berti, F.; Costantino, P.; Fragai, M.; Luchinat, C. *Biophys. J.* **2004**, *86*, 3–9.
32. Di Pasquale, A.; Preiss, S.; Da Silva, F. T.; Garçon, N. *Vaccines (Basel)* **2015**, *3*, 320–343.
33. Marrack, P.; McKee, A. S.; Munks, M. W. *Nat. Rev. Immunol.* **2009**, *9*, 287–293.
34. The European Medicines Agency. *Guideline on Adjuvants in Vaccines for Human Use*, January 2005. http://www.ema.europa.eu/docs/en_GB/document_library/Scientific_guideline/2009/09/ WC500003809.pdf (accessed Mar. 9, 2018).
35. Søgaard, O. *Human Vaccines* **2011**, *7*, 276–280.
36. Sogaard, O. S.; Lohse, N.; Harboe, Z. B.; Offersen, R.; Bukh, A. R.; Davis, H. L.; Schonheyder, L.; Ostergaard, L. *Clin. Infect. Dis.* **2010**, *51*, 42–50.
37. Vernacchio, L.; Bernstein, H.; Pelton, S.; Allen, C.; MacDonald, K.; Dunn, J.; Duncan, D. D.; Tsao, G.; LaPosta, V.; Eldridge, J.; Laussucq, S.; Ambrosino, D. M.; Molrine, D. C. *Vaccine* **2002**, *20*, 3658–3667.
38. Rutgers. *Pneumococcal Vaccine Induces Potent Immune Response in the Elderly*. http://sci.rutgers.edu/forum/showthread.php?t=4298 (accessed May 5, 2010).
39. Garcon, N.; Pascquale *Hum. Vaccin. Immunother.* **2017**, *13*, 19–33.
40. Rowe, R. C.; Sheskey, P. J.; Owen, S. C. *Handbook of Pharmaceutical Excipients*fifth ed.; Pharmaceutical Press: Grayslake, IL, 2006. www.gmpua.com/RD/RD/HandbookPharmaceutical%20Excipients.pdf (accessed May 1, 2018).
41. Lindblad, E. B. *Vaccine* **2004**, 3658–3668.
42. Hem, S. L.; White, J. L. *Pharm. Biotechnol.* **1995**, 249–276.
43. Hem, S. L.; Hogenesch, H. *Expert Rev. Vaccines* **2007**, 685–698.
44. Olatomirin, O.; Kolade, W. J.; Tengroth, C.; Green, K. D.; Bracewell, D. G. *J. Pharm. Sci.* **2015**, *104*, 378–387.
45. Muthurania, K.; Ignatius, A. A.; Jin, Z.; Williams, J.; Ohtake, S. *J Pharm Sci.* **2015**, *104*, 3770–3781.
46. Maa, Y. F.; Hsu, C. C. *Biotechnol. Bioeng.* **1996**, *51*, 458–465.

47. Centers for Disease Control and Prevention. *Vaccine Storage & Handling Toolkit*, January 2018. https://www.cdc.gov/vaccines/hcp/admin/storage/toolkit/storage-handling-toolkit.pdf (accessed Apr. 8, 2018).
48. World Health Organization. *WHO Policy Statement: Multi-dose Vial Policy (MDVP)*, Revision 2014. http://apps.who.int/iris/bitstream/handle/10665/135972/WHO_IVB_14.07_eng.pdf;jsessionid=6E9551CE96BBEBF0B6B746B7CD31F273?sequence=1 (accessed Apr. 22, 2018).
49. World Health Organization. *Meningococcal A Conjugate 10 Dose Presentation*. http://www.who.int/immunization_standards/vaccine_quality/PQ_197_MenAconjugate_10dose_SII/en/ (accessed Nov. 6, 2015).
50. Khandke, L.; Yang, C.; Krylova, K.; Jansen, K. U.; Rashidbaigi, A. *Vaccine* **2011**, *22* (29), 7144–7153.
51. Khandke, L.; Rashidbaigi, A. U.S. Patent 9095567 B2, 2015.
52. Cirefice, G. In *Alternatives to Thiomersal as Preservatives for Vaccines*; Proceedings of the UNEP-convened Intergovernmental Negotiating Committee Meeting 4, Geneva, Switzerland, April 3–4, 2012. http://www.who.int/immunization/sage/meetings/2012/april/Alternatives_thiomersal_preservatives_vaccines.pdf (accessed Jan. 31, 2018).
53. Gavi, The Vaccine Alliance. *AMC Supply Agreements*. https://www.gavi.org/funding/pneumococcal-amc/manufacturers/supply-agreements/ (accessed Feb. 2, 2018).
54. Bharat Biotech International Ltd. *Typbar-TCV Pack Insert*. https://www.bharatbiotech.com/images/typbartcv/Typbar-TCVPackInsert.pdf (accessed Feb. 11, 2018).
55. *International Conference on Harmonisation of Technical Requirements for Registration of Pharmaceuticals for Human Use*. Stability Testing of Biotechnological/Biological Products, Q5C, Nov. 1995. https://www.ich.org/fileadmin/Public_Web_Site/ICH_Products/Guidelines/Quality/Q5C/Step4/Q5C_Guideline.pdf (accessed May 1, 2018).
56. World Health Organization. Guidelines on stability evaluation of vaccines. In WHO Expert Committee on Biological Standardization, Fifty-seventh report. Geneva, Switzerland, 2011, *WHO Technical Report Series*, No. 962, Annex 3.
57. World Health Organization. Recommendations to assure the quality, safety and efficacy of pneumococcal conjugate vaccines. In WHO Expert Committee on Biological Standardization, Fifty-fourth report, Geneva, Switzerland, 2005; *WHO Technical Report Series*, No. 927, Annex 2.
58. World Health Organization. Recommendations to assure the quality, safety and efficacy of Group A Meningococcal Conjugate vaccines. In WHO Expert Committee on Biological Standardization, Fifty-fifth report. Geneva, Switzerland, 2006; *WHO Technical Report Series*, No. 932, Annex 3.
59. World Health Organization. Recommendations for the production and control of group C meningococcal conjugate vaccines. In WHO Expert Committee on Biological Standardization, Fifty-third report. Geneva, Switzerland, 2003; *WHO Technical Report Series*, No. 926, Annex 3.
60. Knezevic, I. *Biologicals* **2009**, *37*, 357–359.

61. Dobbelaer, R.; Pfleiderer, M.; Haase, M.; Griffiths, E.; Knezevic, I.; Merkle, A.; Hongzhang, Y.; Candrian, U.; Castillo, M. A.; Wood, D.; Daviaud, J.; Dellepiane, N.; Hernandez, C. A.; Lambert, S.; Shin, J.; Knezevic, I.; Daviaud, J.; Fournier-Caruna, J.; Kopp, S.; Zhou, T.; Zaffran, M.; Bektimirov, T. A.; Cooper, G.; da Silveira, S. C.; Egan, W.; Medveczky, N.; Morris, T.; Griffiths, E.; Nunez, Y. H.; Horiuchi, Y.; Jivapaisarnpong, T.; Krause, P.; Martin, J.; Southern, J.; Tyas Utami, A. R.; Jadhav, S.; Susanti, I.; Yamaguchi, I. K.; Duchene, M.; Laschi, A.; Schofield, T. L. *Biologicals* **2009**, *37*, 424–434.
62. Pfleiderer, M. *Biol.: J. Int. Assoc. Biol. Stand.* **2009**, *37*, 364–368.
63. Griffiths, E.; Knezevic, I. *Methods Mol Med.* **2003**, *87*, 353–376.
64. World Health Organization. *Guidelines on the international packaging and shipping of vaccine*, Geneva, Switzerland, 2005. whqlibdoc.who.int/hq/2005/WHO_IVB_05.23_eng.pdf (accessed Feb. 24, 2018).
65. PATH. *Summary of stability data for licensed vaccines Produced by Working in Tandem Ltd for the PATH Vaccine and Pharmaceutical Technologies Group*, Nov. 2012.
66. World Health Organization. *Controlled temperature chain (CTC): Beyond the traditional cold chain.* http://www.who.int/immunization/programmes_systems/supply_chain/ctc/en/ (accessed Apr. 8, 2018).
67. World Health Organization. *Controlled temperature chain (CTC): Background material and other CTC resources.* http://www.who.int/immunization/programmes_systems/supply_chain/ctc/en/index3.html (accessed Mar. 18, 2018).
68. World Health Organization. *Meningococcal meningitis.* http://www.who.int/immunization/diseases/meningitis/en/ (accessed Jan. 31, 2018).
69. PATH. *Increasing access to lifesaving vaccines.* http://www.path.org/publications/files/ER_vax_aof_fs.pdf (accessed June 20, 2018).
70. PATH. *Partnering with the US government: Government agencies played key roles in advancing MenAfriVac®.* https://www.path.org/menafrivac/government-partners.php (accessed Mar. 3, 2018).
71. *FDA Vaccines, Blood & Biologics, Vaccines, Approved Products.* http://www.who.int/biologicals/publications/trs/ areas/vaccines/meningococcal/MenA%20Final%20BS204102.Nov.06.pdf?ua=1 (accessed Nov. 6, 2017).
72. *Pfizer Receives World Health Organization Prequalification for Multi-Dose Vial Presentation of Prevenar 13®.* http://press.pfizer.com/press-release/pfizer-receives-world-health-organization-prequalification-multi-dose-vial-presentatio (accessed Jul. 19, 2016).
73. Gavi, The Vaccine Alliance. *How the pneumococcal AMC works.* https://www.gavi.org/funding/pneumococcal-amc/how-the-pneumococcal-amc-works/ (accessed Jan 11. 2018).
74. *Pfizer Vaccines in the Developing World.* https://www.pfizer.com/health/vaccines/developing_world (accessed Feb. 1, 2018).
75. World Health Organization. *Target Product Profile (TPP) for the Advance Market Commitment (AMC) for Pneumococcal Conjugate Vaccines.* http://www.who.int/immunization/sage/target_product_profile.pdf (accessed May 18, 2018).

Chapter 10

Preclinical Assessment of Glycoconjugate Vaccines

Ingrid L. Scully, Kena A. Swanson, Isis Kanevsky, A. Krishna Prasad, and Annaliesa S. Anderson*

Pfizer Vaccines Research and Development, 401 N. Middletown Rd., Pearl River, New York 10965, United States
***E-mail: Annaliesa.anderson@pfizer.com.**

Glycoconjugate vaccines are complex biological entities made from combining a polysaccharide with a carrier protein. The resulting vaccine must retain the biological appearance of the pathogen's polysaccharide to drive an immune response that can recognize and facilitate killing of the pathogen. Polysaccharde structures vary across serotypes and even among strains within a serotype. Thus, there is not a single conjugation approach that can be applied to all polysaccharide conjugate vaccines, and the ability of candidate vaccines to elicit functional immune responses is often empirically determined. To identify an optimally immunogenic vaccine candidate, both *in vivo* and *in vitro* models can be useful. Important considerations for preclinical glycoconjugate vaccine evaluation include: (a) selection of appropriate *in vivo* models to measure immunogenicity, (b) development of *in vitro* assays that measure immune responses that facilitate killing of the pathogen, and (c) identification of glycoconjugate features/epitopes that are critical for eliciting functional immune responses. These considerations are illustrated with case studies.

Introduction

Capsular polysaccharides (CP), expressed on the cell surface of most bacterial pathogens, have evolved to serve as an immune evasion mechanism, effectively masking the bacterial cell from recognition by the host innate immune system. When presented in the context of the whole bacterium they are poorly immunogenic. It is the repetitive and dense antigenic structure of these molecules that convert these structures from the bacteria's armour to their Achilles heel, as vaccines made from these structures have proven highly effective at generating immune responses that can facilitate the killing of the pathogen upon entry into the blood stream. CP alone vaccines have been licensed since the 1970s for the prevention of disease caused by *Neisseria meningitidis,* eg quadrivalent meningococcal polysaccharide vaccine (MSPV4), and *Streptococcus pneumoniae*, eg pneumococcal polysaccharide vaccine – 23 valent (PPSV23). However, unconjugated capsular polysaccharide vaccines have been shown to be poorly immunogenic in infants, and may even induce blunted booster responses (*1*, *2*), limiting their application in preventing disease in infants and young children, the age groups that are at greatest risk for poor outcomes of infectious disease.

Early in the 20th century, as polysaccharides from *S. pneumoniae* were first being discovered, Oswald Avery and colleagues at the Rockefeller Institute for Medical Research used preclinical studies to demonstrate that purified polysaccharides were unable to generate protective immune responses in animal experiments, whereas vaccines made from whole bacteria could induce serotype-specific responses and confer protection in vaccinated animals. This observation led them to conjugate the polysaccharides to proteins after which they observed improved protection in animal models (*3*). Protection observed in humans but not animals with unconjugated CP vaccines could be due to the fact that humans have prior exposure to the pathogen and thus CP vaccination elicits an anamnestic response. In contrast, animals, especially those housed in specific pathogen free facilities, commonly lack any pre-existing immunity, and thus cannot mount memory responses to the CP vaccine. The immunization schedules used in early animal experiments did not address this concept. Likewise, for young children (under two years of age) who do not have fully developed immune systems and have not had the opportunity to build natural immunity, CP vaccines alone were not effective (*4*). The addition of the protein carrier to the CP shifts the immune response from a T cell-independent response to a T cell-dependent response, resulting in both enhanced immunogenicity, especially in young children under the age of two, and enhanced immunological memory (*5*).

Although the concept of the glycoconjugate (CP conjugated to a protein carrier) is similar across vaccines, the details of the immune response to each glycoconjugate vaccine is unique, and requires preclinical evaluation. Predictive in silico modeling techniques do not yet exist that can accurately determine which polysaccharide conjugate attributes, such as size, substituents, or degree of cross-linking, result in functional, protective immune responses. As described in other chapters in this volume, many chemical approaches can be tried to design and develop a glycoconjugate. Each approach may or may not impact the presence and/or accessibility of important immunogenic epitopes on the glycoconjugate. In examining the variables associated with conjugate production (Figure 1), the preclinical assessments can broadly distinguish whether vaccines can still induce antibodies that recognize and facilitate killing of the pathogen.

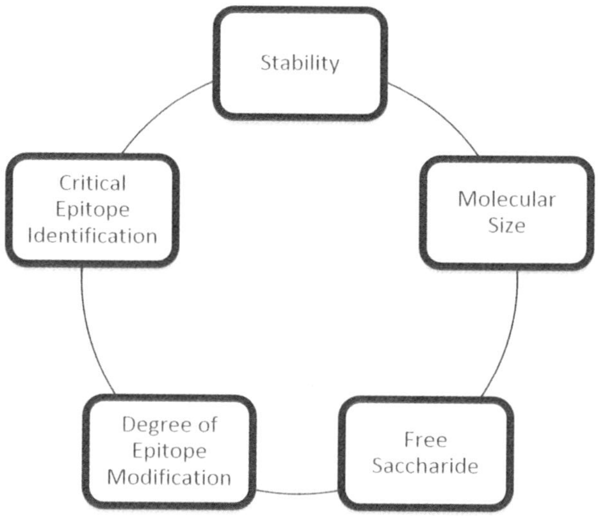

Figure 1. Critical quality attributes for CP-conjugate vaccine design and the role of preclinical assessments.

Ideally, both an animal model that recapitulates human disease and an *in vitro* assay that can accurately measure functional antibody responses would be available to test candidate conjugate vaccines. In practice, the perfect pair of animal model that accurately predicts clinical efficacy and an in vitro assay that perfectly monitors the functional immune response rarely exists. This truism requires that preclinical animal model data be married with *in vitro* assay data to identify the appropriate path forward for vaccine development. In practice, this approach can be used to demonstrate broad differences between vaccine formulations. However, the approach is marred by the intrinsic difference between humans and preclinical animal models, most notably the inability of many animals to differentiate between CP and CP-conjugate vaccines.

Rodent and rabbit models are often selected for initial assessment of candidate vaccine immunogenicity, due to cost and handling considerations. While small animal models are powerful tools and can rapidly screen large numbers of candidates, they do have their limitations, as highlighted in the following case studies. Nonhuman primates are often used to evaluate the immunogenicity of more advanced candidates, as they are more closely phylogenetically related to humans. However, even nonhuman primates cannot always completely predict immune responses in humans, especially in terms of the magnitude of responses.

Some considerations for the use of animal models are:

- Is the dose and route of immunization appropriate for the animal model?
- Is the age of the animals appropriate? Immune responses vary in young vs older mammals, due to the development of the immune system.
- Is the species appropriate? Does this species respond to the glycoconjugate?
- Does the species become infected with the pathogen from which the glycoconjugate is derived? How similar is the infectious process in the preclinical animal model and in the human vaccine target population?
- If challenge models are to be used, are clinically relevant strain(s) being tested in the model or conventional lab-adapted strains?

The impact of animal model selection is illustrated in the case studies presented later in this chapter.

As highlighted above, selection of appropriate in vitro assays to complement preclinical in vivo models is important for successful development of candidate conjugate vaccines. Levels of total immunoglobulin generated by conjugate vaccines are often measured in ligand binding assays, such as enzyme-linked immunosorbent assays (ELISAs) or Luminex-based immunoassays (LIAs). The concept of ELISA and LIA is similar. In the case of ELISAs, the readout is the product of an enzymatic reaction that is proportional to the amount of antibody bound; while in the case of LIAs, the readout is fluorescence that is proportional to the amount of antibody bound to the target antigen. LIAs are powerful as the technology allows for assessment of antibody responses to multiple antigens in a single sample reaction, termed multiplexing. These assays can be run in different formats, such as direct-binding LIA (dLIA), capture LIA (caLIA), or competitive LIA (cLIA). All of these have the benefit of providing information on the magnitude of the antibody response. They do not provide direct assessment of the functional activity (quality) of the antibodies.

The prototypical serological assays used for evaluation of functional antibody responses are the serum bactericidal assay (SBA) and the opsonophagocytic killing assay (OPA/OPK). The SBA measures the ability of antibodies to kill bacteria by the classical and alternate complement pathways: specifically antibodies bind to their target and recruit complement, which is activated and induces the membrane attack complex to kill the pathogen. The SBA measures this by mixing of target bacteria with serial dilutions of test antibody containing serum and a source of complement. After an incubation period, aliquots of the assay reaction are plated, and the surviving bacteria are enumerated. The reciprocal of the serum dilution where half of the bacteria survive, compared to an input value, is reported as the SBA titer. As the SBA measures direct bactericidal activity of serum mixed with exogenous complement, the source of complement can have a strong outcome on the titers obtained. Therefore, the species source of complement must be carefully considered (e.g. human versus animal) and complement must be prescreened to ensure nonspecific background killing activity is not present.

The OPA measures the ability of antibodies to induce phagocytic killing of the pathogen. It is similar to the SBA, except for the addition of phagocytes, which engulf and kill the pathogen in the presence of functional antibodies, as opposed to direct bactericidal activity monitored in the SBA. Like the SBA, bacteria are mixed with serum in the OPA, allowing antibody to bind to and "opsonize" the bacteria. A source of complement and phagocytes are then added to the bacteria-serum mixture, and the reaction is allowed to proceed for a period of time, generally 30 minutes to 2 hours. The phagocytes included in the OPA should be relevant for the pathogen being tested. A commonly used phagocytic cell source is HL-60 cells, a human promyelocytic cell line that can be differentiated into neutrophil-like cells. Surviving bacteria are enumerated at the end of the assay period by plating aliquots of the reaction mixture onto agar plates or onto filter plates (microcolony assay). Colonies are counted the next day, either through direct visualization or through staining and counting by means of an immunospot analyzer. OPA titers are generally reported as the inverse of the serum dilution that results in killing of 50% of the input bacteria. By using bacterial strains that have been genetically modified to contain antibiotic resistance genes, the OPA can be multiplexed. It should be noted that the term OPA can also be used to refer to a bacterial uptake assay, as opposed to a true assessment of bacterial killing; for this reason some members of the field prefer the term OPK to distinguish the killing assay from the uptake assay. As OPAs measure serum-induced bacterial killing, they are widely used in the assessment of glycoconjugate vaccine responses. It is critical to note that binding antibody levels are not always indicative of a functional serological response. There are many clinical examples where polysaccharide antigens have induced high binding responses without subsequently demonstrating functional responses. For example, binding responses to pneumococcal serotype 19A following immunization with serotype 19F polysaccharide do not correlate with OPA responses to 19A (*6*). Determining the quality of the antibody response is important when assessing glycoconjugate vaccine candidates. Therefore, it is critical to incorporate in vitro assays that can assess both binding and functional antibody responses.

Case studies are presented in this chapter, to highlight the following considerations:

- Selection of appropriate models to evaluate immunogenicity of conjugate variants.
- Development of in vitro models that assess functional immune responses, related to interruption of microbial pathogenesis.
- Identification of critical epitopes that must be preserved to ensure a conjugate elicits the desired functional response.

Preclinical Evaluation of Conjugate Vaccines – *Haemophilus influenzae* b

Before effective vaccines became available, *Haemophilus influenzae* type b (Hib) was the most prevalent cause of invasive bacterial infection in children under 5, a rate of ~1 in 250 live births; approximately 2/3 of cases were in children under 18 months of age. Over half of Hib invasive disease in the prevaccine era presented as meningitis, with a 3-6% case fatality rate, despite antimicrobial therapy. Up to 30% of survivors of Hib meningitis experienced neurologic sequelae and/or hearing loss (*7*). Natural history studies of Hib infection in the 1930s (*8*) noted that the risk of Hib-induced meningitis was greatest in infants aged 9-12 months, representing the time between when antibody titers transferred from the mother waned and when the child gained their own protective antibodies. It was later shown in the 1960s that the protective Hib antibodies generated in older children and adults reacted to the Hib capsular polysaccharide. These findings highlighted the need to identify a method of generating protective antibodies to pathogen capsular polysaccharides in infants and young children.

Seminal work by John Robbins, Rachel Schneerson and colleagues showed that free polysaccharide was poorly immunogenic in mice and nonhuman primates, while glycoconjugates induced robust protective immune responses (*9*, *10*) (Figure 2).

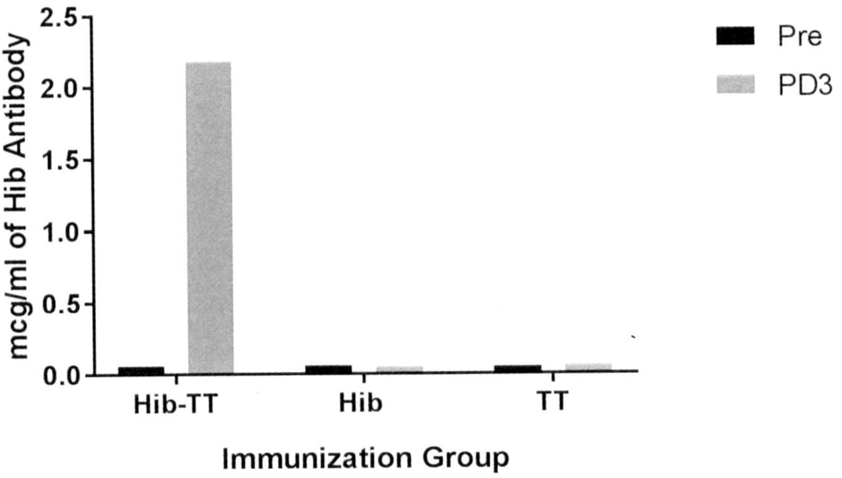

Figure 2. Responses of Juvenile Rhesus Macaques to Polysaccharide vs Conjugate Vaccination. Juvenile rhesus macaques were immunized three times subcutaneously with 50 mcg of Hib-tetanus toxoid conjugate (Hib-TT), Hib alone or TT alone, and anti-Hib antibody titers were measured by ELISA. (Adapted from (10). Copyright 1984, American Society for Microbiology).

Likewise, immunogenicity of glycoconjugates in human adults was not predictive of responses in infants (*11*, *12*). Infants and young children do not have fully developed immune systems, and so the immune responses elicited in adults are not always replicated in infants and young children. Importantly, it was noted that not all proteins and conjugation methods were able to act as efficient polysaccharide carriers to elicit functional antibody responses. Porter Anderson and colleagues found that cross-reactive material 197 (CRM_{197}), a genetically detoxified variant of diphtheria toxin, was an effective carrier protein to induce high-titer antibody responses. This was reduced to practice by David Smith and colleagues at Praxis Biologics, resulting in a successful vaccine against Hib that could protect infants and young children from invasive Hib disease (*7*). For this lifesaving work, John Robbins, Rachel Schneerson, Porter Anderson and David Smith received the Lasker Award in 1996. Today, vaccine-mediated protection against Hib prevents Hib-mediated meningitis in infants and young children.

Of Mice and Men: A Pneumococcal Case Study

Streptococcus pneumoniae is a major cause of pneumonia, bacteremia, and acute otitis media and is associated with significant morbidity and mortality worldwide in children less than 2 years of age and elderly adults (*13*, *14*). *S. pneumoniae* is a gram-positive bacterium containing a capsular polysaccharide (CP). Composition of the CP varies across the more than 90 known pneumococcal serotypes and is the major component in currently licensed vaccines against the prevalent disease-causing serotypes. The 23-valent CP alone vaccine (PPSV23, PneumoVax®), is licensed globally for the prevention of pnemococcal disease in adults, and two glycoconjugate vaccines, the13-valent PS conjugate vaccine (PCV13, Prevnar13®), licensed globally for use in all ages, and another 10-valent vaccine, PCV10, licensed outside the US for use in children 6 months to 5 years of age (*15–18*). Conjugation of the CPs to the protein carrier CRM_{197} (non-toxic mutant form of diphtheria toxin) in PCV13 drives a T-cell-dependent immune response and subsequent improved memory B-cell and antibody response, in contrast to the T-independent response elicited by unconjugated vaccines whose protection has been observed to wane following vaccination (*19*).

The power of a conjugate vaccine to elicit protective immune responses was demonstrated in 1929, when Avery and Goebel showed that immunization with a bacterial polysaccharide conjugated to a protein carrier induced robust protective antibody responses against the bacterium (*20*). In 1931, Avery and Goebel expanded their work with a *Streptococcus pneumoniae* serotype III polysaccharide conjugated to a protein carrier, which induced anti-serotype III antibodies in rabbits, and these antibodies, when passively transferred to mice, could protect mice from lethal infection with *S. pneumonia* (*21*).Development of pneumococcal PS conjugate vaccines has relied on many animal models for evaluation of immunogenicity and protection against bacterial challenge or colonization, e.g. mice, rabbits, nonhuman primates, infant rat bacteremia model, and the chinchilla otitis media model. Mice are routinely used (adult, infant, and aged mice) for assessment of conjugate immunogenicity in part due to ease of access to animals and immunological reagents. However, in some cases, mice do not mount immune responses against all *S. pneumoniae* serotypes, making rabbits a reliable model to discriminate between conjugated and unconjugated PSs. Some reports suggest responses to PS and PS-conjugate vaccines in mice and rabbits may predict efficacy in humans (*22–24*). As described earlier, similar to humans, young mice do not respond well to unconjugated PS vaccines, and priming with unconjugated PS can result in a suboptimal response following subsequent PS conjugate boosting (*3*, *25*).

A comparison of responses in humans to those observed in mice has not been directly tested. However, a recent study by Caro-Aguilar et al found varying responses to a 15-valent PS conjugate vaccine (PCV15) depending on the strain of mouse and route of immunization (*26*). This vaccine is currently in clinical development by Merck & Co. IgG responses were observed for all serotypes in all mouse strains tested: two inbred (Balb/c, Swiss Webster) and two outbred (C3H, and CD1). Serotypes 6B, 23F and 33F were less immunogenic in all strains except CD1 mice. Previous studies have shown that some serotypes, including 23F (*27*) and 33F (*26*), are poorly immunogenic in mice. CD1 mice, being outbred, may have sufficient genetic diversity to engender an improved immune response to a more diverse array of PS serotypes compared to inbred mice but further study would be warranted. CD1 mice were also indifferent to an intramuscular (IM) or intraperitoneal (IP) vaccination route, showing similar antibody responses compared to other mouse strains that displayed increased responses with IP vaccination.

Because of their genetic relatedness to humans, nonhuman primates (NHPs) have been used to assess PS conjugate vaccine responses. Recently published observations in adult cynomolgus macaques suggest NHPs, like human adults, show a superior immune response to conjugated PS vs. PS alone. NHPs vaccinated with a 7-valent PS conjugate vaccine and boosted five years later with either a 23-valent PS vaccine (PPSV23) or 13-valent PS conjugate vaccine (13vPnC) displayed an increased breadth and diversity of the antigen-specific memory B cell response against the 13vPnC booster compared to animals that received 23vPS booster (*28*).

An infant macaque immunogenicity model has been used to evaluate whether it is predictive of responses in human infants. Infant macaques possess immature immune systems similar to humans and are susceptible to some infectious pathogens making them useful to test in the context of vaccine immunogenicity and efficacy (*29*). For pneumococcal conjugate vaccines, infant rhesus monkeys have been used to assess potential immunological interference with increasing valency in conjugate vaccines. As the disease epidemiology has evolved post-introduction of PPV23 and PCV13, non-vaccine serotypes are emerging (*30*), suggesting more broadly protective vaccines may be needed. The PCV15 vaccine in clinical development is attempting to address the expanding medical need by including two new disease-causing serotypes (22F and 33F) together with the existing 13 serotypes in PCV13. Prior to studies in humans, antibody responses to PCV15 compared to Prevnar 7 were compared in infant rhesus macaques. In infant monkeys 2 to 3 months of age, IgG and functional opsonophagocytic (OPA, described earlier) antibody responses following three doses of either PCV15 or Prevnar 7 (7-valent PS conjugate vaccine) were comparable for the 7 serotypes common to both vaccines suggesting no immunological interference with PCV15 (*31*). Post-vaccination responses to PCV15 were >10-fold higher than baseline for the 8 additional serotypes. The PCV15 vaccine has been tested in human adults, toddlers, and infants, as described in Chapter 2 in this volume (*32*). In contrast to data in infant rhesus monkeys, IgG responses in human infants showed reduced serum IgG levels for serotypes 6A and 19A. These observations suggest the infant rhesus model insufficiently predicts signs of immunological interference or other differences in immunity in human infants when comparing pneumococcal conjugate vaccines with increasing numbers of serotypes. It is unknown whether this would apply to other glycoconjugate vaccines or alternative vaccine platforms.

Immunogenicity studies can also be complemented with challenge models to further assess the functionality of immune responses. In the case of pneumococcal vaccines, the infant rat passive protection model has been used extensively to assess functional serological responses. Pneumonia models in rats and nonhuman primates have been employed to assess preclinical efficacy of vaccine candidates. In addition, a chinchilla otitis media model is sometimes employed to assess preclinical responses.

There are many parallels with the immune response in rodents and NHPs to those observed in humans. Mice and rabbits are simple models that allow for preclinical screening of PS conjugate immunogenicity. In cases where mice do not mount an immune response against a specific *S. pneumoniae* serotype, it is important to consider tailoring the PS conjugate dose and/or inclusion of adjuvant to overcome these apparent challenges. Sufficient evidence supports the use of adult NHPs for characterization of the antibody and cellular immune response. With the disease landscape evolving and some evidence for the emergence of antibiotic-resistant *S. pneumoniae* strains, these underscore the continuing importance of improving animal models in support of advancing new vaccines against pneumococcal disease.

A Tale of Two O-Acetyls: A Case Study of *Staphylococcus aureus* and *Neisseria meningitidis* Glycoconjugates

Staphylococcus aureus is carried asymptomatically in the nares of 20-50% of the general population (*33*). Colonization increases the risk of infection, ranging from relatively mild skin infections, such as impetigo, to life-threatening invasive disease. *S. aureus* is recognized as a leading cause of morbidity and mortality in both healthcare-associated and community settings. In particular, infections in surgical patients carry high mortality rates and survivors of *S. aureus* surgical infections require an additional 13-17 days in the hospital, significantly increasing healthcare costs (*34*). The burden of *S. aureus* disease is exacerbated by the emergence of *S. aureus* isolates that are resistant to new classes of antibiotics, highlighting the need for alternative approaches such as a prophylactic vaccine.

One of the *S. aureus* vaccines that was not successful in the clinic was comprised of capsular polysaccharide conjugates. Capsular polysaccharides help bacteria evade immune-mediated killing through inhibiting phagocytosis (*35*, *36*). Vaccine-induced antibodies against capsular polysaccharides can overcome this virulence mechanism by enabling the organism to be opsonized and subsequently phagocytosed. All invasive human *S. aureus* isolates encode the genes required to express either type 5 or type 8 capsule (denoted CP5 and CP8, respectively), and most adults have anticapsular antibodies, demonstrating that the capsule is expressed in vivo. Due to its highly repetitive nature, capsular antigens have high epitope density, allowing multiple antibodies to bind, and thus are attractive candidates for prophylactic vaccines. Bacterial polysaccharides often contain an array of substituents, such as O-acetyl, phosphate, and sialic acid (*37*), which may constitute an important part of the immunodominant epitopes.

Both *S. aureus* CP5 and CP8 are comprised of 2-acetamido-2-deoxy-D-mannuronic acid (ManNAcA), 2-acetamido-2-deoxy-L-fucose (L-FucNAc), and 2-acetamido-2-deoxy-D-fucose (Figure 3). Both capsules are O-acetylated and differ in the stereochemical nature glycosidic linkages between the sugars and the site of O-acetylation. The sites of O-acetylation are the 3′OH moiety of L-FucNAc for CP5 and 4′OH substituent of ManNAcA for CP5 (*38*).

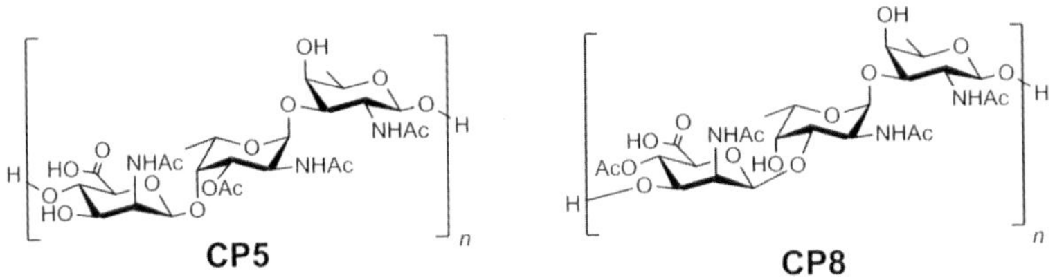

Figure 3. Structures of staphylococcal polysaccharide serotypes 5 and 8.

The presence of O-acetyl groups in capsular polysaccharides of pathogenic bacteria including *Escherichia coli* K1 (*39*), *N. meningitidis* groups A (*40*), C, H, I, K, W, and Y, *Salmonella enterica* serovar Typhi (*41*) Paratyphi A (*42*), *S. aureus* serotypes 5 and 8 (*38, 41*) *S. pneumoniae type* 9V (*43*) and group B streptococcus serotypes Ia, Ib, II, III, V and VI (*44*) has been widely observed. However, the role of O-acetylation in the immunogenicity and pathogenicity of microorganisms cannot be generalized (*39, 45, 46*).

In the case of *S. aureus*, O-acetylation of the capsular polysaccharide was shown to be critical for the elicitation of functional antibodies that can kill the organism in vivo and in vitro (*47*). *S. aureus* causes a wide range of disease in a variety of host microenvironments. Therefore the preclinical development of an effective vaccine targeting *S. aureus* must involve the use of multiple preclinical in vivo models which represent different infection modalities. In the case of the O-acetylation assessment, the end-organ infection model, pyelonephritis, was used. In this model, immunization with de-O-acetylated conjugates were poorly able to reduce bacterial load, while immunization with fully O-acetylated conjugates were able to reduce bacterial load by an additional two logs, a highly significant reduction ($p<0.0001$, Figure 4).

This in vivo finding was recapitulated by measuring functional antibody titers in OPAs. As described earlier, OPAs are in vitro assays that measure the ability of antibodies to induce the uptake and killing of bacteria by phagocytes. In the case of the *S. aureus* vaccine cited above, serum from mice immunized with O-acetylated *S. aureus* capsular polysaccharide conjugates had high OPA titers, while serum from mice immunized with de-O-acetylated *S. aureus* capsular polysaccharide conjugates had low or no detectable OPA titers.

Like *S. aureus*, *Neisseria meningitidis* is carried in the nasopharynx, and carriage is the predecessor of infection in the subset of individuals affected. *N. meningitidis* infection is notable for its sporadic and episodic nature, its rapid onset and poor outcome. *N. meningitidis* mostly strikes children and young adults, and *N. meningitidis* serotgroup A is endemic in the "meningitis belt" in Africa.

While capsular polysaccharide O-acetylation induces functional responses against *S. aureus*, as shown above, the presence of O-acetyl groups mask epitopes important for protective immune responses against *N. meningitidis* C, Y, and W (*48*). *N. meningitidis* serogroup W isolates produce heterogeneous polysaccharides that either possess or lack O-acetylation. Immunization of mice with O-Ac-negative conjugates induced higher serum bactericidal responses than immunization with O-Ac-positive conjugates (Table 1). Serum from humans immunized with O-Ac-negative *N. meningitidis* serogroup W conjugate containing vaccines also has been shown to have bactericidal activity against both O-Ac-positive and O-Ac-negative isolates (*49*).

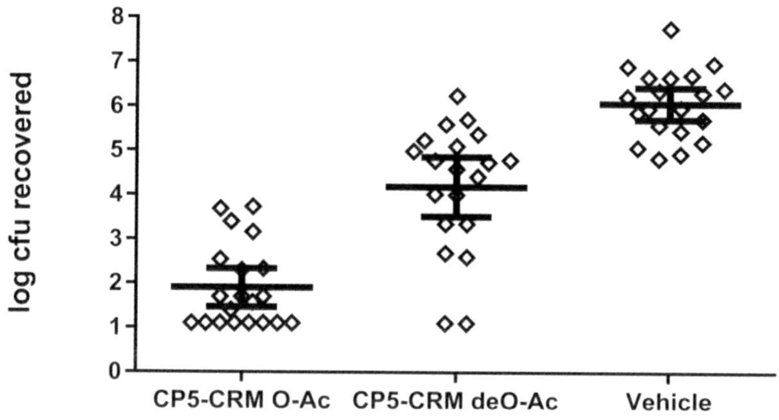

Figure 4. O-acetylation of S. aureus Capsular Polysaccahride is Necessary for an Effective Protective Vaccine Response. Female CD-1 mice were immunized three times with O-acetylated (native) CP5-CRM$_{197}$ with aluminum phosphate ($AlPO_4$), de-O-acetylated CP5-CRM$_{197}$ with $AlPO_4$, or $AlPO_4$ alone, then challenged with S. aureus. Bacterial burden in the kidneys was determined two days after challenge. (Adapted from (47), Copyright 2018, Pfizer Inc.).

In contrast to *N. meningitidis* serogroup W, O-acetylation could be important for inducing functional responses to *N. meningitidis* serogroup A (*50*).

The understanding that O-acetylation is important for the elicitation of functional antibodies to *S. aureus* and *N. meningitidis* serogroup A, but not *N. meningitidis* serogroup W, highlights the need to both identify and monitor critical immunogenic epitopes throughout glycoconjugate vaccine development.

Table 1. Impact of MnW CP O-Acetylation Status on Immune Responses. Mice were immunized with MnW CP conjugates prepared from strains that either do or do not O-acetylate CP. Resulting sera was evaluated both for binding antibody against both OAc+ and OAc- CP by ELISA, and for killing activity by SBA.

		OAc+ Results		*OAc- Results*	
		ELISA Titers vs OAc + CP	*SBA vs. M00242012 (OAc+ Strain)*	*ELISA titers vs OAc CP*	*SBA vs. ATCC35559 (OAc- Strain)*
Immunogen OAc Status + Dose	**OAc+ 0.1 µg**	96,187	100	48,389	100
	OAc+ 1.0 µg	146,973	400	34,314	100
	OAc- 0.1 µg	99,626	1,600	245,002	800
	OAc- 1.0 µg	191,264	800	391,569	800

Looking to the Future: Prospects for Candida and Pseudomonas Glyconjugate Vaccines for Immunocompromised Populations

Fungal infections are associated with high mortality and are on the rise due to emergence of drug-resistant strains. These opportunistic infections afflict the immunocompromised, lending to the assumption that a healthy immune response protects against infection. Furthermore, use of attenuated vaccines is a special challenge against these pathogens due to their intended use in individuals with suboptimal immune status. Similarly, *Pseudomonas sp* infections are most common in individuals with primary immunological syndromes.

Challenges in Designing Vaccines for Fungal Infections

Similar to many bacterial pathogens, successful vaccination against fungal infections, must take into account the multitude of antigens produced and suggests the need for a multivalent vaccine. Key targets in development of an antifungal vaccine include conserved polysaccharides and cell wall components such as agglutinin-like sequences (Als)1, Als3, Sap2, Hsp90 etc as protein targets (*51*, *52*) and B-glucans, chitin, chitosan, glycocalyx and mannans, (reviewed in (*53*, *54*)) for glycoconjugate vaccines. B-glucans are of particular interest as they are highly conserved amongst pathogenic strains (reviewed in (*55*)). One of the main hurdles in developing an antifungal vaccine is evaluation of immunogenicity to carbohydrate antigens. While functional assays such as OPA and SBA have been well established for bacterial pathogens, analogous functional assays have not yet been developed for fungal pathogens. Use of functional assays provides readouts of immunogenicity and potential correlates of protection for assessing vaccine efficacy. In the absence of a defined correlate of protection (*56*) for fungal infections, there are currently no definitive functional assays for assessment of candidate fungal antigens.

Immunogenicity evaluation requires selection of not only a relevant in vitro immune activation measure, but selection of an appropriate in vivo animal model. Since both neutrophils and macrophage functions are related to host susceptibility to infection, there is a critical role for antibody in opsonization, neutralization, and phagocytosis (*57*). However, susceptibility and immune activation varies significantly across species. More interestingly, within mouse species, there are significant differences in susceptibility to Candida infection that are strain-dependent, with certain strains being resistant (Balb/c) and others sensitive (CBA and C57BL/6) (*57*). Relevant models include animals with compromised immune systems to mimic a susceptible human. For example, treatment of mice with corticosteroids and immune suppressants such as cyclosporine can suppress the immune system. Cyclophosphamide induces neutropenia thereby altering susceptibility to infection and responses to vaccination (*58*, *59*). Use of knockout mice allows for study of cellular immunology and a critical role for certain cytokines (IL-17) and T cell lineages (Th17) (*60*) in protection from infection. Consideration should be made for dose and adjuvant as both can significantly impact sensitivity of the model and the ability to detect differences in immunity to protein and conjugate antigens. In the case of Candida infection, inhibition of adhesion, invasion, and susceptibility to immune killing are key mechanisms to target and evaluate during vaccine development (*51*, *57*, *58*).Important readouts from clinical trials include evaluation of humoral (antibody titers), and cell-mediated immunity (B and T cell function and cytokine levels) (*52*, *60*).

Research since 2010 has utilized conjugation to better study fungal carbohydrate components with a focus on B-glucans as antigens. Conjugation of laminarin, a non-fungal B-glycan (1,3) glycan from *Laminaria digitata* to CRM_{197} (Lam-CRM) allowed for evaluation of immunogenicity without concern for contaminating fungal proteins (*61*). The Lam-CRM conjugate induced IgM and IgG titers that correlated with 70% protection and a reduction in kidney colony forming units (CFU) in a murine model of Candida infection with complete Freund's adjuvant (CFA) as an immunogen. Lam-CRM increased clearance of infection in an estrogen-dependent vaginal Candida rat model and provided cross-protection against *A. fumigatus* in an antibody dependent fashion. Conjugation of B-glucans to keyhole limpet hemocyanin (KLH) indicated an impact on immunogenicity when conjugated to different branches at central sugar unit 6 – *O*-position of the B-1,3,-nonaglucan backbone. Immunization of mice with conjugates and CFA induced antibody titers that were lowest when conjugation was at the B-1,3-tetraglucose rather than B-1,6 –tetraglucose or B-1,3-diglucose; immunization with the two latter conjugates induced 40 % survival, compared with 0% in PBS controls, against a lethal dose of *C. albicans* (*55*). Conjugation of B-glucan oligosaccharides of varied length to KLH and subsequent vaccination of mice using Titermax Gold adjuvant demonstrated immunogenicity and similar survival at 40% in a *C. albicans* lethal dose challenge model (*62*). These and similar studies using additional antigens and conjugate combinations (tetanus toxoid (*63*, *64*); mannans/BSA (*65–67*)) allow for evaluation of conjugation, glycosylation efficiency, and oligosaccharide length on immunogenicity in animal models.

To date, only protein-based (Sap2 and rAls3p-N/NDV-3) vaccines have been tested in a phase I clinical trial (*52*) to evaluate IgA1 and IgG antibody titers and antigen specific T cell responses (IFNγ and IL-17A production) (*52*, *68*).

Challenges in Designing Vaccines for Gram-Negative Opportunistic Pathogens

Pseudomonas sp. are gram negative pathogens to which antibiotic resistance is on the rise; a concern both in aquaculture and human health. Similar to human vaccine strategies, the development of multivalent vaccines is necessary to have sufficient coverage of the diversity of bacterial species causing disease. Generating recombinant outer membrane proteins selected from *Vibrio sp.* strains for a polyvalent vaccine suggests an option for increasing coverage to include multiple genera of bacteria (*Vibrio sp.*, *Pseudomonas sp.*, and *Aeromonas sp.* (*69*) Antigen targets to date for *Pseudomonas sp.* include lipopolysaccharide (LPS) O antigen, outer membrane proteins (F and I), and type III secretions system components (PCrV).

Similar to other gram negative pathogens, Pseudomonas use flagellum for dissemination. In addition to providing function to the pathogen, flagellin is recognized by the host and characterized as a pathogen-associated molecular pattern (PAMP), also known as "danger signals." Flagellin elicits both T and B cell responses and contributes to pathogenesis of lung infections, urinary tract infection (UTI), burn wound infections, and keratitis. Since flagellin can modulate the immune response, it can serve as a self-adjuvanting vaccine antigen. Behrouz et al evaluated a bivalent (serotype A and B) flagellin vaccine in murine models that elicited protective T cell responses (CD4+IL-17+ Th17 cells) and conferred protection against intranasal challenge (*70*). The adjuvant attribute of certain proteins and carbohydrates is also considered when evaluating carrier proteins for vaccine development. In the case of typhoid fever, where immunogenicity in children under two does not occur, the Vi capsular polysaccharide can be coupled with recombinant expoprotein A (rEPA) from *P. aeruginosa* to produce a conjugate that elicits protective antibodies (*71*, *72*). Immunization of mice with alginate (mucoid exopolysaccharide), found in biofilm of pseudomonas, conjugated to flagellin elicits IgG antibodies that mediate killing in an OPA (*73*). Animal studies indicate that flagella interact with Toll-like receptor 5 (TLR5) to initiate an immune response that provides protection in lung (*74*) and burn infection (*75*) mouse models as measured by bacterial load in liver, spleen and blood, IgG titers, and OPA activity (*73*, *75*). Regardless of antigen or adjuvant used, a role for a cellular response, especially in light of the lung infections caused by *Pseudomonas sp.*, remain under investigation.

Existing correlates of protection for pseudomonas vaccines focus on antigen-specific binding antibody titers but not functional titers. However, this may need to be re-evaluated in light of a recent failure to meet the primary endpoint in a phase 2 clinical trial for IC43, a fusion protein-based vaccine developed by Valneva, despite detectable OprF/I IgG titers (*76*). Future studies should evaluate correlates of protection that include measurement of functional antibodies by OPA (*75*, *75*, *77*). Additional hurdles and considerations include extrapolation from animal models (*78*) into immunocompromised humans at risk of fungal disease. Due to the compromised immune status of the target population, a highly immunogenic conjugate likely will require pairing with a strong adjuvant, such as (De-*O*-acylated Lipooligosaccharide (dLOS) (*79*), to induce robust protective immune responses (overcoming immunodeficiency without exacerbating disease, as in the case of HIV (*54*)). Novel delivery systems, including liposomal antigen delivery (*80*); delivery by dendritic cells (*59*), and outer membrane vesicles (OMV) (*77*), may provide solutions to overcome common problems in vaccination of immunocompromised populations (*52*).

Summary

The advent of glycoconjugate vaccines revolutionized the approach to acheiving protection against serious pathogens, especially in young children. The case studies presented above illustrate important considerations in the preclinical evaluation of conjugate vaccines: namely, selection and appropriate interpretation of *in vivo* and *in vitro* models. The key difference between CP-alone and CP-conjugate vaccines is the ability to induce T-cell memory responses in humans. This is something that preclinical assessments have not been able to fully mimic, despite efforts in many animal species at different ages. However, the field continues to evolve, with the advent of advanced antibody engineering and characterization techniques, such as the impact of glycosylation on antibody function (*81*). These emerging approaches provide an opportunity to better characterize preclinical CP-conjugate responses in the future.

References

1. Gold, R.; Lepow, M. L.; Goldschneider, I.; Gotschlich, E. C. *J. Infect. Dis.* **1977**, *136* (Suppl), S31–S35.
2. Peltola, H.; Kayhty, H.; Sivonen, A.; Makela, H. *Pediatrics* **1977**, *60*, 730–737.
3. Goebel, W. F.; Avery, O. T. *J. Exp. Med.* **1929**, *50*, 521–31.
4. Douglas, R. M.; Paton, J. C.; Duncan, S. J.; Hansman, D. J. *J. Infect. Dis.* **1983**, *148*, 131–137.
5. O'Brien, K. L.; Steinhoff, M. C.; Edwards, K.; Keyserling, H.; Thoms, M. L.; Madore, D. *Pediatric Infect. Dis. J.* **1996**, *15*, 425–430.
6. Song, J. Y.; Moseley, M. A.; Burton, R. L.; Nahm, M. H. *J. Infect. Chemother.* **2013**, *19*, 412–425.
7. Robbins, J. B.; Schneerson, R.; Anderson, P.; Smith, D. H. *JAMA: J. Am. Med. Assoc.* **1996**, *276*, 1181–1185.
8. Fothergill, L. D. a. W. J. *J. Immunol.* **1933**, *24*, 273.
9. Chu, C.; Schneerson, R.; Robbins, J. B.; Rastogi, S. C. *Infect. Immun.* **1983**, *40*, 245–256.
10. Schneerson, R.; Robbins, J. B.; Chu, C.; Sutton, A.; Vann, W.; Vickers, J. C.; London, W. T.; Curfman, B.; Hardegree, M. C.; Shiloach, J.; Rastogi, S. C. *Infect. Immun.* **1984**, *45*, 582–591.

11. Anderson, P.; Pichichero, M. E.; Insel, R. A. *J. Clin. Invest.* **1985**, *76*, 52–59.
12. Robbins, J. B.; Schneerson, R. *N. Engl. J. Med.* **1990**, *323*, 1415–1416.
13. Melegaro, A.; Edmunds, W. J.; Pebody, R.; Miller, E.; George, R. *J. Infect.* **2006**, *52*, 37–48.
14. Nuorti, J. P.; Whitney, C. G. Centers for Disease Control and Prevention. *MMWR: Morbidity and Mortality Weekly Report* **2010**, *59*, 1–18.
15. Centers for Disease Control and Prevention. *MMWR. Morbidity and Mortality Weekly Report* **2010**, *59*, 258–261.
16. Tomczyk, S.; Bennett, N. M.; Stoecker, C.; Gierke, R.; Moore, M. R.; Whitney, C. G.; Hadler, S.; Pilishvili, T. *MMWR: Morbidity and Mortality Weekly Report* **2014**, *63*, 822–825.
17. WHO. Pneumococcal vaccines WHO position paper--2012. *Weekly Epidemiological Record: Health Section of the Secretariat of the League of Nations* **2012**, *87*, 129–144.
18. Centers for Disease Control and Prevention. *MMWR. Morbidity and Mortality Weekly Report* **2013**, *62*, 521–524.
19. Shapiro, E. D.; Berg, A. T.; Austrian, R.; Schroeder, D.; Parcells, V.; Margolis, A.; Adair, R. K.; Clemens, J. D. *N. Engl. J. Med.* **1991**, *325*, 1453–1460.
20. Avery, O. T.; Goebel, W. F. *J. Exp. Med.* **1929**, *50*, 533–550.
21. Avery, O. T.; Goebel, W. F. *J. Exp. Med.* **1931**, *54*, 437–447.
22. Steinhoff, M. C. *Vaccine* **2007**, *25*, 2465–2470.
23. Shriner, A. K.; Smithson, S. L.; Prinz, D. M.; Rabquer, B.; Khuder, S.; Goomber, R.; Westerink, M. A. *Vaccine* **2006**, *24*, 7197–7203.
24. Sabirov, A.; Metzger, D. W. *Vaccine* **2006**, *24*, 5584–5592.
25. Gonzalez-Fernandez, A.; Faro, J.; Fernandez, C. *Vaccine* **2008**, *26*, 292–300.
26. Caro-Aguilar, I.; Indrawati, L.; Kaufhold, R. M.; Gaunt, C.; Zhang, Y.; Nawrocki, D. K.; Giovarelli, C.; Winters, M. A.; Smith, W. J.; Heinrichs, J.; Skinner, J. M. *Vaccine* **2017**, *35*, 865–872.
27. McCool, T. L.; Harding, C. V.; Greenspan, N. S.; Schreiber, J. R. *Infect. Immun.* **1999**, *67*, 4862–4869.
28. Jia, B.; McNeil, L. K.; Dupont, C. D.; Tsioris, K.; Barry, R. M.; Scully, I. L.; Ogunniyi, A. O.; Gonzalez, C.; Pride, M. W.; Gierahn, T. M.; Liberator, P. A.; Jansen, K. U.; Love, J. C. *PloS One* **2017**, *12*, e0183738.
29. Jensen, K.; Dela Pena-Ponce, M. G.; Piatak, M., Jr.; Shoemaker, R.; Oswald, K.; Jacobs, W. R., Jr.; Fennelly, G.; Lucero, C.; Mollan, K. R.; Hudgens, M. G.; Amedee, A.; Kozlowski, P. A.; Estes, J. D.; Lifson, J. D.; Van Rompay, K. K.; Larsen, M.; De Paris, K. *Clin. Vaccine Immunol.* **2017**, *24*.
30. Rokney, A.; Ben-Shimol, S.; Korenman, Z.; Porat, N.; Gorodnitzky, Z.; Givon-Lavi, N.; Ron, M.; Agmon, V.; Dagan, R.; Valinsky, L. *Emerging Infect. Dis.* **2018**, *24*, 453–461.
31. Skinner, J. M.; Indrawati, L.; Cannon, J.; Blue, J.; Winters, M.; Macnair, J.; Pujar, N.; Manger, W.; Zhang, Y.; Antonello, J.; Shiver, J.; Caulfield, M.; Heinrichs, J. H. *Vaccine* **2011**, *29*, 8870–8876.
32. Lockhart, S. P.; Scott, D. A.; Jansen, K. U.; Anderson, A. S.; Gruber, W. C. Glycoconjugate Vaccines: The Clinical Journey. In *Carbohydrate-Based*

Vaccines: From Concept to Clinic; Prasad, A. K., Ed.; ACS Symposium Series 1290; American Chemical Society: Washington, DC, 2018; Chapter 2.
33. Kluytmans, J.; van Belkum, A.; Verbrugh, H. *Clin. Microbiol. Rev.* **1997**, *10*, 505–520.
34. Noskin, G. A.; Rubin, R. J.; Schentag, J. J.; Kluytmans, J.; Hedblom, E. C.; Jacobson, C.; Smulders, M.; Gemmen, E.; Bharmal, M. *Clin. Infect. Dis.* **2007**, *45*, 1132–1140.
35. Nanra, J. S.; Buitrago, S. M.; Crawford, S.; Ng, J.; Fink, P. S.; Hawkins, J.; Scully, I. L.; McNeil, L. K.; Aste-Amezaga, J. M.; Cooper, D.; Jansen, K. U.; Anderson, A. S. *Hum. Vaccines Immunother.* **2012**, *9*.
36. Thakker, M.; Park, J. S.; Carey, V.; Lee, J. C. *Infect. Immun.* **1998**, *66*, 5183–5189.
37. Geno, K. A.; Gilbert, G. L.; Song, J. Y.; Skovsted, I. C.; Klugman, K. P.; Jones, C.; Konradsen, H. B.; Nahm, M. H. *Clin. Microbiol. Rev.* **2015**, *28*, 871–899.
38. Jones, C. *Carbohydr. Res.* **2005**, *340*, 1097–1106.
39. Orskov, F.; Orskov, I.; Sutton, A.; Schneerson, R.; Lin, W.; Egan, W.; Hoff, G. E.; Robbins, J. B. *J. Exp. Med.* **1979**, *149*, 669–685.
40. Jennings, H. J.; Bhattacharjee, A. K.; Bundle, D. R.; Kenny, C. P.; Martin, A.; Smith, I. C. *J. Infect. Dis.* **1977**, *136* (Suppl), S78–S83.
41. Lemercinier, X.; Martinez-Cabrera, I.; Jones, C. *Biologicals* **2000**, *28*, 17–24.
42. Konadu, E.; Shiloach, J.; Bryla, D. A.; Robbins, J. B.; Szu, S. C. *Infect. Immun.* **1996**, *64*, 2709–2715.
43. McNeely, T. B.; Staub, J. M.; Rusk, C. M.; Blum, M. J.; Donnelly, J. J. *Infect. Immun.* **1998**, *66*, 3705–3710.
44. Lewis, A. L.; Nizet, V.; Varki, A. *Proc. Natl. Acad. Sci. U. S. A.* **2004**, *101*, 11123–11128.
45. Arakere, G.; Frasch, C. E. *Infect. Immun.* **1991**, *59*, 4349–4356.
46. Glode, M. P.; Lewin, E. B.; Sutton, A.; Le, C. T.; Gotschlich, E. C.; Robbins, J. B. *J. Infect. Dis.* **1979**, *139*, 52–59.
47. Scully, I. L.; Pavliak, V.; Timofeyeva, Y.; Liu, Y.; Singer, C.; Anderson, A. S. *Hum. Vaccines Immunother.* **2018**, *14*, 81–84.
48. Fusco, P. C.; Farley, E. K.; Huang, C. H.; Moore, S.; Michon, F. *Clin. Vaccine Immunol.* **2007**, *14*, 577–584.
49. Giardina, P. C.; Longworth, E.; Evans-Johnson, R. E.; Bessette, M. L.; Zhang, H.; Borrow, R.; Madore, D.; Fernsten, P. *Clin. Diagn. Lab. Immunol.* **2005**, *12*, 586–592.
50. Berry, D. S.; Lynn, F.; Lee, C. H.; Frasch, C. E.; Bash, M. C. *Infect. Immun.* **2002**, *70*, 3707–3713.
51. Sui, X.; Yan, L.; Jiang, Y. Y. *Vaccine* **2017**, *35*, 5786–5793.
52. Wang, X. J.; Sui, X.; Yan, L.; Wang, Y.; Cao, Y. B.; Jiang, Y. Y. *Virulence* **2015**, *6*, 309–315.
53. Levitz, S. M.; Huang, H.; Ostroff, G. R.; Specht, C. A. *Semin. Iimmunopathol.* **2015**, *37*, 199–207.
54. Scriven, J. E.; Tenforde, M. W.; Levitz, S. M.; Jarvis, J. N. *Curr. Opin. Microbiol.* **2017**, *40*, 95–103.

55. Liao, G.; Zhou, Z.; Liao, J.; Zu, L.; Wu, Q.; Guo, Z. *ACS Infect. Dis.* **2016**, *2*, 123–131.
56. Carvalho, A.; Cunha, C.; Iannitti, R. G.; Casagrande, A.; Bistoni, F.; Aversa, F.; Romani, L. *Front. Microbiol.* **2012**, *3*, 176.
57. Li, W.; Hu, Y. *Methods Mol. Biol.* **2017**, *1625*, 313–326.
58. Segal, E. *Methods Mol. Biol.* **2017**, *1625*, 343–353.
59. Xin, H. *Vaccine* **2016**, *34*, 245–251.
60. Whibley, N.; Gaffen, S. L. *Cytokine* **2015**, *76*, 42–52.
61. Torosantucci, A.; Bromuro, C.; Chiani, P.; De Bernardis, F.; Berti, F.; Galli, C.; Norelli, F.; Bellucci, C.; Polonelli, L.; Costantino, P.; Rappuoli, R.; Cassone, A. *J. Exp. Med.* **2005**, *202*, 597–606.
62. Liao, G.; Zhou, Z.; Burgula, S.; Liao, J.; Yuan, C.; Wu, Q.; Guo, Z. *Bioconjugate Chem.* **2015**, *26*, 466–476.
63. Cartmell, J.; Paszkiewicz, E.; Dziadek, S.; Tam, P. H.; Luu, T.; Sarkar, S.; Lipinski, T.; Bundle, D. R. *Carbohydr. Res.* **2015**, *403*, 123–134.
64. Lipinski, T.; Fitieh, A.; St Pierre, J.; Ostergaard, H. L.; Bundle, D. R.; Touret, N. *J. Immunol.* **2013**, *190*, 4116–4128.
65. Paulovicova, L.; Paulovicova, E.; Karelin, A. A.; Tsvetkov, Y. E.; Nifantiev, N. E.; Bystricky, S. *Int. Immunopharmacol.* **2012**, *14*, 179–187.
66. Paulovicova, L.; Paulovicova, E.; Karelin, A. A.; Tsvetkov, Y. E.; Nifantiev, N. E.; Bystricky, S. *J. Microbiol., Immunol., Infect.* **2015**, *48*, 9–19.
67. Liu, M.; Machova, E.; Nescakova, Z.; Medovarska, I.; Clemons, K. V.; Martinez, M.; Chen, V.; Bystricky, S.; Stevens, D. A. *Med. Mycol.* **2012**, *50*, 818–828.
68. Moriyama, B.; Gordon, L. A.; McCarthy, M.; Henning, S. A.; Walsh, T. J.; Penzak, S. R. *Mycoses* **2014**, *57*, 718–733.
69. Peng, B.; Lin, X. P.; Wang, S. N.; Yang, M. J.; Peng, X. X.; Li, H. *Fish Shellfish Immunol.* **2018**, *72*, 104–110.
70. Behrouz, B.; Hashemi, F. B.; Fatemi, M. J.; Naghavi, S.; Irajian, G.; Halabian, R.; Imani Fooladi, A. A. *J. Immunol. Res.* **2017**, *2017*, 5689709.
71. Thompson, D. K.; Chourey, K.; Wickham, G. S.; Thieman, S. B.; VerBerkmoes, N. C.; Zhang, B.; McCarthy, A. T.; Rudisill, M. A.; Shah, M.; Hettich, R. L. *BMC Genomics* **2010**, *11*, 311.
72. Pan, J.; Li, C.; Ye, Z. *Methods Mol. Biol.* **2016**, *1404*, 519–528.
73. Faezi, S.; Bahrmand, A. R.; Mahdavi, M.; Siadat, S. D.; Sardari, S.; Nikokar, I.; Khanaki, K.; Mirzajani, E.; Goudarzi, G. *Biologicals* **2017**, *47*, 11–17.
74. Saha, S.; Takeshita, F.; Matsuda, T.; Jounai, N.; Kobiyama, K.; Matsumoto, T.; Sasaki, S.; Yoshida, A.; Xin, K. Q.; Klinman, D. M.; Uematsu, S.; Ishii, K. J.; Akira, S.; Okuda, K. *J. Immunol.* **2007**, *179*, 1147–1154.
75. Korpi, F.; Hashemi, F. B.; Irajian, G.; Fatemi, M. J.; Laghaei, P.; Behrouz, B. *Immunol. Lett.* **2016**, *176*, 8–17.
76. Rello, J.; Krenn, C. G.; Locker, G.; Pilger, E.; Madl, C.; Balica, L.; Dugernier, T.; Laterre, P. F.; Spapen, H.; Depuydt, P.; Vincent, J. L.; Bogar, L.; Szabo, Z.; Volgyes, B.; Manez, R.; Cakar, N.; Ramazanoglu, A.;

Topeli, A.; Mastruzzo, M. A.; Jasovich, A.; Remolif, C. G.; Del Carmen Soria, L.; Andresen Hernandez, M. A.; Ruiz Balart, C.; Kremer, I.; Molnar, Z.; von Sonnenburg, F.; Lyons, A.; Joannidis, M.; Burgmann, H.; Welte, T.; Klingler, A.; Hochreiter, R.; Westritschnig, K. *Crit. Care* **2017**, *21*, 22.
77. Zhang, X.; Yang, F.; Zou, J.; Wu, W.; Jing, H.; Gou, Q.; Li, H.; Gu, J.; Zou, Q.; Zhang, J. *Vaccine* **2018**, *36*, 1047–1054.
78. Bielen, K.; s Jongers, B.; Malhotra-Kumar, S.; Jorens, P. G.; Goossens, H.; Kumar-Singh, S. *Ann. Transl. Med.* **2017**, *5*, 132.
79. Ryu, J. I.; Wui, S. R.; Ko, A.; Lee, Y. J.; Do, H.; Kim, H. J.; Rhee, I. M.; Park, S. A.; Kim, K. S.; Cho, Y. J.; Lee, N. G. *J. Microbiol. Biotechnol.* **2017**, *27*, 1539–1548.
80. Carneiro, C.; Correia, A.; Lima, T.; Vilanova, M.; Pais, C.; Gomes, A. C.; Real Oliveira, M.; Sampaio, P. *Acta Biomater.* **2016**, *39*, 133–145.
81. Ackerman, M. E.; Moldt, B.; Wyatt, R. T.; Dugast, A. S.; McAndrew, E.; Tsoukas, S.; Jost, S.; Berger, C. T.; Sciaranghella, G.; Liu, Q.; Irvine, D. J.; Burton, D. R.; Alter, G. *J. Immunol. Methods* **2011**, *366*, 8–19.

Chapter 11

Campylobacter jejuni Capsule Polysaccharide Conjugate Vaccine

Mario A. Monteiro,[*,1] Alexander Noll,[2] Renee M. Laird,[2] Brittany Pequegnat,[1] Zuchao Ma,[1] Lisa Bertolo,[1] Christina DePass,[1] Eman Omari,[1] Pawel Gabryelski,[1] Olena Redkyna,[1] Yuening Jiao,[1] Silvia Borrelli,[1] Frederic Poly,[2] and Patricia Guerry[2]

[1]Department of Chemistry, University of Guelph, Guelph, Ontario N1G 2W1, Canada
[2]Enteric Diseases Department, Naval Medical Research Center, Silver Spring, Maryland 20910, United States
[*]E-mail: monteiro@uoguelph.ca.

Campylobacter jejuni is one of the most common causes of human diarrheal disease worldwide. Campylobacteriosis cases are primarily sporadic, but outbreaks are not uncommon. In developed countries Campylobacteriosis cases are about 50/100,000 per year, but in developing countries, particularly in Southeast Asia, the incidence is estimated to be at least 10 times higher. In these endemic regions, this high incidence represents a life threatening risk, especially towards the pediatric population. *C. jejuni* is considered a zoonotic disease in which the major source of contamination is through the consumption of poultry. In humans, the infectious dose can be as low as 500-1,000 bacteria with symptoms being variable and dependent on the bacterial strain and host factors. Some patients only present mild abdominal pain and mild to no diarrhea, while the most severe cases are associated with severe abdominal cramping, accompanied with fever, headaches, myalgia, and large volumes of mucous and bloody diarrhea that can last for several days. If left untreated, severe *C. jejuni* cases can be lethal. *C. jejuni* infections have also been associated with the development of Guillain-Barré syndrome (an autoimmune disease triggered by *C. jejuni* lipooligosaccharide sub-structures that mimic glanglioside forms) inflammatory bowel syndrome, reactive arthritis and stunting in children from developing countries. In this piece, we describe the development of a *C. jejuni* vaccine candidate in which the protective antigens are the cell-surface capsular polysaccharides (CPSs), with focus on (i) CPS discovery highlighting the two key structural features of *C. jejuni* CPSs, 6-deoxy-heptoses of unsual configurations and variably linked *O*-methyl-phosphoramidate moities; (ii) conjugation of *C. jejuni* CPSs to carrier proteins using new methodology for CPS activation (TEMPO-mediated oxidation); and (iii) protection studies in a non-human primate model. Collectively, the data obtained has allowed us to advance our prototype *C. jejuni* CPS conjugate vaccine through cGMP production and onto an ongoing phase I human clinical trial.

Campylobacter jejuni Overview

Campylobacter is one of the most common causes of human diarrheal disease worldwide. However, popular awareness has remained limited compared to other enteric pathogens like *Salmonella* spp., *Vibrio cholerae*, or *Shigella* spp. This may be due to its relative recent discovery: it was not until the early 1960's that the genus *Campylobacter* was identified and in the mid 1970's that its isolation methods and growing media were established (*1*, *2*). Since then, this genus has been given increasing attention and data has shown that *C. jejuni* is the most common *Campylobacter* sp. to be associated with human diarrheal diseases (*3*).

In developed countries Campylobacteriosis cases range from 14 to about 57 cases/100,000/year in US and Europe, respectively (*4*, *5*). Campylobacteriosis cases are primarly sporadic but outbreaks are not uncommon. In developing countries, in particular in Southeast Asia where *C. jejuni* is endemic, the incidence is estimated to be at least 10 times higher (*6*, *7*). This high incidence represents a concerning life threatening risk, especially towards the pediatric population in these endemic regions.

C. jeuni is considered a zoonotic disease. The major source of contamination is through the consumption of poultry, but other sources like unpasteurized milk or water have been referenced (*8*). *C. jejuni* is considered a chicken commensal. The bacteria asymptomatically colonize the mucus of the intestinal epithelium of the intestine and preferentially the ceca (*9*). The contamination of the chicken meat occurs during the slaughtering process (*10*).

The infectious dose can be as low as 500-1,000 bacteria. Following ingestion, 24h to 96h incubation are necessary before the appearance of Campylobacteriosis symptoms (*11*). Campylobacteriosis symptoms are quite variable and depend on the bacterial strain and host factors. Some patients will present mild abdominal pain and mild to no diarrhea, whilst the most severe cases are associated with severe abdominal cramping, accompanied with fever, headaches, myalgia, and large volumes of mucous and bloody diarrhea that can last for several days (*12*). Most cases are self-limiting but severe cases require hospitalization and are treated with re-hydratation fluids and antibiotics (*13*). If left untreated, severe cases can be lethal.

C. jejuni infections have also been linked autoimmune pathologies, such as the development of Guillain-Barré syndrome (GBS) and Miller-Fisher syndrome, an autoimmune disease triggered by specific lipooligosaccharide (LOS) structures present at the surface of the bacteria (*14*). It is believed that one out of a thousand patients, one will develop GBS (*14*). More recently, *C. jejuni* has been associated with inflammatory bowel syndrome (IBS) (*15*), reactive arthritis (*16*), a rare form of mucosa-associated lymphoid tissue (MALT) lymphoma called immunoproliferative small intestinal disease (IPSID) (*17*), and stunting in childs from developing countries (*18*).

Campylobacter jejuni Polysaccharide Capsule Discovery

The first evidence pointing towards the presence of polysaccharides (PS) akin to capsular polysaccharides (capsule, CPS) in *Campylobacter* was obtained in the early 1990s, when structural analysis revealed surface PSs in *C. lari* that were not attached to lipid A or LOS regions (*19*). Following this, the first publically available *C. jejuni* sequenced genome (clinical isolate NCTC 11168) was published in 2000 and also suggested the production of CPS by *Campylobacter* (*20*). However, the *C. jejuni* genome sequence did not clearly identify genetic determinants of pathogenicity. No common or known virulence factors were identified in the *C. jejuni* genome and, compared to other enteropathogens, no obvious acquired pathogenic genetic features (ex-pathogenicity islands) were found. Nevertheless, other whole genome sequencing revealed gene sequences presenting similarities with genes (*kps* family) involved in biosynthesis and transport of CPS to the cell surface. Bacterial CPS is considered a virulence factor due to its role in survival and evasion of the host immune system. The physical location and composition of CPS also prevent host recognition/access of antibody toward the bacterial cell surface proteins (*21*).

One PS moiety was later demonstrated to be attached to a phospholipid anchor (*22, 23*). The genes involved in PS transport and attachment to the phospholipid anchor are genetically conserved between the different capsule types, while the locus responsible for the biosynthesis of the capsule (nestled between the *kpsC* and *kpsF* genes) is variable (*24–26*).

The high molecular weight CPS is the main serodeterminant of the Penner serotyping scheme. This serological typing method has been recognized as the gold standard *C. jejuni* isolate typing methodology for over 3 decades. This method requires antiserum generated by boiling a *C. jejuni* bacterial culture in water. Following centrifugation, the soluble, heat-stable (HS) fraction is used to vaccinate rabbits for serum production and establishing an HS antiserum reference bank. The HS antisera are used to type strains by monitoring passive agglutination on a glass microscope slide (*27, 28*). The method recognizes a total of 47 *C. jejuni* serotypes. Due to the limitation of serum bank maintenance, the Penner serotyping method has been gradually replaced by more adapted typing methods such as multiplex PCR methods (*26, 29*).

As suggested by the number of Penner serotypes identified, C. *jejuni* sp. present several capsule types. Based on sequenced capsule biosynthesis loci comparison and cross-reactions of Penner sera, the total of 47 capsule serotypes can be collapsed in 35 groups (*26*). The capsule loci compositions reflect and thus support this phenotypic complexity, whereknown C. *jejuni* CPS biosynthesis loci genes vary from 11 to more than 30 genes (*26*). Not surprisingly, this CPS serological and genetic diversity results in CPS structural diversity.

Structural Features of *Campylobacter jejuni* Capsules

C. jejuni produces structurally variable structures, reflecting the serological and genetic differences between these different CPS groups described in the previous section. These structural differences are further widened due to the presence of phase variation. Phase variation is a process where genes are turned on and off at high frequency, and this process occurs in *C. jejuni* due to the presence of homopolymeric tracts of 8-10 cytosine nucleotide residues in specific genes that cause slipped-strand mispairing (*30*). Slipped-strand mispairing is a high frequency, reversible, erroneous addition or subtraction of C residues during DNA replication at these poly(C) tracts, resulting in a frameshift mutation and ultimately conversion from a functional to non-functional protein product and vice-versa (*31*, *32*). Phase variation results in CPS biosynthetic enzymes being turned on and off, which ultimately affects expression of CPS structural modifications and even cell surface CPS expression itself (*15*, *33*). Phase variation allows structurally related CPS structures to be expressed by strains within a given serotype complex. For this reason, phase variation mechanisms most likely evolved to provide a mechanism of host immune system and predatory phage escape.

Even though the genes involved in *C. jejuni* CPS biosynthesis are variable in numbers and functions, some genetic and structural similarities are still observed between different CPS serotypes. A unique feature of many C. *jejuni* CPS serotypes is modification with O-methyl-phosphoramidate (MeOPN). The *cjj1415-1418* operon represents the core MeOPN biosynthesis genes, which is present in and highly conserved between *C. jejuni* serotypes expressing MeOPN (*34*). The functions of these four gene products are currently unclear, as is the mechanism of MeOPN biosynthesis. However, a recent study showed that *cjj1418* encodes an L-glutamine kinase enzyme that attaches a phosphate group from ATP onto the side chain amido nitrogen group, forming the P-N bond (*35*), although the enzymatic mechanism has not been experimentally defined. In contrast to the MeOPN biosynthesis genes, the MeOPN transferase genes are not as well conserved between strains, which would be expected given that the MeOPN group is attached to different monosaccharide hydroxyl groups or monosaccharide residues. Phosphoramidate (i.e.-MeOPN without O-methylation) is also sometimes detected on *C. jejuni* CPS (*34*, *36*), but at this point it is unclear whether phosphoramidate is a true CPS modification, an intermediate product in MeOPN biosynthesis, or an MeOPN extraction artifact. Indeed, the MeOPN linkage itself is highly acid- and base-labile and hence will hydrolyze during methods of CPS extraction or monosaccharide analysis methods (*37*). An understanding of the MeOPN biosynthetic pathway may help verify whether phosphoramidate is of biological significance.

The MeOPN modification is not known to be naturally produced by any other living organisms. Therefore, the MeOPN modification is not only unique from a natural standpoint but also from the immune system's perspective, so a *C. jejuni* CPS glycoconjugate vaccine should include the MeOPN modifications to ensure that this rare and unique modification is available for recognition and memory by the immune system.

Other CPS modifications occur in specific CPS serotypes including methylation, acetylation, and dehydration. An interesting feature of all these modifications, including the MeOPN modification, is that they tend to occur non-stoichiometrically. One potential reason for this is phase variation of the genes involved in these modifications, which would cause the fraction of phase-on cells within a *C. jejuni* population to only express the CPS modification while the phase-off cells do not. However, other factors such as slow enzyme kinetics, low substrate concentrations, or possibly even artefactual loss of these modifications during CPS extraction may play a role.

Besides the unique MeOPN modification, the presence of heptose residues in unusual configuration is relatively unique to C. *jejuni* CPS. All strains of *C. jejuni* that contain heptoses in their CPS share four highly conserved enzymes that convert D-sedoheptulose-7-P to GDP-D-*glycero*-D-mannoheptose. With the exception of HS2, the strains that are capable of expressing a heptose in their CPS contain a gene *dmhA* that encodes a putative heptose-6-dehydratase enzyme to generate 6-deoxyheptose units. In addition, known *C. jejuni* CPS structures contain multiple unique isomeric forms of heptoses (*gluco*, *manno*, talo, *ido*, *gulo*, *altro*) as well as L configurate heptoses. CPS loci of HS8, HS10, HS42 and HS44 harbor two distinct copies of putative C4 reductases (MlghC/DdahC homologues) annotated as *fcl* and *fcl2* (*24*). Future sequencing and CPS structural analysis may reveal additional heptose isomeric forms and their responsible enzymes. Future studies of other *C. jejuni* CPS structures are hence expected to further uncover unique CPS structural components.

Known *Campylobacter jejuni* Capsular Polysaccharides

Serotype Complex HS1/HS44

The HS1/HS44 complex consists of Penner serotypes HS1 and HS44 as well as strains typed HS1 and HS44 (HS1/44), all of which contain genetically related bioshyntesis loci (*26*). The first structural studies on a HS1 CPS identified a linear teichoic-acid like CPS [→4)-α-D-Gal*p*-(1→2)-D-Gro-(1-P→]$_n$ (*38*, *39*). At this time, gangliosides in the core regions of C. jejuni were also detected (*40*). Further analysis confirmed the aforementioned teichoic-acid backbone sequence in HS1/44 seroclass CPS and in the HS1 type strain ATCC 43429, but also revealed the presence of fructofuranose branch [→4)-[β-D-Fru*f*-(2→2)]-α-D-Gal*p*-(1→2)-D-Gro-(1-P→]$_n$ (*41*). The D-glycerol-1-phosphate linkage is a phosphodiester rather than a glycosidic linkage, such connection has also been observed in the

HS53 CPS structure (described below). In some cases, the D-Gal*p* residue can be non-stoichiometrically modified at the 3-position with another β-Fru*f* residue. Additionally, both the stoichiometric and non-stoichiometric β-Fru*f* residues can be non-stoichiometrically modified at the 3-position with MeOPN. Exploratory studies on seroclass HS44 have revealed that it produces two distinct CPSs, a teichoic-acid CPS similar to that in HS1 and HS1/44, and a heptan CPS composed of 6-deoxy-*galacto*-heptose, 6-deoxy-*altro*-heptose and its methylated derivative 6-deoxy-3-*O*-methyl-*altro*-heptose (*42*). This HS44 heptan CPS is also decorated with MeOPN units (*42*).

Serotype Complex HS2

The structure of the HS2 type strain NCTC11168 is [→2)-β-D-Rib*f*-(1→5)-β-D-Gal*f*NAc-(1→4)-α-D-Glc*p*A6EtN-(1→]$_n$ (*43*). NCTC 11168, as described above, was both the first *C. jejuni* strain whose gene sequence and *kps* locus was uncovered. *C. jejuni* HS2 (NCTC 11168), along with strains of HS23/36 and HS19 (*44*, *45*), was one of the first strain discovered to harbor the MeOPN modification, which specifically occurs non-stoichiometrically on the 3-OH group of Gal*f*NAc.This CPS structure is further decorated with other non-stoichiometric modifications including a phase-variable CH_2OH modification at C7 of the α-D-GlcpA6EtN residue forming α-D-GlcpA6NGro, D-*glycero*-α-L-*gluco*-heptose at the 3-OH of the Glc*p*A residue (which was originally thought to be stoichiometric but later found to be non-stoichiometric), 3-O-methyation of the non-stoichiometric heptose, and 6-dehydration of the non-stoichiometric heptose.

Serotype Complex HS3

The CPS structure of the HS3 type strain ATCC 43431 (TGH9011) is [→3)-L-*glycero*-α-D-*ido*-Hep*p*-(1→4)-α-D-Gal*p*-(1→]$_n$ (*46*). The L-*glycero*-α-D-*ido*-Hep residue occurs non-stoichiometrically in the 6-deoxy form with MeOPN units non-stoichiometric present its position 2 (*47*). The Gal residue of HS3 CPS may also be substituted at position 3 with 3-hydroxypropanoyl. Genetic and biochemical evidence have shown that the genes *dmhA* and *dmhB* encode a heptose-C6-dehydratase and heptose-C4-reductase enzyme, respectively, required for 6-deoxyheptose formation. Putative *dmhA* and *dmhB* genes have been identified in the TGH9011 genome, and the *dmhA* gene has a poly(C) tract present that can cause slipped-strand mispairing and hence phase-variation in 6-deoxyheptose expression, which likely explains the non-stoichiometric 6-deoxyheptose modification in the HS3 CPS structure.

Serotype Complex HS4

The HS4 complex consists of multiple Penner serotypes including the HS4, HS13, HS16, HS43, HS50, HS62, HS64, and HS65 serotypes that contain genetically identical capsule biosynthetic loci (*25*). Some strains, such as the HS4/13/64 strain CG8486, cross-react with multiple

Penner antisera. The HS4/13/64 strain CG8486 strain has the CPS structure [→4)-β-D-GlcpNAc-(1→3)-L-*glycero*-β-D-*ido*-Hepp-(1→]$_n$ or [→4)-β-D-GlcpNAc-(1→3)-6-deoxy-β-D-*ido*-Hepp-(1→]$_n$ (*48*). A number of the *ido*-heptopyranose units were observed to carry O-methyl phosphoramidate moieties at the O-2 or O-7 position (*48*). An HS13 strain has been investigated in our laboratories, and found to contain a Glc unit (in the place of the GlcNAc residue seen in HS4 strains) [→4)-β-D-Glcp-(1→3)-6-deoxy-β-D-*ido*-Hepp-(1→]$_n$.

Serotype Complex HS5

The CPS structure of HS5 consists of the main structure [→6)-[α-3,6-dideoxy-*ribo*-Hepp-(1→2,6)]-α-D-*glycero*-D-*manno*-Hepp-(1→3)-[α-3,6-dideoxy-*ribo*-Hepp-(1→2)]-glucitol-(6→P]$_n$, with non-stoichiometric MeOPN modification at the 7-OH position of the α-3,6-dideoxy-*ribo*-Hep*p* residues (Figure 1). The HS5 CPS is the first *C. jejuni* serotype reported to contain either the α-3,6-dideoxy-*ribo*-Hep*p* or glucitol residue (*49*). In addition to the main structure, there are also three less abundant variations of the CPS structure that can be observed: [→6)-α-D-*glycero*-D-*manno*-Hepp-(1→2)-glucitol-(6→P]$_n$, [→6)-[α-3,6-dideoxy-*ribo*-Hepp-(1→2)]-α-D-*glycero*-D-*manno*-Hepp-(1→2)-glucitol-(6→P]$_n$, and [→6)-[α-3,6-dideoxy-*ribo*-Hepp-(1→2)]-α-D-*glycero*-D-*manno*-Hepp-(1→3)-[α-3,6-dideoxy-*ribo*-Hepp-(1→2)]-glucitol-(6→P]$_n$, all with non-stoichiometric MeOPN modification at the 7-OH position of the α-3,6-dideoxy-*ribo*-Hep*p* residues (*49*).

Serotype Complex HS6/7

In the HS6/7 complex, two carbohydrate structures were identified termed Structure A and Structure B in the HS6 type strain 81116 (RM1863) and occurring in a molar ratio of about 3:1, A:B, respectively. Structure A has the structure [→3)-β-D-Glcp-(1→2)-α-D-GlcpA-(1→3)-α-D-Manp-(1→3)-α-D-Glcp(1→]$_n$, with non-stoichiometric O-acetylation occurring at the 3-OH of the α-D-Glc*p*A residue and 6-OH position of the α-D-Glc*p* residue; Structure B was determined to be [→3)-β-D-GlcpNAc-(1→6)-α-D-Glcp-(1→4)-(β-D-GlcpNAc-(1→3))-α-D-Galp-(1→]$_n$ (*50*). Structure B may be associated with Lipid A.

Serotype Complex HS10

The CPS of HS10 seroclass is composed of a GalNAc backbone with branches of 6-deoxy-L-*galacto*-heptofuranose: [→3)-[6d-α-L-*gal*-Hepf-(1→4)-α-D-GalNAc-(1→]$_n$ (*51*, *52*). Some of the 6-deoxy-L-*galacto*-heptofuranose units carry a MeOPN at position 3 (*52*).

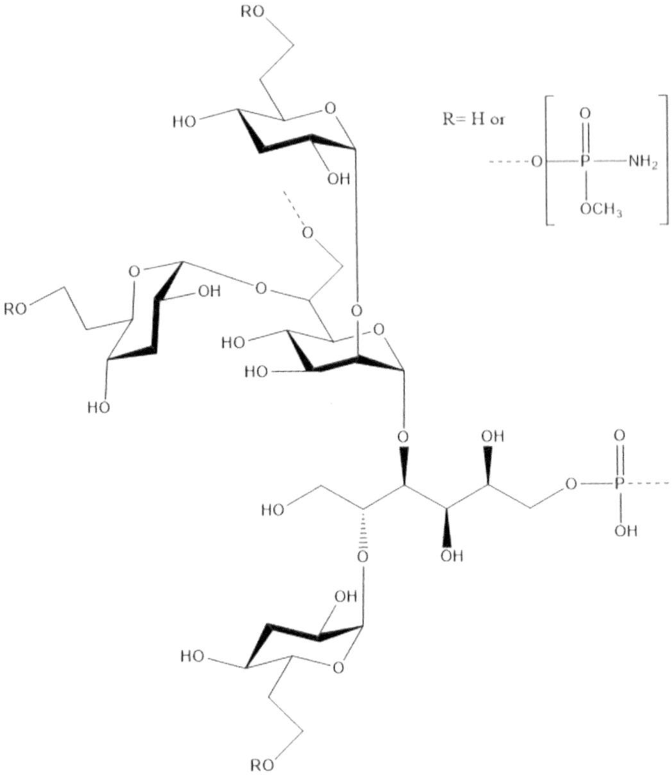

Figure 1. A structural variant of the oligosaccharide-phosphate repeating block of C. jejuni HS5 (49).

Serotype Complex HS15

The HS15 type strain ATCC 43442 core CPS structure is [→3)-6d-α-L-*gulo*-Hep*p*-(1→3)-α-L-Ara*f*-(1→]$_n$. This CPS structure has a non-stoichiometric α-L-Ara*f* residue non-reducing end (*53*), but the mechanism by which this α-L-Ara*f* is added to the HS15 core CPS unit remains unknown. No MeOPN was detected during structural analysis of the ATCC 43442 CPS, but the MeOPN biosynthesis and transferase genes are annotated to be present in the HS15 CPS bioshyntesis locus. Thus, the ATCC 43442 strain used for structural analysis likely contained a phase-off MeOPN transferase gene. For this reason, the location(s) of the MeOPN modification(s) on HS15 CPS are unknown at this time. The 6-deoxy group of the heptose appears was stoichiometric during structural studies. However, since the *dmhA* heptose dehydratase gene is phase-variable and hence non-stoichiometric in most C. *jejuni* strains containing *dmhA* and 6-dexoxyheptose, it may be possible that this 6-dehydration of *glycero*-β-D-altro-Hep*f* is also phase variable but is 100% phase-on in the ATCC 43442 strain used in these studies.

Serotype Complex HS19

The HS19 core CPS structure of strain ATCC 42446 (MK104) is [→4)-β-D-Glc*p*A6NGro-(1→3)-β-D-Glc*p*NAc-(1→]$_n$ (*54*). While the 6NGro group is stoichiometrically present on the β-D-Glc*p*A6NGro residue, a non-stoichiometric α-L-sorbofuranose (α-L-Sor*f*) ketohexose group occurs at the 2-OH of the β-D-Glc*p*A6NGro residue. Similarly to the β-Fru*f* ketohexose residues of HS1/44 CPS, the α-L-Sor*f* ketohexose residue is also highly acid-labile. Additionally, a non-stoichiometric MeOPN modification was identified, which was shown to occur on the 4-OH group of the β-D-Glc*p*NAc residue. Interestingly, structural analysis also revealed an appreciable amount of phosphoramidate in place of MeOPN at the 4-OH of GlcNAc. As described above, the biological significance of phosphoramidate on *C. jejuni* CPS remains to be determined.

Serotype Complex HS23/36

The main CPS structures of the HS23/36 complex are [→3)-α-D-Gal*p*-(1→2)-6d-3-*O*-Me-α-D-*altro*-Hep*p*-(1→3)-β-D-Glc*p*NAc-(1]$_n$ or [→3)-β-D-Glc*p*NAc-(1→3)-α-D-Gal*p*-(1→2)-D-*glycero*-α-D-*altro*-Hep*p*-(1→]$_n$ (*44*, *55*). The HS23/36 seroclass consists of strains that are type HS23, HS36, and both HS23 and HS36 (HS23/36) by the Penner serotyping scheme, and sequencing of the CPS biosynthesis loci has revealed >97% nucleotide identity between the HS23 type strain, the HS36 type strain, and the HS23/36 strain 81-176. The trisaccharide repeat structure of HS23/36 also contains multiple non-stiochiometric modifications. The heptose residue can be methylated at the O-3 position and/or dehydrated at the 6-position. Moreover, three phase-variable MeOPN modifications can occur on the Gal residue at the 2-OH, 4-OH, or 6-OH group. The *cjj1420* MeOPN transferase was recently shown to transfer MeOPN to the 4-OH of Gal, while the *cjj1435* gene encodes a bifunctional MeOPN transferase that transfers MeOPN to both the 2-OH and 6-OH positions. The *cjj1435* MeOPN transferase preferentially adds the MeOPN to the 2-OH position vs. 6-OH position of Gal, but the mechanistic reason for this preference is currently unclear. Moreover, whether multiple MeOPN modifications can occur on a single Gal residue is unclear at this time. Interestingly, the *waaC* gene encoding a heptosyltransferase outside the CPS locus is required for 3-O-methylation of heptose (*44*) and may also play a role in the MeOPN modification of HS23/36 CPS. However, whether this enzyme acts as a methyltransferase enzyme or if this gene is indirectly required for heptose methylation and MeOPN biosynthesis has yet to be experimentally determined.

Serotype Complex HS41

The HS41 strain 176.83 (a non-type strain) core CPS structure is [→2)-β-L-Ara*f*-(1→2)-6d-β-D-*altro*-Hep*f*-(1→2)-α-D-Fuc*f*-(1→]$_n$ and of [→2)-β-L-Ara*f*-(1→2)-6d-β-D-*altro*-Hep*f*-(1→2)-β-6d-L-Alt*f*-(1→]$_n$ (*56*). The only difference between the two structures is that the α-D-Fuc*f* is substituted with 6d-β-L-Alt*f*. α-D-Fuc*f* and 6d-β-L-Alt*f* only vary from one another in that they are C5 epimers, indicating that a non-stoichiometric C5 epimerase enzyme is responsible for these two different residues. To date, this C5 epimerase enzyme has yet to be identified and, until this enzyme is identified, it is unclear which monosaccharide residue represents the "core" monosaccharide residue of the HS41 CPS repeat unit. The HS41 CPS biosynthesis locus lacks genes involved in MeOPN biosynthesis, highly suggesting that HS41 CPS does not contain the MeOPN modification. The HS41 CPS structure is unique among known *C. jejuni* CPS in that the heptose is in the furanose rather than pyranose ring form. The 6-deoxy form of this heptofuranose was stoichiometric in these structural studies (*11*, *28*). Nvertheless, as described above for HS15 CPS structure, it cannot be ruled out that this modification is actually phase-variable and merely 100% phase-on in the 176.83 strain used in these studies.

Serotype Complex HS53

The HS53 type strain RM1221 has a relatively unique poly-heptose core CPS structure of [→3)-β-6d-D-*manno*-Hep*p*-(1→3)-α-6d-D-*manno*-Hep*p*-(1→3)-α-6d-D-*manno*-Hep*p*-(1-P→]$_n$ (*57*). The HS53 core CPS structure can be non-stoichiometrically modified with the ketohexose D-xylulose (D-*threo*-pent-2-ulose) in the furanose form (D-Xlu*f*) at the 2- and/or 4-OH positions of the β-6d-D-*manno*-Hep*p* residue, although the two D-Xlu*f* anomeric configurations are still unclear. Similarly to the non-stoichiometric ketohexose groups in HS1/44 and HS19 CPS, the α-D-Xlu*f* linkage in HS53 is highly acid-labile. Similarly to HS41 CPS biosynthesis locus, the HS53 locus lacks genes responsible for MeOPN biosynthesis. The HS53 CPS biosynthesis locus contains a putative *dmhA* gene, which correlate to the presence of the 6-deoxy form of D-*manno*-Hep*p* in the HS53 CPS structure. As described above for the HS15 and HS41 CPS structures, it cannot be ruled out that the 6-deoxy form of 6d-D-*manno*-Hep*p* is actually phase-variable and merely 100% phase-on in the RM1221 strain used in these studies.

Campylobacter jejuni Glycoconjugate Vaccine

C. jejuni CPS present a myriad of unique structural and conformational features representing unique targets for an anti-*C. jejuni* vaccine. Still, the range of unique CPS structures (albeit similar within serotype clusters), numerous phase-variable modifications, repeated *C. jejuni* infections observed in children in endemic areas, and repeated *C. jejuni* infection by the same serotype indicate that further study is required to fully understand the specificity of the immune response after *C. jejuni* exposure. (*18*, *58*). Nevertheless, conjugate vaccine development and manufacturing efforts are intricate and require optimization of multiple factors, from CPS discovery to CPS conjugation methodology.

Valency

Based on Penner serotyping, at least 35 different CPS structures exist within the *C. jejuni* sp. Due to cost, an efficient *C. jejuni* conjugate vaccine needs to target the most prevalent CPS types responsible for campylobacteriosis. A systematic review of the last 30 years of Penner serotyping analysis revealed that the HS4 complex, HS1/44, and HS2 CPS serotypes accounted for more than half of the sporadic Campylobacteriosis cases. Unfortunately, the majority of the typing was performed on cases from developed countries. Data is lacking from other regions of the world where the vaccine is needed the most. To address this deficiencyof serotyping data in endemic regions, a PCR based typing methods was developed that is currently able to characterize the 35 *C. jejuni* CPS groups. This typing system was employed to type over 2,100 *C. jejuni* human isolates collected from Africa (*59*), Middle East , Southeast Asia (*24*, *25*), and South America. Results presented in Figure 2 demonstrate a CPS distribution quite different than the one observed in developed countries. Nevertheless, the results suggest that the HS4 complex is the most prevalent serotype and that there are regional differences of CPS distribution. This preliminary study demonstrates that a CPS conjugate vaccine targeting about 10 types would cover ~70% of Campylobacteriosis cases within these endemic regions.

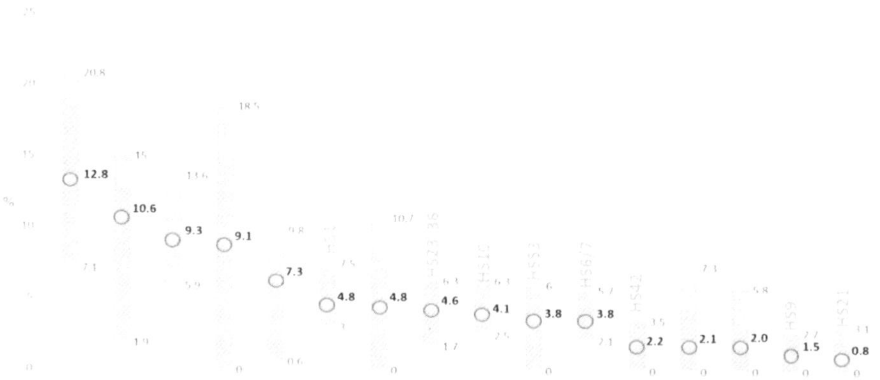

Figure 2. C. jejuni serotype distribution from the analysis of 2332 clinical isolates by CPS multiplex PCR (996 from Southeast Asia, 531 from Peru. 367 from Bangladesh, 120 from Israel and 318 from Egypt). Other serotypes constituted 15%.

Choice of Protein Carrier

The prototype *C. jejuni* glycoconjugate vaccine utilized a non-toxic diphtheria toxin protein, CRM_{197}, as the carrier (*60*). CRM_{197} has the advantage of being commercial available in large quantities and high purity as well as containing a large number of "privileged" lysine groups available for conjugation to activated *C. jejuni* CPS via reductive amination or carbodiimide chemistry (*60*, *61*). A shortcoming of CRM_{197} as the protein carrier is that humans are now routinely immunized against diphtheria toxin in diphtheria vaccines and other glycoconjugate vaccines containing CRM_{197} as the protein carrier (e.g.-the pneumococcal Prevnar13® vaccine). Therefore, the immune system may not have sufficient time to generate a sufficiently strong *C. jejuni* CPS-specific immune response before pre-existing immunity against the diphtheria toxin carrier clears the glycoconjugate from the body. Also, CRM_{197} does not generate any additional immune coverage against *C. jejuni* or other important pathogens to which humans are not routinely immunized. Therefore, future studies on *C. jejuni* vaccines will evaluate cell surface *C. jejuni* proteins as potential protein carriers in the *C. jejuni* glycoconjugate vaccine, which may help broaden vaccine coverage against *C. jejuni* by targeting multiple cell surface antigens in addition to CPS. The protein carriers may also be vaccine candidates against other enteric pathogens besides *C. jejuni* to generate a multivalent diarrheal disease vaccine, such as enterotoxigenic *Escherichia coli* fimbriae proteins.

Campylobacter jejuni Capsule Conjugate Vaccine Syntheses and Immunogenicity

Conjugate vaccines have been widely used and have been safe and efficacious in preventing disease against a number of encapsulated bacteria including *Hemophilus influenzae* type B, *Streptococcus pneumoniae*, and *Neisseria meningiditis*. Furthermore, multivalent glycoconjugate vaccine platforms have been successfully developed and marketed to include a quadrivalent *N. meningiditis* vaccine, 7- and 13-valent *S. pneumoniae* vaccines, and a multipathogen combination vaccine including *H. influenzae* type B and *N. meningiditis* conjugates. Because there are 35 CPS types, an efficacious *C. jejuni* conjugate vaccine will be multivalent. The valency required to provide broad protection is yet uncertain, however, recent data discussed above suggest that a final vaccine formulation containing ~10 CPS types may provide adequate coverage. Immunization of mice with escalating doses of purified *C. jejuni* CPS delivered alone or adsorbed to aluminum hydroxide demonstrated that the *C. jejuni* polysaccharide alone is poorly immunogenic (*60*). These data are consistent with a T cell-independent immune response induced by unconjugated polysaccharides. However, polysaccharide conjugation to a carrier protein has been shown to induce a T cell-dependent immune response and enhance immunogenicity of many bacterial capsule polysaccharides. Conjugation strategies have been developed by us several key *C. jejuni* CPS types (HS1, HS2, HS3, HS4/13/64, HS5, HS10, HS15, HS23/36 and HS53) to the diphtheria toxin protein CRM_{197}, and the CPS-CRM_{197} conjugates have proven immunogenic (*53*, *60*). Stoichiometric oxidation of the CPSs, to generate aldehydes or carboxylic acid moieties for conjugation to amines in the protein, were carried out by the well known periodate oxidation method (*60*) or by oxidation with 2,2,6,6-tetramethylpiperidine-1-oxyl (TEMPO) (*61*). Below, we describe key experiements carried out with some of the representative aforementioned conjugate vaccines.

The prototype CPSHS23/36-CRM_{197} vaccine has been the most extensively tested of the three conjugates (*60*). Monteiro and colleagues compared vaccination intervals and conjugate dose by immunizing mice subcutaneously at 0-2-4, 0-4-8 or 0-6-12 week intervals with 1, 5 or 25 μg of CPSHS23/36-CRM$_{197}$ (dosed by total conjugate weight) delivered without an adjuvant. Immunization with CPSHS23/36-CRM$_{197}$ primarily generated anti-CPSHS23/36 IgG responses and only the 25 μg dose at 4- and 6-week intervals induced significant levels of anti-CPSHS23/36 IgM compared to baseline. Anti-CPSHS23/36 IgG responses increased as a function of both antigen dose and increasing intervals between vaccinations (Figure 3A). At 2-week intervals, higher doses of CPSHS23/36-CRM$_{197}$ in the 5 and 25 μg groups induced a significant increase in IgG responses compared to baseline titers. Regardless of the dose of CPSHS23/36-CRM$_{197}$, all mice immunized at 4- or 6-week intervals seroconverted by a 4-fold rise criteria and this IgG response persisted >26 weeks after vaccination (Figure 3A).

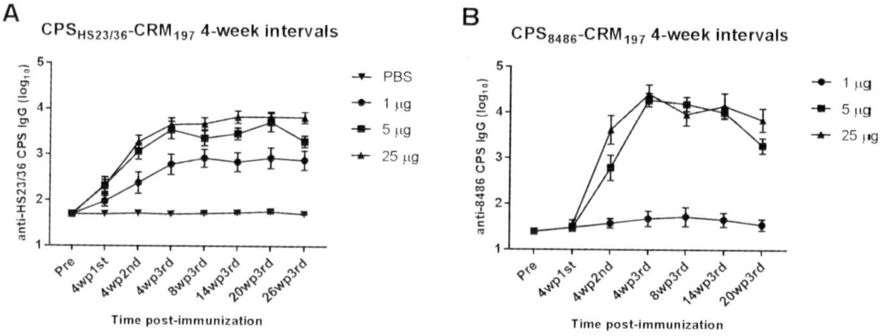

Figure 3. Immunogenicity of CPS$_{HS23/36}$-CRM$_{197}$ (A) and CPS$_{8486}$-CRM$_{197}$ (B) conjugates in mice. Mice were immunized subcutaneously three times with the respective conjugate or PBS at 4-week intervals with the indicated doses. Data represent mean log$_{10}$ anti-CPS titer ± SEM for each group at the indicated time point post 1st, 2nd or 3rd vaccination. Pre, pre-vaccination; 4wp1st, 4-weeks post 1st dose; 4wp2nd, 4-weeks post 2nd dose; 4wp3rd, 4-weeks post 3rd. Data are adapted from Monteiro et al. (60).

The pattern of response is consistent with boosting of immunological memory that can be best elicited only after an interval of 4 or more weeks have elapsed between prime and boost vaccinations. Only a modest induction of CPS-specific serum IgA was observed in mice immunized with 5 or 25 μg at 4-week intervals. Importantly, mice immunized with 5 or 25 μg CPSHS23/36-CRM$_{197}$ at either 2-, 4-, or 6-week intervals showed significant protection following intranasal challenge with *C. jejuni* strain 81-176. All CPSHS23/36-CRM197 vaccinated animals showed a significant reduction in illness index score after challenge compared to the sham immunized mice. These data suggest that anti-CPSHS23/36 IgG responses may be sufficient to mediate protection from *C. jejuni* infection in mice although the mechanism that affords this protection remains unknown.

There are currently no reliable small animal models of *C. jejuni* enteric infections and intestinal pathogenesis. While piglets, ferrets, rats, chickens and insects have been used as models, mouse models have required genetic manipulations or have failed to exhibit enteric disease. Indeed,

CPSHS23/36-CRM_{197} vaccine efficacy in the mouse intranasal challenge model described above may not translate into efficacy in a true enteric *C. jejuni* challenge. To examine conjugate vaccine efficacy in an enteric model, Monteiro and colleagues utilized a New World non-human primate *C. jejuni* enteric challenge model where infection of *Aotus nanycmaae* with high inoculum doses of 1010-1012 CFU of strain 81-176 induce diarrheal disease (*62*). The CPSHS23/36-CRM_{197} vaccine was administered to *A. nancymaae* subcutaneously at 1, 5 and 25 μg (total conjugate weight) with aluminum hydroxide in three doses at 6-week intervals (*60*). A dose dependent anti-CPS IgG response was observed where increasing doses of CPSHS23/36-CRM_{197} induced higher levels of anti-CPSHS23/36 IgG and only the 25 μg group had significantly higher IgG titers compared to the sham vaccinated animals (Table 1).

Animals were challenged with 1011 CFU of 81-176 nine weeks following the last immunization and a statistically significant trend towards protection from diarrhea was observed with increasing vaccine dose. Diarrhea attack rates were 40%, 20% and 0% in the 1, 5 and 25 μg groups, respectively, compared to a 60% attack rate in the sham vaccinated control animals (Table 1). A follow-on study was conducted with two new batches of CPSHS23/36-CRM_{197} to assess lot-to-lot reproducibility. Three doses of 25 μg of each conjugate plus aluminum hydroxide were administered at 6-week intervals. Immunized monkeys showed significantly higher anti-CPSHS23/36 IgG responses compared to control animals and all vaccinated animals were protected from 81-176 challenge. Interestingly, all monkeys were colonized with the challenge strain regardless of the vaccine dose administered demonstrating that anti-CPSHS23/36 IgG responses generated by the CPSHS23/36-CRM_{197} vaccine delivered with aluminum hydroxide are sufficient to protect against diarrheal disease, but not against *C. jejuni* colonization.

There were no significant differences in serum anti- CPSHS23/36 IgA responses between vaccinated and control animals and fecal IgA responses were not measured. The mechanism by which IgG affords protection from diarrhea but not colonization remains unclear, however, preliminary analysis indicates that vaccination with CPSHS23/36-CRM_{197} induces functional serum bactericidal IgG titers and we are currently exploring whether these responses contribute to protection in this model.

A second *C. jejuni* CPS conjugate vaccine that has been developed and evaluated is based on the CPS of strain CG8486, described above that types as HS4/13/64 (*60*), conjugated to CRM_{197}. The CPS8486-CRM_{19}7 vaccine was delivered to mice subcutaneously at 1, 5 and 25 μg (dosed by total conjugate weight) at 4-week intervals. While the low 1 μg dose failed to induce anti-CPS8486 IgG, 5 and 25 μg of CPS8486-CRM_{197} induced high levels of anti-CPS8486 IgG and 90% of the animals seroconverted (Figure 3B). In addition, animals in the 5 and 25 μg groups were protected from intranasal challenge with CG8486 which is similar to the efficacy observed with the CPSHS23/36-CRM_{197} vaccine in this model.

Table 1. Immunogenicity and protective efficacy of three lots of $CPS_{HS23/36}$-CRM_{197} against diarrheal disease in *Aotus nancymaae*. Data are adapted from Monteiro *et al.* (*60*).

Study	*Vaccine and dose*	*N*	*Mean anti-CPS IgG titer (log_{10})*	*Diarrhea attack rate (%) (no. of positive animals/total no.)*
1	PBS	5	2.54	60 (3/5)
	$CPSCPS_{HS23/36}$-CRM 1 μg	5	3.57	40 (2/5)
	$CPSCPS_{HS23/36}$-CRM 5 μg	5	5.38	20 (1/5)
	$CPSCPS_{HS23/36}$-CRM 25 μg	5	6.47	0 (0/5)
2	PBS	10	2.30	70 (7/10)
	$CPSCPS_{HS23/36}$-CRM Batch A 25 μg	9	5.09	0 (0/9)
	$CPSCPS_{HS23/36}$-CRM Batch B 25 μg	5	5.24	0 (0/5)

Another *C. jejuni* conjugate vaccine synthesized and characterized is based on the HS15 strain ATCC 43442 conjugated to CRM_{197} (*53*). A dose of 25 µg of the CPSHS15-CRM_{197} vaccine was administered to mice with aluminum hydroxide at 4-week intervals and anti-HS15 CPS antibodies were detected in the serum two-weeks after the third immunization. Although the antibody response to the CPSHS15-CRM_{197} vaccine was not well defined in this study due to a difficulties using the HS15 CPS in standard immunologic assays.

Conjugation methods using periodate oxidation (Figure 4) or TEMPO-mediated oxidation (Figure 5) for the HS53 CPS to CRM_{197} and other proteins have also been developed (*49*) and immunogenicity of this vaccine has recently been tested in mice. Gel-electrophoresis (Figure 6) has become the tool of choice to confirm and analyze the products of conjugations.

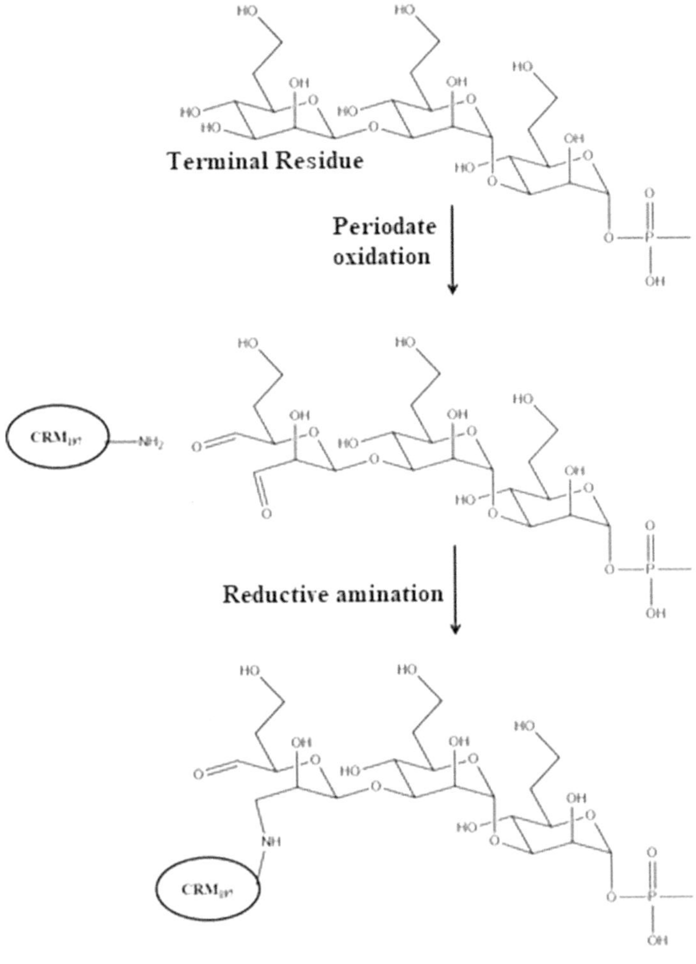

Figure 4. Conjugation of C. jejuni HS53 CPS to protein carrier CRM_{197} by periodate oxidation/reductive amination method. The monosaccharide at the non-reducing end was first oxidized with periodate and then conjugated coupled to a protein by reductive amination (49).

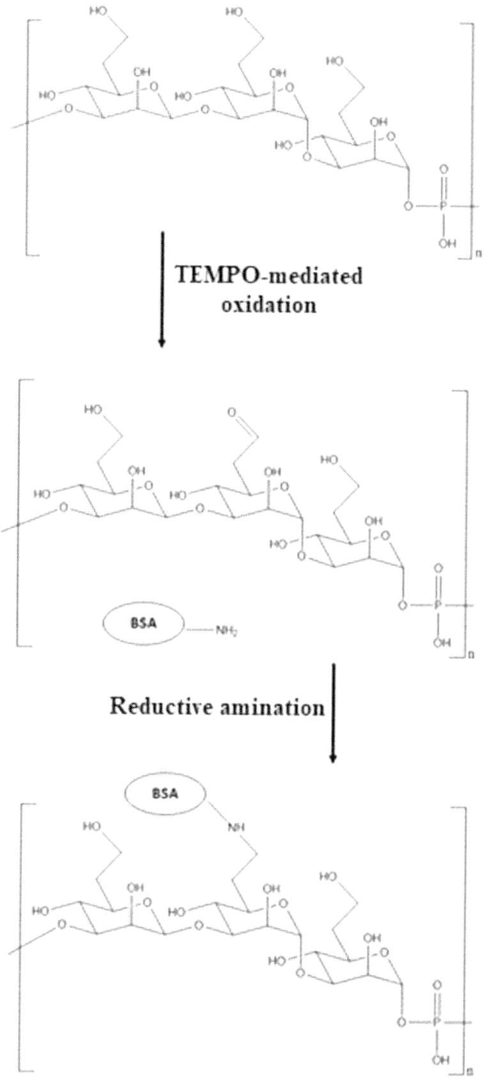

Figure 5. Activation of C. jejuni HS53 CPS by TEMPO-mediated oxidation. 10% of the primary hydroxyls (C-7 of heptoses) in the CPS were first stoichiometrically oxidized with TEMPO/bleach to aldehyde or carboxylic acid, and then coupled to a protein by reductive amination (in case of aldehyde) or carbodiimide chemistry (in case of carboxylic acid) (61).

Animals were dosed with increasing amounts of the CPSHS53-CRM_{197} conjugate at 0.5, 3.5 and 10 µg (dosed by total CPS weight, not total conjugate weight) at 4-week intervals without an adjuvant. Lower doses of 0.5 and 3.5 µg were not immunogenic after three doses, however, 10 µg induced anti-HS53 CPS IgG responses at levels comparable to titers observed with the CPSHS23/36-CRM_{197} vaccine (Figure 7). Conjugates containing the CPSs of

HS1, HS2, HS3, HS5 and HS10 have also been observed to raise antibodies against the corresponding CPSs (*47*, *49*, *52*, *63*). Upcoming experiments aim to optimize conjugation methods and analytical methods to evaluate these and other new CPS-CRM_{197} conjugate vaccines.

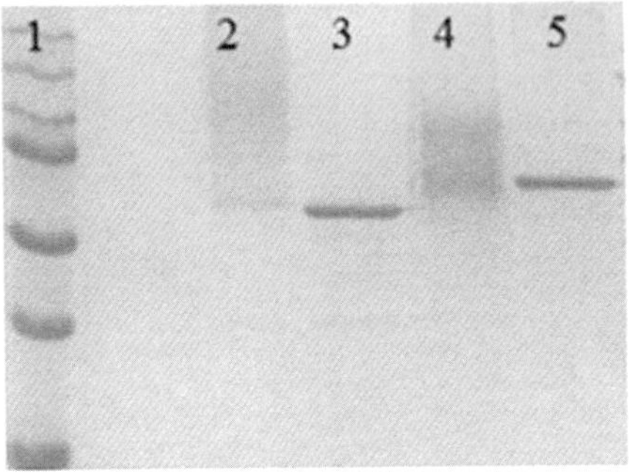

Figure 6. Analysis of CPSHS53 conjugates by gel-electrophoresi0s. Lane 1: molecular weight marker; Lane 2: CPSHS53-CRM197 conjugate obtained via periodate oxidation/reductive amination method; Lane 3: CRM197; Lane 4: CPSHS53-BSA conjugate generated by TEMPO-mediated oxidation/carbodiimide method; Lane 5: BSA (49).

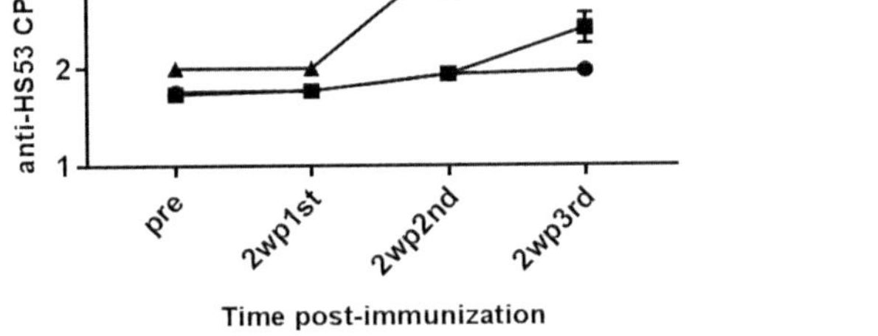

Figure 7. Immunogenicity of CPS_{HS53}-CRM_{197} conjugate vaccine in mice. Mice were immunized subcutaneously three times with the CPS_{HS53}-CRM_{197} conjugate at 4-week intervals with the indicated doses. Data represent mean log_{10} anti-CPS_{HS53} titer ± SEM for each group at the indicated time point post 1st, 2nd or 3rd vaccination. Pre, pre-vaccination; 2wp1st, 2-weeks post 1st dose; 2wp2nd, 2-weeks post 2nd dose; 2wp3rd, 2-weeks post 3rd.

The prototype HS23/36-CRM_{197} conjugate vaccine has been produced under cGMP and is now being evaluated in a phase I human clinical trial. Future studies must also target combination of monovalent *C. jejuni* conjugate vaccines into multivalent formulations. As conjugation methods for additional CPS types are established, monovalent and multivalent formulations will be examined. Preliminary studies have indicated that two monovalent CPS conjugates can be successfully combined into a bivalent formulation where immunogenicity to each CPS type is maintained. In addition, pathogen-relevant proteins are being explored as carriers for *C. jejuni* CPS conjugate vaccines. *C. jejuni* proteins and Enterotoxigenic *Escherichia coli* (ETEC) fimbriae proteins have been successfully conjugated to *C. jejuni* CPS in our laboratories and show promise of developing a conjugate vaccine that induce not just CPS-specific antibody responses, but also the potential for antibody and T cell-mediated responses against pathogen-relevant carrier proteins (*49*, *63*, *64*).

Conclusions

C. jejuni represents a threat to global health and a vaccine would be welcomed. *C. jejuni* exposes sero specific CPSs, whose main characteristic is the presence of 6-deoxy-heptoses and MeOPN moieties. Each serotype complex generates CPSs with specific 6-deoxy-heptoses (many of unusual configuration) and MeOPN decorations at selected linkage sites. Strains that fall within a serotype complex share structural characteristics, in that only a single CPS conjugate, containing only a CPS from a single strain, affords protection against the other strains of the same serotype family.

Immunogenic CPS-CRM_{197} conjugates have been made for all significant serotypes. So far, the CPS conjugate of serotype complex HS23/36 (strain 81-176) has been the most studied, and protection experiments in a non-human primate model have shown full efficacy in preventing diarrhea against HS23/36 strains. The prototype HS23/36 CPS conjugate vaccine was manufactured under cGMP and it is now in a phase I human clinical trial. Eventually, a *C. jejuni* vaccine will have to be multivalent. Recent trials have shown that a divalent vaccine, composed of HS23/36 and HS3 CPS conjugates, is indeed capable of equally raising antibodies against both serotypes. This encouraging result opens the door to a future multivalent *C. jejuni* vaccine that may be used globally.

Acknowledgments

Funding for this work was received through the Natural Sciences and Engineering Research Council (Canada) and Naval Medical Reseacrch Centre (USA).

References

1. Dekeyser, P.; Gossuin-Detrain, M.; Butzler, J. P.; Sternon, J. *J. Infect. Dis.* **1972**, *125*, 390–392.
2. Butzler, J. P. *Clin. Microbiol. Infect.* **2004**, *10*, 868–876.
3. Klena, J. D.; Parker, C. T.; Knibb, K.; Ibbitt, J. C.; Devane, P. M.; Horn, S. T.; Miller, W. G.; Konkel, M. E. *J. Clin. Microbiol.* **2004**, *42*, 5549–5557.
4. Preliminary FoodNet data on the incidence of infection with pathogens transmitted commonly through food-selected sites, United States, 2003. *MMWR Morb. Mortal. Wkly. Rep.* **2004**, *53*, 338–343.
5. EU summary report on zoonoses, zoonotic agents and food-borne outbreaks 2015. *EFSA J.* **2016**, *14*, 4634. https://doi.org/10.2903/j.efsa.2016.4634.
6. Riddle, M. S.; Sanders, J. W.; Putnam, S. D.; Tribble, D. R. *Am. J. Trop. Med. Hyg.* **2006**, *74*, 891–900.
7. Premarathne, J.; Satharasinghe, D. A.; Huat, J. T. Y.; Basri, D. F.; Rukayadi, Y.; Nakaguchi, Y.; Nishibuchi, M.; Radu, S. *Crit. Rev. Food. Sci. Nutr.* **2017**, *57*, 3971–3986.
8. Guerry, P.; Poly, F.; Riddle, M.; Maue, A. C.; Chen, Y. H.; Monteiro, M. A. *Front. Cell. Infect. Microbiol.* **2012**, *2*, 7.
9. Beery, J. T.; Hugdahl, M. B.; Doyle, M. P. *Appl. Environ. Microbiol.* **1988**, *54*, 2365–2370.
10. Rouger, A.; Tresse, O.; Zagorec, M. *Microorganisms* **2017**, *5*, 50.
11. Black, R. E.; Levine, M. M.; Clements, M. L.; Hughes, T. P.; Blaser, M. J. *J. Infect. Dis.* **1988**, *157*, 472–479.
12. Janssen, R.; Krogfelt, K. A.; Cawthraw, S. A.; van Pelt, W.; Wagenaar, J. A.; Owen, R. J. *Clin. Microbiol. Rev.* **2008**, 505–518.
13. Pike, B. L.; Guerry, P.; Poly, F. *PloS One* **2013**, *8*, e67375.
14. Nachamkin, I.; Allos, B. M.; Ho, T. *Clin. Microbiol. Rev.* **1998**, *11*, 555–567.
15. Pequegnat, B.; Laird, R. M.; Ewing, C. P.; Hill, C. L.; Omari, E.; Poly, F.; Monteiro, M. A.; Guerry, P. *J. Bacteriol.* **2017**, *199*, e00027. https://doi.org/10.1128/JB.00027-17.
16. Hannu, T.; Kauppi, M.; Tuomala, M.; Laaksonen, I.; Klemets, P.; Kuusi, M. *J. Rheumatol.* **2004**, *31*, 528–530.
17. Lecuit, M.; Abachin, E.; Martin, A.; Poyart, C.; Pochart, P.; Suarez, F.; Bengoufa, D.; Feuillard, J.; Lavergne, A.; Gordon, J. I.; Berche, P.; Guillevin, L.; Lortholary, O. *N. Engl. J. Med.* **2004**, *350*, 239–248.
18. Lee, G.; Pan, W.; Penataro, Y. P.; Paredes, O. M.; Tilley, D.; Gregory, M.; Oberhelman, R.; Burga, R.; Chavez, C. B.; Kosek, M. *PLoS Neglected Trop. Dis.* **2013**, *7*, e2036.
19. Aspinall, G. O.; Monteiro, M. A.; Pang, H. *Carbohydr. Res.* **1995**, *279*, 245–264.
20. Parkhill, J.; Wren, B. W.; Mungall, K.; Ketley, J. M.; Churcher, C.; Basham, D.; Chillingworth, T.; Davies, R. M.; Feltwell, T.; Holroyd, S.; Jagels, K.; Karlyshev, A. V.; Moule, S.; Pallen, M. J.; Penn, C. W.; Quail, M. A.; Rajandream, M. A.; Rutherford, K. M.; van Vliet, A. H.; Whitehead, S.; Barrell, B. G. *Nature.* **2000**, *403*, 665–668.

21. Cress, B. F.; Englaender, J. A.; He, W.; Kasper, D.; Linhardt, R. J.; Koffas, M. A. *FEMS Microbiol. Rev.* **2014**, *38*, 660–697.
22. Karlyshev, A. V.; Linton, D.; Gregson, N. A.; Lastovica, A. J.; Wren, B. W. *Mol. Microbiol.* **2000**, *35*, 529–541.
23. Corcoran, A. T.; Annuk, H.; Moran, A. P. *FEMS Microbiol. Lett.* **2006**, *257*, 228–235.
24. Karlyshev, A. V.; Champion, O. L.; Churcher, C.; Brisson, J. R.; Jarrell, H. C.; Gilbert, M.; Brochu, D.; St. Michael, F.; Li, J.; Wakarchuk, W. W.; Goodhead, I.; Sanders, M.; Stevens, K.; White, B.; Parkhill, J.; Wren, B. W.; Szymanski, C. M. *Mol. Microbiol.* **2005**, *55*, 90–103.
25. Poly, F.; Serichatalergs, O.; Schulman, M.; Ju, J.; Cates, C. N.; Kanipes, M.; Mason, C.; Guerry, P. *J. Clin. Microbiol.* **2011**, *49*, 1750–1757.
26. Poly, F.; Serichantalergs, O.; Kuroiwa, J.; Pootong, P.; Mason, C.; Guerry, P.; Parker, C. T. *PloS one.* **2015**, *10*, e0144349doi:10.1371/journal.pone.0144349.
27. Penner, J. L.; Hennessy, J. N. *J. Clin. Microbiol.* **1980**, *12*, 732–737.
28. Mills, S. D.; Congi, R. V.; Hennessy, J. N.; Penner, J. L. *J. Clin. Microbiol.* **1991**, *29*, 2093–2098.
29. Penner, J. L.; Hennessy, J. N.; Congi, R. V. *Eur. J. Clin. Microbiol.* **1983**, *2*, 378–283.
30. Poly, F.; Serichantalergs, O.; Kuroiwa, J.; Pootong, P.; Mason, C.; Guerry, P.; Parker, C. T. *PLoS One.* **2016**, *11*, e0151410DOI: 10.1371/journal.pone.0151410.
31. Dorrell, N.; Mangan, J. A.; Laing, K. G.; Hinds, J.; Linton, D.; Al-Ghusein, H.; Barrell, B. G.; Parkhill, J.; Stoker, N. G.; Karlyshev, A. V.; Butcher, P. D.; Wren, B. W. *Genome. Res.* **2001**, *11*, 1706–1715.
32. van Belkum, A.; Scherer, S.; van Alphen, L.; Verbrugh, H. *Microbiol. Mol. Biol. Rev.* **1998**, *62*, 275–293.
33. Bacon, D. J.; Szymanski, C. M.; Burr, D. H.; Silver, R. P.; Alm, R. A.; Guerry, P. *Mol. Microbiol.* **2001**, *40*, 769–777.
34. McNally, D. J.; Lamoureux, M. P.; Karlyshev, A. V.; Fiori, L. M.; Li, J.; Thacker, G.; Coleman, R. A.; Khieu, N. H.; Wren, B. W.; Brisson, J. R.; Jarrell, H. C.; Szymanski, C. M. *J. Biol. Chem.* **2007**, *282*, 28566–28576.
35. Taylor, Z. W.; Brown, H. A.; Narindoshvili, T.; Wenzel, C. Q.; Szymanski, C. M.; Holden, H. M.; Raushel, F. M. *J. Am. Chem. Soc.* **2017**, *139*, 9463–9466.
36. McNally, D. J.; Jarrell, H. C.; Khieu, N. H.; Li, J.; Vinogradov, E.; Whitfield, D. M.; Szymanski, C. M.; Brisson, J. R. *FEBS J.* **2006**, *273*, 3975–3989.
37. Szymanski, C. M.; St. Michael, F.; Jarrell, H. C.; Li, J.; Gilbert, M.; Larocque, S.; Vinogradov, E.; Brisson, J. R. *J. Biol. Chem.* **2003**, *278*, 24509–24520.
38. MacDonald, A. G. Ph.D. thesis, York University, Toronto, ON, Canada, 1993.
39. Aspinall, G. O. *Carbohydr. Eur.* **1998**, 2124–2129.
40. Aspinall, G. O.; McDonald, A. G.; Raju, T. S.; Pang, H.; Moran, A. P.; Penner, J. L. *Eur. J. Biochem.* **1993**, *213*, 1017–1027.

41. McNally, D. J.; Jarrell, H. C.; Li, J.; Khieu, N. H.; Vinogradov, E.; Szymanski, C. M.; Brisson, J. R. *FEBS J.* **2005**, *272*, 4407–4422.
42. Gabryelski, P. M.Sc. thesis, University of Guelph, Guelph, ON, Canada, 2016. http://hdl.handle.net/10214/9505 (accessed 15/4/2018).
43. St. Michael, F.; Szymanski, C. M.; Li, J.; Chan, K. H.; Khieu, N. H.; Larocque, S.; Wakarchuk, W. W.; Brisson, J. R.; Monteiro, M. A. *Eur. J. Biochem.* **2002**, *269*, 5119–5136.
44. Kanipes, M. I.; Papp-Szabo, E.; Guerry, P.; Monteiro, M. A. *J. Bacteriol.* **2006**, *188*, 3273–3279.
45. McNally, D. J.; Jarrell, H. C.; Khieu, N. H.; Li, J.; Vinogradov, E.; Whitfield, D. M.; Szymanski, C. M.; Brisson, J. R. *FEBS J.* **2006**, *273*, 3975–3989.
46. Aspinall, G. O.; Lynch, C. M.; Pang, H.; Shaver, R. T.; Moran, A. P. *Eur. J. Biochem.* **1995**, *231*, 570–578.
47. Chen, Y.-H. Ph.D. thesis, University of Guelph, Guelph, ON, Canada, 2011.
48. Chen, Y,-H.; Poly, F.; Pakulski, Z.; Guerry, P.; Monteiro, M. A. *Carbohydr. Res.* **2008**, *343*, 1034–1040.
49. Pequegnat, B. M. Ph.D. thesis, University of Guelph, Guelph, ON, Canada, 2016. http://hdl.handle.net/10214/9737 (accessed 15/4/2018).
50. Muldoon, J.; Shashkov, A. S.; Moran, A. P.; Ferris, J. A.; Senchenkova, S. N.; Savage, A. V. *Carbohydr. Res.* **2002**, *337*, 2223–2229.
51. Shin, J. E.; Ackloo, S.; Mainkar, A. S.; Monteiro, M. A.; Pang, H.; Penner, J. L.; Aspinall, G. O. *Carbohydr. Res.* **1997**, *305*, 223–232.
52. DePass, M.Sc. thesis, University of Guelph, Guelph, ON, Canada, 2011. http://hdl.handle.net/10214/3192 (accessed 15/4/2018).
53. Bertolo, L.; Ewing, C. P.; Maue, A.; Poly, F.; Guerry, P.; Monteiro, M. A. *Carbohydr Res.* **2013**, *366*, 45–49.
54. Aspinall, G. O.; McDonald, A. G.; Pang, H. *Biochemistry* **1994**, *33*, 250–255.
55. Aspinall, G. O.; McDonald, A. G.; Pang, H. *Carbohydr. Res.* **1992**, *231*, 13–30.
56. Hanniffy, O. M.; Shashkov, A. S.; Moran, A. P.; Prendergast, M. M.; Senchenkova, S. N.; Knirel, Y. A.; Savage, A. V. *Carbohydr. Res.* **1999**, *319*, 124–132.
57. Gilbert, M.; Mandrell, R. E.; Parker, C. T.; Li, J.; Vinogradov, E. *ChemBioChem* **2007**, *8*, 625–631.
58. Kirkpatrick, B. D.; Lyon, C. E.; Porter, C. K.; Maue, A. C.; Guerry, P.; Pierce, K. K.; Carmolli, M. P.; Riddle, M. S.; Larsson, C. J.; Hawk, D.; Dill, E. A.; Fingar, A.; Poly, F.; Fimlaid, K. A.; Hoq, F.; Tribble, D. R. *Clin. Infect. Dis.* **2013**, *57*, 1106–1113.
59. Sainato, R.; ElGendy, A.; Poly, F.; Kuroiwa, J.; Guerry, P.; Riddle, M. S.; Porter, C. K. *Am. J. Trop. Med. Hyg.* **2018**, *98*, 581–585.
60. Monteiro, M. A.; Baqar, S.; Hall, E. R.; Chen, Y.-H.; Porter, C. K.; Bentzel, D E.; Applebee, L.; Guerry, P. *Infect. Immun.* **2009**, *77*, 1128–1136.
61. Ma, Z.; Bertolo, L.; Arar, S.; Monteiro, M. A. *Carbohydr. Res.* **2011**, *346*, 343–347.

62. Jones, F. R.; Baqar, S.; Gozalo, A.; Nunez, G.; Espinoza, N.; Reyes, S. M.; Salazar, M.; Meza, R.; Porter, C. K.; Walz, S. E. *Infect. Immun.* **2006**, *74*, 790–793.
63. Omari, E. M.Sc. Thesis, University of Guelph, Guelph, ON, Canada, 2015. http://hdl.handle.net/10214/9130 (accessed 15/4/2018).
64. Guerry, P.; Monteiro, M. A.; Savarino, S. U.S. Patent US20170266300A9, 2017.

Chapter 12

Chemistry Manufacturing, Control, and Licensure for Carbohydrate-Based Vaccines

Christopher Jones*

St Albans, Herts, United Kingdom AL4 0DW
***E-mail: chrisjones715@btinternet.com.**

Current glycoconjugate vaccines are a group of products with similar gross structures, are manufactured by common processes, under common regulatory requirements and with similar critical quality attributes (CQAs). The analytical methods used to characterise and control the quality of this family of products are similar. Due to their relatively simple structures, high purity and the limitations of *in vivo* approaches, physicochemical methods have been widely used to characterise these vaccines and predominate for the control of these products. Product control typically uses a subset of the methods used for characterisation. This chapter attempts to provide a structured guide to the manufacturing processes, regulatory requirements and analytical methods applied to current products, and in the expectation that similar requirements will exist for future novel products.

Introduction

Antibodies directed against cell surface polysaccharides allow the human immune system to attack and kill invasive bacteria, and provide immune protection against a range of infections (*1*, *2*). Covalent conjugation of the polysaccharide, or oligosaccharide fragments derived from it, to a suitable carrier protein stimulates an enhanced immune response compared to unconjugated polysaccharides. The response is elicited through a different molecular mechanism than that for unconjugated polysaccharides, stimulates a protective response in infants, and, in many cases reduces carriage of and transmission of the bacterial pathogens. Reduced transmission establishes a degree of protection for non-vaccinated individuals, called "herd immunity", both for other infants and adults who are in contact with the infants (*3*). Glycoconjugate vaccines (*4*) against four pathogens are currently available. Those against *Haemophilus influenzae* Type b were the first to be introduced as monovalent products. With time, this immunogen has most often been used as a component of complex combination paediatric vaccines with diphtheria and tetanus toxoids and cellular or acellular pertussis immunogens. Meningococcal Group C vaccines were first introduced as monovalent products, but are frequently available now as tetravalent products, including Group A, W and Y components. A monovalent Group A vaccine was specifically developed to protect against epidemic meningococcal Group A infection in sub-Saharan Africa. These epidemics had previously appeared on a 5 to 12-year cycle, leading to many deaths (*5*). Pneumococcal conjugate vaccines were first introduced as a 7-valent product, but 10- and 13-valent products are now marketed and higher valency vaccines are under development. Typhoid Vi conjugates are the newest in the family, developed to fight the disease in developing countries. Only one such vaccine is currently licensed, but many are in development. Many other bacterial vaccines utilising the conjugate approach are in development (*2*).

Structures of Immunogens

Whilst the common feature of glycoconjugate immunogens is the attachment of an oligo- or poly-saccharide to a carrier protein, differences in conjugation approach can lead to structural variants. Two structural types are common, one has been used in a single licensed product and the fourth approach is known to give rise to immunogenic conjugates. These four structural families are shown in Figure 1.

"Fuzzy balls" [Figure 1a] arise from coupling of mono- or bi-functional oligosaccharides to a carrier, usually CRM_{197}. These are relatively low molecular weight, have a polysaccharide: protein ratio of ca. 0.4 to 1, and usually minimal crosslinking. Molecular weights in the range 94 -146 kDa have been reported for meningococcal-CRM_{197} glyco-conjugates (*6*). "Crosslinked network" immunogens arise from the conjugation of high molecular weight polysaccharides with multiple activation sites with carrier proteins also capable of multi-site attachment [Figure 1b]. Molecular masses in the range 0.62 to 12.2 MDa have been reported (*7*) for pneumococcal-CRM_{197} conjugates prepared by reductive amination, with a median value of ca. 3.5 MDa, determined using SEC-MALLS methods. The molecular weights of the starting polysaccharides were in the range 92 to 866 kDa. Weight average molecular weights (M_w) of approximately 10MDa have been reported for each of the four components of the tetravalent meningococcal-tetanus toxoid vaccine Nimenrix, determined using hydrodynamic methods (*8*). Merck conjugated size-reduced Hib PRP to a meningococcal outer membrane protein vesicle to produce a very high mass immunogen [Figure 1c] which elicited somewhat different immunological responses compared with other vaccines types, with a strong antibody response after the first dose (*9*). The "apples-on-a-branch" structural class [Figure 1d] arises by reaction of a high mass polysaccharide with multiple activation sites with carrier proteins containing only a single complementary reaction site, denying the possibility of network formation through the carrier protein.

Current glycoconjugate vaccines are heterogeneous products, and many of the analytical methods for characterisation and quality control return average values, rather than profiles of the distribution of forms. Sometimes an average value is insufficient to define fully the product, and the distribution of forms may need to be taken into account (*10*).

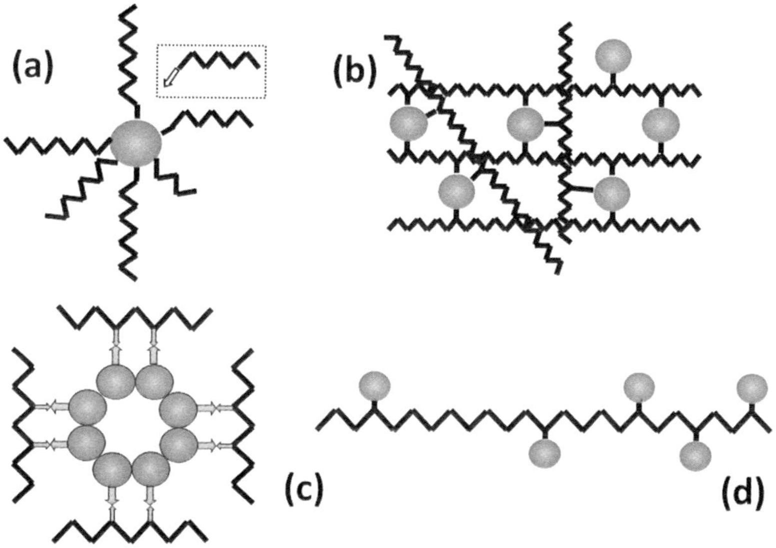

Figure 1. Cartoon structures of the four families of glycoconjugate immunogens (a) "Fuzzy ball", (b) Cross linked network immunogen, (c) Vesicle immunogen, and (d) "apples on a branch".

Compared to other products, polysaccharide and glycoconjugate vaccines depend very heavily on physicochemical methods for quality control and release (*11*). This is partly a consequence of the lack of accessible *in vivo* models which correlate with responses in humans, the relative simplicity of the individual components (intermediates, prior to conjugation) of these vaccines and their extensive physicochemical characterisation. Many of the key characterisation approaches have translated into quality control procedures after validation.

However, product licensing obviously still requires extensive *in vivo* preclinical analysis and well-controlled clinical trials in adults and infants to assess product immunogenicity, protection and safety. Increasingly, the effects of immunisation on pathogen carriage and transmission are being studied. The multistage manu-facturing process for glycoconjugate vaccines, with key intermediates of bulk purified polysaccharides and carrier proteins, is shown in Figure 2.

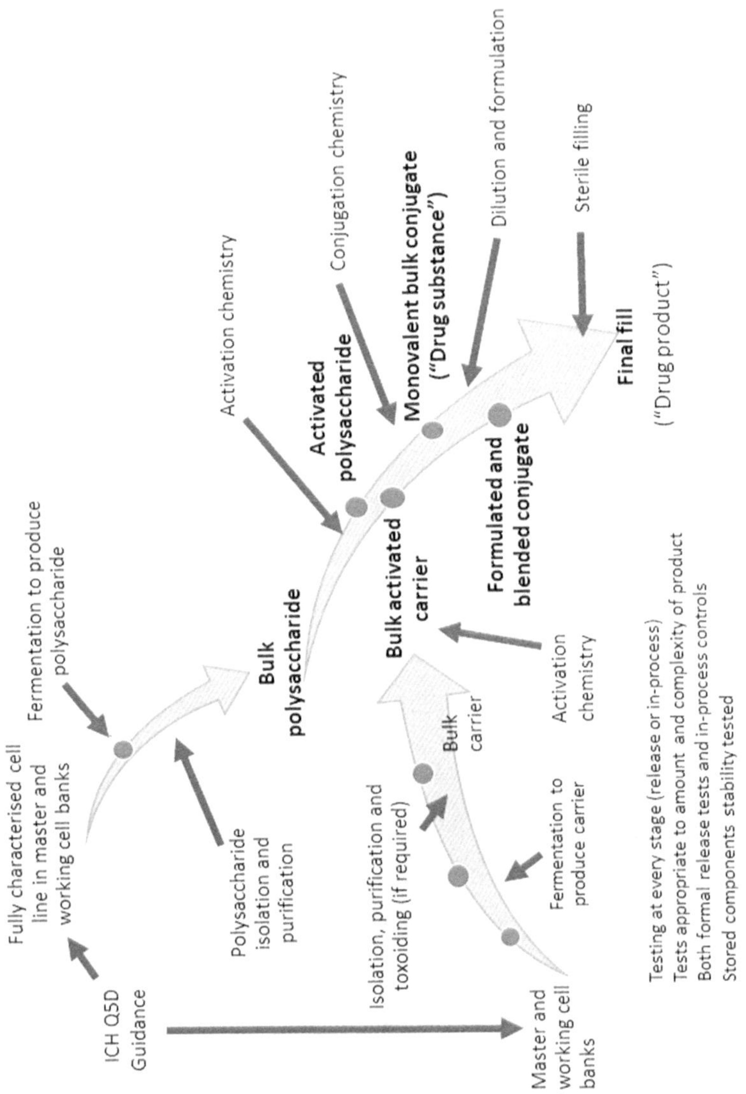

Figure 2. Process flow diagram of the manufacture of glycoconjugate vaccines, with progress from fully characterised cell lines to final product. Intermediate stages at which control may be carried out are highlighted with bold font.

Quality control occurs throughout the process, from the choice of and qualification of raw materials used in fermentation and bacterial strain characterisation through to ensuring consistency of filling and the quality of the final filled vials. The manufacturing facility must be licensed, and the facilities, equipment and records are inspected by regulatory authorities. Within this chapter, I will attempt to mention all of the key stages, but those aspects which are common across all biological and vaccine products will be only described briefly, and with references to available guidance. Coverage will focus on those vaccines which are currently marketed, but the principles can generally be expected to be applicable to novel products of this type. Vaccines licensed in the US are listed in Table 1. The United States Pharmacopeia has recognised this by considering these vaccines as a "Product Class", with common quality expectation and Critical Quality Attributes (CQAs).

Final product and in-process control and release testing is only one part of a fully validated manufacturing system that includes raw materials selection and testing, a validated stability testing programme, validated manufacturing facilities, including equipment, environmental control, cleaning and staff training, operating under Good Manufacturing Practices (GMP) (*12*).

Table 1. Licensed glycoconjugate vaccines (US, Dec 2017)

Product Name	*Material*	*Format*	*Manufacturer*
Pentacel	DTaP[a]-IPV[b]-Hi[b]	Liquid	Sanofi Pasteur
PedvaxHib	Hib-OMPC	Liquid	Merck & Co. Inc.
ActHIB	Hib-TTx	Lyophilised	Sanofi Pasteur
Hiberix	Hib-TTx	Lyophilised	GSK
Menveo	4Men[c] - CRM197	Liquid	GSK (Novartis)
MenHibrix	Men C/Y-TTx + Hib-TTx	Lyophilised	GSK
Menactra	4Men DTx	Liquid	Sanofi Pasteur
Nimenrix	4Men-TTx	Liquid	Pfizer (GSK)
Prevnar 13	13 valent PnC[d]-CRM_{197}	Liquid	Pfizer
Synflorix	10 valent PnC[d]-various	Liquid	GSK

[a] DTaP = diphtheria, tetanus and acellular pertussis. [b] IPV = inactivated polio virus. [c] 4Men = tetravalent meningococcal – Men A, Men C, Men Y and Men W. [d] PnC = pneumococcal polysaccharides.

Sources of Regulatory Information

There are many organisations providing regulatory iguidance, including the World Health Organisation (WHO), National Regulatory Authorities (NRAs), the *International Conference on Harmonisation* of Technical Requirements for Registration of Pharmaceuticals for Human Use (ICH), and Pharmacopoeias. The documentary guidance covers different aspects of the manufacturing and licensing procedures and explicit product-specific information. ICH Guidelines are frequently incorporated into national regulatory guidance and pharmacopoeial requirements, so that expectations are consistent.

The same property of an intermediate, bulk conjugate and final product may be measured at several stages during manufacture, for different purposes, by different methods, against different reference standards and requiring different extents of validation. Ensuring consistent and comparable results is a challenge. In general, CQAs which define the product require the highest degree of validation. Table 2 below highlights this, for quantification of the polysaccharide component of pneumococcal conjugate vaccines. The degree of validation required depends upon the developmental stage for the product (*13*): "fit for purpose" assays can be used in early product characterisation but more comprehensive studies are required for licensed products. The ICH Guideline ICH Q2(R1) "Validation of Analytical Procedures: Text and Methodology" describes (*14*) appropriate validation for different types of test – identity tests, impurity limit tests, impurity quantity tests, quantification/potency tests etc. This document defines key concepts, and terminology are defined. Validation for accuracy, precision, specificity, relative accuracy and limits of detection (LOD) and of quantification (LOQ) are likely to be required for quantitative tests.

In addition, the analyst is expected to be aware of the robustness of the assay: the effects of small errors in defined parameters or sample matrix on the quality of the final result, A comparison between tests based on different physicochemical properties is valuable for method validation. Whilst product-specific pharmacopoeial methods are generally considered fully validated methods when used with the supplied reference standard, they require verification for use in the manufacturer's laboratory (*15*). "Generic" pharmacopoeial methods which use a heterologous reference standard, such as dye-binding assays for proteins, are likely to require more detailed verification prior to use, or choice of an alternative, more appropriate reference standard, to ensure that they give accurate results for the specific material being analysed. Objective acceptance criteria derived from data on a number of final process lots, using the final form of the assay, and numerical outcomes are increasingly expected for all validated assays.

Table 2. Use of different methodologies for the same measurand at different manufacturing stages

Manufacturing stage	*Assay methodology*	*Outcome*	*Purpose*
Bulk PS	Colorimetric assay or degradation/ HPAEC vs monosaccharide or polysaccharide std.	Amount of material for next manufacturing step Purity assessment	Internal: Progress to next stage
Monovalent bulk	Colorimetric assay or degradation/ HPAEC vs monosaccharide or polysaccharide std.	Conjugation yield PS/protein ratio Free saccharide Amount of bulk material to blend for final product	Internal Specification Specification Internal
Blended bulks	Immunochemical: Rate nephelometry vs reference conjugate std.	Dilution for filling	Internal
Final fill	Immunochemical: Rate nephelometry vs reference conjugate	Dosage determination	Final release

Product Characterisation

Full characterisation of the product, and several early manufacturing batches, is a regulatory requirement. This will be for both the product and key intermediates, assessing composition, structure, purity and impurity profiles, stability and biological activities. Those character-isation tests which address what are considered key aspects of product quality, the safety, potency or stability of the material being tested, are typically those further developed into lot release assays.

Consistency of Manufacture

At the heart of the licensing procedure for any pharmaceutical product is the Phase III clinical trial: the proof that the chosen batch of material tested is safe and efficacious. Repeating this level of assessment is clearly unreasonable on a routine basis, and assays are chosen to ensure that the composition and properties of all subsequent manufact-uring batches are consistent with the batch(es) used in the clinical trial(s). During product development and subsequent manufacturing, many parameters are tracked to ensure a consistent and well-controlled process. Action limits are established which warn that process parameters differ from historical values, and that additional action may be required.

Where a product characteristic has been shown to be consistently within the specifications of the process, continued analytical testing on a routine basis may no longer be required – the analysis has been "validated out". Alternatively, results from in-process tests found predictive of final product quality may be incorporated into release criteria.

Stability

ICH Guideline Q5C "Stability testing of Biotechnological/Biological Products", is an international guideline (*16*) outlining stability testing requirements. It is expected that stability is assessed for key intermediates to set storage lifetimes, and for final fills to set shelf lives. For most glycoconjugate vaccines a shelf life of two to three years for the final product is usually approved by licensing authorities, if supported by sufficient experimental evidence.

Specifications and Control Values

Specifications are defined as "a list of tests, references to analytical procedures, and appropriate acceptance criteria which are numerical limits, ranges, or other criteria for the tests described (*17*). It establishes the set of criteria to which a drug substance, drug product or materials at other stages of its manufacture should conform to be considered acceptable for its intended use". These may be for final release of for an intermediate to be processed to the next stage. The ICH Guideline Q6B "Specifications: Test Procedures and Acceptance Criteria for Biotechnol-ogical/ Biological Products" provides guidance on setting of specific-ations. Action limits, which indicate compliant results, but which deviate from the range of values historically obtained, and which trigger further investigation are also likely to be established.

Polysaccharide Production

Polysaccharides are isolated and purified from bacterial cultures in a well-established process. Many manufacturers and contract manu-facturing organisations (CMOs) have this capability. In Gram negative organisms, such as Hib, the meningococcus and *S.* Typhi, the capsular polysaccharide is attached to the cell membrane through a lipid tail on the glycan. In the case of Hib and the meningococcus the linkage between the lipid and the CPS is very labile (*18*). The lipid anchor on the *S.* Typhi Vi CPS has a lipid anchor that resembles the LPS lipid A moiety (*19*). In Gram positive bacteria the CPS is covalently attached to peptide-glycan in the cell wall (*20*). Most bacterial polysaccharides are anionic in character, some are uncharged [for example, the pneumococcal types 7F and 14] and a small number are zwitterionic [for example, the pneumococcal Type 1].

Strain Characterisation and Cell Banking

Regulatory guidance for the characterisation and banking of non-recombinant cell lines, such as those used to produce the poly-saccharide components of glycoconjugate vaccines, are covered in the ICH Guideline Q5D "Derivation and Characterisation of Cell Substrates used for Production of Biotechnological/

Biological Products (*21*)." The expectation is that a master- and working-cell bank [MCB and WCB] approach will be used, and that the cell will be "grown and stored in a medium free of material of animal origin (*22*)." The WCB is produced by expansion of a vial from the MCB, and cells from a WCB are then further expanded to support manufacturing-scale fermentation. Manufacturers should consider the size of the MCB to ensure that it is sufficiently large to support the product throughout its lifetime.

Growth and Harvesting

Cells should be grown in media free of animal-based materials: soy-based medium has been used for the production of pneumococcal polysaccharides (*23*, *24*). Pathogenic organisms are typically killed (eg. with formaldehyde (*25*), 1% phenol for 2-12 hr at ambient temperature (*26*), or heating to 56 °C for 10-30 min. (*27*, *28*)) before harvesting the product. Thermal treatment may promote release of the polysaccharide from the cell debris by cleaving the labile pyrophosphate linkage. Cell material is removed by tangential flow centrifugation.

Polysaccharide Purification

CPS is typically purified by a combination of precipitation with increasing concentrations of isopropyl alcohol (*26*) or other organic solvents and, for anionic polysaccharides, co-precipitation with cetyltrimethyl-ammonium bromide (CTAB), a hydrophobic counterion. Fractional CTAB precipitation can also be used to remove impurities (*26*). Residual nucleic acid may be removed by nuclease digestion and diafiltration (*26*). Other techniques such as anion-exchange and gel permeation chromatography can be used. Removal of small molecule process impurities is typically by diafiltration. Often the bulk PS is stored as a powder of defined moisture content after a final precipitation with organic solvent. These polysaccharides tend to be extremely hygro-scopic and can contain between 5-30% by weight of volatiles (*26*).

Characterisation and Quality Control of Polysaccharide Bulks

In general terms, polysaccharides used in the manufacture of glycoconjugate vaccines should meet the requirements for those used in purified polysaccharide vaccines. The polysaccharides are, in general, moderately or highly flexible chains, as evidenced by relatively narrow line widths in NMR spectroscopy (*29*) and hydrodynamic measurement (*8*). For example, the persistence length in 10 pneumococcal poly-saccharides used in the manufacture of GSK's Synflorix were 4 to 9 nm (*30*). This facilitates the use of NMR spectroscopy in characterisation and quality control.

Table 3 below lists the release criteria for the intermediates and final fill for the 10-valent pneumococcal conjugate vaccine Synflorix (*22*). As an important intermediate the bulk polysaccharide will be subject to a formal release regime before it can be used as the next stage in the manufacturing process.

Table 3. Synflorix: Release tests carried out on intermediates and final fill (*22*)

Manufacturing stage	*Release tests*
Bulk Polysaccharides	Water content, alcohol content, identity, molecular size distribution, residual protein and nucleic acid content, phosphorus content, nitrogen content, O-acetyl content, hexosamines content, methylpentoses content, uronic acid content, CPS content and endotoxin.
Protein D carrier	Includes tests such as identity, purity, sterility, protein content and endotoxin content.
Tetanus toxoid carrier	Complying with WHO (TRS n° 800) and Ph. Eur. requirements on bulk Tetanus Toxoid (Ph. Eur. 0452).
Diphtheria toxoid carrier	Complying with WHO (TRS n° 800) and Ph. Eur. requirements on bulk Diphtheria Toxoid (Ph. Eur. 0443).
Bulk conjugate	Identity, sterility, molecular size distribution, protein content, PS content, PS/carrier ratio, free PS content, free carrier content and endotoxin content.
Final fill	Description, identity of serotype, sterility, pH, endotoxin content, volume, aluminium content and polysaccharide content for each serotype.

Polysaccharide intermediate bulks are often stored as dried powders, and optimal moisture and residual solvent levels will have been determined from stability studies and developed as specifications for internal company use.

Polysaccharide Identity

A useful definition of the expectations of an identity test in 21CFR 610.14 is that it "shall be specific for each product in a manner that will adequately identify it as the product designated on final container and package labels and circulars, and distinguish it from any other product being processed in the same laboratory" (*31*). Classically, before reliable structural information was available, the identity of the polysaccharides was determined by a combination of immunochemical approaches and chemical composition. For example, the (draft) WHO requirements for pneumococcal polysaccharides, incorporated into the European Pharmacopoeia (*32*), was based on the content of different sugar types (pentoses, 6-methyl pentoses, amino sugars, uronic acids) and related values (total nitrogen, total phosphorus, O-acetyl group) per unit mass of dried polysaccharide. Sugar content specifications were based on manufacturing experience, rather than structural data, and reflected differences in response factors as well as composition. More recent WHO requirements have used compositions based on structural data (*33*). O-Acetyl specifications are absence for some pneumococcal polysaccharides now known to be O-acetylated, and low specifications for total nitrogen and phosphorus in polysaccharides not containing these elements represent limit specifications on the pneumococcal C-polysaccharide content [see below].

Bacterial polysaccharides are repeating polymers, with a repeat unit of between one and eight sugars. The pneumococcal polysaccharides usually have more complex repeat unit structures than others used to make vaccines. The repeat structure is tightly controlled, with the only heterogeneity arising from O-acetylation. The structure determination (or confirmation and, sometimes, revision) of these polysaccharides, largely by the use of high-field multidimensional NMR (*34–42*), has allowed more modern approaches to be used. The use of NMR identity testing has become the norm, with comparison of the spectra of the native or de-*O*-acetylated polysaccharide with that of a reference sample being used to confirm identity (Figure 3). Spectral comparison can be by "visual comparison," by comparing the chemical shifts of five peaks found in the spectrum of the test sample with those of reference data (*43*) or by calculation of a correlation coefficient between the anomeric region of the test and reference spectra (*44*). Full assignment of the NMR spectrum is a key element in the assay validation. Spectra from different manufacturers are all essentially the same, although variation in the degree of O-acetylation does occur (*45*).

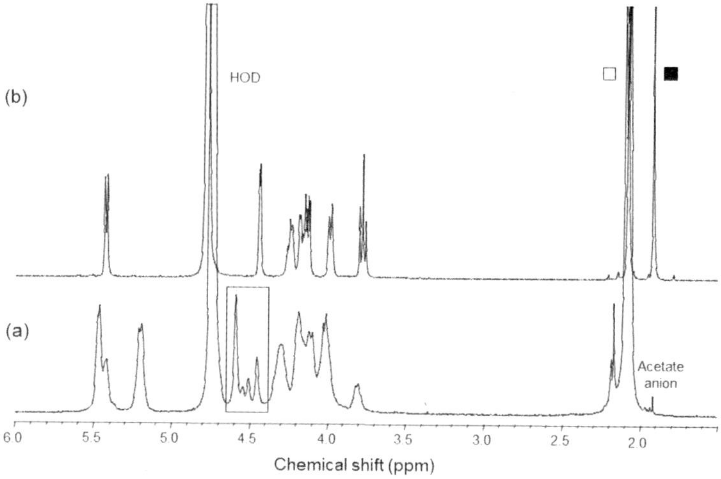

Figure 3. Partial 500MHz ^{1}H NMR spectra of (a) native, and (b) de-O-acetylated meningococcal Group A polysaccharide.

Degree of O-Acetylation

The criticality of polysaccharide O-acetylation in obtaining a protective immune response has been much debated. Its importance for Vi clear (*46*), whilst there is conflicting information for MenA (*47*, *48*). O-acetylation is not required for other meningococcal polysaccharides (*49*). There is much less data in the pneumococcal field. Deliberate de-O-acetylation of the PS [specifically MenC (*50*), Pn18C] has been used to simplify manufacturing processes, and the vaccines containing these components are highly protective. Quantitative *O*-acetyl specifications are in place for the four Men CPS, some pneumococcals CPSs and the *S.* Typhi Vi CPS. Quantitation of the degree of O-acetylation can be through the use of a Hestrin colorimetric assay (*51*), using acetylcholine as a reference standard, and providing a result in mmol O-acetyl per gram dry weight of CPS, or NMR based assays, either by direct integration of peak areas in the spectrum of the native polysaccharide, or samples in which the O-acetyl is released as acetate by base treatment and quantification by peak integration (*42*). In Figure 3, the boxed area around 4.5 ppm in (a) arises from the ManNAc H-2 in variously O-acetylated forms, allowing the degree of O-acetyl-ation to be estimated. In Figure 3(b) the black square arises from acetate anion and the white square from the N-acetyl residue. Comparison of the integrals is a measure of the O-acetyl content of the original polysaccharide.

This leads to a value related to the mean number of O-acetyl groups per saccharide repeat unit: these specifications have been correlated.

Polysaccharide Purity

There are of two types of impurities present in the CPSs, residual small molecule impurities from the polysaccharide purification and macro-molecular residuals from the polysaccharide fermentation process.

Process Impurities

These include residual cetyltrimethylammonium bromide (CTAB) and organic solvents from precipitation steps. Control of these materials can individually by specific chromatographic approaches or by quantitative NMR. Practically, since the glycoconjugates will be purified by ultra-filtration after production there is another opportunity to remove small molecule impurities.

Other Impurities

Bacteria may elaborate more than one cell surface polysaccharide. For example, Group B Streptococcus have both a Type-specific CPS and a Group-specific antigen (the “Lancefield antigen”), or Gram negative bacteria can have both capsular- and lipo-polysaccharides. The purified target capsular polysaccharide may be contaminated with another polysaccharide. The best studied case is pneumococcal C-poly-saccharide, which is the common glycan chain of the teichoic and lipoteichoic acids of *S. pneumoniae* and a ubiquitous contaminant in pneumococcal CPSs (*52*). There are three structural variants of C-polysaccharide (*53*, *54*). There are no direct formal specifications, but C-polysaccharide is phosphorus- and nitrogen-rich and is limited by compositional assays, and by HPAEC methods (*55*). Quantitative NMR methods have also been used (*56*). The C-polysaccharide content is serotype dependent (from one manufacturers) and content varies from very low to ca. 10% of the CPS, based on repeat units (*45*). Traces of lipid may be present in Hib, meningococcal and *S.* Typhi Vi CPSs (*18*, *19*), and fragments of peptidoglycan in *S. aureus* CPSs (*40*) and pneumococcal CPSs. Residual antifoam agents may also fail to be completely removed by the purification process, especially for the hydrophobic S. Typhi Vi CPS (*45*).

Residual Protein and Residual DNA

Limits for these contaminants are based on colorimetric or spectro-scopic analysis [Table 4]. In the specific case of the pneumococcal type 5 a higher limit for protein is set (5%) due to interference of the polysaccharide in the Lowry assay. If the cells used in CPS manufacture have been exposed to materials containing blood group substances, then absence of these materials should be confirmed.

Table 4. Protein and nucleic acid content specification for different polysaccharides used in conjugate vaccine manufacture. Data is taken from relevant European Pharmacopoeial monographs or WHO Recommendations, unless otherwise referenced.

	Hib PRP	*Meningococcal A/C*	*Pneumococcal*	*S. Typhi Vi*
Protein[a]	NMT[d] 1% w/w	NMT 1% w/w	NMT 3% w/w[e]	NMT 1% w/w
Nucleic acid[b]	NMT 1% w/w	NMT 1% w/w	NMT 2% w/w	NMT 2% w/w
Endotoxin[c]	NMT 10 IU per mcg CPS	NMT 100 IU per mcg CPS	NMT 0.1 IU per mcg	NMT 150 IU per mcg (*57*)

[a] Estimated by the Lowry method vs BSA. [b] Estimated by UV absorption spectroscopy at 260 nm. [c] Estimated using a LAL assay. [d] NMT – Not more than. [e] Variable for different serotypes.

Molecular Size Analysis

Whilst polysaccharide molecular size is a CQA for purified poly-saccharide vaccines, because immunogenicity requires a minimum molecular size, its relevance for glycoconjugate vaccines is less clear: the native polysaccharides are frequently size-reduced (see below) or chemically depolymerised to oligosaccharides prior to conjugation (see below). However, requirements for polysaccharides used in glyco-conjugate manufacture typically match those for purified poly-saccharide vaccines. The classical method, and public specifications, relate to chromatography on a Sepharose CL-2B or CL-4B soft gel column (of dimensions 0.9 x 90 cm). The specification can be either that the peak maximum elutes before a specified K_d value (Hib PRP and pneumococcal CPSs) or the proportion of material eluting before a specified K_d (meningococcal and *S.* Typhi Vi). Soft gel chromatography has largely been replaced by HPSEC on rigid column matrices. A series of sized dextran calibrants has been developed which supports the translation of specifications between these methods (*58*). Merck reported the molecular weights of pneumococcal polysaccharides, using HPSEC with coupled refractive index (RI) and multiple angle laser light scattering (MALLS) (*59*), and used controlled depolymerisation through sonication, linked molecular size specifications from soft gel analysis with molecular weights (*60*). The advantage of this approach is that it becomes essentially independent of the separation matrix (soft gel or HPSEC, matrix type, matrix batch), but does require an estimate of the refractive index increment (*dn/dc*) for the specific polysaccharide-eluent combination used. Hydrodynamic approaches such as analytical ultracentrifugation are generally too slow to support routine use (*61*), but they are valuable absolute approaches to support validation of alternative approaches.

Quantification Assays

The two physicochemical approaches are commonly applied for bulk polysaccharides are (i) colorimetric and (ii) combined degradation and chromatographic HPAEC quantification of the fragments. Some colorimetric assays are (largely) specific for different types of sugars (*51*) – the orcinol assay for pentoses, the resorcinol assay for sialic acids, and anthrone assays for total sugar (*62*). Many such generic assays are listed in the European Pharmacopoeia: the key issue is the choice of reference material to create a standard curve. If a monosaccharide is used, equal response factor cannot be assumed and validation of assay's accuracy is required (*14*). Alternatively, a quantitative sample of the homologous polysaccharide can be used (*26*). For the *S.* Typhi Vi CPS quantification of the O-acetyl content has been used as a surrogate for saccharide content, as the degree of O-acetylation is typically close to 100% and that the O-acetyl group is regarded as critical for the establishment of protective immunity. Many polysaccharides can be degraded to monosaccharide or disaccharide components separable by HPAEC, by dilute acid (*63*), dilute base (*64*) or strong base hydrolysis (*65*). Complete and specific degradation of the polysaccharide cannot be assumed, and only some components may be quantified by HPAEC (*64*). Methods which quantify the saccharide component (eg. HPAEC of Neu5Ac derived from MenC vs Neu5Ac standard) require knowledge of the degree of O-acetylation to convert micromole estimates of saccharide into milligram estimates of the amount of polysaccharide. Quantitation of poly-saccharide bulks is required to allow the correct amount of material to be released for the conjugation step.

If an NMR identity assay is being used, addition of a suitable internal standard can allow polysaccharide quantification: this approach can be as accurate and precise as the alternative methods.

Rocket immunoelectrophoresis is a quantification method that has been widely used (*66*) for both proteins and polysaccharides, particularly in polysaccharide mixtures, and is still included in some guidelines (*25*). The use of defined monospecific antisera created using allows specificity for individual serotypes in complex mixtures.

Polysaccharide Stability

The stability of a polysaccharide intermediate is a key factor defining optimal storage conditions and the allowable storage time, before further processing. Typically, internal release specifications will be defined, on the basis of manufacturing experience or existing pharmacopoeial specifications.

Depolymerisation

Some polysaccharides, notable Hib PRP and meningococcal Groups A, C, W and Y, have limited stability in solution, due to spontaneous hydrolysis. This process is illustrated for the Hib PRP in Figure 4.

Instability can be monitored by a reduction in molecular size using HPSEC chromatography or by NMR spectroscopy. Molecular size monitoring is simple and easily implemented. The NMR approach, using either 1D ^{1}H NMR or 2D ^{1}H-^{13}C correlation methods, to observe new resonances arising from newly formed end groups is less sensitive to low levels of depolymerisation, but allows the degradation pathways to be defined.

O-Acetylation Changes

Spontaneous migration of O-acetyl groups occurs from the original position, defined by the biosynthetic pathway, to one of greater thermodynamic stability, or for entropic reasons. This has been shown for the meningococcal Group C polysaccharide (*37*) and for the pneumococcal Type 1 (with ca. 50% acetylation on each of O-2 and O-3, the two available hydroxyl groups in the GalA residue) (*67*). This process can be monitored by NMR spectroscopy. In general, it has not been considered that the location of the O-acetyl groups is a CQA, and specifications are based on total O-acetylation, rather than specific locations.

Carrier Proteins and Activated Carrier Proteins

The function of the carrier protein in the final glycoconjugate is to modify the immune response to the vaccine to a T cell-dependent B cell-modulated antibody response, which is more effective in at pathogen killing and reduction of carriage than the T cell-independent Type 2 response typical of purified polysaccharide immunogens. Two major families of carrier proteins have been employed (*68*): bacterial toxoids [tetanus or diphtheria toxoid] or (usually bacterial) proteins which are non-toxic or cell surface protein. The non-toxoided proteins may be recombinant.

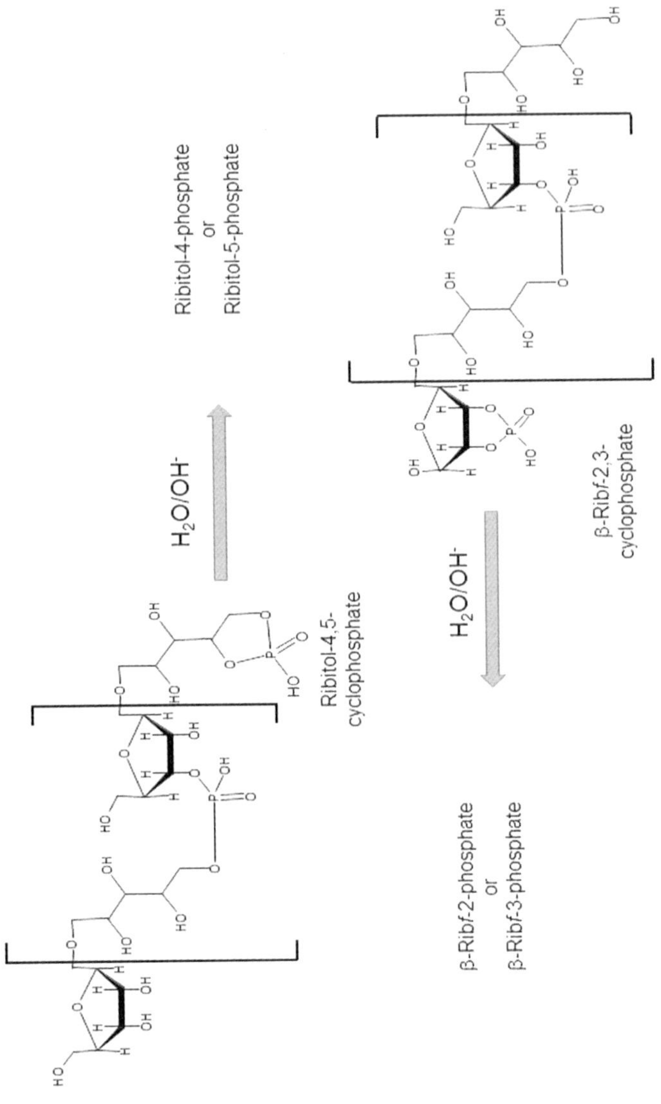

Figure 4. Degradation pathway of Hib PRP in aqueous solution, with initial formation of either the ribofuranose-2,3-cyclophosphate or the ribitol-4,5-cyclophosphate in a ratio of approximately 9:1.

Toxoid Carrier Proteins

Toxoids used as carrier proteins should meet the requirements for their use in other vaccines. Identity is confirmed by immunochemical methods, such as immunoprecipitation (flocculation, radial immuno-diffusion, and nephelometry), immunoelectrophoretic methods (rocket immunoelectrophoresis), or immunoenzymatic methods (immunoblots and ELISA). The manufacturing consistency of diphtheria and tetanus toxoids is monitored by determining the content of monomers vs. dimers and other aggregates using HPSEC coupled to a static light scattering detector. Typically, antigenic purity for both tetanus and diphtheria toxoids, as determined by the flocculation test should be at least 1500 Lf units/mg of protein: higher than otherwise required for tetanus toxoid immunogens. Additional characterisation of the product may be appropriate (such as residual available amino groups) which have a bearing on their use as carrier proteins. Toxoids are available as commodity products, and may not need to be manufactured in-house. Compared to toxoid vaccines, mass quantification of carrier proteins (rather than by Lf) is the key measure of quantity.

Recombinant Non-Toxoided Carriers and Cell Surface Proteins

CRM_{197} is available both as a "native" protein, produced by a strain of *Clostridium diphtheriae* strain C7 (β197) and, increasingly as recom-binant proteins produced in *Pseudomonas fluorescens* (*69*) or *E. coli* (*70*). CRM_{197} contains an exposed loop of three arginine residues that is clipped by proteases present in the culture medium, resulting in a so-called nicked form, and domain-swapped dimers are also formed (*71*). The manufacturing process should demonstrably be able to regularly produce CRM_{197} with a consistently low degree of nicking. In the presence of a reducing agent like dithiothreitol, the nicked form breaks down into two distinct polypeptides called fragments A and B that can be easily detected by SDS-PAGE, which accordingly is a suitable method to determine the degree of nicking (*72*). The required purity of CRM_{197} is not less than (NLT) 90% (*68*).

The extracellular domains of recombinant *Haemophilus influenzae* protein D is used by GSK in Synflorix (*73*), (required purity NLT 95%) and recombinant exoprotein A from *Pseudomonas aeruginosa* (rEPA) has been used in developmental bivalent *Staphylococcus aureus* (*74*) and a Typhoid Vi vaccine (*75*). In practice, these materials can all be controlled as though they are recombinant materials. Purity should be monitored with an appropriate test such as HPLC, SDS-PAGE, or capillary electrophoresis (CE) (*72*).

Meningococcal OMPC carrier should be monitored for consistent composition by SDS-PAGE or by another suitable method, and the lipopolysaccharide (LPS) content should not exceed 8% by weight. Suitable methods for LPS determination include HPLC, colorimetric analyses, SDS-PAGE, and GC-MS (*72*).

Control of Activated Carrier Protein

Some manufacturing procedures require activation of the protein carrier. This process step introduces functional groups onto the protein that react with the poly- or oligo-saccharide intermediates with complementary reactivity. Glutamic or aspartic acid are functionalised with a bifunctional reagent such as adipic acid dihydrazide or hydrazine: the nucleophilic hydrazide group is available for coupling with the polysaccharide. In other manufacturing strategies, the lysine side chains are derivatised with bromo-acyl, thiol or maleimido groups: however, chemical toxoiding destroys many of the reactive amino groups which are often favoured as conjugation sites on the carrier protein. Methods to quantify the newly formed chemical functions should be in place. In a validated process where production consistency has been established, and depending on the conjugation chemistry used and the results of clinical trials, testing may be used as an in-process control. In some scenarios, such as immediate conjugation after activation, consistency in degree of carrier protein activation may be demonstrated as part of process validation or reflected by characteristics of the final conjugate bulk (*72*).

Conjugation Chemistries

Conjugation is the coupling of the saccharide to the carrier protein. As there are no complementary reactive groups on the two components in their native states to allow a controlled coupling, the saccharide and, sometime the carrier protein, are activated prior to reaction. Often a bifunctional linker is included to allow alternative chemistries to be applied. The groups available in the CPS which can be activated are hydroxyl groups, *vicinal* diols and carboxylate groups. Amino and phosphodiester groups are sometimes available, and have been used in developmental vaccines. Depolymerisation with dilute acid leads to the uncovering of the hidden aldehyde group of the anomeric centre and provides an alternative reaction site for conjugation chemistry. The carrier protein has carboxylic acids and amino groups, and, sometimes free thiols. In toxoid carrier proteins many of the amino groups in lysine residues have been destroyed by reaction with formaldehyde during the toxoiding process.

The choice of appropriate conjugation chemistry for the preparation of any given immunogen will depend principally on the structure of the repeat unit of the polysaccharide, and the reactive groups it contains. In some cases, chemical de-O-acetylation may be required to reveal *vic*-diols required for periodate oxidation. The activated polysaccharide is not always isolated, and therefore, characterisation and quality control are hence not relevant. Consistency in the creation of the activated polysaccharide needs to be deduced from the consistency of the resulting conjugate product.

Preparation of Polysaccharide for Conjugation

In many manufacturing processes (*26*) there is a controlled size reduction of the polysaccharide to create a material of lower polydispersity and consistent size to support reproducibility in the conjugation step. Confirmation of the continuing quality of the size reduced PS is required. Microfluidisation methods, in which the polysaccharide is forced through a narrow opening at high pressure and depolymerised as a result of high shearing forces, are widely used. Harding et al. report (*8*) quantitative data on the molecular weights on native and size-reduced meningococcal polysaccharides used in the manufacture of Nimenrix [Table 5].

Table 5. Molecular weights[b] (in kDa) of meningococcal polysaccharides used in manufacture of Nimenrix

	Men A[a]	*Men C*	*Men W*	*Men Y*
Native	710 ± 35	1950 ± 100	1350 ± 70	1370 ± 70
Size-reduced	195 ± 10	185 ± 10	275 ± 15	110 ± 5

[a] sodium salt. [b] molecular weights are M_X. The paper also includes M_Z data.

Controlled Periodate Oxidation of Vicinal Diols

This results in the formation of two aldehyde groups. The rate of reaction will depend on the relative orientation of the hydroxyl groups and the conformational flexibility of the carbon-carbon bond between the two. When the *vic*-diol is part of an acyclic system forming part of the polysaccharide backbone, such as the ribitol residue in the Hib PRP or the sidechain in MenC, concomitant depolymerisation will occur. The degree of polymerisation of the resultant oligosaccharides is controlled through the relative ratios of polysaccharide and periodate and careful choice of reaction conditions. The size of the resulting oligosaccharide is assessed by a combination of chemical colorimetric assays [total saccharide and aldehydic groups] or chromatographic approaches (*76*). The detailed chemistry for this process for Hib PRP is shown in Figure 5.

The revealed aldehyde groups can react with free amino groups on the carrier protein (usually CRM_{197}) to form a Schiff's base, that is locked by controlled reduction with sodium cyanoborohydride. Residual free aldehydic groups are removed ["capped"] by reduction with sodium borohydride.

Unreacted periodate can be quenched with butan-2,3-diol or removed by diafiltration (*7*). Conjugation chemistry of this type will result in "fuzzy ball"/ "neoglycoconjugate" type conjugates, with some degree of oligomerisation due to the presence of a reaction of the bifunctional linker at both ends. Typically, conjugates contain 6-8 glycan chains per carrier protein (*77*). After conjugation, residual small molecules and unreacted oligosaccharides are removed by diafiltration using, for example, a 100 kDa MWCO ultrafiltration membrane.

Where the *vic*-diol is part of a cyclic system, the oxidation step does not necessarily result in depolymerisation. Site selectivity is realised through the differential reaction rates of various *vic*-diols in the polysaccharide repeat unit. The number of attachment sites and the degree of crosslinking between the polysaccharide and the carrier protein can be controlled by the degree of oxidation of the CPS. This is the case of all the pneumococcal polysaccharide components of the 7- or 13-valent vaccines, and leads to crosslinked network type immunogens [Figure 1b]. Periodate oxidation of pneumococcal C-polysaccharide contaminant in the CPSs does result in degradation down to small oligosaccharides. These are lost in diafiltration of the crude activated CPS and not incorporated into the glycoconjugate. the phosphocholine methyl resonance, prominent on the spectrum of the starting polysaccharide, is absent in NMR analysis of bulk CRM_{197}-pneumocococal immunogens (*45*).

As Schiff base formation is reversible by hydrolysis, reaction in non-aqueous solvents such as DMSO or DMF is preferred (*7*, *78*). Some patents describe the use of lyophilised CRM_{197} re-suspended in DMSO (or 90% DMSO/10% water) and lyophilised activated polysaccharide (with sucrose as a cryoprotectant) as the reactants (*7*, *24*).

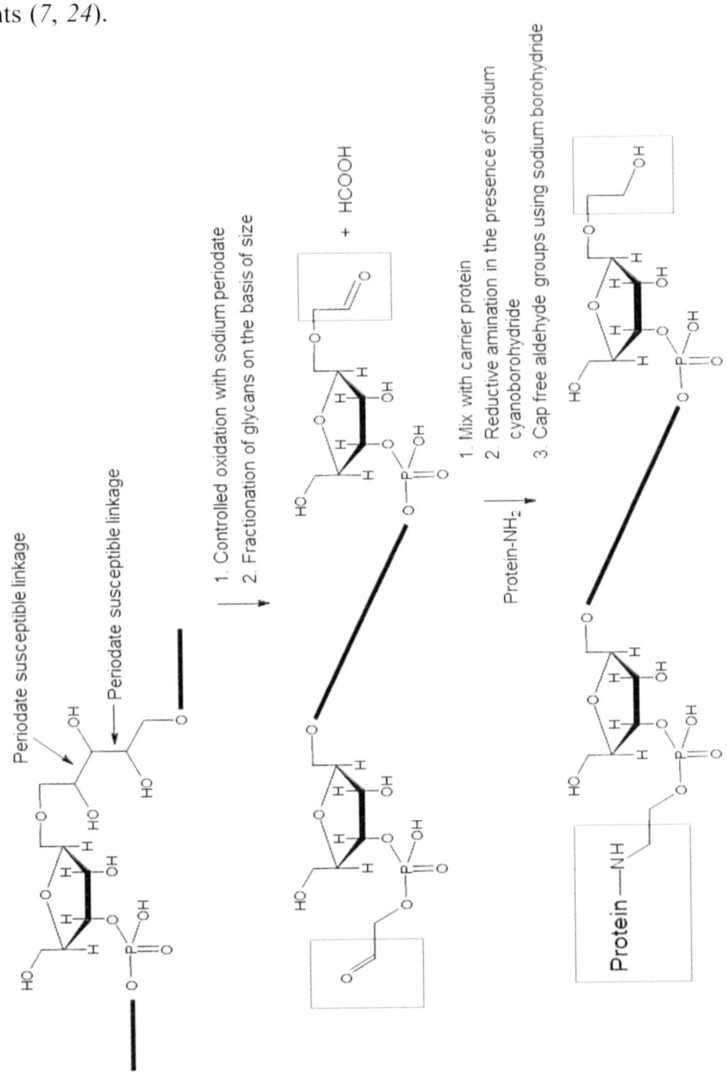

Figure 5. Schematic showing the periodate oxidation of Hib PRP and conjugation to a protein (typically CRM_{197}). The heavy straight line represents an undefined number of repeat units.

Frasch *et al.* developed (*79*, *80*) an approach to increase conjugation yields with toxoid carriers. Treatment of the toxoid with hydrazine or ADH in the presence of a water-soluble carbodiimide converts some of the carboxylic acids into hydrazides. With more, and more reactive, groups on to the protein (compared to the number of residual amino groups), yields of reductive amination products are significantly improved [Figure 6]. This approach is used in the manufacture of MenAfrivac, a low-cost MenA-TTx vaccine developed specifically for use in sub-Saharan Africa (*81*).

In some developmental vaccines, controlled oxidation with TEMPO/ N-chloro-succinimide has been used to convert hydroxymethyl groups into aldehydes, prior to conjugation (*82*).

Protein-COOH

↓ ADH (or H_2NNH_2) / EDC

Protein-$CONHNHCO(CH_2)_4CONHNH_2$

↓ Sugar-CH=O / $NaCNBH_3$

Protein-$CONHNHCO(CH_2)_4CONHNHCH_2$-Sugar

Figure 6. Schematic showing the activation of a carrier protein (typically TTx) using hydrazine or ADH and EDC, as developed by Frasch et al. Creation of a hydrazide functionality increases the reactivity of the carrier protein in reductive amination reactions.

Acid-Catalyzed Depolymerisation

Susceptible glycosidic linkages in polysaccharides, such as that of the ribofuranose residue in Hib PRP, the neuraminic acids in meningococcal Groups C, W and Y, or the anomeric phosphodiester in the meningo-coccal Group A CPSs allow controlled dilute acid depolymerisation of these polysaccharides, as shown in Figure 7. Progress of Hib PRP hydrolysis can be monitored through the changing optical rotation of the solution, and terminated at an appropriate point. A consistent molecular size fraction can be then purified from the crude hydrolysis mixture by anion exchange chromatography (*83*). The molecular size of the resulting oligo-saccharides can be confirmed by HPLC-SEC, NMR (*84*) or a combination of colorimetric assays for total saccharide and reducing end group (*85*, *86*). The aldehydic (or ketosidic, in the case of neuraminic cids) group provides a site for conjugation.

Direct reductive amination coupling of the reducing terminal of these oligosaccharides to carrier proteins has not been used, due to the relatively low proportion of uncovered aldehyde in solution. Reductive amination in the presence of high concentrations of ammonium salts produces an aminated sugar, which is trapped by reaction with a large excess of a bifunctionalised carboxylic acid - adipic acid di-*N*-hydroxysuccinimide ester has been used.

The remaining activated carboxylic acid can then be reacted with a free amino groups on a carrier protein (typically CRM_{197}) in a mixed aqueous/nonaqueous solvent mix. Again this procedure results in the formation of "fuzzy ball" type neoglycoconjugates [Figure 1a], without cross linking, and typically 8-10 glycan units per carrier protein (*6*).

Several groups have used peptide mapping and HPLC-MS-MS approaches to determine the preferred sites of attachment of the glycan chains onto CRM_{197} (*87*, *88*): glycan attachment at the N-terminal amino group is favoured (*89*).

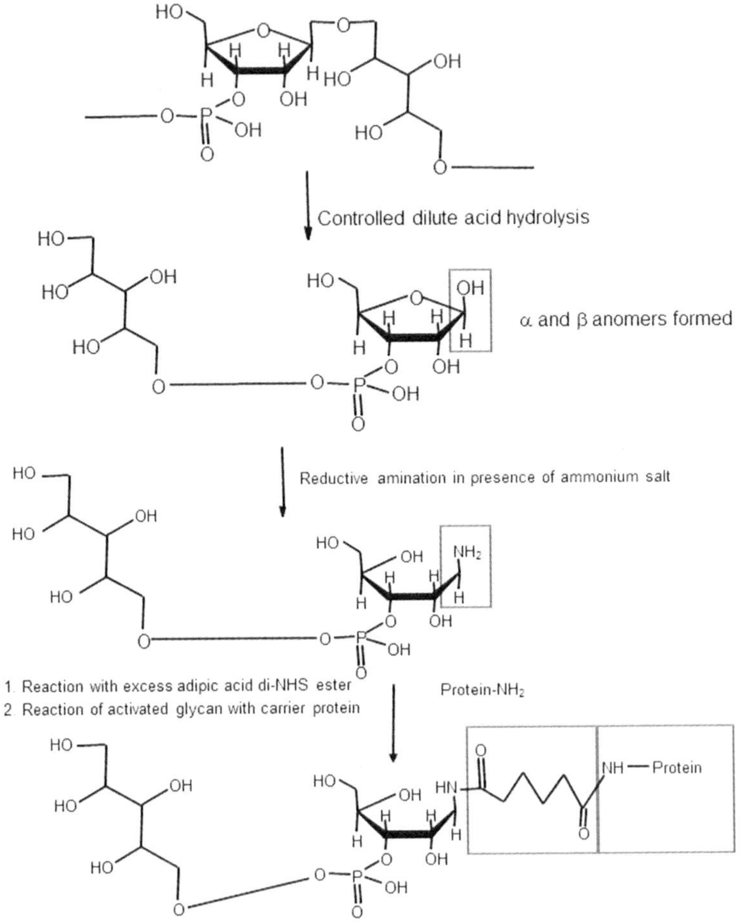

Figure 7. Schematic showing the dilute acid hydrolysis, reductive amination and coupling of the resulting activated oligosaccharide to a carrier protein (typically CRM_{197}). This chemistry was developed by Chiron. The heavy straight line represents an undefined number of repeat units.

Cyanogen Bromide or CDAP Activation of Polysaccharide

Hydroxyl groups on polysaccharides can be activated through reaction with a variety of reagents. Cyanogen bromide has long been used to activate Sepharose column matrices for coupling of ligands and use in affinity chromatography. It has also been widely used in glycoconjugate vaccine manufacture. 1-Cyano-4-aminopyridine [CDAP] is a crystalline reagent able to perform the same chemistry whilst easier and safer to handle (*90*, *91*). Cyanogen bromide or CDAP activation of hydroxyl groups initially results in a low stability cyanate which is trapped *in situ* with an excess of adipic acid dihydrazide (ADH) or 1,6-diaminohexane. Detailed NMR analysis of ADH-activated Hib PRP showed an essentially random activation of hydroxyl groups (*45*). After removal of excess ADH or 1,6-diaminohexane this material is stable enough for storage, characterisation and control, although conjugation may be performed immediately. For high molecular weight polysaccharides activated in this manner it is important to ensure a consistent, known degree of activation, to ensure a consistent degree of crosslinking in the final immunogen. The number of hydrazide groups per unit of saccharide can be determined using colorimetric or fluorophore labelling. Harding's group reported slight depolymerisation of Men A and Men C when using CDAP and ADH for activation in the manufacture of Nimenrix (*8*). The activated polysaccharide is coupled to carboxyl groups on carrier proteins, using N-ethyl-N′-(dimethyl-aminopropyl) carbodiimide (EDC) to create hydrazide or amide linkages [Figure 8]. Toxoid carriers have been widely used with this type of conjugation. The reaction of multiply activated high molecular weight polysaccharides with carriers with multiple reactive sites results in crosslinked network type conjugate matrices [Figure 1b]. Residual reactive groups can be quenched with glycine (*78*) and excess reagents are removed by diafiltration, with particular concern for N-ethyl-N′-(dimethylaminopropyl) urea (EDU). The number of uncapped hydrazide groups in the final conjugate should be determined (*32*, *92*, *93*) unless their absence has been validated.

Sugar-OH

↓ CNBr or CDAP

Sugar-O-C≡N

↓ ADH or 1,6-diaminohexane

Sugar-O-C(=N).NHNHCO$(CH_2)_4$CONHNH_2

↓ Protein + EDC

Sugar-O-C(=N).NHNHCO$(CH_2)_4$CONHNHCO-Protein

Figure 8. Schematic showing the activation of hydroxyl groups on a polysaccharide using cyanogen bromide (or CDAP) and trapping of the unstable activated polysaccharide with ADH or 1,6-diaminohexane. The linker is suitable for attachment to carboxyl groups on a carrier protein (typically TTx) through EDC-mediated coupling.

Linkers through Uronic Acids

This approach has proven especially valuable for conjugates based on the *S.* Typhi Vi polysaccharide (or the polysaccharide from *Citrobacter freundii* WR7011 which has an essentially identical structure) (*94*), as this polysaccharide lacks reactive groups apart from the uronic acid carboxylate.

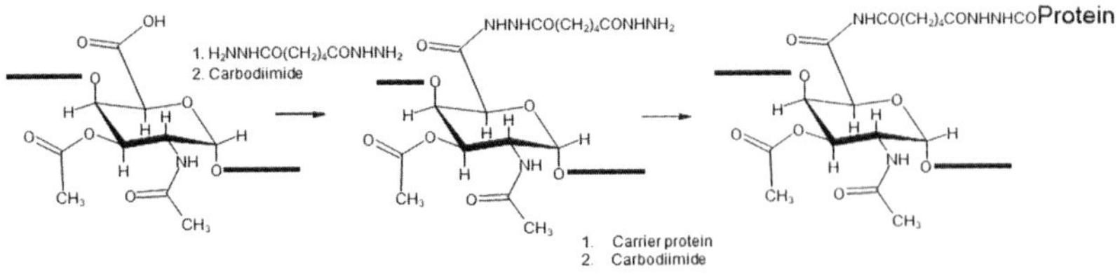

Figure 9. Schematic showing the conjugation of Vi CPS to a carrier protein. The Vi CPS is activated by attachment of a bifunctional linker to the carboxyl group of the Vi CPS and forms functionality for attachment to the carrier protein.

Reaction of the polysaccharide with sub-stoichiometric amounts of ADH in the presence of EDC creates an activated polysaccharide, as shown in Figure 9, which can be further reacted with carboxylic acids on the carrier protein (rEPA, DTx and CRM197 have been used) (*95–98*) and producing a cross-linked network immunogen [Figure 1b]. This linker has good stability. Control tests focus on ensuring a consistent degree of activation of the polysaccharide.

Activation with 1,1′-Carbonyldiimidazole

Merck developed the use of 1,1′-carbonyldiimidazole to activate Hib PRP through a hydroxyl group, followed by attachment of a 1,4-diaminobutane linker, as shown in Figure 10. This stable intermediate can be stored and characterised. A qNMR method was developed (*99*) to define the degree of activation of the Hib PRP , quantifying both the degree of initial activation and trapping of the initially formed intermediate. More detailed NMR analysis was consistent with random activation at all available hydroxyl groups. The carrier proteins, a mixture of LPS-depleted meningococcal outer membrane proteins (OMPC) in a vesicle were activated by treatment with *N*-acetyl-homocystine, EDTA and dithiothreitol. Prior to conjugation the activated polysaccharide is reacted with *p*-nitrophenyl bromoacetate, and coupling to the activated OMPC is by reaction of the bromoacetyl function with the free thiol on the protein.

This coupling step is a rapid process and requires careful control of experimental conditions. Residual free thiol on the activated protein after conjugation is capped with N-ethyl maleimide. This combination of components and conjugation chemistry results in vesicle-type immunogens [Figure 1c].

Attachment of Synthetic Oligosaccharides through Maleimide Chemistry

In the Hib vaccine manufactured in Cuba, the oligosaccharide chain (averaging about 7 repeat units) is produced by oligomerisation of synthetically produced repeat units, with a maleimide functionality at the "reducing terminal" to allow for conjugation to a thiol group on an activated tetanus toxoid carrier protein (*100*). The structure of the resulting chains and their linkage to the carrier is shown in Figure 11. The resulting immunogen has a PS:protein ratio of 1:2.6, which suggests an average of approximately 30-35 glycan chains per carrier protein.

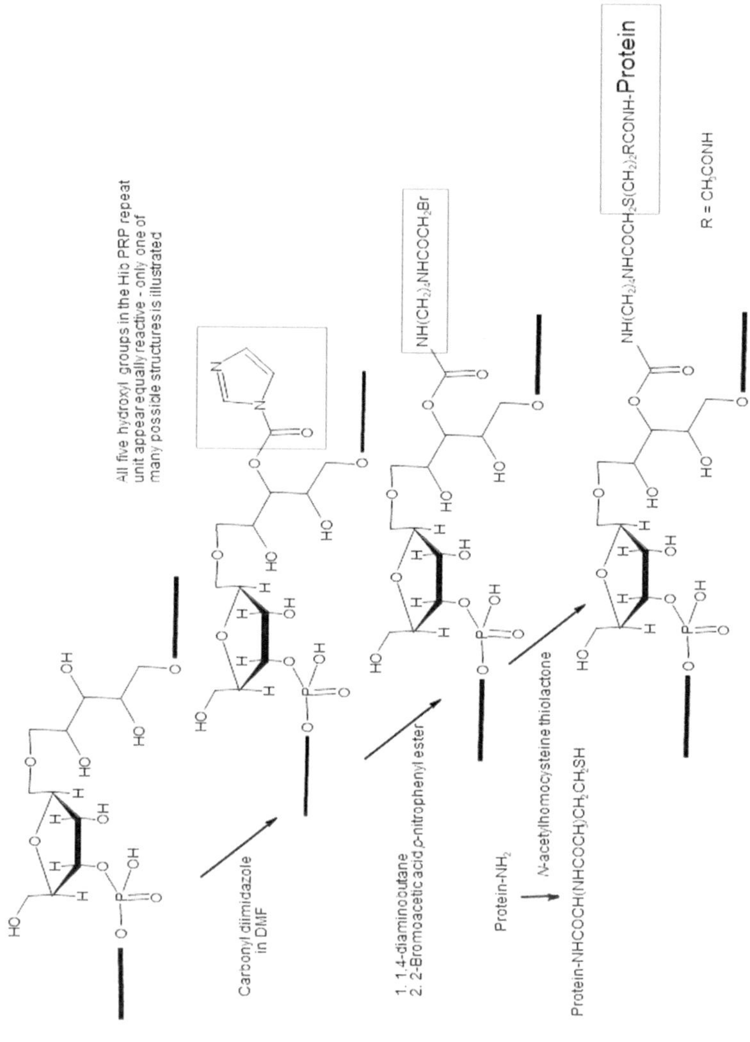

Figure 10. Schematic showing the chemistry developed by Merck for conjugation of size-reduced Hib PRP to meningococcal OMPCs.

Other Conjugation Chemistries

A large number of other approaches to attach the polysaccharide to a carrier protein have been applied in developmental vaccines, but have not yet been applied in licensed products. These include: coupling of synthetic oligosaccharides through squaric acid (*101*) or the Pawlowski approach (*102*), activation of carboxylic acids on the polysaccharide and protein and formation of disulphide linkages (*103*), deamination of GlcN to create aldehydic groups suitable for reductive amination coupling (*104*), the use of oxime chemistry (*105*), or the Huisgen 1,3-dipolar cyclo-addition (*106*) and Staudinger ligation (*107*). Many of these are best suited to coupling synthetic oligosaccharides where suitable reactive groups can be incorporated into the glycan chain when it is constructed. The potential use of synthetic, relatively small, oligosaccharide chains to manufacture glycoconjugate vaccines has been greatly advanced by the work of the group of Seeberger, which has used scalable, automated synthesis of complex oligosaccharides (*108*) and developed suitable glycoconjugate immunogens for protection against a number of pneumococcal serotypes (*109–112*). Access to these synthetic glycoconjugates provides a means to explore the critical protective epitopes in the polysaccharides (*113*).

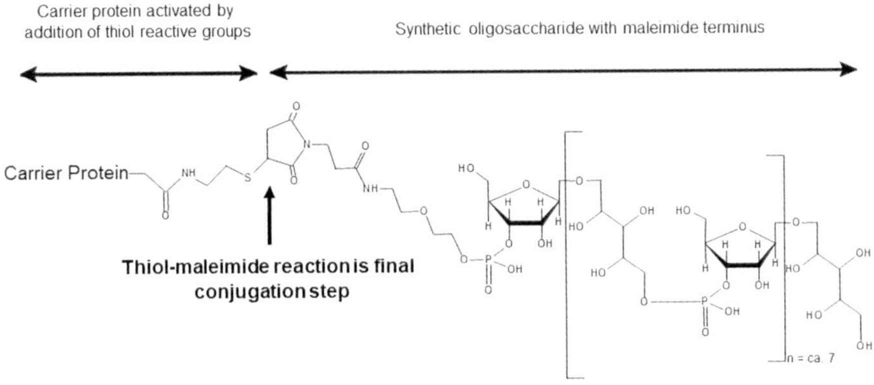

Figure 11. Partial structures of the Hib-TTx immunogen employing synthetic Hib PRP oligosaccharides, and highlighting the structure of the linker and the TTx carrier protein.

Control of Monovalent and Polyvalent Bulk Conjugates (or Drug Substance)

Many of the control assays applied to the monovalent bulk conjugates are those expected for any biopharmaceutical product – visual inspection, identity, composition and the purity/impurity profile – but complicated by the need to consider both the polysaccharide and carrier protein components. The structural integrity of the poly-saccharide and carrier proteins in the conjugate will probably have been confirmed in characterisation studies and validated out, although it is likely to be an aspect of stability studies. The polysaccharide: protein ratio, typically a CQA, is likely derived from individual quantification assays of the two components, and estimation of molecular size is a valuable indicator of the consistency of the manufacturing process, and a stability-indicating parameter in those studies. Process related impurities such as residual conjugation reagents should be shown to be within acceptable limits, but quantification of unconjugated "free" polysaccharide and carrier protein are specific for glycoconjugate vaccines. Quantification of the saccharide in monovalent bulks is essential to inform the next manufacturing step, either blending to produce a polyvalent bulk, a combination vaccine bulk or filling of a monovalent product.

Polysaccharide Identification

This assay confirms that the correct antigen was used during the manufacturing process and that no critical epitope was lost during conjugation. Polysaccharide identity should be confirmed using immunological methods such as ELISA, immunoblot analysis, and rate nephelometry. Alternatively, the identity of the polysaccharide can be confirmed using chemical or physical methods such as HPAEC-PAD, GC, or NMR if acceptable specificity can be demonstrated and it can be shown that the carrier protein does not substantially interfere with the identification of the polysaccharide.

Polysaccharide Quantification

The colorimetric and degradation/HPAEC methods described above [see above] usually remain viable assays for determination of the saccharide content of monovalent bulk conjugates and some simple multivalent bulks (*114*). For other multivalent bulks, and especially those for pneumococcal vaccines, rate nephelometry is widely used (*115*). Reaction of bivalent antibodies with high mass conjugates gives rise to very high mass aggregates which scatter light and can be detected by nephelometry using widely available clinical auto-analyser instruments. The rate of formation of these aggregates (rate of change of light scattering) can be related to the amount of material present. The use of antibodies (and typically antiserum is used, rather than monoclonal antibodies) provides the specificity so that the individual components can be quantified in a complex mix. For pneumococcal conjugates adsorbed onto aluminium phosphate, the adjuvant is solubilised by addition of 1M NaOH and then immediately neutralised with 1M citric acid prior to rate nephelometry (*23*). Similarly, conjugates adsorbed onto an aluminium hydroxide adjuvant can be solubilised by dialysis against 3% sodium citrate for 6h at room temperature prior to analysis (*26*).

Carrier Protein Identification

Depending on the nature of the manufacturing process and the manufacturing controls, it may be necessary to confirm the identity of the carrier protein, e.g. during a manufacturing process for a multivalent product in which different antigens are conjugated to different carrier proteins within the same facility. Carrier protein identification can be performed using an immunological method such as ELISA or, if possible, an appropriate chemical method such as peptide mapping.

Carrier Protein Quantification

The concentration of the carrier protein is confirmed for all lots of monovalent conjugate. Analysts should select a test method that is specific for the carrier protein and does not suffer from interference from the polysaccharide components. Suitable methods may include amino acid analysis, colorimetric protein tests such as the bicinchoninic acid assay or UV absorbance. The accuracy of colorimetric depends crucially on the choice of protein used to create the standard curve, and may also be compromised by the glycan chains (*116*). UV absorption at 280 nm has been widely used, but the author's experience has been that there is light scattering from high molecular size complexes. The EP chapter on protein quantification has a suitable procedure to correct for light scattering effect (*117*).

Polysaccharide:Protein Ratio

This is usually determined by calculation from the values from determined by individual assays (see above) and the accuracy of this value depends on that of the other assays. Using the data in Table 6 below, values are typically in the range of 0.4 to 1. Direct determination of the ratio is possible in some cases: this may be valuable for validation of the calculated number. One-dimensional ^{1}H NMR of denatured – either after addition of chaotropic agents such as guanidinium hydrochloride, or heat or both – conjugates based on CRM_{197} carriers allows both the saccharide and carrier protein components to be observed. Integration of resonances arising from the saccharide chain (typically resonances from the sugar anomerics) and the carrier protein (typically the sidechains of aromatic amino acids) provides a direct comparison after allowance is made for the molecular weights of the components (*84*). In the unusual case of a Vi-DTx conjugate, where the polysaccharide has a spectrum of comparable intensity to that of the carrier protein, as shown in Figure 12, deconvolution of the far UV CD spectrum was a surrogate measurement of polysaccharide-protein ratio, to help validate other approaches (*45*).

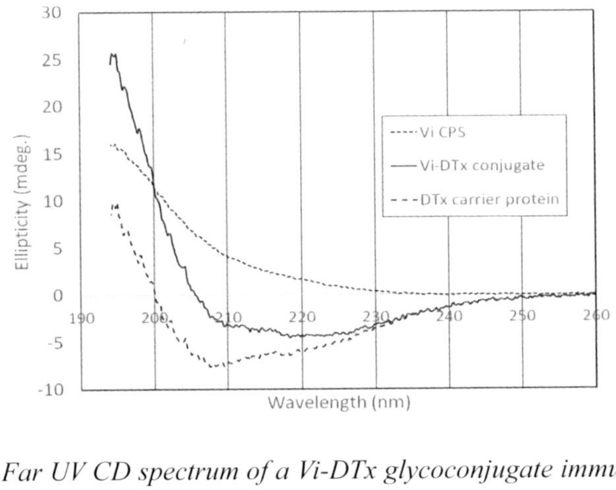

Figure 12. Far UV CD spectrum of a Vi-DTx glycoconjugate immunogen, with component spectra for the polysaccharide and DTx. The component spectra allow the PS:protein ratio to be estimated.

Degree of Substitution of Carrier Protein

This parameter is the number of glycan chains attached per carrier protein, and is most easily determined when there is a stable linkage through a lysine residue. In favourable cases, for an immunogen prepared by reductive amination, after acid hydrolysis of the CRM197 carrier a defined variant of lysine, N-hydroxymethyllysine, is observed and quantified by amino acid analysis (AAA) (*77*). If no single well-defined lysine variant is present, the loss of unsubstituted Lys can be assessed by qAAA. With the Merck chemistry with activation of the OMPC carriers, S-carboxymethylhomo-cysteine or S-carboxymethyl cysteamine are released as markers of the degree of substitution in AAA analysis (*26*).

Molecular Size

The molecular size of conjugates is a valuable marker of consistency of manufacture, and may be linked to immunogenicity (*118*), but there are many confounding factors for such studies. As for polysaccharide sizing, the original approach of soft gel chromatography, with the gel type appropriate for the individual conjugates, has largely been replaced by HPSEC using columns with rigid matrices. The size differences between neoglycoconjugate and cross-linked network immunogens necessitates different column matrices. To transition from molecular size to molecular weight specifications, HPSEC-MALLS has been used. However, a validated basis to estimate the refractive index increment, *dn/dc*, for complex heterogeneous glycoconjugates is unclear, and this has a direct impact on the accuracy of the figures reported. Hydro-dynamic methods provide a complementary means to align molecular sizing and molecular weight analyses, without a dependence on an estimate of *dn/dc*. As for polysaccharides, changes in molecular sizing during storage is important as a stability-indicating assay (see below).

Unreacted Functional Groups

Even after capping, residual unreacted functional groups on the polysaccharide or carrier protein that may react with host tissue may remain on the immunogen. "Each batch should be shown to be free of activated functional groups on either the chemically modified polysaccharide or the carrier protein. Alternatively, the product of the capping reaction can be monitored or the capping reaction can be validated to show removal of unreacted functional groups. Validation of the manufacturing process during vaccine development can eliminate the need to perform this analysis for routine control" (*119*). Appropriate methods may include gas chromatography, HPLC with fluorescence, or UV detection following hydrolysis. With the Merck chemistry, AAA is used to measure S-carboxymethylhomocysteine (SCMHC) which marks the covalent linkages between PS and protein and S-carboxymethyl-cysteamine (SCMC) which marks the number of capped active BrAc groups (*26*).

Unconjugated "Free" Polysaccharide

A specification for the content of unconjugated (or "free") poly-saccharide is based on findings that its presence can reduce the immune response to the glycoconjugate (*120*). The general approach is to separate conjugated and unconjugated saccharide, based on differences in size (especially for glycoconjugates based on oligosaccharides), hydrophobicity (*121*), or precipitation. This has proven difficult in some cases, especially, Vi conjugates (*122*), and non-clinical studies where the immunogen has been spiked with Vi polysaccharide have been used to re-assess the role of free saccharide in Vi conjugate vaccines (*123*). Hib conjugate vaccines containing between <10% and <40% free PRP have been found effective in clinical trials (*124*). For Typbar TCV a specification of NMT 20% has been established (*98*). In glyco-conjugates containing low stability polysaccharides [eg. Hib, MenA, MenC, Pn6A, Pn6B], determination of free saccharide during storage is an important stability indicating assay complementing changes in molecular size (see below).

Unconjugated "Free" Carrier Protein

There is less information in the potential clinical implications of free carrier protein, compared to that for free polysaccharide. In some combination vaccines the glycoconjugate is formulated with DTx or TTx as immunogens in their own right. A method using micellar electrokinetic chromatography (MEKC), using SDS to create micelles, was developed to quantify DTx in meningococcal glycoconjugates (*125*). Published specifications for free carrier protein are typically much lower than for free saccharide: eg. NMT 1% in Typbar TCV (*98*).

Process Residuals

Limit specifications may exist for impurities in the final vaccine, such as formaldehyde or glutaraldehyde (from the carrier protein toxoiding process), antibiotic residues, and conjugation reagents. Measurement in the bulk is generally simpler, and acceptance can be based on the calculated dilutions. An HPLC-MS assay has been developed for EDU, the product from EDC conjugation (*126*).

Glycoconjugate Stability and Integrity

As with other stored intermediates, the stability of the bulk conjugate should be assessed to support a maximum hold time. Molecular sizing is a simple approach to monitor depolymerisation of the glycan chains, or cleavage of the linkage between the polysaccharide and the carrier protein. Increasing free/unconjugated saccharide will provide similar information. 1D ^{1}H and 2D ^{1}H-^{1}H NMR analysis of bulk conjugate immunogens shows the appearance of resonances indicative of poly-saccharide degradation (*127*). In those cases where the polysaccharide does not interfere, circular dichroism has been used to demonstrate that the carrier protein in the conjugate has a comparable structure to the starting material (ie. the protein has not been degraded during conjugation) (*128*). Denaturation of the carrier protein can be monitored by far-UV or near-UV circular dichroism studies in accelerated degradation experiments (*23*, *129*), and may result in glycoconjugate aggregation and precipitation (*23*). The formation of soluble aggregates can be monitored by size exclusion chromatography. ^{1}H NMR monitoring of the acetate anion resonance showed in a sample of Vi-DTx conjugate demonstrated the stability of O-acetylation in that system under long term the storage conditions (*45*).

Formulation and Filling Steps

Few glycoconjugate vaccines are now marketed as monovalent products, and there is a requirement to quantitatively blend multiple monovalent bulks – typically four for the meningococcal and up to 13 for pneumococcal vaccines – or combine glycoconjugate components into complex combination vaccines (which is the case for most Hib conjugates). These multivalent bulks will then require dilution with approved excipients and buffers, such as saline, Tris or succinate buffers (*9*, *23*, *24*). Many products also contain an adjuvant. Multidose presentations include a preservative such as 2-phenoxyethanol, thimerosal or phenol. Formulated bulks are sterile filled into vials and capped, or single use syringes. Bulking agents such as sucrose, lactose or mannitol are added to bulks destined for lyophilised products, and a sterile diluent provided: the adjuvant can be present in the diluent.

Degree of Adsorption to Adjuvant

The role of an adjuvant is to boost and direct the immune response to the vaccine, and aluminium phosphate has proven to be the adjuvant of choice. The degree of adsorption should be determined. In monovalent formulated bulks this must be performed as a release test if it cannot be performed on the final product. After consistent production has been demonstrated, the test can be omitted. Adjuvant and adsorbed material is removed by centrifugation and non-adsorbed antigen is quantified in the supernatant. The degree of adsorption of a CRM_{197}-based pneumococcal conjugate to aluminium phosphate adjuvant is typically 85% or more, although this varies with serotype – Pn4 reportedly has a relatively high adsorption and Pn6B relatively low adsorption (*23*). In complex combination vaccines (eg. DTaP-IPV-Hib) the adjuvant may be present to support the immunogenicity of one or more of the other immunogens. Adsorption onto an adjuvant may improve stability by inhibiting aggregation (*23*). The adjuvant content should be determined: the amount of aluminium should not exceed 1.25 mg per single human dose or as otherwise required by the governing agency (*130*).

Merck reported the degradation of the Hib PRP chain of an OMPC-based glyco-conjugate immunogen when in contact with the surface of the aluminium hydroxide adjuvant, as evidenced by an increase in free saccharide. This arose from hydrolysis at the phosphodiester linkage catalyzed by the adjuvant particle surface. The large size of the conjugate and the tight adsorption meant that only a proportion of the glycan chains could access the surface of the adjuvant (*131*).

Product Presentation

These vaccines are often available in several different presentation formats, including pre-filled syringes and single- and multi- dose vials. The quality requirements for vials and closures are specified in the United States Pharmacopeia (*132–135*) and FDA documentation (*136*).

Control of Final Fill (Drug Product)

As the material which will be delivered to the patient, tests for the final product in the vial focus on ensuring its identity and quantity, and the absence of material which may be harmful. The characteristics of many glycoconjugate vaccines are listed in Table 6. Potentially harmful material which have not been controlled at an earlier stage includes particulates formed during final filling and storage, and confirming vaccine pH and isotonicity. Demonstration of product sterility is a requirement (*137*). There are also regulatory requirements for the labels on the vials and package insert containing information for the patient and person delivering the vaccine. Many other companies in developing countries, members of the Developing Countries Vaccine Manufacturers' Network (DCVMN) (*138*) are producing Hib conjugate vaccines and, increasingly, other glycoconjugate vaccines.

Polysaccharide Identity

This assay confirms that the correct antigen was used during the manufacturing process and that antigenic integrity was maintained during conjugation. For multivalent products this assay should confirm that all of the stated immunogens are indeed present. Immunological methods including rate nephelometry, have proven suitable. Alternatively, the identity of the polysaccharide can be confirmed using a validated method such as HPAEC-PAD.

Table 6. Characteristics of licensed glycoconjugate vaccines

Product Name	*Saccharide*	*Saccharide content*	*Carrier protein*	*Carrier content*	*Adjuvant and other ingredients*
PedVaxHib (*139*)	Hib PRP	7.5 mcg	OMPCs	125 mcg	Amorphous aluminium hydroxyphosphate sulfate
ActHib (*140*)	Hib PRP	10 mcg	TTx	24 mcg	None
Hiberix (*141*)	Hib PRP	10 mcg	TTx	25 mcg	None
HibTiter	Hib PRP oligo	10 mcg	CRM_{197}	25 mcg	None
Vaxem-Hib (*142*)	Hib PRP oligo	10 mcg	CRM_{197}	25 mcg	Aluminium phosphate
QuimiHib	Synthetic Hib PRP	10 mcg	TTx	26 mcg	Aluminium phosphate
Pentacel (*143*)	Hib PRP	10 mcg	TTx	24 mcg	Aluminium phosphate
MenHibrix (*144*)	Men C, Men Y Hib PRP	5 mcg each 2.5 mcg	TTx	17.75 mcg	None
Meningitec (*145*)	Men C oligo	10 mcg	CRM_{197}	15 mcg	Aluminium phosphate
Menveo (*146*)	Men A oligo Men C, Men W, Men Y oligos	10 mcg 5 mcg each	CRM_{197}	32.7 to 64.1 mcg	None
Menactra (*147*)	Men A, Men C Men W, Men Y	4 mcg each	DTx	48 mcg	None
Nimenrix (*148*)	Men A, Men C Men W, Men Y	5 mcg each	TTx	44 mcg	None
MenAfriVac (*149*)	Men A	10 mcg	TTx	10-33 mcg	Aluminium phosphate (in diluent)

Continued on next page.

Table 6. (Continued). Characteristics of licensed glycoconjugate vaccines

Product Name	*Saccharide*	*Saccharide content*	*Carrier protein*	*Carrier content*	*Adjuvant and other ingredients*
Prevnar 13 (*150*)	12 Pn serotypes Serotype 6B	2.2 mcg each 4.4 mcg	CRM_{197}	34 mcg	Aluminium phosphate
Synflorix (*22*, *151*)	Pn1, Pn5, Pn6B, Pn7F, Pn9V, Pn14, Pn23F Pn4, Pn18C, Pn19F	1 mcg 1mcg 3 mcg	Protein D, TTx or DTx	9-16 mcg 5-10 mcg 3-6 mcg	Aluminium phosphate
Typbar-TCV (*98*)	Vi CPS	25 mcg	TTx	n.r.[a]	None

[a] Not reported.

Polysaccharide Quantity

The amount of material, typically in terms of the saccharide component, in the final vial should be determined [Table 7]. This is the key assay that confirms that the correct dosage is being delivered. Pharmacopoeial acceptance criteria differ between vaccines.

The acceptance limits on the specifications are designed to allow for analytical uncertainty and unavoidable variations in manufacturing and formulation. The methods used are physicochemical or rate nephelo-metry as an immunochemical method. Typically, the point estimate of the content of the vial should be 80-120% of the label for the product to be compliant [Table 7], although there is no upper limit for Men C conjugates and wider limits for the adsorbed pneumococcal conjugates.

Table 7. Acceptance criteria for different glycoconjugate vaccines – saccharide content

Vaccine	*Acceptance criteria*
Hib conjugates (*92*)	• Content is within the range 80.0 to 120.0% of the label content
Men C conjugates (*93*)	• Content is NLT 80.0% of the label content
Pn conjugates (*32*)	• Content is within the range 70.0 to 130.0% of the label content • 95% confidence range of estimated content is within the range 80-120% of the estimated content
Vi conjugates (*57*)	• Content is within the range 80.0 to 120.0% of the label content • 95% confidence range of estimated content is within the range 80-120% of the estimated content

The criterion of a 95% CI within the range 80-120% of the estimate is a limit on the imprecision of the immunochemical assay being used. Often a quantitative physico-chemical or immunochemical assay will also provide the information to confirm the identity of the polysaccharide.

O-Acetylation

Uniquely, for Vi conjugate vaccines there is a specification for the O-acetylation in the final fill, with an anomaly in the requirements for the Vi CPS (*152*) and conjugate vaccines (*57*). These require a higher degree of O-acetylation in the final fill (NLT 0.085 ± 25% µmol per 25 mg s.h.d, or NLT 3.4 µmol mg^{-1}) than in the bulk polysaccharide (NLT 2 µmol mg^{-1}) from which it is produced.

Pyrogenicity Test or Endotoxin Content

During development and "wherever revalidation is required (*32*, *92*, *93*)" a rabbit pyrogenicity test (*153*) is required, but during routine manufacturing the endotoxin content is measured using a LAL endotoxin assay (*154*). Typical values are NMT 12.5 IU per s.h.d (Pn conjugates) (*32*) or NMT 25 IU per s.h.d (Men C conjugate) (*93*).

Preservative

Multidose vaccines may contain antimicrobial preservatives such as 2-phenoxyethanol, thimersal, benzethonium chloride or phenol (*134*). The content in the vial should be determined. For example, Typbar TCV multi-dose vials contains 5 mg per s.h.d of 2-phenoxyethanol, with an allowed variation of ±15% (*155*). Preservatives should meet pharmacopeial expectations of antimicrobial effectiveness (*156*).

Common Regulatory Requirements

There are a series of standard tests applied to all parenteral products to minimise pain and other adverse events at the site of injection. These include sample osmolarity, pH, sterility and freedom from particulate matter. A visual inspection of the colour and appearance of the sample, and freedom from visible particles are quick probabilistic tests to ensure product consistency. These are described in the US Pharmacopeia (*157*) The average moisture content should be determined for a lyophilised product (*158*): values should typically be NMT 2.5%, and no vial should be found to have a residual moisture content of more than 3%. Vaccine product labelling is regulated in compliance with 21 CFR 201 and 610. Requirements are set for container labelling and package labelling. The information content of the package insert is regulated in 21CFR201.57. The requirement is summarised in an FDA Guidance Note (*159*). The FDA website (*160*) on licensed vaccine products leads through to the package inserts for products licensed in the US. Previously, biological products that are intended for administration to humans were subject to a "general safety test" to detect any extraneous toxic contaminants (*161*). Whilst this is no longer required for newly licensed, it remains a requirement for legacy products.

Licensing

A full description of the licensing process for vaccines is beyond the scope of this chapter. For biological products the process is essentially the same, but there are some aspects which are unusual.

Preclinical Evaluation and Phase 1 and 2 Studies

Preclinical evaluation of the potential vaccines will include a full physico-chemical evaluation of the composition, structure and stability of the immunogen and an impurity profile established. The immuno-genicity of the development conjugates will be characterised in animal models, and the quality of the immune response analysed. This will include antibody isotype, avidity and functional activity. The composition and manufacturing approach can be changed to maximise immunogenicity and minimise impurities and residuals. The T-cell dependent nature of the immune response should be demonstrated. Assay validation will take place, manufacturing consistency assessed and GMP and GCP quality systems put into place. Phase 1 and 2 studies are typically small and medium size studies to evaluate safety and immunogenicity of these vaccines.

Phase 3

The Phase 3 clinical trial, which is the critical study for licensure, is usually an efficacy study, showing reduced incidence of disease in the vaccinated compared to the control arms of the study. Because in developed countries disease incidence is low, Phase 3 clinical trials for vaccines often requires large numbers of participants. Even then they and may not produce clear-cut outcomes. In the Prevnar 7 trial in North California (*162*), the small number of cases caused by some serotypes in the control arm of the study meant that no efficacy estimate could be made, and where only a small number of cases were observed the statistical analysis will have wide uncertainty limits. A key outcome from these clinical trials is to establish an antibody level which best correlates with protection from disease (*163*). Follow-on products are generally licensed on the basis of the non-inferiority of the antibody levels following immunisation, although there can be a balancing act for regulators between increased serotype type coverage and reduced antibody responses against some serotypes. In the case of the MenC products in the UK, vaccines were licensed on the basis of immuno-genicity rather than efficacy, as there was a clear understanding between the relationship between antibody levels and reductions in disease incidence (*164*). Only after licensing and the start of mass paediatric immunisation was an estimate of vaccine effectiveness produced, confirming the correctness of the decision. As disease cases caused by other meningococcal serogroups are generally rarer than for Men C, a similar approach would be appropriate for vaccines with expanded valencies.

Post-Marketing Surveillance

Regulatory approval for a vaccine product will typically require the manufacturer to continue to monitor the product for adverse incidents too rare to show up in clinical trials. Also, in some cases, the estimates of vaccine effectiveness arise from post licensing surveillance: this was particularly the case for the MenC vaccine, which had been licensed on immunogenicity rather than efficacy (*165*). The complex nature of protection offered by current glycoconjugate vaccines – direct immunological defence against invading organisms, reduction in pathogen carriage reducing the frequency of host invasion by commensal bacteria, and a concomitant reduction in transmission between carrier and naïve individuals – makes assessment of the cost-benefit performance of these vaccines complex. It requires extended monitoring of disease levels in the entire population, not just vaccinated individuals (*166*). Introduction of PnC7 paediatric immunisation in California resulted in major reductions in disease caused by serotypes contained in the vaccine in the unvaccinated over-65s population (*3*). In addition, the infecting serotype in cases of disease need to be identified to account for and understand the extent of serotype replacement (*167*). In the US, this work was carried out by the Active Bacterial Surveillance Core group (*168*), now the National Notified Disease Surveillance System. These data help define the additional serotypes to be included in a next-generation vaccine with expanded coverage.

Post-marketing surveillance also provides information on changes in the disease profile which might indicate that a different immunisation strategy/ timetable is optimal.

Reference Standards

When they are available, manufacturers should use the appropriate pharmacopoeial standards to support release testing. Reference standards can be qualitative, as used to confirm identity, or quantitative materials and system suitability standards. Whereas reference materials for common materials, such as the polysaccharides or carrier protein components, used in more than one product are feasible pharmacopoeial targets, manufacturer-specific materials – such as individual vaccine products – are not. There are few reference materials for the bacterial polysaccharides, and those for toxoid carriers are focussed on toxoid vaccines. It is expected that the manufacturer will have fully characterised in-house glycoconjugate reference standards prepared from a lot (or lots) representative of the manufacturing process for use in specific assays (eg. rate nephelometry assays). International and pharmacopoeial reference standards exist for endotoxin content, but, at present, not for most other impurities.

For a reference standard to support a quantitative specification the uncertainty of measurement [UoM, a combination of analytical bias determined in validation studies, and method imprecision] on the reference standard should be "negligible", as outlined in the "pharmacopoeial exemption" in ISO 34 (*169*) and exemplified in the European Pharmacopoeia (*170*) with respect to the allowable uncertainty on the specification [eg. ± 20% for Hib glycoconjugates: see Table 7].

The choice of an appropriate and validated reference standard is critical. For example, a recent study of the determined saccharide content of Hib conjugate vaccines differed by 10% depending on whether the reference standard was ribitol or the 1st WHO International Standard for Hib PRP (*171*).

System suitability standards are deigned to challenge the performance of an analytical method and to challenge its ability to deliver quality results. This might be the limit of detection of the assay, or the chromatographic resolution of an analytical separation.

Manufacturing Variations

Should a manufacturing process be altered, for economic reasons or due to the unavailability of a starting material, manufacturers should demonstrate the comparability of their product before and after the manufacturing change. This situation is covered in ICH Guideline Q5E (*172*). Similarly, manufacturing of a licensed vaccine through an established process at a new site cannot be assumed to result in a clinically equivalent product. Comparability testing and re-validation for intermediates and final fill products, perhaps including possibly immunogenicity testing in humans, should be expected. In general, for well characterised materials, which in this case typically includes the bulk polysaccharide intermediate(s), detailed physicochemical comparability is usually sufficient to establish comparability. More complex products, such as the final glycoconjugates, are likely to require some type of biological or immunological testing, perhaps even in humans, to confirm comparability of the new product with the existing material (*173*). The European Pharmacopoeia highlights the need to demonstrate through tests in animals that the re-validated product consistently elicits a T-cell-dependent B-cell immune response (*32*, *92*, *93*).

Concluding Remarks

The current family of glycoconjugate immunogens are structurally related, produced by a common series of manufacturing processes, are controlled through largely common analytical approaches and with similar regulatory requirements. Beyond those currently licensed a number of other glycoconjugate vaccines are in development, to add protection against additional serotypes or to protect against pathogens for which no vaccines currently exist. Monovalent Vi conjugate vaccine is seen as a stepping stone to a bivalent material (*123*), which also includes a conjugate prepared from the LPS O-chain from *Salmonella* paratyphi A (*174*), to also immunise against paratyphoid. Some developmental vaccines are intended for high-risk populations, rather than mass paediatric use. Tri-, penta- and hexa-valent CRM_{197}-based Group B Streptococcus vaccines, intended for the protection of the newborn through vaccination of the pregnant mother are in development (*175*, *176*). A multivalent vaccine containing both two CRM_{197}-based glyco-conjugates and two recombinant proteins, designed to offer protection against *Staphylococcus aureus* infections in patients undergoing elective surgery, is moving towards large scale clinical trials (*177*, *178*). It would seem probable that quality expectations for existing vaccines are likely to extend to those products. Different conjugation chemistries are appropriate in different cases: the choice is constrained principally by the structure of the repeat units of the polysaccharide, and secondarily by the choice of carrier protein (a native protein or a chemically toxoided protein). It is this choice which leads to the different families of immunogen structures highlighted in Figure 1.

References

1. The chapter is based upon the best understanding of the author on current expectations, but these may differ between countries, and with time, and for individual products.
2. Jones, C. *An. Acad. Bras. Cienc.* **2005**, *77*, 293–324.
3. Pilishvili, T.; Lexau, C.; Farley, M. M.; Hadler, J.; Harrison, L. H.; Bennett, N. M.; Reingold, A.; Thomas, A.; Schaffner, W.; Craig, A. S.; Smith, P. J.; Beall, B. W.; Whitney, C. G.; Moore, M. R. Active Bacterial Core Surveillance/Emerging Infections Program Network. *J. Infect. Dis.* **2010**, *201*, 32–41.
4. In this chapter, the term "vaccine" refers to the formulated, filled final product ready for use. Bulk glycoconjugates are called "immunogens", and carrier proteins and saccharides prior to conjugation are "components" or "antigens".
5. Teyssou, R.; Muros-Le Rouzic, E. *Vaccine* **2007**, *25*, A3–A7.
6. Berti, F.; Costantino, P.; Fragai, M.; Luchinat, C. *Biophys. J.* **2004**, *86*, 3–9.
7. Emini, E. A.; Watson, W. J.; Prasad, A. K.; Han, M.; Kim, J. H.; Gu, J.; Yang, Y.-y.; Kainthan, R. K.; Cooper, D.; Pride, M. W.; Jansen, K. U. *Immunogenic compositions comprising conjugated capsular antigens and uses thereof.* U.S. Patent 9,492,559 B2, November 15, 2016.
8. Abdelhameed, A. S.; Morris, G. A.; Almutairi, F.; Adams, G. G.; Duvivier, P.; Conrath, K.; Harding, S. E. *Sci. Rep.* **2016**, *26* (6), 35588.
9. Shehab, Z. M.; Azimi, P.; Asmar, B. I.; Dunphy, M. G.; Ayoub, E. M.; Dajani, A. S. *Scand. J. Infect. Dis.* **1991**, *23*, 763–769.
10. Egan, W.; Frasch, C. E.; Anthony, B. F. *J. Am. Med. Assoc., JAMA* **1995**, *273*, 888–889.
11. Jones, C. Glycoconjugate Vaccines: The Regulatory Framework. In *Carbohydrate-Based Vaccines: Methods and Protocols*; Lepenies, B., Ed.; Springer: New York, 2015; Vol. 1331, pp 229–251.
12. *Good Manufacturing Practice Guide for Active Pharmaceutical Ingredients Q7.* http://www.ich.org/fileadmin/Public_Web_Site/ICH_Products/Guidelines/Quality/Q7/Step4/Q7_Guideline.pdf (accessed 18 Dec. 2017).
13. Investigational New Drug Application, *Code of Federal Regulations*, Part 312.23(a)(7), Title 21, 2017.
14. *Validation of Analytical Procedures: Text and Methodology Q2(R1).* http://www.ich.org/fileadmin/Public_Web_Site/ICH_Products/Guidelines/Quality/Q2_R1/Step4/Q2_R1__Guideline.pdf (accessed 09 Oct. 2017).
15. "Also, according to these regulations [21 CFR 211.194(a)(2)], users of analytical methods described in USP–NF are not required to validate the accuracy and reliability of these methods, but merely verify their suitability under actual conditions of use." See 1226 Verification of Compendial Procedures. *United States Pharmacopeia 40*, United States Pharmacopeial Convention, Rockville, MD, 2017.
16. *Quality of Biotechmological Products: Stability Testing of Biotechnological/Biological Products Q5C.* http://www.ich.org/fileadmin/Public_Web_Site/

ICH_Products/Guidelines/Quality/Q5C/Step4/Q5C_Guideline.pdf (accessed 18 Dec. 2017).
17. *Specifications: Test Procedures and Acceptance Criteria for Biotechnological/Biological Products Q6B*. http://www.ich.org/fileadmin/Public_Web_Site/ICH_Products/Guidelines/Quality/Q6B/Step4/Q6B_Guideline.pdf (accessed 11 Oct. 2017).
18. Willis, L. M.; Stupak, J.; Richards, M. R.; Lowary, T. L.; Li, J.; Whitfield, C. *Proc. Natl. Acad. Sci. U. S. A.* **2013**, *110*, 7868–7873.
19. Liston, S. D.; Ovchinnikova, O. G.; Whitfield, C. *Proc. Natl. Acad. Sci. U. S. A.* **2016**, *113*, 6719–6724.
20. Larson, T. R.; Yother, J. *Proc. Natl. Acad. Sci. U. S. A.* **2017**, *114*, 5695–5700.
21. *Derivation and Characterisation of Cell Substrates Used for Production of Biotechnological/Biological Products Q5D*. http://www.ich.org/fileadmin/Public_Web_Site/ICH_Products/Guidelines/Quality/Q5D/Step4/Q5D_Guideline.pdf (accessed 18 Dec. 2017).
22. *Assessment Report for Synflorix*. http://www.ema.europa.eu/docs/en_GB/document_library/EPAR_-_Public_assessment_report/human/000973/WC500054349.pdf (accessed 18 Dec. 2017).
23. Khandke, L.; Malone, R.; Yang, C. X.; Han, H.; Look, J. L.; Jin, Z.; Seid, R. C. Jr.; Chen, Y. *Formulations which stabilize and inhibit precipitation of immunogenic compositions*. U.S. Patent 8,562,999 B2, October 22, 2013.
24. Park, M.-H., Kim, H.; Yang, J.-H.; Yang, S.-Y.; Noh, M.-J.; Park, S.-J.; Shin, J.-H. *Multivalent Pneumococcal Polysaccharide-Protein Conjugate Composition*. U.S. Patent Application 2015/0,343,076 A1, December 3, 2015.
25. *Guidelines on the quality, safety and efficacy of typhoid conjugate vaccines*. http://www.who.int/biologicals/areas/vaccines/TYPHOID_BS2215_doc_v1.14_WEB_VERSION.pdf (accessed 17 Jan. 2018).
26. Kniskern, P. J.; Ip, C. C.; Hagopian, A.; Hennessey Jr, J. P.; Miller, W. J.; Kubek, D. J.; Burke, P. D.; Marburg, S.; Tolman, R. L. *Pneumo-coccal Polysaccharide Conjugate Vaccine*, Eur. Pat. Appl. 0 497 525 A2, January 27, 1992.
27. *Requirements for Meningococcal Polysaccharide Vaccine*. http://www.who.int/biologicals/publications/trs/areas/vaccines/meningococcal/WHO_TRS_594_MeningA2.pdf (accessed 18 Dec. 2017).
28. *Requirements for Meningococcal Polysaccharide Vaccine (addendum 1980)*. http://www.who.int/biologicals/publications/trs/areas/vaccines/meningococcal/WHO_TRS_658_(part2)_MeningA6.pdf (accessed 18 Dec. 2017).
29. Jones, C.; Lemercinier, X.; Crane, D. T.; Gee, C. K.; Austin, S. *Dev. Biologicals* **2000**, *103*, 121–136.
30. Harding, S. E.; Abdelhameed, A. S.; Morris, G. A.; Adams, G.; Laloux, O.; Cerny, L.; Bonnier, B.; Duvivier, P.; Conrath, K.; Lenfant, C. *Carbohydr. Polym.* **2012**, *90*, 237–242.

31. *PART 610: General Biological Products Standards: Subpart-General Provisions: Sec. 610.14 Identity*. https://www.accessdata.fda.gov/scripts/cdrh/cfdocs/cfcfr/CFRSearch.cfm?fr=610.14 (accessed 06 Oct. 2017).
32. Ph.Eur. Monograph 2150 Pneumococcal polysaccharide conjugate vaccines, *European Pharmacopoeia*, EDQM, Strasbourg, 2017.
33. *Recommendations to assure the quality, safety and efficacy of pneumococcal conjugate vaccines*. http://www.who.int/biologicals/areas/vaccines/pneumo/Pneumo_final_23APRIL_2010.pdf?ua=1 (accessed 18 Jan 2018).
34. Jones, C.; Mulloy, B.; Wilson, A.; Dell, A.; Oates, J. E. *J. Chem. Soc. Perkin Trans. 1* **1985**, 1665–1673.
35. Jones, C.; Currie, F. *Carbohydr. Res.* **1988**, *184*, 279–284.
36. Jones, C. *Carbohydr. Res.* **1995**, *269*, 175–182.
37. Lemercinier, X.; Jones, C. *Carbohydr. Res.* **1996**, *296*, 83–96.
38. Lemercinier, X.; Whitley, C.; Jones, C. *Carbohydr. Res.* **2000**, *325*, 192–201.
39. Jones, C.; Lemercinier, X. *Carbohydr. Res.* **2005**, *340*, 403–409.
40. Jones, C. *Carbohydr. Res.* **2005**, *340*, 1097–1106.
41. Lemercinier, X.; Jones, C. *Carbohydr. Res.* **2006**, *341*, 68–74.
42. Jones, C.; Lemercinier, X. *J. Pharm. Biomed. Anal.* **2002**, *30*, 1233–1247.
43. 198 Nuclear Magnetic Resonance Spectroscopy Identity Testing of Bacterial Polysaccharides Used In Vaccine Manufacture. *Pharmacopeial Forum* **2017**, *43* (2).
44. Abeygunawardana, C.; Williams, T. C.; Sumner, J. S.; Hennessey, J. P., Jr. *Anal. Biochem.* **2000**, *279*, 226–240.
45. Jones, C.; *NIBSC*, UK, Unpublished data collected on various projects, 1982–2015.
46. Szu, S. C.; Li, X. R.; Stone, A. L.; Robbins, J. B. *Infect. Immun.* **1991**, *59*, 4555–4561.
47. Berry, D. S.; Lynn, F.; Lee, C. H.; Frasch, C. E.; Bash, M. C. *Infect. Immun.* **2002**, *70*, 3707–3713.
48. Lupisan, S.; Limkittikul, K.; Sosa, N.; Chanthavanich, P.; Bianco, V.; Baine, Y.; Van der Wielen, M.; Miller, J. M. *Clin. Vaccine Immunol.* **2013**, *20*, 1499–1507.
49. Fusco, P. C.; Farley, E. K.; Huang, C. H.; Moore, S.; Michon, F. *Clin. Vaccine Immunol.* **2007**, *14*, 577–584.
50. Michon, F.; Huang, C. H.; Farley, E. K.; Hronowski, L.; Di, J.; Fusco, P. C. *Dev. Biol. (Basel, Switz.)* **2000**, *103*, 151–160.
51. Composition of polysaccharide vaccines, *British Pharmacopoeia*, BPC, London, 2017; pp V-A479−481
52. Fischer, W.; Behr, T.; Hartmann, R.; Peter-Katalinić, J.; Egge, H. *Eur. J. Biochem.* **1993**, *215*, 851–857.
53. Karlsson, C.; Jansson, P. E.; Skov Sørensen, U. B. *Eur. J. Biochem.* **1999**, *265*, 1091–1097.
54. Vialle, S.; Sepulcri, P.; Dubayle, J.; Talaga, P. *Carbohydr. Res.* **2005**, *340*, 91–96.
55. Talaga, P.; Bellamy, L.; Moreau, M. *Vaccine* **2001**, *19*, 2987–94.

56. Xu, Q.; Abeygunawardana, C.; Ng, A. S.; Sturgess, A. W.; Harmon, B. J.; Hennessey, J. P., Jr. *Anal. Biochem.* **2005**, *336*, 262–272.
57. Typhoid Vi Conjugate Vaccine, *Indian Pharmacopoeia*, Indian Pharmacopoeia Commission, Delhi, 2017; pp 4277−4281.
58. Clément, G.; Dierick, J.-F.; Lenfant, C.; Giffroy, D. *Pharmeuropa* **2014**, 40–59.
59. Bednar, B.; Hennessey, J. P., Jr. *Carbohydr. Res.* **1993**, *243*, 115–130.
60. MacNair, J. E.; Desai, T.; Teyral, J.; Abeygunawardana, C.; Hennessey, J. P., Jr. *Biologicals* **2005**, *33*, 49–58.
61. Harding, S. E.; Vårum, K. M.; Stokke, B. T.; Smidsrød, O. Molecular Weight Determineation of Polysaccharides. In *Advances in Carbohydrate Analysis*, Vol. 1; White, C. A., Ed.; JAI Press: London, 1991; pp 63−144.
62. Turula, V. E., Jr.; Gore, T.; Singh, S.; Arumugham, R. G. *Anal. Chem.* **2010**, *82*, 1786–1792.
63. Cook, M. C.; Bliu, A.; Kunkel, J. P. *Vaccine* **2013**, *31*, 3702–3711.
64. Tsai, C. M.; Gu, X. X.; Byrd, R. A. *Vaccine* **1994**, *12*, 700–706.
65. Micoli, F.; Rondini, S.; Pisoni, I.; Proietti, D.; Berti, F.; Costantino, P.; Rappuoli, R.; Szu, S.; Saul, A.; Martin, L. B. *Vaccine* **2011**, *29*, 712–720.
66. Walker, J. M. Rocket immunoelectrophoresis. In *Proteins*; Walker, J. M., Ed.; Methods Mol. Biol. Vol. 1; Humana Press: New York, 1984**;** pp 317-323.
67. Stroop, C. J.; Xu, Q.; Retzlaff, M.; Abeygunawardana, C.; Bush, C. A. *Carbohydr. Res.* **2002**, *337*, 335–344.
68. Ph.Eur. Chapter 5.2.11 Carrier proteins for the production of conjugated polysaccharide vaccines for human use. *European Pharmacopoeia*, EDQM, Strasbourg, 2017.
69. CRM197. https://www.reagentproteins.com/product_detail/CRM197_Lyophilized.html (accessed 29 Dec. 2017).
70. Stefan, A.; Conti, M.; Rubboli, D.; Ravagli, L.; Presta, E.; Hochkoeppler, A. *J. Biotechnol.* **2011**, *156*, 245–252.
71. Steere, B.; Eisenberg, D. *Biochemistry* **2000**, *39*, 15901–15909.
72. 1234 Vaccines for Human Use – Polysaccharide and Glycoconjugate Vaccines. *United States Pharmacopeia USP 40*, United States Pharmacopeial Convention, Rockville, MD, 2017.
73. Forsgren, A.; Riesbeck, K.; Janson, H. *Clin. Infect. Dis.* **2008**, *46*, 726–731.
74. Welch, P. G.; Fattom, A.; Moore, J., Jr.; Schneerson, R.; Shiloach, J.; Bryla, D. A.; Li, X.; Robbins, J. B. *J. Am. Soc. Nephrol.* **1996**, *7*, 247–253.
75. Szu, S. C.; Taylor, D. N.; Trofa, A. C.; Clements, J. D.; Shiloach, J.; Sadoff, J. C.; Bryla, D. A.; Robbins, J. B. *Infect. Immun.* **1994**, *62*, 4440–4444.
76. D'Ambra, A.; Baugher, J. E.; Concannon, P. E.; Pon, R. A.; Michon, F. *Anal. Biochem.* **1997**, *250*, 228–236.
77. Seid, R. C., Jr.; Boykins, R. A.; Liu, D. F.; Kimbrough, K. W.; Hsieh, C. L.; Eby, R. *Glycoconj. J.* **1989**, *6*, 489–498.
78. Biemans, R. L.; Duviivier, P. *Conjugation Process*. U. S. Patent Appl. US 2013/0344103 A1, December 26, 2013.
79. *FDA-Patented Invention Earns 2016 Patents for Humanity Award for Impact on Global Public Health.* https://blogs.fda.gov/fdavoice/index.php/2017/01/

fda-patented-invention-earns-2016-patents-for-humanity-award-for-impact-on-global-public-health/ (accessed 18 Jan. 2018).
80. *A Method with Increased Yield for Production of Polysaccharide-Protein Conjugate Vaccines Using Hydrazide Chemistry*, https://www.fda.gov/ScienceResearch/CollaborativeOpportunities/Inventions/ucm464876.htm (accessed 18 Jan. 2018).
81. Frasch, C. E.; Kapre, S. V.; Lee, C. H.; Préaud, J. M. *Clin. Infect. Dis.* **2015**, *61* (Suppl 5), S404–409.
82. Ma, Z.; Bertolo, L.; Arar, S.; Monteiro, M. A. *Carbohydr. Res.* **2011**, *346*, 343–347.
83. Costantino, P.; Norelli, F.; Giannozzi, A.; D'Ascenzi, S.; Bartoloni, A.; Kaur, S.; Tang, D.; Seid, R.; Viti, S.; Paffetti, R.; Bigio, M.; Pennatini, C.; Averani, G.; Guarnieri, V.; Gallo, E.; Ravenscroft, N.; Lazzeroni, C.; Rappuoli, R.; Ceccarini, C. *Vaccine* **1999**, *17*, 1251–1263.
84. Ravenscroft, N. The application of NMR spectroscopy to track the industrial preparation of polysaccharide and derived glycoconjugate vaccines. In *Biologicals Beyond 2000: Challenges for Quality Standards in an Evolving Field*; EDQM: Strasbourg, France, 1999; pp 131–144.
85. Bardotti, A.; Averani, G.; Berti, F.; Berti, S.; Galli, C.; Giannini, S.; Fabbri, B.; Proietti, D.; Ravenscroft, N.; Ricci, S. *Vaccine* **2005**, *23*, 1887–1899.
86. Ravenscroft, N.; Averani, G.; Bartoloni, A.; Berti, S.; Bigio, M.; Carinci, V.; Costantino, P.; D'Ascenzi, S.; Giannozzi, A.; Norelli, F.; Pennatini, C.; Proietti, D.; Ceccarini, C.; Cescutti, P. *Vaccine* **1999**, *17*, 2802–2816.
87. Möginger, U.; Resemann, A.; Martin, C. E.; Parameswarappa, S.; Govindan, S.; Wamhoff, E. C.; Broecker, F.; Suckau, D.; Pereira, C. L.; Anish, C.; Seeberger, P. H.; Kolarich, D. *Sci. Rep.* **2016**, *6*, 20488.
88. Bardotti, A.; Averani, G.; Berti, F.; Berti, S.; Carinci, V.; D'Ascenzi, S.; Fabbri, B.; Giannini, S.; Giannozzi, A.; Magagnoli, C.; Proietti, D.; Norelli, F.; Rappuoli, R.; Ricci, S.; Costantino, P. *Vaccine* **2008**, *26*, 2284–2296.
89. Crotti, S.; Zhai, H.; Zhou, J.; Allan, M.; Proietti, D.; Pansegrau, W.; Hu, Q. Y.; Berti, F.; Adamo, R. *ChemBioChem* **2014**, *15*, 836–843.
90. Lees, A.; Nelson, B. L.; Mond, J. J. *Vaccine* **1996**, *14*, 190–198.
91. Shafer, D. E.; Toll, B.; Schuman, R. F.; Nelson, B. L.; Mond, J. J.; Lees, A. *Vaccine* **2000**, *18*, 1273–1281.
92. Ph. Eur. Monograph 1219 *Haemophilus* Type b conjugate vaccine. In *European Pharmacopoeia*; EDQM: Strasbourg, 2017.
93. Ph. Eur. Monograph 3112 Meningococcal Group C conjugate vaccine. In *European Pharmacopoeia*; EDQM: Strasbourg, 2017.
94. Rondini, S.; Micoli, F.; Lanzilao, L.; Hale, C.; Saul, A. J.; Martin, L. B. *Clin. Vaccine Immunol.* **2011**, *18*, 460–468.
95. Szu, S. C. *Exp. Rev. Vaccines* **2013**, *12*, 1273–1286.
96. Micoli, F.; Rondini, S.; Pisoni, I.; Giannelli, C.; Di Cioccio, V.; Costantino, P.; Saul, A.; Martin, L. B. *Vaccine,* **2012**, *30*, 853–861.
97. Micoli, F.; Rondini, S.; Pisoni, I.; Proietti, D.; Berti, F.; Costantino, P.; Rappuoli, R.; Szu, S.; Saul, A.; Martin, L. B. *Vaccine* **2011**, *29*, 712–720.

98. *Typbar-TCV: Summary of Product Characteristic.* http://cdsco.nic.in/writereaddata/Bharat%20Biotech%20Typbar-TCV.pdf (accessed 15 Dec. 2017).
99. Xu, Q.; Klees, J.; Teyral, J.; Capen, R.; Huang, M.; Sturgess, A. W.; Hennessey, J. P., Jr.; Washabaugh, M.; Sitrin, R.; Abeygunawardana, C. *Anal. Biochem.* **2005**, *337*, 235–245.
100. Verez-Bencomo, V.; Fernández-Santana, V.; Hardy, E.; Toledo, M. E.; Rodríguez, M. C.; Heynngnezz, L.; Rodriguez, A.; Baly, A.; Herrera, L.; Izquierdo, M.; Villar, A.; Valdés, Y.; Cosme, K.; Deler, M. L.; Montane, M.; Garcia, E.; Ramos, A.; Aguilar, A.; Medina, E.; Toraño, G.; Sosa, I.; Hernandez, I.; Martínez, R.; Muzachio, A.; Carmenates, A.; Costa, L.; Cardoso, F.; Campa, C.; Diaz, M.; Roy, R. *Science* **2004**, *305*, 522–525.
101. Safari, D.; Dekker, H. A.; Joosten, J. A.; Michalik, D.; de Souza, A. C.; Adamo, R.; Lahmann, M.; Sundgren, A.; Oscarson, S.; Kamerling, J. P.; Snippe, H. *Infect. Immun.* **2008**, *76*, 4615–4623.
102. Pawlowski, A.; Källenius, G.; Svenson, S. B. *Vaccine* **1999**, *17*, 1474–1483.
103. Fattom, A.; Vann, W. F.; Szu, S. C.; Sutton, A.; Li, X.; Bryla, D.; Schiffman, G.; Robbins, J. B.; Schneerson, R. *Infect. Immun.* **1988**, *56*, 2292–2298.
104. Laferrière, C. A.; Sood, R. K.; de Muys, J. M.; Michon, F.; Jennings, H. J. *Vaccine* **1997**, *15*, 179–186.
105. Kubler-Kielb, J. Conjugation of LPS-derived oligosaccharides to proteins using oxime chemistry. In *Bioconjugation Protocols*; Mark, S. S., Ed; Methods Mol. Biol. Vol. 751; Humana Press: New York, 2011; pp 317–327.
106. Wan, Q.; Chen, J.; Chen, G.; Danishefsky, S. J. *J. Org. Chem.* **2006**, *71*, 8244–8249.
107. Grandjean, C.; Boutonnier, A.; Guerreiro, C.; Fournier, J. M.; Mulard, L. A. *J. Org. Chem.* **2005**, *70*, 7123–7132.
108. Hahm, H. S.; Schlegel, M. K.; Hurevich, M.; Eller, S.; Schuhmacher, F.; Hofmann, J.; Pagel, K.; Seeberger, P. H. *Proc. Natl. Acad. Sci. U. S. A.* **2017**, *114*, E3385–E3389.
109. Parameswarappa, S. G.; Reppe, K.; Geissner, A.; Ménová, P.; Govindan, S.; Calow, A. D.; Wahlbrink, A.; Weishaupt, M. W.; Monnanda, B. P.; Bell, R. L.; Pirofski, L. A.; Suttorp, N.; Sander, L. E.; Witzenrath, M.; Pereira, C. L.; Anish, C.; Seeberger, P. H. *Cell. Chem. Biol.* **2016**, *23*, 1407–1416.
110. Seeberger, P. H.; Pereira, C. L.; Govindan, S. *Beilstein J. Org. Chem.* **2017**, *13*, 164–173.
111. Lisboa, M. P.; Khan, N.; Martin, C.; Xu, F. F.; Reppe, K.; Geissner, A.; Govindan, S.; Witzenrath, M.; Pereira, C. L.; Seeberger, P. H. *Proc. Natl. Acad. Sci. U. S. A.* **2017**, *114*, 11063–11068.
112. Emmadi, M.; Khan, N.; Lykke, L.; Reppe, K.; Parameswarappa, S.; Lisboa, M. P.; Wienhold, S. M.; Witzenrath, M.; Pereira, C. L.; Seeberger, P. H. *J. Am. Chem. Soc.* **2017**, *139*, 14783–14791.
113. For example, see Geissner, A.; Pereira, C. L.; Leddermann, M.; Anish, C.; Seeberger, P. H. *ACS Chem. Biol.* **2016**, *11*, 335–44.
114. Cook, M. C.; Kunkel, J. P. *Anal. Chem.* **2015**, *87*, 5009–5011.
115. Lee, C. J. *J. Biol. Stand.* **1983**, *11*, 55–64.

116. Noble, J. E.; Knight, A. E.; Reason, A. J.; Di Matola, A.; Bailey, M. J. *Mol. Biotechnol.* **2007**, *37*, 99–111.
117. Ph. Eur. Method Total protein. In *European Pharmacopoeia*; EDQM: Strasbourg, 2017.
118. An, S. J.; Yoon, Y. K.; Kothari, S.; Kothari, N.; Kim, J. A.; Lee, E.; Kim, D. R.; Park, T. H.; Smith, G. W.; Carbis, R. *Vaccine* **2011**, *29*, 7618–7623.
119. *Recommendations to assure the quality, safety and efficacy of pneumococcal conjugate vaccines*. Replacement of WHO Technical Report Series, No. 927, Annex 2. http://www.who.int/biologicals/vaccines/TRS_977_Annex_3.pdf (accessed 16 Dec. 2017).
120. Peeters, C. C.; Tenbergen-Meekes, A. M.; Poolman, J. T.; Zegers, B. J.; Rijkers, G. T. *Vaccine* **1992**, *10*, 833–840.
121. Lei, Q. P.; Lamb, D. H.; Heller, R.; Pietrobon, P. *J. Pharm. Biomed. Anal.* **2000**, *21*, 1087–1091.
122. Jones, C.; Lee, C. K.; Ahn, C.; Shin, J.; Knezevic, I. WHO Working Group on Quality, Safety and Efficacy of Typhoid Vi Capsular Polysaccharide Conjugate Vaccines. *Vaccine* **2013**, *31*, 4466–4469.
123. Giannelli, C.; Cappelletti, E.; Di Benedetto, R.; Pippi, F.; Arcuri, M.; Di Cioccio, V.; Martin, L. B.; Saul, A.; Micoli, F. *J. Pharm. Biomed. Anal.* **2017**, *139*, 143–147.
124. *Recommendations for the Production and Control of Haemophilus influenzae Type b conjugate vaccines*. http://www.who.int/biologicals/publications/trs/areas/vaccines/influenza/WHO_TRS_897_A1.pdf?ua=1 (accessed 23 May 2018).
125. Lamb, D. H.; Summa, L.; Lei, Q. P.; Duval, G.; Adam, O. *J. Chromatogr. A* **2000**, *894*, 311–318.
126. Lei, Q. P.; Lamb, D. H.; Shannon, A. G.; Cai, X.; Heller, R. K.; Huang, M.; Zablackis, E.; Ryall, R.; Cash, P. *J. Chromatogr. B: Anal. Technol. Biomed. Life Sci.* **2004**, *813*, 103–112.
127. Jones, C.; Lemercinier, X.; Crane, D. T.; Gee, C. K.; Austin, S. *Dev. Biol. (Basel, Switz.)* **2000**, *103*, 121–136.
128. Crane, D. T.; Bolgiano, B.; Jones, C. *Eur. J. Biochem.* **1997**, *246*, 320–327.
129. Ho, M. M.; Lemercinier, X.; Bolgiano, B.; Crane, D.; Corbel, M. J. *Biotechnol. Appl. Biochem.* **2001**, *33*, 91–98.
130. 1235 Vaccines for Human Use – General Considerations. In *United States Pharmaocopeia 40*; United States Pharmacopeial Convention, Rockville, MD, 2017.
131. Sturgess, A. W.; Rush, K.; Charbonneau, R. J.; Lee, J. I.; West, D. J.; Sitrin, R. D.; Hennessy, J. P., Jr. *Vaccine* **1999**, *17*, 1169–1178.
132. USP 1 Injections. In *United States Pharmaocopeia 40*; United States Pharmacopeial Convention, Rockville, MD, 2017.
133. USP 660 Containers – Glass. In *United States Pharmaocopeia 40*; United States Pharmacopeial Convention, Rockville, MD, 2017.
134. USP 381 Elastomer Components Used in Injectable Pharmaceutical Packaging/Delivery Systems. In *United States Pharmaocopeia 40*; United States Pharmacopeial Convention, Rockville, MD, 2017.

135. 1207 Package Integrity Evaluation—Sterile Products. In *United States Pharmaocopeia 40*; United States Pharmacopeial Convention, Rockville, MD, 2017.
136. *Guidance for Industry. Container Closure Systems for Packaging Human Drugs and Biologics. Chemistry, Manufacturing, and Controls Documentation.* https://www.fda.gov/downloads/drugs/guidances/ucm070551.pdf (accessed 01 Jan. 2018).
137. 71 Sterility tests. In *United States Pharmacopeia 40*; United States Pharmacopeial Convention, Rockville, MD, 2017.
138. *DCVMN Developing Countries Vaccine Manufaturing Network.* http://www.dcvmn.org/ (accessed 01 Jan. 2018).
139. *Liquid PedvaxHIB® [Haemophilus b Conjugate Vaccine (Meningococcal Protein Conjugate)].* https://www.fda.gov/downloads/BiologicsBloodVaccines/Vaccines/ApprovedProducts/UCM253652.pdf (accessed 15 Dec. 2017).
140. *Sanofi Pasteur - ActHIB®.* https://www.fda.gov/downloads/BiologicsBloodVaccines/Vaccines/ApprovedProducts/UCM109841.pdf (accessed 15 Dec. 2017).
141. *Package insert.* https://www.fda.gov/downloads/BiologicsBloodVaccines/Vaccines/ApprovedProducts/UCM179530.pdf. (accessed 15 Dec. 2017).
142. *VAXEN-Hib – Haemophilus influenzae Type b conjugate vaccine.* http://www.who.int/immunization_standards/vaccine_quality/VaxemHib_product_insert.pdf. (accessed 15 Dec. 2017).
143. *sanofi pasteur 242 – Pentacel®.* https://www.fda.gov/downloads/BiologicsBloodVaccines/Vaccines/ApprovedProducts/UCM109810.pdf (accessed 15 Dec. 2017).
144. *Package insert.* https://www.fda.gov/downloads/BiologicsBloodVaccines/Vaccines/ApprovedProducts/UCM308577.pdf (accessed 15 Dec. 2017).
145. *Product Information: Meningitec 0.5 mL. Meningococcal Serogroup C Conjugate Vaccine.* https://gp2u.com.au/static/pdf/M/MENINGITEC-PI.pdf (accessed 15 Dec. 2017).
146. *Assessment Report for Menveo.* http://www.ema.europa.eu/docs/en_GB/document_library/EPAR_-_Public_assessment_report/human/001095/WC500090150.pdf (accessed 15 Dec. 2017).
147. *Sanofi Pasteur Inc. 284 Menactra®.* https://www.fda.gov/downloads/BiologicsBloodVaccines/Vaccines/ApprovedProducts/UCM131170.pdf (accessed 15 Dec. 2017).
148. *Assessment report Nimenrix.* http://www.ema.europa.eu/docs/en_GB/document_library/EPAR_-_Public_assessment_report/human/002226/WC500127664.pdf (accessed 15 Dec. 2017).
149. *Meningococcal A conjugate vaccine lyophilised.* http://www.who.int/immunization_standards/vaccine_quality/pq_197_menAconj_SII_PI_624-2.pdf (accessed 15 Dec. 2017).
150. *Package insert.* https://www.fda.gov/downloads/BiologicsBloodVaccines/Vaccines/ApprovedProducts/UCM574852.pdf (accessed 15 Dec. 2017).
151. *Synflorix suspension for injection in pre-filled syringe.* https://www.medicines.org.uk/emc/medicine/22743 (accessed 14 Dec. 2017).

152. Ph.Eur. Monograph 1160 Typhoid polysaccharide vaccine. *European Pharmacopoeia*; EDQM: Strasbourg, 2017.
153. 151 Pyrogen Test. In *United States Pharmacopeia 40*; United States Pharmacopeial Convention, Rockville, MD, 2017.
154. Bacterial Endotoxins Test. In *United States Pharmacopeia 40*; United States Pharmacopeial Convention, Rockville, MD, 2017.
155. *Typhoid (Vi capsular polysaccharide) – tetanus toxoid conjugate vaccine Typbar TCV®.* http://www.vaxinpoint.in/wp-content/uploads/2014/02/TCV1-A4_merged...pdf (accessed 17 Jan. 2018).
156. 51 Antimicrobial Effectiveness Testing. In *United States Pharmacopeia 40*; United States Pharmacopeial Convention, Rockville, MD, 2017.
157. 1790 Visual Inspection of Injections. In *United States Pharmacopeia 40*; United States Pharmacopeial Convention, Rockville, MD, 2017.
158. 921 Water Determination. In *United States Pharmacopeia 40*; United States Pharmacopeial Convention, Rockville, MD, 2017.
159. *Guidance for Industry: FDA Review of Vaccine Labeling: Requirements for Warnings, Use Instructions, and Precautionary Information.* https://www.fda.gov/downloads/biologicsbloodvaccines/guidancecomplianceregulatoryinformation/guidances/vaccines/ucm092196.pdf (accessed 19 Oct. 2017).
160. *Vaccines Licensed for Use in the United States.* https://www.fda.gov/biologicsbloodvaccines/vaccines/approvedproducts/ucm093833.htm (accessed 19 Oct. 2017).
161. General Safety, *Code of Federal Regulations*, Part 610.11, Title 21, 2011.
162. Black, S.; Shinefield, H.; Fireman, B.; Lewis, E.; Ray, P.; Hansen, J. R.; Elvin, L.; Ensor, K. M.; Hackell, J.; Siber, G.; Malinoski, F.; Madore, D.; Chang, I.; Kohberger, R.; Watson, W.; Austrian, R.; Edwards, K. *Pediatr. Infect. Dis. J.* **2000**, *19*, 187–195.
163. Siber, G. R.; Chang, I.; Baker, S.; Fernsten, P.; O'Brien, K. L.; Santosham, M.; Klugman, K. P.; Madhi, S. A.; Paradiso, P.; Kohberger, R. *Vaccine* **2007**, *25*, 3816–3826.
164. Maclennan, J. *Arch. Dis. Child.* **2001**, *84*, 383–386.
165. Campbell, H.; Andrews, N.; Borrow, R.; Trotter, C.; Miller, E. *Clin. Vaccine Immunol.* **2010**, *17*, 840–847.
166. Ramsay, M. E.; Andrews, N. J.; Trotter, C. L.; Kaczmarski, E. B.; Miller, E. *BMJ* **2003**, *326*, 365–366.
167. Weinberger, D. M.; Malley, R.; Lipsitch, M. *Lancet* **2011**, *378*, 1962–1973.
168. *Surveillance Reports.* https://www.cdc.gov/abcs/reports-findings/surv-reports.html (accessed 17 Jan. 2018).
169. *ISO Guide 34:2009. General requirements for the competence of reference material producers.* https://www.iso.org/standard/50174.html (accessed 15 Jan. 2018).
170. Ph.Eur. General text 5.12 Reference standards. In *European Pharmacopoeia*; EDQM: Strasbourg, 2017.
171. Rosskopf, U.; Daas, A.; Terao, E.; von Hunolstein, C. *Pharmeuropa* **2017**, 44–68.

172. *Comparability of Biotechnological/Bological Products Subject to Changes in Their Manufacturing Process Q5E.* http://www.ich.org/fileadmin/Public_Web_Site/ICH_Products/Guidelines/Quality/Q5E/Step4/Q5E_Guideline.pdf (accessed 21 Oct. 2017).
173. Holliday, M. R.; Jones, C. *Meeting Report: Biologicals* **1999**, *27*, 51–53.
174. Ravenscroft, N.; Cescutti, P.; Gavini, M.; Stefanetti, G.; MacLennan, C. A.; Martin, L. B.; Micoli, F. *Carbohydr. Res.* **2015**, *404*, 108–116.
175. Fabbrini, M.; Rigat, F.; Tuscano, G.; Chiarot, E.; Donders, G.; Devlieger, R.; Filippini, S.; Frigimelica, E.; Forte, P.; Wittke, F.; Halperin, S. A.; Slobod, K.; Grandi, G.; Margarit, I. *J. Infect.* **2018**pii: S0163-4453(18)30026-4. DOI: 10.1016/j.jinf.2018.01.006.
176. *A Phase 1/2, Randomized, Placebo-Controlled, Observer-Blinded Trial To Evaluate the Safety, Tolerability, and Immunogenicity of a Multivalent Group B Streptococcus Vaccine in Healthy Adults 18 to 49 Years of Age.* https://clinicaltrials.gov/ct2/show/study/NCT03170609?term=Streptococcus&lead=Pfizer&gndr=Female&age=1&rank=1 (aAccessed 28 March 2018)
177. *Staphylococcus aureus 4-Antigen Vaccine (SA4Ag). Vaccines and Related Biological Products Advisory Committee (VRBPAC) Meeting.* Briefing Document. Meeting Date: 07 November 2017. https://www.fda.gov/downloads/AdvisoryCommittees/CommitteesMeetingMaterials/BloodVaccinesandOtherBiologics/VaccinesandRelatedBiologicalProductsAdvisoryCommittee/UCM583915.pdf (accessed 28 March 2018).
178. Begier, E.; Seiden, D. J.; Patton, M.; Zito, E.; Severs, J.; Cooper, D.; Eiden, J.; Gruber, W. C.; Jansen, K. U.; Anderson, A. S.; Gurtman, A. *Vaccine* **2016**, *35*, 1132–1139.

Chapter 13

Lessons Learned and Future Challenges in the Design and Manufacture of Glycoconjugate Vaccines

John P. Hennessey, Jr.,*,[1] Paolo Costantino,[2] Philippe Talaga,[3] Michel Beurret,[4] Neil Ravenscroft,[5] Mark R. Alderson,[6] Earl Zablackis,[7] A. Krishna Prasad,[8] and Carl Frasch[9]

[1]Consultant, Lower Gwynedd, Pennsylvania 19002, United States
[2]GSK, Siena 53100, Italy
[3]Department of Analytical Research and Development, Sanofi Pasteur, Marcy l'Etoile 69280, France
[4]Janssen Vaccines & Prevention B.V., Leiden, 2301 CA, The Netherlands
[5]Department of Chemistry, University of Cape Town, Rondebosch 7701, South Africa
[6]PATH, Seattle, Washington 98121, United States
[7]Analytical Process Technology, Sanofi Pasteur, Swiftwater, Pennsylvania 18370, United States
[8]Pfizer Vaccines Research and Development, Pearl River, New York 10965, United States
[9]Consultant, Martinsburg, West Virginia 25402, United States
*E-mail: johnphennessey@gmail.com.

Glycoconjugate vaccines have proven to be highly effective in preventing a variety of bacterial diseases. Publicly available data provides a partial understanding of best practices to design and manufacture new vaccines of this type without undue risk of technical failure (e.g. consistent quality, yield, stability, cost of goods, supply) or immunological insufficiency in providing a suitable protective response (e.g. inadequate antibody and/or T cell response). Not surprisingly, a great trove of knowledge exists within companies that have made substantial investments in glycoconjugate vaccine development and commercialization, with much of that knowledge currently unpublished. In this chapter, veterans from the "trenches" of glycoconjugate vaccine development outline the key concepts that underlie successful design and manufacture of new glycoconjugate vaccines and provide general guidance from "lessons learned" with the intent of improving the chances of success for new glycoconjugate vaccines.

Licensed Polysaccharide-Based Vaccine Historical Overview

Bacterial capsular polysaccharides (Ps) have been recognized as key virulence factors for more than a century (*1*) and have been explored as potential vaccine antigens from the outset (*2*). This led to licensed vaccines for adults based on purified Ps for *Neisseria meningitidis* (Nm; 2-valent 1975 and 4-valent in 1981), *Streptococcus pneumoniae* (Sp; 14-valent in 1972, 22-valent in 1980, then 23-valent in 1983), *Haemophilus influenzae* type b (Hib) in 1985 and *Salmonella typhi* (St) in 1988.

Even in the early days of research on pneumococcal vaccines, Avery and Goebel found that vaccines containing purified Ps were not effective in infants, but showed that conjugation of Ps to a carrier protein resulted in a protective immunological response even in infants (*3*). From this auspicious foundation, it took 58 years to license the first Ps vaccine (Meningo A & C) and another twelve years to license the first glycoconjugate vaccine, the Hib Ps-diphtheria toxoid conjugate vaccine (ProHIBiT) made by Connaught (see Table 1). Subsequently, at least twelve more monovalent Hib conjugate vaccines have been licensed. Similarly, first to market Ps-based vaccines for pneumococcal (Pn), meningococcal (Mn) and *Salmonella typhi* (St) vaccines have produced numerous follow-on competitors. Subsequently, Ps vaccines have consistently been displaced in the market by glycoconjugate vaccines showing that medical and market forces bring pressure to support the most effective and cost-efficient vaccines (e.g. via WHO prequalification) resulting in technically sound products failing in the market place if they aren't viewed as competitive or meeting the demands of the medical community and of global subsidizers of high-volume vaccines for the developing world. As examples, three highly effective Hib-conjugate vaccines were displaced from the market in large part due to their failure to be incorporated into pediatric combination vaccines with DTP components. Pneumococcal conjugate vaccines show a similar dynamic, with licensed products being displaced by vaccines containing more serotypes and/or better technology (i.e. conjugates vs Ps). While there may be some geographic regions where a tailored (lower valence) product would be acceptable, the global commercial market tends to be dominated byproducts that have the most serotypes included in the vaccine, even when the scientific evidence to support the higher valency may be marginal.

This history makes clear the future of Ps and glycoconjugate vaccines, with glycoconjugates succeeding Ps vaccines and higher-valent vaccines dominating lower-valent competitors. Time will tell whether this latter dynamic will apply to regional vaccine markets (e.g. for India or Africa) where the epidemiology and/or cost may support a lesser-valent product, with a slightly different vaccine composition, being medically justified and potentially more affordable and cost-effective.

Table 1. History of Ps and glycoconjugate vaccines

Vaccine name	*Manufacture and marketing*	*First Approval date*	*Active Components/Description/Serogroups/Serotypes*
b-CAPSA Hib-VAX Hib-IMUNE	Praxis Biologics (now Pfizer) Connaught Laboratories (now Sanofi Pasteur) Lederle (now Pfizer)	1985	Purified Hib Ps. All withdrawn from the market in 1988.
ProHIBiT	Connaught Laboratories (now Sanofi Pasteur)	1987	Hib Ps conjugated to diphtheria toxoid (DT). First Hib conjugate vaccine to be approved by FDA. Withdrawn from the market in 2000.
PedvaxHIB	Merck & Co.	1989	Hib Ps conjugated to meningococcal outer membrane protein complex.
HibTITER	Wyeth Pharmaceuticals (now Pfizer)	1990	Hib oligosaccharide (Os)* conjugated to diphtheria toxin mutant (CRM_{197}). Transferred to Nuron Biotech in 2007.
OmniHIB	Pasteur Mérieux Vaccins (Vaccine sold to GSK)	1993	Hib Ps conjugated to tetanus toxoid (TT). No longer available in the United States.
ActHIB	Sanofi Pasteur	1993	Hib Ps conjugated to tetanus toxoid (TT).
VaxemHib	Chiron (now GSK)	1995	Purified Hib Os conjugated to CRM_{197}
Hiberix	GSK	1996	Hib Ps conjugated to tetanus toxoid (TT). Replaced OmniHIB.
QuimiHib	Center for Genetic Engineering & Biotechnology (Cuba)	2004	Chemically synthesized Hib Os conjugated to TT
SII HibPro	Serum Institute of India	2007	Hib Ps conjugated to tetanus toxoid (TT).
Pneumovax14	Merck & Co.	1977	Purified pneumococcal (Pn) Ps for 14 serotypes. Displaced by 22-valent product.

Continued on next page.

Table 1. (Continued). History of Ps and glycoconjugate vaccines

Vaccine name	*Manufacture and marketing*	*First Approval date*	*Active Components/Description/Serogroups/Serotypes*
Pneumovax22	Merck & Co.	1980	Purified pneumococcal (Pn) Ps for 22 serotypes. Displaced by 23-valent product.
PNEUMO23	Sanofi Pasteur	1981	Purified Pn Ps for 23 serotypes. Retired 2017
Pneumovax23	Merck & Co.	1983	Purified Pn Ps for 23 serotypes.
Pnu-imune	Wyeth Pharmaceuticals (now Pfizer)	1983	Purified Pn Ps for 23 serotypes. Withdrawn in 2004.
Prevnar	Wyeth Pharmaceuticals (now Pfizer)	2000	Purified Pn Ps for 7 serotypes, each conjugated to CRM_{197}. Displaced by Prevnar13.
Synflorix	GSK	2008	Purified Pn Ps for 10 serotypes, each conjugated to DT, TT or non-typable *H. influenzae* protein D.
Prevnar13	Pfizer	2010	Purified Pn Ps for 13 serotypes, each conjugated to CRM_{197}.
Meningo A and C	Sanofi Pasteur	1975	Purified meningococcal (Mn) Ps for 2 serogroups. Still available through WHO prequalification program.
Menomune-ACYW-135	Sanofi Pasteur	1981	Purified meningococcal (Mn) Ps for 4 serogroups. Still available through WHO prequalification program.
NmVac4-A/C/Y/W135	JN Internat. Medical Corp.	ND	Purified Mn Ps for 4 serogroups. Displaced by 4-valent conjugate product.
MenAfriVac	Serum Institute of India	2010	Purified Mn A Ps for 1 serogroup conjugated to TT.
Meningitec	Pfizer (sold to Neuron Biotech)	1999	Purified Mn C Osconjugated to CRM_{197}

Vaccine name	*Manufacture and marketing*	*First Approval date*	*Active Components/Description/Serogroups/Serotypes*
Menjugate	Chiron (now GSK)	2000	Purified Mn C OS conjugated to CRM_{197}
NeisVac-C	North American Vaccine (now Pfizer)	2000	Purified Mn C Os (de-OAc) conjugated to TT.
MenHibrix	GSK	2012	Purified Mn C,Y and Hib Ps conjugated to TT
Menactra	Sanofi Pasteur	2005	Purified Mn A,C,Y,W135 Os conjugated to DT
Menveo	Novartis V&D (now GSK)	2010	Purified Mn A,C,W,Y Os conjugated to CRM_{197}
Nimenrix	GSK (transferred to Pfizer)	2012	Purified Mn A,C,Y,W135 Ps conjugated to TT
Typhim VI	Sanofi Pasteur	1988	Purified St Vi Ps
Typherix	GSK	2008	Purified St Vi Ps
Typbar	Bharat Biotech	ND	Purified St Vi Ps
Typbar-TVC	Bharat Biotech	2017	Purified St Vi Ps conjugated to TT

[*] According to IUPAC nomenclature, sequences of more than ten repeating units are defined as polysaccharides, but there is a general consensus to use the term Os for products where the Ps has been substantially reduced in size like HibTITER, VaxemHib, Meningitec, Menjugate, Menactra and Menveo. ND – not determined.

A number of glycoconjugate vaccines candidates that advanced to clinical development to address a variety of bacterial and fungal disease targets, including *Eschericia coli* (*4*), *S. aureus* (*5*), Shigella (*6*) and Candida (*7*), fell short of licensure. Development programs are commonly terminated pre-maturely or delayed for business reasons (e.g. program reprioritization) or due totechnical challenges. However, it is often not a matter of public knowledge what that technical issue was, leaving us to decipher historical events rather than utilize data to establish guidances to improve the prospects for successful commercial launch of new glycoconjugate vaccines. Despite the challenges associated with publishing negative data in target journals, there is a need to address this gap to document these valuable learnings. These terminated/delayed development programs, along with the successful programs, provide us with a "road map" of lessons learned, which highlight technical areas that can make or break a glycoconjugate development program. We provide an overview of the major technical areas essential to development of glycoconjugate vaccines, with specific focus on some of the key areas where lessons from the past can inform the development of new conjugate vaccines.

Polysaccharides

Bacterial Ps are high molecular weight polymers with the primary structure of their repeating unit (RU) defined by the constituent monosaccharides, the sequence of these monosaccharides, the anomeric configuration of the glycosidic linkages and the presence of non-saccharide substituents, such as phosphate, O-acetyl groups and phosphoglycerol residues. Many Ps are anionic due to the presence of phosphate, uronic acids or sialic acids residues in their RU, while some Ps are neutral or zwitterionic. The structure and composition of the Ps RU, along with their molecular size and size distribution, dictate many of their physico-chemical characteristics as well as their inherent stability profiles. Purposeful or incidental (e.g. during processing) modification of that RU from its native form (i.e. as produced by the bacterium) is a topic of concern throughout development of any Ps-based vaccine.

A limited amount of basic structural information of saccharide antigens is available based on biosynthetic patterns and genomic sequencing. With the continued application of historical immunological and chemical analytical methods and development of advanced structural technologies, in particular modern NMR approaches, definitive determination of the structure of a Ps is well-supported, which is an essential starting point for any Ps- or Os-based vaccine development program to ensure that Ps or Os with the correct RU is being used. Many Ps have structures that are very similar, e.g. in the pneumococcal field some serotypes only differ by a single glycosidic linkage (Ps 6A and 6B and Ps 19F and 19A), so definitive identification is essential prior to early development efforts. This is made challenging as a continuum of NMR-based corrections of published structures of pneumococcal, staphylococcal and Group B Streptococcal (GBS) Ps continue to be published (see below for examples). Meanwhile advanced genomic analyses reveal further complexities in providing definitive typing of bacterial strains that make the Ps. As an example, two "new" pneumococcal serotypes in serogroups 6 and 19 were identified by genetic analysis. The worldwide distribution of what appeared to be a new multidrug-resistant Pn serotype, designated 6E, was shown by NMR and chemical analysis to produce the serotype 6B Ps (*8*). Likewise what appeared to be "new" Pn19A strains recently were shown to still produce 19A Ps (*9*). These examples illustrate the difficulties of assigning new bacterial serotypes based solely on genetic or immunological findings alone, making essential the application of physico-chemical techniques as well.

While the chemical structure of a Ps RU has become the definitive evaluation of Ps being used in modern day Ps-based vaccines, it is critical to remember that the historical differentiation between Ps of different bacteria as well as serogroups/serotypes within the same species of bacteria were originally based on serological evaluation, i.e. carefully prepared antisera or monoclonal antibodies that can differentiate one Ps from another. It is essential here to distinguish between immunogenicity (the ability of a hapten to elicit antibodies to be produced in a living system) and antigenicity (the detection of a hapten by a specific antisera). Of relevance here is the concern that Ps of similar structures can induce an immunogenic responses (e.g. a spectrum of antibodies) that can cross-react with other bacterial Ps that share specific epitopes and both Ps can be detected by antisera when it may not be intended. As such, just as the structure of the Ps RU must be well characterized, so must the immunological reagents that are brought into play to measure antigen content, in particular in a multivalent vaccine.

Ps Identity

Capsular Ps of bacteria present unique and defining markers for the diverse pathogens that impact human health. Historically the identity of bacterial organisms with capsular Ps has relied on serogroup- or serotype-specific assessment using immunological reagents. This is possible because of the highly specific reagents that bind only to the unique chemical structures of the Ps repeating units. As one example, the Statens Serum Institut provides "factored antisera" that are capable of identifying the serogroup and/or serotype of up to 95 serologically distinct types of pneumococcus. These reagents, which are made by inducing an immune response to killed whole bacteria and removing antibodies that cross-react with specific epitopes, are in fact the basis for defining these distinct serotypes. However, slight structural variants within a serogroup and sometimes within a serotype, previously referred to as "chemotypes", can challenge the specificity of these reagents (e.g. see (*10*, *11*). This has contributed to inadvertent use of a similar but incorrect serotype for a period of time in a licensed pneumococcal vaccine product.

As the Ps structure of licensed vaccines is known, official pharmacopoeial methods include composition analysis using a combination of colorimetric methods for different saccharide types (sialic acid, uronic acid, hexosamine, methylpentose) and for non-saccharide substituents such as O-acetyl groups and phosphorus. Colorimetric assays are also employed for antigen quantification and purity, with expected limits based on dry weight described in WHO and Pharmacopoeia guidelines. As shown by analysis of Pn Ps, this ensemble of chemical methods is unable to differentiate antigens only composed of hexoses (e.g. Pn 33F) and cannot differentiate between several pairs of Ps within serogroups (e.g. 19A and 19F) and between serogroups (e.g. 5 from 9V; 6B from 18C; 10A from 20 and 23 from 17F) (*12*). This lack of specificity and sensitivity of colorimetric assays has led to the application of more sophisticated physicochemical methods for defining and testing antigen identity. Chromatographic methods that involve depolymerization using acid/base and profiling of the constituent sugar monomers and non-carbohydrate components offer greater specificity and have been applied to carbohydrate antigen identification and quantification. The two main techniques are based on hydrolysis followed by either liquid chromatography or gas-liquid chromatography, after derivatization. As described for Pn Ps, the most common methods are high-performance anion-exchange chromatography with pulsed amperometric detection (HPAEC-PAD) (*13*, *14*) and gas chromatography with mass spectrometric detection (GC-MS), performed after methanolysis, re-N-acetylation and trimethylsilylation (*15*). Additionally, ionic components such as uronic acids, phosphate and acetate can also be analyzed by HPAEC with conductivity detection (HPAEC-CD) (*16*). Depolymerization conditions are product-specific and may not be quantitative for all components, and several monomers such deoxy sugars (FucNAcN in Pn 1) are destroyed during analysis. Additionally, even when achieved quantitatively, composition analysis still cannot differentiate between serotypes that differ only in linkage positions such as Pn 6A/6B and Pn 19A/19F.

Nuclear magnetic resonance (NMR) spectroscopy has long been a definitive method for defining composition and structure of bacterial Ps (*17*). The repetitive series of saccharides and other entities found in Ps RUs provide unique NMR spectra for each Ps type, clearly distinguishing even those Ps that differ by only a single linkage on a saccharide in the repeating unit. NMR spectra contain characteristic signals depending on the electronic environment of the nuclei (protons, carbons and phosphorus) present in the repeating unit and their assignment follows from the use of 1D and 2D NMR correlation experiments (*18*). The sensitivity and selectivity of 1D proton NMR permits fingerprinting of Ps antigens and the relative peak intensities can confirm the proportions of different sugar residue types and substituents (such as N- or O-acetyl) present (*19*). Quantification of Ps and NMR-sensitive residual process contaminants (Pn cell wall Ps (C-Ps), buffers and solvents) can be achieved by comparison of these intensities to that of an added internal standard (*18*, *20*). The unique proton NMR spectra obtained can differentiate between two closely related Ps that differ due to a single linkage such as Pn 6A/6B (→3)-D-Rib-ol/→4)-D-Rib-ol) and Pn 19A/19F (→3)-α-L-Rha*p*/→2)-α-L-Rha*p*), or only in configuration at a single carbon as between Mn W and Y (→6)-α-D-Gal*p*/→6)-α-D-Glc*p*). 1D NMR identity tests for Hib (*21*), Mn (*22*), Pn (*12*), St (*23*) and GBS (*24*) Ps have been published. NMR played a leading role in the correction of published structures for Pn serotypes Pn 4 (*25*), 15B (*26*), 17F (*26*) and 33B (*27*)., *S. aureus* Types 5 and 8 (*28*) and GBS type III (*29*) and the USP is currently developing a bacterial Ps NMR identity test for future implementation.

The specificity of NMR typically exceeds that of the immunological assessment discussed above whereby now the definitive basis of Ps identity has become NMR, particularly proton NMR; in a recent example proton NMR has been explored as a basis for developing an identity test for GBS Ps (*24*) In fact, we have entered an era where NMR is defining new chemotypes within a serotype, meaning that the chemical variants can be distinguished by NMR but not by currently available immunological reagents (e.g. see (*22*)). Although not recognized by typing sera, the use of capsule-specific monoclonal antibodies revealed that the apparent increase in 6A disease after the introduction of Prevnar7 was due to a new serotype 6C (*30*). The structure of the 6C Ps, which is as 6A, but with the 2-linked α-Gal replaced by α-Glc, was fully characterized by NMR spectroscopy (*31*). Genomics revealed the worldwide distribution of a new multidrug-resistant serotype designated 6E, however, NMR and chemical analysis showed that it produces the serotype 6B Ps (*8*). Similarly, different capsular serotype 19A subtypes identified were shown to still produce 19A Ps (*9*). These examples illustrate the difficulties of identify new bacterial antigens based on genetic findings alone and the importance of NMR spectroscopy in carbohydrate antigen identification and vaccine control.

The quantification of Ps antigens by NMR, first described for the multivalent Pn vaccine (*12*), has been further applied as a quantitative NMR (qNMR) method. Secondary comparative methods were used to assign a value to WHO international MnC and Hib Ps standards; however, qNMR has been adopted as the primary method for determining the Ps content of MnA and MnX candidate standards (*32*) and future Ps standards in preparation.

In addition to the NMR spectroscopic analysis of Ps in solution, the introduction of high-resolution magic-angle sample spinning (HR-MAS) probes has extended use of this technology to semi-solids. This means that high resolution NMR can also be performed directly on bacterial cells, not possible using liquid state NMR (*33*). This provides structural information on the surface carbohydrates produced by the pathogen and can be used to confirm that the fermentation, isolation and purification method employed did not alter the structural integrity of the antigen. HR-MAS tracking of the O-antigen produced by living *Salmonella enterica* serovar *Typhimurium* cells showed that it was O-acetylated at O2 of Abe but a rise in pH during the stationary phase resulted in de-O-acetylation (*34*). Some examples include the use of HR-MAS to confirm the O3 and O4 O-acetylation of Mn A (*35*) and to demonstrate that O-acetylation on O2 and O3 of Rha of the *S. paratyphi* A O-antigen was not an artefact from purification (*36*). HR-MAS has also been used to identify new surface antigens that could be potential vaccine targets e.g. the surface glycans (*37*) and lipoteichoic acid (*38*) of *Clostridium difficile*.

Production of Ps for use in vaccines can result in the repeating unit produced being different than that seen in the native Ps structure. This can be purposeful or an oversight in the design of the production process. As an example, during GBS capsular Ps purification a base treatment is often used which removes O-acetylation from sialic acid residues (*39, 40*). Additionally, the O-acetylation level and position may also be affected by different media composition and fermentations conditions (*41*). The decision to accept such structural alterations needs to be carefully evaluated as minor changes can have significant immunological consequences (Pn 9V (*42*), MnC (*43*) and GBS III (*40*)), while seemingly significant changes may have no impact on functional activity of the resulting antisera (e.g. GBS III (*40*)). The decision on whether such alterations of the natural structure are acceptable require extensive supportive data.

Production of Capsular Ps

Production of bacterial capsular Ps is typically based on biological processes whereby a clone of the bacterium of interest is isolated, methodically grown by fermentation processes, and the capsular Ps purified using a variety of precipitation, chromatographic and/or diafiltration processes. Earlier methods for fermentation involved casein and soy hydrolysates, tryptic soy broth as media containing supplement such as yeast extract for nitrogen and carbon sources. Chemically defined synthetic fermentation media were later introduced, allowing easier approaches for extraction and purification of capsular polysaccharides. The standards for acceptability for Ps to be used in vaccines is typically well-defined by precedent as reflected in World Health Organization (WHO) technical report series (TRS) guidelines for the production of Pn and Mn Ps (*44*, *45*) and elsewhere, e.g. (*46*, *47*). However, for new Ps-based vaccines, such as current efforts to develop vaccines targeting Group B Streptococcus (GBS), the specifics of the chemical makeup of the RU may still be a work in progress. That said, there are other considerations for what constitutes a "preferred" vs "acceptable" Ps preparation. A few of these specific considerations are touched on below.

Ps Purity

The Ps must be tested for moisture (if dried) and impurities such as protein, nucleic acid and process residuals using pharmacopoeial methods. The levels of process-specific reagents must be determined and meet the limits specified. Anionic Ps are typically precipitated and purified using cetyltrimethylammonium bromide (CTAB), followed by cation exchange to result in the calcium or sodium salt of the Ps. Incomplete exchange results in residual CTA ionically associated to the Ps which cannot be removed by diafiltration against water. CTA results in characteristic peaks in NMR spectroscopy, which are frequently present in published spectra of Mn and Vi Ps. An ion-exchange HPLC method for CTAB has been published, but it detects residual bromide, not CTA itself (*35*). Another isolation approach uses the detergent sodium deoxycholate (DOC); HPLC methods for residual DOC have been published with detection using refractive index (HPLC-RI) or the more sensitive tandem mass spectrometry (HPLC-MS/MS) (*48*). Common solvent residuals can be quantified by NMR spectroscopy (*12*) or gas chromatography. Preparations of gram negative PsPs may contain trace amounts of lipids from the reducing end of the PsPs and co-purified endotoxin; the latter is measured using the *Limulus amoebocyte* lysate (LAL) test. The lipids present can be characterized by their fatty acid content by use of reversed-phase liquid chromatography of their fluorescent derivatives (*49*) or as fatty acid methyl esters by GC-MS (*50*).

Ps may be covalently bound to the bacterial cell-wall (such as Pn polysaccharides), attached to the outer membrane via a lipid anchor (such as Hib or St Vi) and / or excreted to the culture media during fermentation. This will define the choice of the Ps first step of isolation after fermentation. For Pn polysaccharides, deoxycholate addition at the end of the fermentation will activate the major Pn autolysin (LytA) and releases the capsular Ps in the supernatant. While the synthesis of capsular polysaccharide (CPS) of most Pn serotypes is wzy enzyme -dependent, the strains of some serotypes, e.g. type 3, synthesize CPS by the synthase-dependent pathway, resulting in a Ps that is not covalently linked to peptidoglycan and can be released during growth (*51*). Therefore, the deoxycholate treatment step, during the recovery stage, prior to purification is not needed for these serotypes. For excreted polysaccharides, pelleting the cells and starting directly from the culture supernatant may be an advantage as cell lysis will release other bacterial components such as proteins, nucleic acids, endotoxins from Gram negative bacteria and also cell wall Ps from Gram positive bacteria.

Ps composition/purity may also be affected by different media composition and fermentations conditions and potentially even by different strains of the same bacterium. As a specific example, Hib Ps has been shown to have a lipid tail under some fermentation conditions, which may require removal of the tail during the purification process to increase the efficiency of the production process (*52*). Such technical solutions are often patented or kept as trade secrets within the manufacturing company.

Similarly, cell wall Ps content is a common concern among process developers producing bacterial capsular Ps from Gram positive bacteria. For pneumococcus, the key concern is the cell wall Ps, referred to as C-Ps, which can be present both as a co-purified impurity or as an impurity covalently attached to the capsular Ps, typically through a peptidoglycan link. The presence of C-Ps in Pn Ps is both serotype-dependent – serotype 3 typically contains little if any C-Ps, whereas serotype 5 presents challenges to keep the level below 10% (wt/wt). For serotype 5 and 15B individual isolates can consistently produce as low as 5 or as high as 15% C-Ps, almost all of which is covalently attached to the Ps by peptidoglycan linkages. The levels of C-Ps can be determined by NMR spectroscopy (*53*) or ribitol content by chromatographic methods (*54*). For GBS the analogous concern is the rhamnose-containing Group B-specific carbohydrate; levels of contamination indicated by Rha content can be determined by NMR spectroscopy or high performance anion-exchange chromatography with pulsed amperometric detection (HPAEC-PAD) (*55*). In neither case is there a regulatory expectation for the amount of this type of impurity, i.e. a requirement to test against a fixed specification, as the impact of these impurities on the immunogenicity of the Ps has not been clearly defined. As such, expectations focus more on consistency of manufacture rather than absolute limits. In both cases, vigilance during strain and clonal isolate selection and process/analytical development is warranted to avoid issues in subsequent quality control and regulatory development activities.

Ps Yield

Acceptable productivity and yield for a Ps production process is relative to how much Ps is needed to support the final product. As an example, for Pn conjugate vaccines the expectations for number of serotypes (equal to or greater than the standard of care product) and the dose level (ranging from 1 to 4μg Ps per serotype per dose) are usually fixed. As such, given considerations for Ps step yields for Ps and for glycoconjugate production, as well as for formulation and filling overages and losses; e.g. it usually takes ~0.7 mL in a single-dose vial to ensure consistent removal of 0.5 mL to administer, and a fixed volume of final formulated material is "lost" in the bottom of the filling tank and lines in the filling system, means that ~100 g of purified Ps for each serotype is needed to provide 10M doses of final vaccine product. While this is not so daunting, increasing the number of serotypes (Pn vaccines contain up to 23 serotypes), the dose level (up to 30 μg/dose for some products) or the number of doses needed (Pn vaccines projected global need in 2008 was >300M doses per year) and the need for purified Ps can become very demanding, requiring either process efficiency or very large, dedicated facilities to keep the cost of goods for the Ps in a reasonable range for large low-margin markets such as in the developing world. In this context, design of experiments (DOE) approach for the media composition, fermentation conditions and purification steps can be a powerful tool for the optimization of Ps production.

Chemically Synthesized Oligosaccharides

Synthetic Os, which are available for an ever-increasing number of structures, may be even more constrained in vaccine applications if they are held to the same dose levels as required for Ps purified from bacteria. However, a potential benefit of chemical synthesis of Os for use in glycoconjugates is that activation of the Os to prepare it for conjugation can potentially be accomplished as a step in the chemical synthesis, thus eliminating the need for subsequent activation. Another potential advantage of chemical synthesis is that it does not require fermentation and biomass processing of the related pathogen thus limiting only to the carrier protein the source of typical process related impurities like endotoxins, C-Ps, peptidoglycan, proteins, nucleic acids, etc. A first example of a conjugate vaccine with fully synthetic carbohydrate moiety is the QuimiHib made in Cuba (*56*). The scalability of oligosaccharide chemical synthesis is highly dependent on the structure and length of the glyco-antigens to be synthesized, although significant progress has been made in recent years (see below). Optimization of the Ps structure for better stability (see MnA example (*57*)) and better immunogenicity is also possible by chemical synthesis, though great care must be taken that structural alterations don't lead to generation of antibodies that don't recognize the native immunogenic epitopes of the Ps. Chemical synthesis will, however, tend to be used to produce Os with a smaller average number of repeating units per molecule than are produced in biological systems due to the expense of synthesis and the repetitive losses in each cycle of synthesis. In other words, smaller synthetic Os will have a lower cost of goods than larger synthetic Os. Subsequently, head-to-head comparisons of the dose-response profiles of Os-based glycoconjugates versus Ps-based glycoconjugates will lead to an understanding of whether moles of saccharide (i.e. number of saccharide molecules) is more important than the total mass of saccharide in an effective dose of vaccine. Unfortunately this is also clouded by a lack of knowledge of the optimal number of repeating units in the conjugated saccharide. Additionally, the impact of higher carrier protein load in a vaccine becomes more in focus with Os vs Ps based glycoconjugates in multivalent vaccines due to the loading density of the saccharide on the carrier protein, which may be limited by number of reactive linkage sites on the protein (see "Number of linkage sites per protein molecule" below).

Ps Stability

Under identical conditions, glycoconjugates can be expected to be no more stable than their component Ps unless the conjugation linkages stabilize the structure. The underlying similarity in the chemical composition and structure of the Ps, which is largely similar in Ps and glycoconjugate samples. As such, it is advantageous and simpler to characterize the stability of the purified Ps under relevant conditions when the Ps become available. Such studies will reveal the "problem children" if any. As an example, for Pn Ps the least stable Ps among 23 Pn Ps used in various vaccines are serotypes 6A (*58*), 19A and 19F (*59*) where hydrolysis of a phosphodiester bond in the Ps backbone leads to size reduction, and 9V where higher pH can lead to de-O-acetylation (*42*). Another phenomenon seen with Mn Ps is the migration of O-acetyl groups at higher pHs, which can lead to unnatural structural alterations as well. This leads to a general expectation that when the structure of the Ps RU is well-defined, potential stability issues may become predictable.

Knowing such Ps stability limitations avoids surprises during development of products composed of these Ps. For example in the case of Mn Ps, MnA is the least stable of the four most common vaccine serogroups, and its poor stability has likely been the cause of the failure of a clinical trial (*46*). In fact, subsequent analyses on samples returned from the field revealed that the MnA Ps was substantially depolymerized, with consequent loss of vaccine potency, even if the product was lyophilized (*60*). This led to the introduction of size analysis as the potency assay in guidelines for Ps vaccines, which is applied together with tests for identity and purity. Subsequent studies identified lactose as an optimal stabilizer for the lyophilized drug production (*61*). This knowledge and consequent concern, together with additional stability studies resulted in the choice of a lyophilized rather than liquid formulation for most of the glycoconjugate vaccines containing MnA to accommodate longer shelf life and reduced risk in case of failure of the cold chain.

Another example of stability issues detectable in the purified Ps that have direct impact on conjugate manufacturing is represented by Hib Ps. Egan et al. (*62*) published the mechanism of Hib Ps hydrolysis in aqueous solution and the nature of base-catalyzed hydrolytic depolymerization; divalent counter ions like Ca^{++} catalyze the hydrolysis which is about 25-fold faster than for the Ps in sodium form. A similar interaction and destabilizing effect has been proposed also for Hib conjugate vaccines adsorbed onto aluminum hydroxide (*63*). These considerations are important to take into account in the design of processes for Ps purification, activation, conjugation and formulation.

Polysaccharide Molecular Size Distribution

By their nature, Ps are polymers of identical repeating units that can range from tens to thousands of repeating units in a chain. Additionally, unlike protein synthesis where there is a specific "stop" codon to ensure that all protein molecules are identical, there is no such signal in bacterial Ps production so Ps molecules in a given preparation cover a broad range of sizes, i.e. are polydisperse, with a rough approximation being that 80% of the mass of a purified Ps sample will fall within a ten-fold size range.

The molecular size distribution of each lot of purified Ps needs to be determined as it reflects manufacturing consistency and indicates the structural integrity of the Ps. Soft gel filtration techniques with collection and colorimetric analysis of fractions have been largely replaced by high performance size-exclusion chromatography (HPSEC) equipped with in-line refractive index (RI), and multi-angle laser light scattering (MALLS) detectors (*64*) or rate nephelometry (RN) detection (*65*). The HPSEC-MALLS/RI method permits the absolute molecular weight to be determined and the system as a whole allows direct evaluation of a variety of hydrodynamic properties for Ps. The HPSEC-RN method provides the benefit that it can be used in multivalent Ps solutions such as the 23-valent Pn Ps vaccines. The advantages of the HPLC method in terms of speed, specificity, and inherent reproducibility was demonstrated in a comparative study performed on Hib Ps (*66*). Depending on the isolation procedure, Hib and Mn Ps contain covalently associated lipid at the reducing end which results in Ps aggregation and interactions with the stationary phase. This problem was addressed by the addition of SDS to the elution buffer which resulted in quantitative recovery and prevented aggregation (*66*). The elution of charged Ps can vary with elution conditions; the retention time of charged Pn Ps was shown to increase with increasing ionic strength and decreasing pH. This was attributed to compaction of the Ps associated with a reduction in the intramolecular electrostatic interactions (*67*). Another study of Pn Ps used orthogonal techniques: analytical ultracentrifugation (sedimentation velocity and sedimentation equilibrium), HPSEC-MALLS and viscometry. Comparable results were obtained with all Ps exhibiting semi-flexible coil shapes (*68*).

Ps size may be directly correlated to viscosity (increased molecular size and increased concentration result in greater viscosity) and thus to chemical reactivity (lower MW Ps tend to be more structurally flexible and reactive). Viscosity is also directly linked to the conformation of the Ps in solution with rod-like Ps having a much higher intrinsic viscosity than random coil Ps. Molecular modeling (see Chapter 7 (*69*)) may predict the conformation of a given Ps in solution that determine its hydrodynamic behavior as well as potential cross-reactivity.

Cross-Reactive Ps

Given the previous discussion on the extensive chemical similarity between some bacterial Ps, particularly between Pn serotypes, staphylococcal serotypes and between GBS serogroups, there is certain to be similar immunogenic and antigenic epitopes between chemically distinct Ps. The strategy for evaluating the immunological endpoints for deciding to include or exclude similar Ps types in a vaccine (e.g. including Pn Ps 6A in a vaccine that includes 6B) is a real-life example in which two manufacturers of Pn conjugate vaccines arrived at different decisions (e.g. Prevnar13 vs Synflorix). However, at a process and analytical level, it is critical that identity of the Ps be definitive and that analytical tools be "sharp" enough to make clear the distinction between the two similar Ps. This can be challenging for differences such as the presence/absence of O-acetyl groups on Pn Ps 9V versus 9A, where it is not clear if the O-acetyl makes a difference in the protective response or not (*42*).

A summary of the important lessons learned from the above wealth of data on design and manufacture Ps components for Ps-based vaccines includes the following:

- It is essential that the bacterial strains and the Ps they produce are critically evaluated to ensure that Ps of the desired type are produced. Physico-chemical analytical methods, including NMR, and immunological methods (with well-defined reagents) are essential to this task.
- The chemical composition, RU structure and molecular size distribution of Ps preparations are all critical parameters for meeting consistency of manufacturing standards, but may also impact on the protective immune response elicited by the Ps.
- Production of Ps must be of suitable and consistent purity (i.e. in compliance with existing pharmacopeial requirements), yield and cost of goods to support commercial manufacture of the vaccine product.

Carrier Proteins

The clinical development of Ps vaccines revealed that while they are immunogenic and protective in adults, they are typically poorly immunogenic and non-efficacious in children below two years of age. Going back to the pioneering studies of Avery and Goebel (*3*, *70*, *71*), a solution was found by covalently coupling saccharides to proteins, forming glycoconjugates which were shown to to be immunogenic in the first months of life. Ps are T-cell independent antigens that directly stimulate B cells to differentiate into plasma cells, which secrete specific antibodies without involvement of T cells. This mechanism is not yet mature in infants where a cooperation between B and T cells is required for efficient immune responses to Ps immunogens. In glycoconjugates the carrier protein and the saccharide moieties complement each other, the first providing the so-called T-cell epitopes which recruit T-cell help and the second the B cell epitopes that elicit functional antibodies. Excellent reviews provide detailed and comprehensive explanation of the current understanding of mechanism of action of carrier proteins in glycoconjugate vaccines (*72–74*).

Five carrier proteins are currently used for licensed conjugate vaccines: diphtheria toxoid (DT) and tetanus toxoid (TT), which are chemically inactivated, usually with formaldehyde, genetically modified diphtheria toxin (CRM_{197}), the outer membrane protein complex of serogroup B meningococcus (OMPC) and non-typeable *Haemophilus influenzae* protein D (PD). Each of these has been shown to effectively engage the needed T cell immune responses essential for functional response of glycoconjugate vaccines and has been produced consistently at commercial scale. Other carrier protein candidates have been evaluated in animal studies or even in clinical studies (e.g. cholera toxin B (*75*), *Pseudomonas aeruginosa* exotoxin A (*76*) and meningococcal B factor H binding protein (*77, 78*)). While introduction of new carrier proteins requires extensive clinical evaluation to establish safety and to demonstrate that the biomarkers of an effective carrier protein are seen, i.e. a booster response to conjugated Ps antigens in an infant population, it is important to acknowledge that the primary measure of success for a new carrier protein will continue to come from non-inferiority studies versus the current standard of care for follow on vaccines with established surrogate markers of protection and efficacy studies for novel vaccines without such surrogate markers.

Choice of Carrier Protein

Some manufacturers selected DT and TT as carrier proteins attracted by the safety record established over decades of vaccination against tetanus and diphtheria, thus facilitating the acceptance of the corresponding conjugates into immunization schedules, and also by their ready availability in commercial-scale production facilities. Modern applications of these two carrier proteins have evolved to use more purified forms of these proteins, typically prepared to increase the content of protein monomers, and thus improve the consistency of resulting glycoconjugates.

In contrast, CRM_{197} is a 58 kDa nontoxic point mutation-containing variant of diphtheria toxin that does not require chemical detoxification and is considered a well-defined carrier protein. This protein is made both as "native" CRM_{197}, using expression in the natural host organism, *Corynebacterium diphtheria*, or as a recombinant protein in heterologous host cells systems (e.g. *Pseudomonas fluorescens* (Pfenex) or *E. coli* (Fina Biosolutions)). There are no data available to date suggesting which form of CRM_{197} is "best", although a recent work showed high level of physicochemical comparability of the recombinant ones with that from *Corynebacterium diphtheria* (*79*) and the recombinant systems tend to be more productive and have a lower cost of goods. So far, all the currently licensed vaccines use CRM_{197} produced from *Corynebacterium diphtheria* with other sources of CRM_{197} still undergoing preclinical and clinical evaluation.

OMPC was used for a licensed Hib conjugate vaccine (*80*) and was explored as a carrier protein in a Pn glycoconjugate development program. In extensive cross-comparison of Hib conjugate vaccines using each of the above carrier proteins (*81*), the two notable influences that stood out were 1) that OMPC only required one priming dose to allow induction of an anamnestic response, whereas the others required two priming doses, and 2) that DT and TT effectiveness were influenced by the prior exposure to these antigens in DTP vaccines.

In a more recent glycoconjugate vaccine development program, protein D, a 40 kDa cell-surface protein from non-typeable *H. influenzae* produced in a recombinant strain of *E. coli*, was introduced as the carrier protein for most of the serotypes in a multivalent Pn conjugate vaccine, Synflorix (e.g. (*82*)). Licensure of this vaccine is proof that new carrier proteins can meet licensure standards and provides a "road map" of sorts in pursuing such efforts.

The choice of the carrier protein is driven by several considerations including safety, manufacturability, solubility, stability, and reactivity towards chemical conjugation with glyco-antigens and, obviously, the proven ability of the protein to elicit the T-dependent response to the Ps antigen essential for a glycoconjugate antigen. In vaccines for infants, that choice is also complicated by the relevance of the immune response to the carrier protein as a protective antigen in the glycoconjugate vaccine as well as in co-administered vaccines, e.g. DTP-containing vaccines.

Carrier Proteins That Are Also Antigens

An emerging concept in carrier protein selection is based on the utilization of carrier proteins which can serve a dual role as both a carrier and as a protective immunogen against a pathogen of concern. Examples of this approach include the use of PD (see above), *Staphylococcus aureus*, *Clostridium difficile* and GBS-derived proteins (*83–89*). In these cases, the carrier protein provides not only T cell epitopes, but also protective B cell epitopes, therefore the design and control of the conjugation and the quality control of the product are of paramount importance in order to avoid loss of protective epitopes on the protein (*90*, *91*). As an example, the TT component of a Mn A glycoconjugate (MenAfriVac) has been shown to boost the protective TT response induced by prior exposure to TT (*92*). Moreover, recent work with a recombinant surface protein from *Candida albicans* shows the potential for protein antigens to elicit strong antigen-specific cytokine responses from T cells that may contribute to protection as well (*93*). Some conjugation approaches might be more suitable than others to maintain protective epitopes of the protein carrier, in particular those based on site directed conjugation, in vitro or in vivo (e.g. bioconjugation).

Availability of protective monoclonal antibodies in conjunction with structural biology tools like the crystal structure of protein/antibody complexes or Hydrogen Deuterium Exchange coupled to Mass Spectrometry could allow epitope mapping and inform the chemist on how to develop specific, critical epitope-preserving and scalable glycosylation methods that don't adversely impact the structure of immunogenic epitopes on the protein (*94*). Likewise, biophysical and immunochemical methods for post-conjugation confirmation of the presence of intact protein and saccharide B epitopes should be in place.

Carrier Protein-Induced Immune Interference

When glycoconjugate vaccines are combined with other antigens to increase the valency among the same microorganism or to cover different pathogens, a potential for interference with the immune response to other conjugates of the formulation or to co-administered antigens exists. Though no conclusive evidence of the mechanism has been demonstrated, two main interference mechanisms have been proposed: a) carrier-induced epitope suppression (CIES) where pre-existing antibodies to the carrier protein can have a suppressive effect on the response to the carbohydrate moiety of the corresponding conjugate, and b) the so called "bystander interference" where a given conjugate co-administered or combined with other antigens generates an immune interference that extends to unrelated antigens due to competition for limited immune response capacity within the lymph nodes.

The theoretical basis of CIES, which has been examined in animal models, is that antibodies directed at the carrier protein, bind to the carrier protein in the glycoconjugate antigen, preventing access of carbohydrate-specific B cells to the Ps epitopes by steric hindrance, thereby reducing the response to the Ps. Additionally, clonal expansion of carrier protein-specific B cells can drive the immune response toward the carrier protein, reducing the response to the Ps epitopes (*95*). Although the preclinical data are intriguing and the topic has been extensively reviewed, analysis of data from clinical trials do not present clear and consistent conclusions of the role of CIES in human immune response (*96*).

It has been reported that CRM_{197} and corresponding conjugates are less likely to induce CIES as compared to TT, while conjugate vaccines containing CRM_{197} have been reported to reduce the anti-Hib response when co-administered with DTaP vaccines due to the bystander effect (*96–98*). Two key studies that exemplify the carrier-induced immune interference by TT are the reduction of Hib antibodies observed following immunization of infants with a tetravalent Pn-TT conjugate concomitantly with DTP-IPV-Hib-TT vaccine (*99*), and the negative impact on the immunogenicity of the 7-valent Pn Ps-TT conjugates or the Pn Ps-D/TT 11-valent vaccine when co-administered with DTaP/IPV/PRP-T (*100*). An example of the bystander interference is the reported reduction of the anti-Hib response when CRM_{197} based conjugates are co-administered with DTaP vaccines containing Hib-TT (*96*). However, the interpretation of these must be viewed in the context that there are multiple variables that may impact immune responses to glycoconjugate vaccines (*101*).

Immunological and biophysical approaches have been applied to investigate CIES for glycoconjugates in animal models. These studies also showed that DT, TT and corresponding conjugates are more susceptible to CIES than CRM_{197}, especially after pre-exposure to high doses of carrier protein. Notably, pre-existing anti-DT antibodies do not have a negative impact on CRM_{197} conjugates (*90, 102–104*). A tendency to reduce risk of CIES by increasing the Ps loading density on the carrier protein was also observed, in agreement with early studies which showed that increasing the density of the hapten on the carrier protein prevented CIES (*95*), probably by favoring hapten B cells in the competition for the conjugate epitopes,, e.g. by masking carrier B epitopes and/or inducing conformational changes which disturb the carrier B epitopes.

The real clinical relevance of most of the interactions observed when glycoconjugates have been co-administered with other antigens is not clear, sometimes being inconcistent or even contradictory (*105*), however the knowledge acquired so far suggests that the possibility of vaccine interference should be considered during the design and development of glycoconjugate vaccines and for co-administration of these vaccines with other vaccines. The lack of a clear mechanistic explanation for the interferences observed to date does not the help the design of conjugate vaccines nor lower the risk of immune interference with other vaccines. Mixed carrier strategy could reduce the risk of carrier-mediated suppression and possible bystander interference with co-administered conjugate vaccines as proposed for PHiD-CV, which contains PD as the carrier eight of the ten Pn serotypes of Synflorix (*89*). Interestingly Serum Institute India and PATH are using both TT and CRM_{197} as carrier proteins in their ACWYX conjugate vaccine (*106*). More research is recommended in order to unravel the precise mechanism of CIES and bystander effect in human vaccines.

As the number of glycoconjugates in a vaccine formulation expands and clinical results show less than expected immune responses, the question arises whether the increasing amounts of specific carrier proteins in a vaccine may induce "tolerance" for the T-cell response resulting in reduced immune response to the carbohydrate antigen. This is an evolving area of research (see (*72*)) that should be directly addressed in the context of any new conjugate vaccine being explored, especially for multi-valent vaccines.

Production of Carrier Proteins

Production methods for the various carrier proteins used in glycoconjugates are varied, but there are some unifying concepts for producing these critical components that facilitate easier the consistent and reliable production of glycoconjugates. Considering that the typical monovalent Ps-protein glycoconjugate bulk drug substance (DS) contains as much if not more mass of protein than it does Ps, the dominant component in a multivalent glycoconjugate vaccine will almost certainly be the carrier protein, and large-scale production of carrier protein will be critical to support such a product. As such, production must be scaled to support the intended product and should not be the rate-limiting factor in production of the vaccine. Specifically, consideration must be made for purity, yield and cost of goods for production of carrier proteins. For large size carrier proteins, attention should be paid to the production systems (native organism produced vs recombinant approaches) for structural integrity and microheterogeneity.

Purity

Protein impurities in the carrier protein batch are exposed to the same activation/conjugation process steps as the carrier protein and therefore may become part of the glycoconjugate product with unknown biological activity. Protein carriers obtained with recombinant DNA technology, such as PD (and CRM_{197} in development), need to comply with additional requirements established for recombinant products including testing for host cell DNA and proteins (*107*).

Yield

Given that carrier protein is the dominant component in a glycoconjugate vaccine, higher yielding production processes are essential to satisfy production demands, particularly if they are designed to need fewer lots or with smaller production equipment trains. In this context, design of experiments (DOE) approach for the media composition, fermentation and purification processes can be a powerful tool for the optimization of carrier protein production.

Cost of Goods

Although the CRM_{197} production processes based on cultures of *Corynebacterium diphtheriae* C7(β197) tox(-) strain is capable to support substantial glycoconjugate vaccines supply, nowadays more efficient processes based on recombinant expression systems using less pathogenic organisms such as *E. coli* (*108–110*) are utilized. Another example is *Pseudomonas fluorescens*, which has been used to produce high levels of CRM_{197} (*107*). The recombinant CRM_{197} was shown to be physicochemically equivalent to a well characterized CRM_{197} reference standard and the derived Ps conjugates elicited comparable immune responses in humans, as indicated by IgG and opsonophagocytic activity (OPA) assays (*111*). Recombinant CRM_{197}-based conjugate vaccines are currently in various stages of clinical trials in the U.S., Europe, and Asia.

Stability

Proteins in solution are subject to time-dependent physico-chemical modifications such as oxidation, deamidation, aggregation, etc., which in turn could impact the success of the conjugation reaction, the stability of the resulting conjugate and/or the immune response to the conjugate antigen(s). Historical use has shown that toxoids have better thermal stability (e.g. DT versus CRM_{197} or TT versus tetanus toxin), suggesting that CRM_{197} from the various sources and PD may be the most susceptible and analyzable of the carrier proteins (*112*, *113*). As an example, CRM_{197} contains an exposed loop of three arginines that can be clipped by proteases present in the culture medium, resulting in a the so-called nicked form. The degree of nicking needs to be controlled (typically <5% of total CRM_{197}) as it may affect the processing and conjugation characteristics of the protein; the impact of nicking on the immunogenicity of the derived conjugate has not been reported. Aside from concerns related to aggregation, it is not clear that other changes in a carrier protein can impede its conjugation proficiency or functional immunological activity.

Aggregation State

Control of the aggregation state of the carrier protein utilized for conjugation is important to control both the reactivity of the protein as well as the size of the conjugate particles. Carrier protein aggregates (e.g. dimers, trimers, multimers) can be generated during (e.g.) chemical detoxification, protein purification or during storage storage, including during freeze/thaw events, and can be determined using methods such as SEC-HPLC coupled with static light scattering detection (*47*). While surfactants can be added to the bulks to limit time-dependent changes in aggregation, such remedies can have downstream consequences if diafiltration is required as it may be hard to prevent concentration of the surfactant if micelles trap the surfactant in the protein solution. Removal of aggregates in the carrier protein starting material becomes necessary as very large conjugate particles can be refractory in downstream processing (e.g. not elute from chromatography columns or precipitate out of solution). Such problems are in fact less likely to be an issue for proteoliposomes such as OMPC as they are large to start with and don't tend to aggregate, thus processes to handle the OMPC and the resulting conjugates need to accommodate large particles anyway.

A summary of the important lessons learned from the above considerations for the choice and manufacture the carrier protein component(s) for glycoconjugate vaccines includes the following:

- Selection of which carrier protein to use requires considerations of the valence of the vaccine to be made, the target population for the vaccine and their anticipated immune response to the carrier protein in the context of the glycoconjugate vaccine as well as other vaccines that may be co-administered. Freedom to operate is also a critical consideration.
- Given that the carrier protein can be the dominant component of the glycoconjugate vaccine, the manufacturing scale needed to support vaccine requirements and the sourcing of the protein production need to be carefully considered.
- Consistency of manufacture is particularly important as the key functional attributes of most carrier proteins are not well-defined so the "process defines the product" mantra is in play.

Development of Glycoconjugates

For glycoconjugates licensed to date, the physical association of Ps and carrier protein for the purposes of engaging T-cell help in the induction of anti-Ps immune responses has relied on 1) chemical activation of Ps, 2) mixing of activated Ps with carrier protein (itself activated or not), and 3) purification of Ps-protein covalent complexes, with particular intent to reduce the presence of non-conjugated Ps. While many products in development follow this general path, some new approaches to association of Ps and carrier protein are being evaluated (see Table 2). Lessons of the past point to some very specific considerations of how to associate Ps and carrier protein to have the desired functional activity and these are addressed below. Specific chemistries and conjugation approaches used in licensed products (e.g. Prevnar13) are commonly utilized in development of new and follow-on glycoconjugate vaccines too numerous to mention here (see Chapter 4 (*114*)).

Ps Size Reduction

With many Ps, e.g. several Pn serotypes, the large size (weight-average molecular mass (M_W) >1000 kDa) interferes with the chemical reactions inherent to the conjugation process, resulting in either poor yields or inconsistent results or both. Starting with such larger size Ps may also result in the formation of very big glycoconjugates, with could be lost during processing using (e.g.) chromatographic purification or on sterile filtration or even result in similar characteristics as seen in biogels.

As such many Ps are typically subjected to molecular size reduction to achieve M_W on the order of <300 kDa. In other cases manufacturers prefer to use smaller fragments defined as oligosaccharides (Os) which are coupled to the carrier protein via their end groups which have been selectively activated (*133*, *134*). Size reduction can be accomplished in a variety of ways, though all are not necessarily suitable to a specific Ps structure and to an industrial scale process (see Table 3).

Thermal hydrolysis, with some challenges, could be made scalable and reproducible, but elevated temperatures can disrupt the chemical structure of the repeating units along with reducing molecular size. Chemical hydrolysis has been used successfully in some conjugate vaccine development programs (e.g. (*124*, *133*, *135*, *136*), but it can come at the risk of chemical side-reactions due to the use of acids or bases (e.g. impact of base on O-acetyl on St Vi Ps). In both of these approaches size reduction is a random event, meaning that the Ps chain can break at any point along the chain with equal probability, sometimes resulting in an increase in sample polydispersity, which may necessitate a follow-up size-based purification step to keep the size distribution from being too broad.

In order to ensure process consistency chemical hydrolysis can be monitored using at-line in-process controls to stop the reaction at a predetermined average chain length and, with subsequent application of size-exclusion chromatographic purification, to restrict the polydispersion. For example a correlation between average degree of polymerization and optical activity was found for Hib Ps (*136*). Additionally, the most thermally labile saccharide-saccharide bond is the primary reaction site of interest, but other reactions, particularly if they impact side-chain structures of the repeating unit, can render these methods unsatisfactory for some Ps. The main application of acid or base hydrolysis for size reduction has been to prepare pools of sized Hib and Mn oligosaccharides for terminal activation and conjugation, discussed under "Creation of Glycoconjugates". These Ps contain a single in-chain labile linkage: Rib-P (Hib), ManNAc-P (Mn A) or NeuNAc (Mn C, Y and W) that is selectively cleaved using mild acid hydrolysis; this is not typical for most multi-residue repeating unit polysaccharides.

Table 2. Chemical, physical, and biological approaches to associate Ps with carrier protein

Method (ref) and user Manufacturer	*Ps Activation Rate*[a]	*Linker between Ps and protein?*	*Carrier Protein Activation?*	*Conjugation Processing time*[b]	*Ps Yield*[c]	*Benefits & Risks*
Covalent						
Periodate oxidation w/ reductive amination (Anderson (*115*) e.g. Pfizer	Slow	No	No	Slow	Medium	• Not suitable for all Ps • May require removal of substituents prior to activation
CDAP (Lees (*116*, *117*)) e.g. GSK	Fast	Yes or no	Can be with or without	Moderate	Medium/ High	• CDAP activated Ps coupled to lysines of proteins or their hydrazido derivatives • CDAP activated Ps derivatized with bifunctional amino-linkers and coupled to protein COOH groups via EDC
CNBr-ADH EDAC (Robbins (*118*, *119*)) e.g. Sanofi and GSK	Fast	6-carbon	Can be with or without	Fast	Medium	• Side reactions on the carrier protein when EDAC is used (e.g. N-acyl urea derivatives)
CDI-BuA2 (Marburg (*120*)) e.g. Merck	Fast	Bigeneric	Thiolated	Fast	Low	• Reduced control • Over activation can happen
Reductive amination w/ protein hydrazide (Frasch (*121*)) e.g. Serum Institute of India	Slow	No	Yes	Fast	High	• Not suitable for all Ps

Continued on next page.

Table 2. (Continued). Chemical, physical, and biological approaches to associate Ps with carrier protein

Method (ref) and user Manufacturer	*Ps Activation Rate*[a]	*Linker between Ps and protein?*	*Carrier Protein Activation?*	*Conjugation Processing time*[b]	*Ps Yield*[c]	*Benefits & Risks*
Terminal Active Ester (Costantino (*122–124*)) e.g. GSK	Moderate	6-carbon	No	ND	Medium	• Largely clinical evaluated • Coupling to terminal sugar preserves structure & epitopes of Os. • Good control of conjugation • Mainly applicable to oligosaccharides • Requires isolation and purification of the activates Ps
N-linked *in vivo* glycosylation of recombinantly expressed proteins (Wacker (*125*)) e.g. LimmaTech/GSK	NA	No	No	NA	NA	• Clinically evaluated • Specific sites of Ps linkage to carrier protein can be designed by genetic engineering • Structure and epitopes on the PS are preserved • Especially suitable for pneumococcus-like capsule; a variety of gram-negative organisms produce such Ps
Dry glycation following periodate oxidation (Gerson (*126*)) e.g. PnuVax	Slow	No	No	Slow	High?	• Others have had difficulty with this approach to conjugation • No specific removal of unconjugated Ps • Not yet evaluated in clinic

Method (ref) and user Manufacturer	*Ps Activation Rate*[a]	*Linker between Ps and protein?*	*Carrier Protein Activation?*	*Conjugation Processing time*[b]	*Ps Yield*[c]	*Benefits & Risks*
Incorporation of unnatural amino acids into specific sites in carrier protein (Murray (*127*, *128*)) e.g. Sutrovax	ND	ND	ND	ND	ND	• Specific azide-activated sites of Ps linkage to carrier protein designerd by genetic engineering • Ps activation/conjugation approach not yet defined • Non-specific antibodies against bulky linkers, a potential concern
Chemical synthesis of Os (Pozsgay (*129*)) e.g. Hilleman Institute, Cuba, Institute Pasteur	NA	Yes	Yes	ND	ND	• Product w/ synthetic Hib Os licened in Cuba. • Success in clinical evaluation in other programs is Ps dependent
Non-Covalent						
Ps biotinylation w/ avidin-like fusion protein (Multi-Antigen Presenting System; MAPS) (Malley (*130*)) e.g. Affinivax	Slow	Biotin-avidin-like peptide association	No	Fast	High?	• High affinity association of linkers • Not yet evaluated in clinic • Non-covalent linkage • Non-specific antibodies against avidin – a potential concern ?

Continued on next page.

Table 2. (Continued). Chemical, physical, and biological approaches to associate Ps with carrier protein

Method (ref) and user Manufacturer	*Ps Activation Rate*[a]	*Linker between Ps and protein?*	*Carrier Protein Activation?*	*Conjugation Processing time*[b]	*Ps Yield*[c]	*Benefits & Risks*
Ps and carrier protein cross-linked by polymer (Protein Capsular Matrix Vaccine; PCMV) (Mekalanos (*131*)) e.g. Matrivax	NA	No	No	ND	High?	• Clinical data not yet public
Encapsulation of Ps and carrier protein in liposome (Liposomal encapsulation of polysaccharides; LEPS) (Jones (*132*)) e.g. Abcombi Biosciences	NA	No	No	ND	ND	• "Encapsulation" approach not yet proven in humans

[a] Ps activation rate (time from addition of activating agent to next processing/storage step): fast = <1 h; moderate = 1-6 h; slow = >6 h. [b] Conjugation processing time (from beginning of reaction to start of purification): fast = <1 h; moderate = 1-6 h; slow = >6 h. [c] Ps yield post purification (Ps mass recovered post-purification vs Ps input into conjugation reaction): low = <20%; medium = 20–40%; high = >40%. NA – not applicable ND – not defined.

Table 3. Methods for reduction of Ps size

Method	*Simple*	*Scalable*	*No Side Rxns*	*Reproducible*
High temperature (auto-hydrolysis due to buffer or Ps charged groups)	++	+	-	+
Chemical hydrolysis (i.e. acidic, alkaline, ozonolytic, free radical)*	++	++	+	++
Sonication	+	-	+	-
Homogenization	++	++	++	+++

* Utility is structure dependent. If well controlled, chemical hydrolysis can meet the needs for size reduction. More "+"s indicates superior performance for this parameter.

Sonication and homogenization both reduce size via cavitation energy (expansion and collapse of micro-bubbles in the solution), one by inputting energy directly into the solution via a probe and the other using high-pressure discharge of the Ps solution through a small-bore orifice, both resulting in physical shearing of the Ps at equidistant nodes on the molecule, inter-node distance being inversely proportional to energy input. In both cases, temperature control is essential to minimize unwanted side reactions. As such, both methods reduce the polydispersity of the Ps molecules in solution – larger molecules are reduced but smaller molecules aren't impacted. Sonication is notoriously difficult to conduct consistently as the efficiency of the probe changes with use and the sonication energy is position-dependent, being highest at the point of the probe. This can result in increased polydispersity and inconsistent size reduction. Scalability is also an issue as the energy-to-volume translation is not exact. Additionally, the sonication probe is degraded during the sonication process, introducing metal contaminants into the Ps solution.

Homogenization is a desirable approach to size-reduction as once the Ps concentration, pressure and number of cycles are determined, the process can be highly consistent. However, homogenization does have its limits as the pressures needed to create very small Ps or Os may not be feasible with commercially available equipment. That being said, each Ps has to be evaluated independently to ensure that no adverse changes in composition and structure of the repeating unit are encountered.

While large Ps can be a process challenge, it is still unclear if "too small" is an issue. Many studies have been published to evaluate the impact of Os or Ps size (e.g. (*115*, *137*, *138*), but there is still lack of clarity on what specific parameters might impact the immunological readout, be it immune response measures or, ultimately, protection. Interestingly, one of the first studies investigating the impact of saccharide chain length on the immunogenicity of conjugate vaccines was a clinical trial where two Hib-DT conjugates containing Hib fragments of 8 and 20 average chain length (Dp8 and Dp20) and an estimated molar carbohydrate to protein ratio of 3.3 and 2.1 respectively, were compared in adults and infants. While in adults the two preparations elicited similar PRP antibody titers, infants responded better to the conjugate containing longer saccharide chains (*115*). In a subsequent study, the same team tested Hib fragments of 6 – 7 average chain length conjugated to CRM_{197} at different Ps loading densities, reflecting the number of Hib Ps/Os molecules per protein molecule, (although with slightly different chemistry). Infants receiving the conjugates at higher Ps loading density developed significantly higher anti-PRP immune response, in the same range of that obtained in other studies using the conventional Hib-CRM_{197} vaccine (*139*). This study suggested a possible interconnection between chain length and Ps loading density in determining the immune response to the carbohydrate moiety, which is also tied into the concerns for glycoconjugate design regarding carbohydrate-to-protein ratio and dose of carbohydrate and protein in the vaccine. Similar studies with other bacterial Ps conjugates have attempted to address the question of the impact of Ps size on immune response, but these studies often do not provide the details needed to clearly understand the differences in the molecular architecture of the conjugates being compared to allow definitive understanding of the critical relationships between molecular size and immunogenicity. Overall, these studies seed the idea that Ps/Os of different sizes may engage the immune system in different ways, which could be serogroup/serotype dependent, suggesting that the question of "size" may be more challenging than we realize.

An issue that might render difficult to compare conjugates containing saccharides of different size is the polydispersity of the saccharide preparations and the degree of overlap of their molecular size distributions, which can become a confounding factor when analyzing the antibody response. In investigational studies, care should be taken to minimize the overlap in the size distribution of the saccharide in the different preparations. The availability of synthetic glyco-antigens with defined chain length can facilitate the understanding of the impact of chain length on glycoconjugates immunogenicity.

An elegant study with conjugates containing synthetic well-defined *Shigella dysenteriae type 1* oligosaccharides confirmed, in an animal model, that chain length and density of saccharide loading on the carrier protein might be interconnected and that shorter chains need higher density than longer ones to elicit the optimal antibody response (*140*). The development of a synthetic *Shighella flexneri* 2a conjugate vaccine, currently in phase 1 clinical trials, offers another example of the use of carbohydrate synthesis to select the optimal chain length and density of loading: in a first study a conjugate of *Shighella flexneri* 2a synthetic pentadecasaccharide, corresponding to 3 O-antigen repeating units (RU), was found to be more immunogenic than conjugates with 2 or 1 RUs at the same density of loading (*141*), in a subsequent study the optimal average density of loading for the selected pentadeca-saccharide conjugate was determined to be around 17 (mole/mole) and conjugates with saccharide loading density in the range of 5 (mole/mole) were found significantly less immunogenic in mice (*6*). See additional related discussion in the section on "Number of linkage sites per protein molecule" below.

One limitation of most studies aimed at investigating the impact of variables such chain length, loading density, conjugation strategy, and carrier protein on the immunogenicity of glycoconjugates is that they are conducted in animal models like mice, rabbits or guinea pigs, for which prediction of the human response is uncertain at best and misleading at worst. The most valuable response that we see from animal models is to confirm that the immune response is boostable and reflective of the expected behavior of a conjugate vaccine.

Chemical Activation of Ps

Chemical modification of Ps to create a covalent link with the carrier protein is the most often used approach in making glycoconjugate vaccine antigens. That said, the choice of chemistry to use for a given Ps preparation comes with many considerations that can directly impact the integrity and functionality of the resulting conjugate vaccine antigen (e.g. (*142*)). Most licensed vaccines are prepared using random activation of the Ps or size-reduced Ps, while some Mn and Hib vaccines employ depolymerization resulting in terminal activation. Consistent production of the activated intermediate in terms of active groups and size distribution is required to ensure reproducible conjugation and formation of the glycoconjugate vaccine. Systematic design and optimization of the activation process, such as by using a "design of experiments" approach, could provide long-term benefit to the product by controlling one of the most critical process steps in making a glycoconjugate.

The activated Ps can be reacted directly with the carrier protein or isolated as the saccharide intermediate and the structural integrity, active groups and size distribution characterized. If stored before use, the stability and structural integrity of the saccharide intermediate and active groups needs to be demonstrated. Variable derivatization of Ps can result in suboptimal lots of conjugate vaccines that might not be easily identifiable using immunogenicity in animal models for lot release. An example is provided by some batches of PedvaxHIB that were found to be less immunogenic than expected in infants (*142*, *143*)). The specific chemistries used and being used must balance several factors:

Activation Specificity

All chemical reactions are driven by the concentration of the reactants and the suitability of the reactions conditions (e.g. pH, temperature, etc.) to favor one reaction over another. As such, in all chemical activations of Ps the preferred reaction must be favored by the physico-chemical environment. This, however, does not preclude secondary reactions from occurring. In the activation of a specific Ps these secondary reactions must be controlled to the extent possible, but must be identified and quantified to best anticipate the final product of the activation and the resulting impact on Ps conjugation and immunogenic potential of these secondary reaction products. NMR has been established as an excellent tool for this purpose (see (*144*)), allowing both initial characterization of the activated Ps, but also revealing the lot-to-lot consistency of the activation process.

Applicability across Ps Structures

The presence on a single acid-labile linkage in the repeating unit of Hib and Mn Ps allows the use of mild acid hydrolysis (which has little impact on O-acetylation) to generate the activated saccharide intermediate with an aldehyde or keto group at the reducing end (*124*). A similar approach can be used for some Pn serotypes such as Pn 3 (*145*) and Pn 6A (*146*). Hydrogen peroxide has also been used to generate Mn oligosaccharides and the reducing end groups reacted with ADH to form the hydrazide-derivatized saccharide intermediate (*147*). This method of depolymerization may have general applicability; it was used to fragment the St Vi Ps considered to be the most stable of all the Ps vaccines (*148*). Though all Ps (excluding Vi) have exposed hydroxyl groups that can be exploited in a variety of activation chemistries, for specific Ps a given activation chemistry may hit too many sites equally, making it hard to produce a highly specific product, or the activation may alter the repeating unit structure in ways that aren't optimal for immunogenic response to the conjugated Ps. The complexity of periodate oxidation of Pn Ps has been described by Kim et al. 2006 (*149*), together with a method for determining the location and order of activation by methylation/linkage analysis. For Ps containing diols that are not available due to the presence of O-acetyl groups (Pn 1) or pyruvate (Pn 4), these substituents can be partially removed prior to periodate treatment (see Chapter 4 (*114*)). Lastly some Ps possess carboxyl groups (e.g. Vi, Pn 1, 3, 5, 9V, and Mn C, Y and W) or phosphate groups (e.g. Hib, Pn 6B, 18C, 23F) that might be used for introduction of an amino linker using a carbodiimide-mediated chemistry (*150*).

Rate of Reaction

Conjugation reaction rates can be highly variable, reaching equilibrium in times ranging from seconds to hours even under the best of conditions. While reaction rates can usually be controlled by reducing the concentration of reactants or adjusting the physico-chemical conditions for the reaction, it becomes more a balancing of competing reactions that must be achieved. Sometimes this is best acomplished by alternate chemistries (see Table 3). In general, fast reactions can be a challenge to control for consistency, whereas very slow reactions tend to not be suitable for routine commercial manufacturing as they may occupy equipment for longer than desirable.

Extent of Modification

Controlling the extent of chemical activation of a Ps, as measured by the average number of repeating units modified in the final product, is perhaps one of the most important aspects of making a glycoconjugate vaccine. The physicochemical characterization methods applied to the saccharide intermediate are similar to those applied to the Ps, with additional tests for the active groups introduced. For randomly activated Ps this can be determined by colorimetric assays for the reactive species introduced (e.g. amino group or aldehyde) per repeating unit Ps or in a single assay by use of NMR spectroscopy as described for a derivatized Hib Ps intermediate (*144*). Terminal activation can be applied effectively to size-reduced Ps or Os. Assuming that all reducing ends are modified, molecular size can be a surrogate of the extent of modification and can be measured as molecular size (measured as a mean value or as a profile), which can then be carried out by size exclusion or anion exchange chromatography (*151*). Alternatively the average values for end group and sugar chain can be determined by colorimetric assays, chromatography (HPAEC-PAD after end group reduction and hydrolysis to monomers) or NMR spectroscopy (*151–153*). Colorimetric assays for active groups such as aldehydes are simple and easy to perform, but care must be taken to use appropriate standards because their reducing power is structure dependent. This was illustrated by a study of size analysis/extent of activation of Hib fragments generated by periodate oxidation, which showed that use of SEC with dextran calibration or colorimetric assays for end group and sugar chain overestimated the extent of modification compared to direct methods such as HPSEC-MALLS/RI and NMR. This demonstrates the value of physico-chemical techniques and highlights the importance of using orthogonal methods during process and assay development.

Modification of Key Ps Immunogenic Epitopes

Most commercial glycoconjugate vaccines utilize activation processes that result in modification of 2-15% of the repeating units of a Ps. A lower percentage can result in a lengthy reaction time to achieve conjugation of activated Ps to carrier protein. Higher activation rates can perturb the immunogenic epitopes (see (*142*)) of the Ps in the conjugate such that the resulting antibodies lack functional activity and therefore may fail to protect the vaccinated individual. As an example, excessive activation of Ps has resulted in suboptimal immune responses to a Hib glycoconjugate vaccine in humans (*143*). Additionally, periodate oxidation of GBS type III results in loss of a native epitope and creation of a new epitope, generating antibodies that now recognize Pn Ps 14 (*154*). This epitope loss does not occur with cyanylation chemistry (*142*). Activation of Os end groups allow immunogenic epitopes of the carbohydrate to be preserved and they often guarantee good control of the conjugation and manufacturing consistency, however Os manufacturing processes might add complexity.

Chemical Modification of Carrier Protein

Some of the above conjugation chemistries require chemical modification of the carrier protein to produce the desired chemical linkage. Modification of carrier protein, like Ps, targets specific chemical entities, e.g. amino groups on Lys or carboxyl groups on Asp or Glu. These sites may not be fully accessible to modification, depending on the chemistry used and the reaction conditions, so will limit the potential number of Ps molecules linked to protein to some portion of those sites that are modified.

Number of Linkage Sites Per Protein Molecule

The average frequency that saccharide (Ps or Os) links to the carrier protein molecules is a "lever" in the design of glycoconjugates that can potentially increase the saccharide to protein ratio. This ratio can be an important consideration, particularly for multi-valent glycoconjugate vaccines, in that a higher ratio means that less protein is present in the final vaccine. This can help abrogate any impacts of carrier immune suppression (see above) as well as reduce the amount of carrier protein that needs to be made, reducing the overall cost of goods for the vaccine. As an example, there are 39 lysines in CRM_{197} and approximately 8-10 are typically involved in conjugation, these logically being the most surface active lysine molecules in the protein (*155*). Diphtheria toxin has the same total number of lysines, but the toxoiding process (with formaldehyde) reduces the number of reactive lysines, again most likely including those most accessible on the protein. This may result in fewer or at least different lysines being involved in conjugation in CRM_{197} versus DT.

Tetanus toxoid (109 lysines/molecule total (*156*)) presents a similar scenario as with DT, with formaldehyde-based toxoiding making the most accessible Lys unreactive in conjugation. Other carrier proteins (e.g. OMPC and Protein D) will have similar concerns about which and how many of the reactive sites on the protein are available as linkage points for the Ps or Os. In all cases the number of lysines involved in conjugation can likely be increased but only through disrupting the structure of the carrier protein. The impact of this, good or bad, is very hard to assess and thus increases the risk that such a vaccine might have altered immunogenicity and/or protection.

Hib Ps-protein conjugates can be used as a theoretical example to illustrate the practical considerations in the integrated evaluation of Ps chain length and Ps loading density on the carrier protein. With a fixed Ps dose level of 10 ug, CRM_{197} as the carrier protein at specified Ps-protein mass ratios (e.g. 0.5. 1.0 and 1.5), Figure 1A defines the issue using a neoglycoprotein structure as the most simple example: as the number of sites where Ps can attach to a protein molecule decrease, the average chain length (or M_W) of the Ps needs to go up to allow the same mass of Ps and protein to be administered per dose. Either that or the amount of protein in a dose needs to increase in linear proportion to the decreased number of attachment sites. The practical consequence of these considerations is shown in Figure 1B, which shows the relationship between Ps average chain length and Ps loading density per protein molecule for a given dose of Ps (10 ug) and specified Ps-protein ratios. The green shaded area represents the nominal working range for these two parameters under these conditions. Of course any conjugate that approached a single Ps attachment site per protein molecule would be at risk of significant amount of unconjugated protein in the formulation, and any conjugate that tied a single large Ps molecule to each carrier protein molecule would be at the risk of the inherent stability of the Ps and/or the linkage, although the latter is generally very stable (to chemical or shear influences) as it would only take a single break to create unconjugated Ps. With the more complex example of a lattice conjugate, a more stable conjugate product can be achieved as it then takes two breaks in the uninterrupted Ps chain to produce unconjugated Ps and shear forces are less likely to do this.

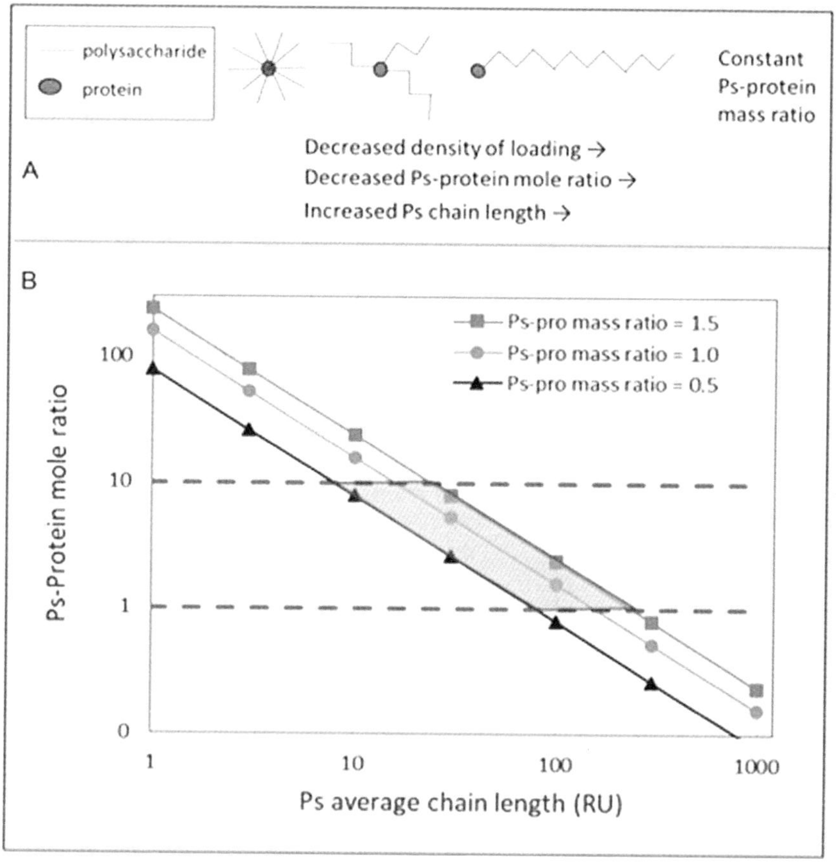

Figure 1. Density of loading of Ps on the carrier protein. A. Carrier protein can, by process or by protein design, have a single point of Ps attachment or multiple points of attachment. B. Balancing Ps and protein content also requires consideration of Ps chain length, Ps loading density on the carrier protein and molar ratio of Ps and protein. At a fixed Ps-protein mass ratio, typically 0.5 to 1.5 in most conjugates, and the limits of Ps molecules bound to a protein molecule ranging between 1 and 10, the average chain length of the Ps conjugated to the protein must increase as the number or attachment points decreases.

For engineered recombinant carrier proteins that control the number of potential linkage sites per protein molecule (e.g. LimmaTech or Sutrovax technologies) the number and placement of Ps attachment sites is a critical design consideration as is the efficiency of their involvement in conjugation to Ps. Too few linkages will impact the balance of Ps chain length and loading density, whereas too many linkages or linkages in the wrong sites in the protein may impact the carrier function of the protein and thus the success of its immunogenic response.

A summary of the important lessons learned from the above considerations for the design and manufacture of the activated Ps and protein components for glycoconjugate vaccines includes the following:

- For some Ps types, purposeful size reduction may be required as viscosity issues may otherwise prevent successful activation/ conjugation processing.
- Selection of which activation/conjugation chemistry to use requires considerations of the Ps structures, the carrier protein to be used and the type of glycoconjugate (e.g. neoglycoconjugate, lattice, etc.) desired.
- Ps activation reactions must be designed and monitored extremely carefully as these reactions have the highest probability of modifying epitopes and/or introducing deleterious properties to the resulting conjugate.
- Choice and production of carrier protein can have a substantial impact on vaccine manufacturability and cost of goods and on the immune response induced by the carrier protein or its homologues used in other co-administered vaccines.
- Ps-to-protein ratio for the glycoconjugate, an extremely important parameter that directly impacts the overall contents of the monovalent conjugate bulk as well as the final vaccine product, is directly influenced by extent of activation of the Ps, the Ps chain length, the number of accessible linkage sites on the carrier protein molecule, and the efficiency of the conjugation reaction in linking Ps to the protein.

Generation of Covalent Glycoconjugates

All commercial glycoconjugate vaccines in current use rely on covalent linkage of Ps to the carrier protein. Combining of the activated Ps and protein components is followed by the actual conjugation reaction, which deserves careful design, evalution and characterization to make sure that the resulting glycoconjugate preparation has suitable characteristics, and allows efficient purification of the glycoconjugate to remove reactants, most importantly unconjugated Ps. Several critical features in the production of glycoconjugates are discussed below, but for more detailed discussion of this topic refer to Chapter 4 (*114*).

The Conjugation Reaction

The actual conjugation reaction can either proceed to completion or be "quenched" at a specified time in the process. Though the chemistry involved can be generally understood, the actual details of such processes are often proprietary in nature. Suffice it to say that this choice should be evaluated carefully and be data-driven. A common motif is often to mix equal masses of Ps and carrier protein to initiate the reaction. When the concentration of activated sites in Ps molecules can be determined, appropriate conjugation stoichiometry can be developed in order to have all the carrier protein glycosylated and therefore simplify the conjugate purification. This addition must be well-mixed, temperature-controlled and under well-controlled physico-chemical conditions to insure lot to lot consistency. The scale of this reaction is also critical in that too small of a scale results in an excessive burden on the production schedule as well as the quality control laboratory, but too large of a scale of conjugation can result in too much of a manufacturing risk in terms of inventory lost if a lot fails release testing. Decisions on manufacturing scale for production of a multivalent glycoconjugate vaccine can have a major impact of the commercial operations of any company, large or small, and should carefully consider a multitude of factors including market demand, market share, inventory management plans, and competing uses for manufacturing facilities.

Simultaneous Conjugation of Multiple Ps

It is worth noting that there have been multiple efforts to conjugate more than one activated Ps to the carrier protein at the same time. Porro (*157*) utilized this technology to simultaneously conjugate Pn Ps 6A and MnC Ps to CRM_{197} and other efforts have been more recently offered by Bongat (*158*) and Rivera (*159*). Although the potential advantage could be to simplify the vaccine formulation (less components in the vial) and these examples offer credible processes, the limitation becomes apparent in the quality control and specifications for the final vaccine product as now the variability in the ratios of the Ps that are simultaneously conjugated adds to the range of the acceptance criteria for Ps content (i.e. potency) in the final container. Even if this variability is limited, this could result in a substantial increase in the targeted dose level for these products that would ultimately increase both their Ps level as well as carrier protein level in a multivalent final product, which can have both immunological implications (see "carrier protein-induced immune interference" above) as well as cost-of-goods impact. A better way to control the conjugation of two antigens to the same carrier could be the use of sequential conjugation, however manufacturing complexity would increase and an additional risk could be the immunodominance of one component of the bivalent conjugate versus the other (*85*).

Purification of Glycoconjugates

Purification of the glycoconjugate can be achieved in several ways, including by use of chromatography, diafiltration or membrane filtration, precipitation or by other means. The primary goal of this process step is to reduce unconjugated saccharide to as low a value as is reasonably possible, preferably $<10\%$ of the total saccharide in the purified conjugate sample. This is a critical consideration as unconjugated Ps does not produce a functional immune response in infants (<2 years old) and has been shown to reduce the immune response to conjugated Ps in animal models (*160*). Although unconjugated Ps is immunogenic in older populations, the relative immunogenicity of conjugated and unconjugated Ps at equal Ps dose levels has not been assessed in these populations.

A key challenge to the physical separation of conjugated and unconjugated saccharide, both for production and analytical applications, are the relative molecular size differences between these two entities and the potential for electrostatic interactions between them. The greater the difference in size of conjugated and unconjugated saccharide, the more effective the purification can be. Similarly, reduction of interactions between conjugated and unconjugated saccharide will facilitate one of the major goals of the purification process. Separation can usually be achieved using dilute solutions and sufficiently high ionic strength, which in turn relies on the validity of the unconjugated saccharide assay to verify the effectiveness of that separation. Accurate measurement of the amount of unconjugated saccharide in the purified glycoconjugate sample is an equally important matter in defining the success of this purification process, with details of that measurement discussed below.

Another important goal of glycoconjugate purification procedure is the removal or reduction within acceptable limits of unconjugated carrier protein. Also in this case classical methods like chromatography and diafiltration can be applied.

The resulting purified glycoconjugate preparation, typically defined as the DS, should be extensively characterized to insure that it reflects the profile of a product that can be successfully incorporated into a commercial vaccine product. Of particular concern are the size distribution of the glycoconjugate particles, proof of covalent association of the Ps and protein, an assessment of yield of the Ps through conjugaton and purification of the conjugate, and eventually stability of the conjugate.

Particle Size/Structure of Conjugates

As shown in Figure 2, glycoconjugates can have a variety of different structures, from the limited interactions seen in simple or neoglycoprotein conjugates, where the emphasis is on a more controlled molecular architecture, to the lattice conjugates, which are inherently larger and more structurally diverse, to the ladder type of conjugate, which reflects the Ps-OMPC conjugates. The structure of a conjugate is specific to the activation-conjugation process used to make the conjugate. For example, activation of the reducing end of a Ps (or Os) will result in a neoglycoprotein conjugate structure, as seen in VaxemHib, Menjugate, Menveo and QuimiHib. Random activation along the Ps molecule, regardless of the frequency of the activation, will result in multiple protein molecules binding to the Ps molecule. If the protein is reactive at a single point, e.g. the N-terminal amino group, then a simple conjugate structure will result. This is a relatively inefficient use of the carrier protein, so few if any commercial conjugate products pursue this structure. Random activation along the Ps molecule and multiple activations on the carrier protein, e.g. accessible amino groups, will result in a lattice conjugate structure of which there can be several variations. This is the more common structural type for glycoconjugates. Ps-OMPC conjugates, due to the unusual size and nature of the proteoliposome complex, form more of a ladder type structure, with multiple Ps molecules attaching to the complex, but little if any link between OMPC particles.

The average size and size distribution of the glycoconjugate particles is a significant attribute that can contribute to the success or failure of a conjugation/purification process. As mentioned above, this can be a critical quality attribute (CQA) for successful removal of unconjugated saccharide, as the greater the size difference between the conjugated Ps and the unconjugated Ps, the more efficient the purification process can be in removing unconjugated Ps. However, too large a conjugate structure can also impose limitations to the successful recovery of conjugated saccharide through the purification process. Very large conjugate particles (e.g. >10 MDa can be refractory to chromatographic purification as the particles may be too large to enter the stationary phase matrix and thus be lost to recovery, resulting in low yields (of Ps) post purification. In contrast, such large particles would be readily recovered by filtration, precipitation or centrigation purification methods where the larger components (glycoconjugates) are retained and the smaller components (unconjugated saccharide) are in the waste stream. Of course, very large particles can be of marginal solubility and contribute to precipitates forming in the purified conjugate bulk, an undesireable attribute that will be discussed more below.

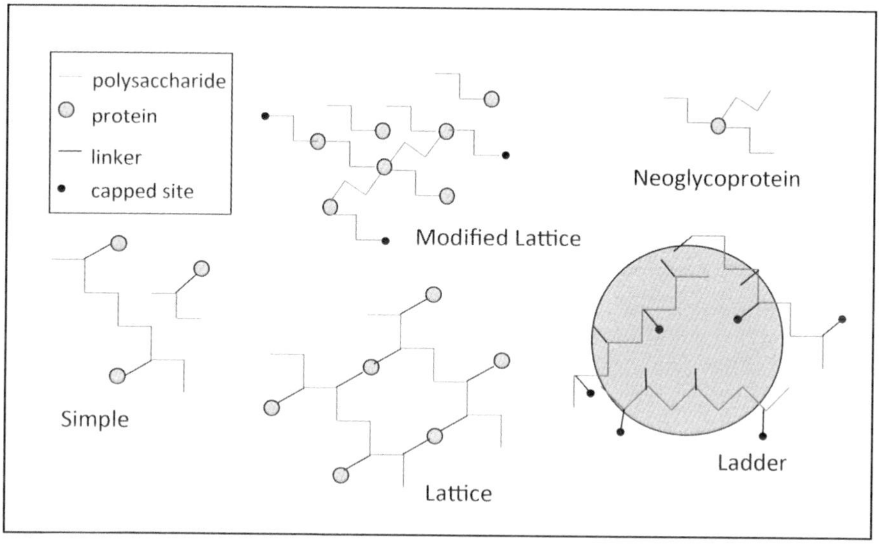

Figure 2. Higher order structures for Ps-protein conjugates.

Reactivity/Yield

A "successful" conjugation reaction requires a balance of reactivity of the Ps and the protein. If the reactivity of the protein is too high (e.g. too many reaction sites exposed) then the resulting glycoconjugate may contain too much or too little Ps (depending on the reactivity of the Ps). If the reactivity of the protein is too low, then the reaction of Ps with protein will be limited and the Ps yield of the conjugation/purification process will be unacceptably low.

Measurement of process yield is deceptively simple; measurement of the Ps mass input into the conjugation reaction and measurement of the Ps mass in the purified glycoconjugate bulk allow calculation of the percent recovery or overall yield of the conjugation/purification process. To be accurate the Ps amount must be measured using the same analytical method for both samples.

As reflected in Table 3, process yields can be specifically influenced by the choice of conjugation chemistry used and, as discussed above, can also be influenced by the conjugate purification method chosen. The range of claimed and demonstrated process yields for various glycoconjugates spans 5 to 85%. The very low numbers reflect inefficient processes that were developed without consideration of process efficiency. The very high numbers may well reflect inadequate analytical methods or inadequate removal of unconjugated saccharide. In practice Ps yields of 25-40% appear to be readily and consistently achievable, though results are dependent on the specific Ps, carrier protein, conjugation chemistry and processes being used, as well as the skill and experience of the process and analytical staff involved. It is worth noting here that altering the ratio of Ps and protein going into the conjugation reaction can influence the Ps yield, but the competing concerns about keeping protein yields high (typically >90%) and not allowing the Ps to protein ratio in the purified conjugate to get too low typically results in at least half of the Ps put into the conjugation reaction being lost.

Alternatives to Traditional Glycoconjugates

Alternatives to the traditional chemical synthesis approach to making glycoconjugates are being explored and are discussed briefly below. In general these approaches are novel, if not promising, but few if any have progressed to the stage of clinical development to establish a full proof of concept, which would be to demonstrate clinical equivalence or superiority to the current licensed standard of care.

In vivo Production of Glycoconjugates

In this approach both the glyco-antigen and the carrier protein are expressed and joined *in vivo* in *E. coli* cells. This *in vivo* "conjugation", referred to as bioconjugation, is directed by genetic design of the *E. coli* glycosylation pathway and of the recombinant carrier protein containing specific glycosylation sites *(161,162)*. The resulting glycoconjugate is then extracted from the bacterial cells and purified using appropriate methods. The possibility of selectively positioning the glycosylation sites on the carrier protein allows for a better exploitation of proteins with the dual carrier/antigen role, as in the example of *S. aureus* α toxin Hla used as carrier for type 5 and 8 CPS (*86*).

Although this technology is an emerging one, Phase 1 clinical trials of monovalent vaccines against *S. dysenteriae O1* and *S. flexneri 2a* (*163*), and of a tetravalent anti-extraintestinal pathogenic *E. coli* (ExPEC) vaccine have been completed (*164*), and Phase 2 trials have been initiated. Development of a multivalent Pn bioconjugate vaccine is also in progress (*165*).

Applicability of bioconjugation technology to a broader range of Ps is possible, especially for pneumococcal-like capsule and a variety of gram negative organisms. Ultimately, later stage clinical studies will establish the suitability of this approach to making glycoconjugates.

An attractive aspect of bioconjugation is the potential to simplify conjugate vaccine manufacturing by avoiding separate production and purification of the Ps and carrier protein moieties and their subsequent activated intermediates, reducing the quality control panel and avoiding handling of pathogenic organisms. This approach is particularly advantageous for O-antigens as it circumvents the challenges associated with removal of lipid A and conjugation of the polydisperse O-polysaccharide (*166*). Many attributes that are measured on the Ps before conjugation (e.g. size, integrity) can be measured directly onto the purified glycoconjugate.

In fact, the glycoconjugates thus obtained are largely defined by the bacterial strains constructed to that effect. After introduction of the Ps biosynthesis locus in the *E. coli* bacterial genome, followed by the plasmids carrying the genes for the glycosylation enzyme (PglB) and the carrier protein, few parameters are available if one wishes to influence the structure of the conjugate, or its production yield. PglB glycosylation enzyme can be selected through screening of a library of plasmids bearing different mutants of the enzyme, leading to increased glycosylation efficiency and therefore production yields (*167*). Similar to convential manufacturing processes the characteristics of the production strain should be defined early in the development of the glycoconjugate antigen to avoid slow down due to regulatory requirements

With chemical conjugation, some variation of (e.g.) the Ps to protein ratio can be obtained by changing the respective proportion of the reagents.

Non-Covalent Association of Ps to CP

One of the first examples of a vaccine candidate consisting of a non-covalent admixture of Ps and protein studied at preclinical and clinical level was based on experimental Hib vaccines. They tested first in animals then in infants a simple mixture of Hib Ps and Hib membrane proteins in comparison to more traditional covalent conjugates of Hib Ps and CRM_{197}, using Hib Ps alone as a control vaccine. In rabbits both the mixture and the conjugate vaccines induced a satisfactory and anamnestic anti-Hib response compared to the Hib Ps alone vaccine. However, in two year old children the mixture vaccine was poorly immunogenic compared with the covalent glycoconjugate (*168*). This study confirmed that a strong covalent association between the Ps and protein might be needed in order to induce the desired anticarbohydrate response in infants.

Despite this precedent, the established dogma that a covalent linkage between Ps and carrier protein is required in order to transform T-independent Ps into T-dependent antigens is still being challenged. The prevailing theory on the immunological mechanism of glycoconjugate vaccines teaches that both the carbohydrate and the protein carrier moieties need to be simultaneously incorporated into the same B cell. New approaches to non-covalent association of Ps and carrier proteins that approach this theory in a different way are being actively investigated. For example

- The Protein Capsular Matrix Vaccine (PCMV; Matrivax) technology is based on the entrapping of Ps antigens in a matrix of polymer crosslinking the Ps and the carrier protein. This association is proposed to allow the Ps and carrier protein to remain associated through entry into T cells and engagement of the immune system. Satisfactory preclinical results have been demonstrated (*131*) and a Phase 1 clinical trial for a typhoid PCMV vaccine (Typhax) has been completed (*169*).
- The Multi-Antigen Presenting System (MAPS; Affinivax) uses biotin bound to the Ps and an avidin-like protein fused to the carrier protein to form an association between Ps and carrier protein (*130*). Though technically not covalent this bond is strong and stable. Satisfactory preclinical results have been published and efforts to develop a new multivalent Pn vaccine are ongoing.
- The Liposomal Encapsulation of Ps (LEPS; Abcombi) system is based on liposomal surface localization of proteins and encapsulation of Ps and has been proposed to develop new Pn vaccines using the LEPS platform (*132*, *170*). The stability in vivo of the encapsulated "native" Ps and their non--covalent but proximate association with the carrier protein is the key attraction of this approach. Preclinical evaluation of a Pn vaccine based on the LEPS technology show promising results (*132*), but clinical evaluation is yet to be obtained.

Given the poor reliability of animal models in predicting the behavior of glycoconjugate vaccines in humans, such as the conflicting results seen in animals and human infants with the non-covalent Hib Ps and protein complex, new approaches for development of Ps-based vaccines based on non-covalent association of Ps and carrier protein have to be evaluated carefully in clinical trials and particularly in young children and infants where alternatives to glycoconjugate vaccines have so far been unsuccessful.

Stability Considerations for Monovalent Glycoconjugate Bulks

The stability of glycoconjugate bulks is primarily a product of the inherent properties of the Ps, the carrier protein and the specific application of the conjugation process technology used. Generally, the latter two can be influenced or controlled such that they needn't become the rate-limiting factors in obtaining a glycoconjugate with suitable stability. On the other hand, one or more of the Ps types being used can prove to be labile such that special considerations are needed for storage of the bulk. WHO produces a number of technical reports providing guidance on monitoring stability of licensed glycoconjugates such as Pn conjugate vaccine (*171*) and Mn conjugate vaccine (*172*). These can serve as good templates for other glycoconjugates as many glycoconjugates share similar stability concerns.

Shelf-Life Considerations

Though the tendency is to have the shelf-life for bulk intermediates be a number of years, for multivalent products the overall design for making all of the various components may provide some helpful options. Whether order of production of components is regimented (e.g. A, B, C, A, B, C, etc.) or campaigned (e.g. A, A, A, B, B, B, C, C, C, etc.) leading up to formulation and filling, the shelf-life of each bulk component need only be enough to cover the time that the bulks need to be stored prior to use (formulation). Whether production is regimented or campaigned, one solution is to make the components is order of decreasing stability so that the least stable bulk need be stored the shortest amount of time. Another solution may be to store bulk components frozen instead of refrigerated as long as there is no negative impact of freeze/thaw.

Additives to Glycoconjugate Bulks

Buffers, stabilizers, surfactants and excipients used in bulk intermediates need to anticipate the compatability of components for final formulated products, both in content and in concentration, to avoid reprocessing of bulks prior to formulation. Aside from achieving a solution that is near isotonic with a buffer that will set a neutral pH with a mild buffer capacity, any other components added need to have a specific reason to be there and concentration should be optimized by statistical design of experiments. Following this simple approach will reduce inquiries from regulators.

Storage Conditions

Glycoconjugate bulks could ideally be stored refrigerated (2-8°C) in liquid form at antigen concentrations suitable to allow formulation with other components to achieve the desired dose levels of all components. In the event that a bulk component is not sufficiently stable under refrigerated conditions, lyophilization or freezing are possible options, but need to be approached carefully. Lyophilization requires specialized equipment and requires processes that are scalable so that all experiments don't need to be done at commercial scale. Freezing requires very specific protocol for both freezing and thawing to prevent cryo-precipitation.

A summary of the important lessons learned from the above considerations for the design and manufacture of the monovalent BDS components for Ps-protein vaccines includes the following:

- The design of the processes for conjugation and purification of the glycoconjugate requires careful consideration of what will be made (e.g. structural type of conjugate) and extensive process characterization (i.e. analysis of intermediates) to understand the sensitivity of product quality and process yields on process variables. Failure to do so can result in an inferior glycoconjugate, e.g. inconsistent, unstable and/or poor yield.

- In choosing the final approach to manufacture of monovalent components, evaluate all technologies involved for feasibility in the organization(s) that will be doing the work, including careful evaluation of contract manufacturing organizations if they will be involved, and make sure that there is freedom to operate with those technologies.
- The complexity of glycoconjugate vaccine production puts a premium on production technologies that are designed, executed and monitored carefully to assure that they produce the expected product and it is produced in a consistent manner (lot to lot and campaign to campaign).
- Evaluation of monovalent bulks must include extensive in-process and product characterization as well as diligent evaluation of stability. Monovalent bulks should be evaluated with the same Ps concentration assay as will be used for final formulation to avoid inter-assay bias from impacting final formulation release specification.
- Evaluation of safety and immunogenicity in animal models is required to proceed to human studies, but demonstration of non-inferiority/efficacy in the target population is required to establish clinical proof of concept that the vaccine is suitable for commercial development.

Formulation

Formulation of glycoconjugate vaccines, particularly multivalent products, requires an ever-increasing set of complex considerations to ensure that the formulation is safe, immunogenic and stable. Much of the premise of how to formulate glycoconjugate vaccines is based on lessons learned from past formulation efforts and the details of formulation efforts are captured in Chapter 9 (*173*), but some of the key learnings will be discussed here.

Formulation of the active components of a glycoconjugate vaccine requires the incorporation into the formulation of excipients (e.g. salts, buffers, stabilizers, bulking agents), often adjuvants and occasionally preservatives to define the final vaccine product that is suitable for commercial use. In all cases, the inclusion of each component in the product formulation must be explained and justified both with data and in relationship to how the vaccine will be used. In deference to Chapter 9 (*173*), only sterility assurance, adjuvants, potency evaluation and stability evaluation will be discussed here.

Sterility Assurance

Historically, many vaccines contained a preservative as a hedge against potential contamination of the final filled container during formulation and filling. As equipment improved and barrier technology became more common, the need for preservatives in single-dose containers became not only unnecessary but not allowed in many parts of the world. Preservatives by their nature are toxic to living organisms so in a sterile, closed system they are not justified. However, with multi-dose containers there is need to guard against a potential breach of sterility due to multiple intrusions into the container. Preservatives are one potential solution to this problem, though for a preservative to provide protection it must meet the pharmacopeial standards for antimicrobial effectiveness as outlined in the USP, BP, JP, etc. Unfortunately there are few preservatives that meet this standard, with mercury-containing thiomersal being one of the most often used.

Another option that has met regulatory acceptance for some licensed multi-dose products is a restricted use period between removal of first and last dose. As an example, the 10-dose presentation of MMR vaccine produced by Serum Institute of India comes with a WHO recommendation that "opened vials of this vaccine should be discarded 6 hours after opening or at the end of the immunization session, whichever comes first" (*174*). This approach is considered sufficient to protect all patients from receiving a dose of vaccine that may be unsafe.

A third option has been recently introduced for multi-dose products, the multi-dose vial adapter: a device which ensures sterility of the vial contents while allowing repeated access to the vial contents.

Adjuvants

It is more common than not for glycoconjugate vaccines to contain an aluminum-based adjuvant. Older vaccines often contained aluminum hydroxide suspensions, whereas more recently the use of aluminum phosphate suspensions has become predominant. While other adjuvants are attractive to potentially establish a "better" immune response, the current correlate of protection has been established as serum serogroup/serotype-specific anti-Ps IgG titre and the enhancement of this titre by aluminum adjuvants, combined with the well-established safety profile of aluminum adjuvants, has held off serious consideration of the use of other adjuvants as evidenced by the lack of late stage clinical studied containing other adjuvants. In fact, the current question with adjuvants tends to focus more on whether the aluminum adjuvants truly boost protection.

Of course this is not a simple issue. Use of aluminum adjuvants typically comes with expectations of very high levels of adsorption of the glycoconjugates to the suspended aluminum particles, both at product release and throughout product shelf-life. This becomes a measure of consistency of manufacture and also ensure effective delivery of the antigens upon vaccination as antigen adsorbed to the suspension cannot be "lost" to other surfaces (e.g. vial walls). Additionally, there is some evidence that for some serotypes the adsorption of the antigen to the aluminum suspension can result in a more stable vaccine (*175*). There are exceptions though as Hib conjugates can actually be destabilized by adsorption to aluminum hydroxide particles, where the particle interaction facilitates catalytic hydrolysis of the Hib Ps, rendering some Hib conjugates to "dissolve" into increasing amounts of unconjugated saccharide (*63*).

Potency Evaluation

Formulation development allows for evaluation of immunological potency and the stability thereof, reflected in both the use of animal models to show any time-dependent changes in the immune response to the vaccine and in establishing a validated measure of potency, i.e. a measure of vaccine content with defined acceptance limits that assures that a vaccine that meets those acceptance limits will elicit the expected immune/protective response in the target human population. In reality most potency assays are a surrogate measure of some vaccine component or activity that translates to a high probability of generating a protective immune response. This protective response must then be correlated to a measured value of immune response, often serum antibody titer, to validate a surrogate marker of protection.

As a simple example, the potency assay for many monovalent Hib conjugate vaccines is a chemical assay that measures the amount of Hib Ps in the vial. Combined with a measure that confirms that the amount of unconjugated saccharide is acceptable (varies by product), the recommended dose regimen for the vaccine (3 or 4 doses depending on the product) will produce >1.0 µg/mL Hib antibody in the vaccinee serum. This is the level of antibody that is considered protective for ~90% of infant vaccinees (*176*, *177*). This "simple" example is used to highlight the fact that a true potency assay is akin to the "holy grail" of vaccines in that a substantial amount of efficacy data is needed to establish that a given dose of vaccine, given as indicated, will be protective. By definition this also requires at least some evidence of vaccine failure at lower doses or with other attributes, which are also essential to defining a surrogate maker of protection, and the level of which will be considered protective (>1.0 µg/mL for Hib conjugates).

These challenges are of course no different for glycoconjugates than for any other vaccine. And the nature of the "potency" challenge is such that animal data can only provide a limited start on the task. However, the large amount of accumulating data has pushed a standard for bacterial glycoconjugate vaccines that accepts serum antibody titer as a reasonable surrogate marker of protection.

The key message here is that a potency assay for a given vaccine is usually an evolving target that starts before a human is ever dosed and can continue to evolve well after product licensure.

Stability

For a formulated glycoconjugate vaccine, the primary stability concern is the measurement of unconjugated saccharide as this material is not potent in immature immune systems as found in human infants. Even in adult populations there is no direct translation of the potency of unconjugated Os or Ps with conjugated Os or Ps. In fact, with most vaccine targets where a Ps-based vaccine and a glycoconjugate vaccine both exist (e.g. Pneumovax23 vs Prevnar13) the dose level of the Ps-based vaccine can be ten-fold higher, though comparative dose-ranging clinical studies have not been done.

The appearance of unconjugated saccharide is dictated both by the inherent chemical stability of each Ps, the suitability of the physico-chemical environment in the formulation. The former cannot be easily altered, at least not without concern for the immunogenic integrity of the Ps, but the latter can and must be addressed.

Other stability concerns for formulated glycoconjugate vaccines will depend in large part on the presence or absence of adjuvant. In the absence of adjuvant, measures of time-dependent changes in aggregation state and adsorption to product contact surfaces become relevant. In the presence of adjuvants, particularly aluminum-containing adjuvants, concerns about adsorption and surface chemistry become a challenge.

A summary of the important lessons learned from the above considerations for formulation development and the final formulated Ps-protein vaccines includes the following:

- Sterility assurance, particularly for a multi-dose presentation, requires early strategy discussions to ensure that the production stream can readily accommodate the demands of this type of product.
- Use of adjuvants comes with the requirement to demonstrate that the adjuvant is needed. Such data almost certainly need to come from clinical evaluation of the vaccine antigens with and without adjuvant.
- A proper measure of vaccine potency requires integration of analytical data, including stability data, animal immune response data and clinical data, and requires a strategy for identifying and measuring potential surrogate markers of protection.
- Establishing multiple stability-indicating assays prior to undertaing extended stability studies will ensure early detection of a poor product stability profile.

Glycoconjugate Quality Control

Release and Stability Test Methods

With the above referenced WHO technical reports serving as templates, quality control testing for glycoconjugate vaccines is fairly well defined. Most of the testing required to assure lot to lot acceptability for identity, purity, potency, safety and stability have ample precedent in the regulatory and scientific literature. Three specific measures have proven to be challenging and most predictive of the suitability of glycoconjugates for commercial use, those being unconjugated Ps, particle size distribution and Ps concentration.

Unconjugated Ps

Generally, there is an expectation that glycoconjugates will contain low unconjugated saccharide at release (e.g. <10% if possible) and that this level will be no higher than <25% at the end of shelflife. There are many reasons that a given glycoconjugate may not meet these expectations, perhaps related to the method of purifying the conjugate or the inherent lability of the Ps.

Determination of unconjugated or "free" Ps is regarded as a critical component in the assessment of potency for glycoconjugate vaccines. This parameter is important because, as mentioned above, only Ps that is covalently bound to the carrier protein, i.e. conjugated Ps, is immunologically effective for clinical protection and the presence of large amounts of unconjugated Ps may suppress the immune response to the Ps antigen. Unconjugated Ps, besides being a unreacted residual of the conjugation reaction, may be present due to incomplete purification of the conjugate after conjugation or may be generated by hydrolysis of sugar linkages in the glycoconjugate. The stability of glycoconjugates based on this assay varies depending on the type of conjugate and the nature of glycosidic bonds present in the repeating unit. Cross-linked lattice conjugates have multiple saccharide-protein linkages and therefore a cleavage in the saccharide chain may not release unconjugated saccharide, whereas it would be detectable for neoglycoprotein conjugates that have only a single covalent link between Ps and protein.

Although the exact method may vary, depending on properties of the conjugate, the unconjugated Ps assay requires a separation of the conjugated Ps from the unconjugated Ps followed by quantification of the unconjugated Ps expressed as a percentage of the total Ps present. Typical published specifications are <10% at product release and <25% over the shelf life of the vaccine. As the vaccines tend to degrade by depolymerisation of the glycan chains, which releases unconjugated saccharide into solution, this is an essential stability-indicating assay for glycoconjugate vaccines. Separation methods that have been used include chromatography (size, ion-exchange or hydrophobic interaction), chemical precipitation of the conjugate with acid, detergents or ammonium sulfate or by adsorption to aluminum particles, capillary electrophoresis, gel filtration, centrifugal ultrafiltration, ultracentrifugation, solid phase extraction or immunoprecipitation. The analytical approach to measuring the content of unconjugated saccharide in a glycoconjugate BDS has to be carefully designed, with validation of the assay to include evaluation of glycoconjugate samples spiked with protein carrier as well as samples spiked with activated saccharide to ensure robustness of the assay. Use of forced degradation studies during vaccine development can be used to demonstrate the suitability of the method chosen (*178*). These forced degradation or accelerated stability studies allow for the formulation to be optimized during development and also indicate the expected stability of the conjugate in the drug product.

Conjugate Size Distribution

As discussed above, the size, or more accurately the size distribution, of conjugate particles is a defining feature of the monovalent bulk production process and if not controlled sufficiently can jeapardize quality and consistency of both the monovalent glycoconjugate and the resulting final product. HPSEC with a variety of detectors has generally been the standard method for evaluating the size distribution of conjugates (*66*, *179*), but is only a suitable method if full recovery of the sample can be documented. Lack of complete recovery can be a problem if either some portion of the conjugate particles are too big to enter the solid phase matrix and therefore won't elute or if interactions between the conjugate and the solid phase matrix are not overcome by the mobile phase, resulting in material adsorbing to the matrix and not eluting. Loss of material on the column can be exacerbated by column load, so column load should be balanced with the sensitivity of the detection method. UV absorbance, particularly at 210 nm, can be a sensitive detection for large protein-containing particles, gaining the additive signal from the protein peptide bond absorbance (maximum at ~190 nm) and the light scattering from larger particles. Light scattering detection (single angle or multi-angle) can be sensitive for larger particles, but may require higher concentrations to detect the smaller size conjugates.

Alternatively, static light scattering instruments or particle size analyzers can provide particle size distribution analysis without influence of a physical means of separating the particles (*66*, *180*, *181*). If samples can be shown to be fully recovered after filtration then the artifacts due to dust particles and fibers can be eliminated, but the burden of proof of recovery is similar to that using chromatographic methods.

One obvious limitation of the measurement of particle size distribution comes with the addition of particulate adjuvants such as aluminum phosphate to either the monovalent bulk or to the final product. The presence of such material in the sample obscures the evaluation of conjugate particle size distribution as most of the material of interest is adsorbed to the adjuvant particles. Release of the conjugates from the adjuvant particles, either directly or by dissolution of the adjuvant particles, can compromise the conjugate particles such that measurements may not reflect useful information about the conjugate particle size distribution in the bulk or final product.

Overall, the measurement of particle size distribution, if rigorously evaluated and validated, becomes a strong support for the consistency of manufacture of a glycoconjugate vaccine. This applies to release testing as well as stability testing of monovalent bulks and final products.

Ps Concentration

For DS lots the measurement of Ps concentration comes with a lot of related considerations. First, the final result is typically the number used to determine the dilution of that DS lot during formulation to set the final concentration of Ps in the final formulated product. As such the precision and accuracy of that number has a direct impact on the final product meeting release and stability specifications for potency (see below). Precision can typically be adjusted by increasing the number of sample replicates (separate samples from the batch), the number of test replicates for each sample and/or the number of repeated tests that are done to define the final result. As an example, a single test of two replicate samples will have less precision that three tests of three replicate samples, assuming that inter-assay variability is well controlled.

Accuracy can be a more challenging parameter to control as often the assay result for the DS lot that is the basis for dilution is a chemical assay that is easy to run and for which precision can be adjusted. Dilution bias and lack of primary reference standards often renders an assessment of accuracy to really only be a reflection of relative accuracy. However, in a multivalent product the chemical assay is unlikely to have the specificity to accurately measure the concentration of that Ps. As such a more specific assay, such as an immunological assay is typically used to confirm the content of that Ps in the final formulated product. Though the immunoassay has the needed specificity, the inherent precision of such assays are rarely as good as for a chemical assay, even with increased test runs and sample replicates. Additionally it must be appreciated that measuring Ps content by a chemical assay is inherently different than measuring Ps antigen content by immunoassay. Though there is going to be a relationship between Ps content and Ps antigen content, the accessibility of the Ps epitope(s) that are recognized by the antibody reagents in the immunoassay can be altered by Ps activation and glycoconjugate formation as well as conjugate particle size distribution. The result is often an assay bias between the chemical assay and the immunoassay that can be a challenge to resolve for each Ps in the formulation.

One seemingly obvious solution to this dilemma is to use the immunoassay to measure Ps antigen content in the DS and use that as the basis of dilution during formulation, but the trade-off of eliminating the bias between two assays for the inherently lower precision of the immunoassay versus the chemical assay is by necessity a data-driven decision, specific for each vaccine product and dependent on the style of the manufacturing organization. See comments below related to potency testing of the final vaccine product.

Setting Specifications

Setting specifications for release and stability evaluation of multivalent glycoconjugate vaccines is similar to other vaccine products with one distinct difference: the total number of release tests for all starting components (Ps and carrier protein), intermediates (DSs) and final product can be many fold higher than for most other vaccine products. As such, the setting of specifications must embrace the need to set well-supported specifications that will minimize quality control bottlenecks in the flow to production and release of final vaccine product. By necessity the concept of setting specifications also encompasses the setting of acceptance criteria for in-process testing and, in the later stages of development and the early stages of full-scale manufacturing, on process and product characterization testing as well. This complexity is recognized in the WHO technical reports for Pn and Mn conjugate vaccines as they stress the importance of interaction with national regulatory authorities (NRA) in setting specifications.

By ICH definition a specification is defined as "… a list of tests, references to analytical procedures, and appropriate acceptance criteria, which are numerical limits, ranges, or other criteria for the tests described." (Q6B) Specifications are part of the overall control strategy for vaccine manufacture and are the critical quality standards that are proposed and justified by the manufacturer to the NRA as conditions for approval of the vaccine lots. Specifications are meant to define and are used to confirm the quality of BDS and final product and should focus on assuring the safety and efficacy of the product that will be administered to humans.

Specifications for the Ps starting materials used in glycoconjugate vaccines are generally based on the WHO requirements for Mn (*45*), Pn (*44*), and Vi (*182*) vaccines, though exceptions are allowed when fully justified. The strategy for control of glycoconjugate vaccines relies heavily on molecular characterization of the Ps starting materials, the Ps intermediates and the glycoconjugates because they are manufactured from purified materials using defined processes.

As an example, the WHO recommended specifications for Mn Ps is presented in Table 4. The recommendations for test methodology have evolved over time with introductions of additional methodologies that are either more sensitive or more specific or both. Manufacturers may have slightly different limits for a specific test, in either case, the limits must be agreed to with the governing health authority. In the case of glycoconjugate vaccines, early discussion with NRA is important as specifications for the Ps preparations that are input to the activation/conjugation processes may vary, in particular if the Ps are reduced in size prior to chemical activation.

Table 4. Meningococcal Ps specifications

Attribute	*Test*	*Purpose*	*Specification*
Protein content	Lowryothers	Impurity	NMT 1% by weight
Nucleic Acid	UV-A260	Impurity	NMT 1% by weight
O-Acetyl content	Hestrin NMR HPAEC-CD	Content, identity	Product/process specific
Sialic Acid content	Svennerholm HPAEC-PAD	Content	Grp C NLT 80% Grp Y, W NLT 56%
Phosphorus content	CHEN HPAEC-PAD	Content	Grp A NLT 8%
Identitification	Serological NMR	Identity	Unique Identification
Moisture*	KF TGA	Impurity	Product/process specific
Endotoxin	Gel clot Chromogenic	Impurity Safety	Product/process specific
Molecular size	Gel Filtration HPSEC HPSEC-MALLS	Manufacturing consistency	Product/process specific

* If lyophilized Ps is to be stored prior to further conjugate manufacture.

Specifications for other Ps (e.g. Pn, St Vi, GBS) may also start with historical specifications, but will be updated to reflect the use of new analytical technology and relevant production technologies specific to the glycoconjugate being made.

Ps intermediates between the native Ps and the monovalent glycoconjugate will likewise require their own specification. A theoretical specification for such intermediates is presented in Table 5.

Table 5. Theoretical Ps intermediate specifications*

Attribute	*Test*	*Purpose*	*Specification*
Depolymerized			
Sialic Acid content	HPAEC-PAD	Content	NLT 10 mg/L NMT 30 mg/L
Molecular size	HPSEC-MALLS	Consistency Efficacy	NLT 20000 Da NMT 50000 Da
Bioburden	Microbial Enumeration	Process consistency Safety	NMT 8CFU/10 mL
Derivatized			
Sialic Acid content	HPAEC-PAD	Conjugation kinetics	Calculation for conjugation load
Free hydrazide	HPSEC-FD	Impurity	NMT 8%
Total hydrazide	HPSEC-FD	Conjugation kinetics	NLT 2000 nmol/mL NMT 5000 nmol/mL
Hydrazide incorporation	HPSEC-FD	Conjugation kinetics Efficacy	
Cyanide	Colorimetry	Safety	NMT 5 ppm
Bioburden	Microbial Enumeration	Process consistency Safety	NMT 4 CFU/10mL

* Based on a process whereby a two step depolymerization followed by derivitization of the Ps is used to create the glycoconjugate using an adipic dihydrazide linker and reductive amination using sodium cyanoborohydride with MnC Ps.

The monovalent DS glycoconjugate using the activated Ps intermediate described above would require a testing specification that covered its CQAs and many of the tests above would be repeated to form that specification. The specification would need additional tests for any conjugation residual impurities (*e.g.* additional conjugation chemicals, free carrier protein), residual processing aids and endotoxin. Additional tests for evaluating conjugation efficiency should be included to ensure the type of conjugate desired was made, for example assessment of unused linker sites or ratio of Ps to protein.

The setting of specifications for glyconjugate DS, Ps intermediates and final product requires understanding of process capability and analytical method capability and evaluation of data generated with a valid method using samples representative of the final production process. Typically initial vaccine specifications are steered by the target product profile but are based on developer experience with the product during development and data from limited process design, scale up and engineering lot production. The CMC Vaccines Working Group published "A Case Study in Using Quality by Design in Vaccines Manufacturing", the so called AVax case study (*183*). The AVax case study offers guidance and examples for designing a glycoconjugate vaccine though perhaps little consideration of the challenges encountered in early product development.

Critical Process Parameters and Critical Quality Attributes

A natural outfall of setting specifications and acceptance criteria during the latter stages of product development, particularly prior to transfer and validation of the production process to the commercial manufacturing facility, is the need to define critical process parameters (CPPs) and critical quality attributes (CQAs). In this regard, glycoconjugate vaccines are no different than any other product. However, given the specific sensitivities (lessons learned) regarding Ps activation and formation of glycoconjugates, selected steps in these processes certainly warrant the designation of "critical" as they can be directly related to the effectiveness of the vaccine. Similarly, the Ps quality attributes of greatest concern are Ps identity (confirming structure) and molecular size distribution. Activated Ps quality attributes include extent of activation and molecular size distribution. DS quality attributes of greatest concern are Ps to protein ratio, unconjugated Ps and conjugate particle size distribution. Other process parameters and quality attributes (e.g. DS Ps loading density on the carrier protein) can warrant the designation of "critical", but those may be more from a process consistency perspective as their tie-in to the effectiveness of the vaccine may be more speculative.

Quality by Design

Quality by Design (Qbd) is defined in ICH Q8 (R2) as "a systematic approach to development that begins with predefined objectives and emphasizes product and process understanding and process control, based on sound science and quality risk management in order to ensure the quality of the product".

This whole chapter discusses the lessons learned from thirty plus years of glycoconjugate vaccine development, so the foundation of knowledge to inform QbD for glycoconjugate vaccines is certainly broad and deep.

The application of QbD principles in designing the manufacturing process allows the possibility of developing robust specifications for the manufacture of glycoconjugates. The same QbD principles can be applied to development of the test methods used in testing glycoconjugates and their intermediates (see (*184*). Using the information and outcome of a QbD process and a QbD method analysis, realistic and justifiable specifications can be developed for glycoconjugate vaccines.

Logically, QbD will focus on the CPPs and CQAs discussed above, the CQAs being a subset of the parameters tested as part of monovalent bulk release. It is important to recognize that the specifications applied to release testing are designed to reflect the realities of a product that has already been made. From a QbD perspective, intentional adjustments in a well-defined process should have known impact on the analytical profile of the product such that adjustments of target process ranges can lead to anticipated adjustments to specification ranges.

As an example, two technologies in use for making glycoconjugates allow design of the sites and number of sites per protein molecule that are intended for linkage to the Ps, those being the carrier protein engineering inherent to the Sutrovax process for introducing non-natural amino acids into the sequence of the recombinant carrier protein to use as linkage points for the Ps, and the recombinant protein design capabilities available in bioconjugation (e.g. Limmatech) that can influence if not define the sites and number of linkage points per protein molecule for Ps.

This leads to the question, "If you can design the architecture of your glycoconjugate, what would you design?". This question comes with the "design" questions centered on our history of development, e.g. how much can we impact Ps to protein ratio? How will this impact vaccine potency? How will this impact process yield and cost of goods? Should we design conjugates that incorporate more than one Ps on a protein molecule? And a host of other questions lie in wait as greater detail comes into focus. The challenge (in industry) will be highly influenced by the cost-time risk assessment, i.e. does use of innovative approaches translate to a superior product? Will the likely increased development time for the innovative product and predicted market share create a better net present value than would be achieved with a follow-on product?

Even as we back off into more familiar territory of using "standard" chemical conjugation there are questions that we know are theoretically relevant, e.g. Does the site of conjugation on the carrier protein impact immunogenicity? What is the optimal Ps loading density with regard to Ps chain length? Which is the "best" conjugation strategy (e.g. random or end group Ps activation) for the program? Which carrier should be used for a particular project? If the carrier protein has to work also as protective antigen which regions should be targeted for conjugation in order to preserve key epitopes? These questions apply not only to conjugate design and practical manufacturability considerations but also to safety, immunogenicity and efficacy considerations.

Questions such as these can appear to make a glycoconjugate development program impossibly complex and never ending. Ideally the optimal specification setting process for a glycoconjugate vaccine should start from understanding the impact of physicochemical parameter variations on safety and efficacy. However, as a practical matter, and in light of the wealth of lessons learned, some of the key variables of concern for a given program can be defined up front and, with increasing product knowledge, re-considered later in the context of life cycle management.

In the context of QbD, epitope mapping on both polysaccharide and carrier protein will be crucial for the design of glycoconjugate vaccines and the strategy for their manufacture. As an example in a recent work on GBS type III, saturation transfer NMR and X-Ray crystallography were used for the exact definition of the epitope mapped by a functional monoclonal antibody, ruling out its conformational nature as opposed to what was previously proposed (*94*).

A summary of the important lessons learned from the above considerations for the quality control of monovalent BDS components for Ps-protein vaccines includes the following:

- Though following the relevant WHO technical reports will provide good guidance on establishing quality control for glycoconjugate products, the measurements of Ps to protein ratio, unconjugated saccharide, particle size distribution and potency for monovalent bulks and, as possible, final products will provide the most relevant measures for characterizing the quality and consistency of glycoconjugates.
- Given the complexity of producing a multivalent glycoconjugate vaccine, it is essential to systematically identify and evaluate the key design variables of concern for making a glycoconjugate. Decide which will be evaluated by experimentation and which will be resolved by selection of precedent historical data. Record this process so that it remains clear where data versus assumptions drove the decision process in the event some of the assumptions need to be revisited.
- Characterization of monovalent BDS lots should include current (new) technologies as appropriate, along with established methods, to provide a comprehensive view of the characteristics of each BDS.
- Measurement of Ps content in BDS lots has complex implications for production and quality control of the final vaccine product and must be approached with an integrated process and analytical strategy.
- Establishment of specifications for release of intermediates and final product, for re-evaluation dating of clinical supplies and for expiry dating of commercial product require an extensive array of data that spans several independently made lots and rigorous attention to the version of the analytical method being used to generate the data. Specifications should also evolve as product knowledge and manufacturing and clinical experience accumulate. This needs a cross-disciplinary approach and should be a project team responsibility, not just an analytical responsibility.

Final Thoughts

A final caution on proceeding with development of glycoconjugate vaccines is in order. Defining an approach to making a glycoconjugate vaccine forces the choice of how to make selections for many variables (e.g. Ps size, conjugation chemistry, etc.). All possible variations do not necessarily lead to a product that "works", meaning you could produce a glycoconjugate that meets all of the physico-chemical and manufacturing expectations that you have, but that conjugate may prove to not be suitable for use in humans for a variety of reasons ranging from the immune response in preclinical models, inconsistent manufacture, poor stability, inadequate safety or inferior immune response in humans. As examples, one early Hib conjugate proved to be quite unstable, resulting in often poor and inconsistent immune response and removal of that product from the market. A redeveloped version of that product fixed the stability issue, resulting in a commercially viable vaccine.

Glycoconjugates have proven to be a very effective platform for vaccines against bacterial diseases. These are complex, semi-synthetic entities that are manufactured to meet ever-rising expectations for their molecular profile and their critical quality attributes. With over 30 years of experience over many different vaccines, there are valuable lessons to guide more efficient and insightful development of future glycoconjugate vaccines, addressing unmet medical needs with very potent tools.

Acknowledgments

We would like to thank Peter Dull of the Bill & Melinda Gates Foundation for instigating the discussion of "lessons learned" for Ps-protein conjugate vaccines and Francis Michon of Clearpath for his insightful review of this article.

References

1. Blake, F. G. *J. Exp. Med.* **1917**, *26*, 67–80.
2. Wright, A.; Parry Morgan, W.; Colebrook, L.; Dodgson, R. W. *Lancet* **1914**, *183*, 1–10.
3. Avery, O. T.; Goebel, W. F. *J. Exp. Med.* **1931**, *54*, 437–447.
4. Cryz, S. J., Jr.; Que, J. O.; Cross, A. S.; Furer, E. *Vaccine* **1995**, *13*, 449–453.
5. Fattom, A.; Schneerson, R.; Watson, D. C.; Karakawa, W. W.; Fitzgerald, D.; Pastan, I.; Li, X.; Shiloach, J.; Bryla, D. A.; Robbins, J. B. *Infect. Immun.* **1993**, *61*, 1023–1032.
6. van der Put, R. M.; Kim, T. H.; Guerreiro, C.; Thouron, F.; Hoogerhout, P.; Sansonetti, P. J.; Westdijk, J.; Stork, M.; Phalipon, A.; Mulard, L. A. *Bioconjug. Chem.* **2016**, *27*, 883–892.
7. Torosantucci, A.; Bromuro, C.; Chiani, P.; De Bernardis, F.; Berti, F.; Galli, C.; Norelli, F.; Bellucci, C.; Polonelli, L.; Costantino, P.; Rappuoli, R.; Cassone, A. *J. Exp. Med.* **2005**, *202*, 597–606.
8. Burton, R. L.; Geno, K. A.; Saad, J. S.; Nahm, M. H. *J. Clin. Microbiol.* **2016**, *54*, 967–971.

9. Brugger, S. D.; Troxler, L. J.; Rufenacht, S.; Frey, P. M.; Morand, B.; Geyer, R.; Muhlemann, K.; Hock, S.; Thormann, W.; Furrer, J.; Christen, S.; Hilty, M. *Infect. Immun.* **2016**, *84*, 3152–3160.
10. Geno, K. A.; Gilbert, G. L.; Song, J. Y.; Skovsted, I. C.; Klugman, K. P.; Jones, C.; Konradsen, H. B.; Nahm, M. H. *Clin. Microbiol. Rev.* **2015**, *28*, 871–899.
11. Yother, J., Bentley, S. D., Hennessey, J. P., Jr. In *Pneumococcal Vaccines*; Siber, G. R.; Klugman, K. P.; Makela, P. H., Eds.; American Society of Microbiology: Washington, DC, 2008; pp 33−46.
12. Abeygunawardana, C.; Williams, T. C.; Sumner, J. S.; Hennessey, J. P., Jr. *Anal. Biochem.* **2000**, *279*, 226–240.
13. Ip, C. C.; Manam, V.; Hepler, R.; Hennessey, J. P., Jr. *Anal. Biochem.* **1992**, *201*, 343–349.
14. Talaga, P.; Vialle, S.; Moreau, M. *Vaccine* **2002**, *20*, 2474–2484.
15. Kim, J. S.; Laskowich, E. R.; Arumugham, R. G.; Kaiser, R. E.; MacMichael, G. J. *Anal. Biochem.* **2005**, *347*, 262–274.
16. Kao, G.; Tsai, C. M. *Vaccine* **2004**, *22*, 335–344.
17. Jones, C., Ravenscroft, N. In *NMR Spectroscopy in Pharmaceutical Analysis*; Holtzgrabe, U.; Wawer, I.; Diehl, B., Eds.; Elsevier: Oxford, U.K., 2008; pp 341−368.
18. Wishart, D. S. *TrAC, Trends Anal. Chem.* **2013**, *48*, 96–111.
19. Berti, F., Ravenscroft, N. In *Carbohydrate-Based Vaccines*, Humana Press: New York, 2015; pp 189−209.
20. Jones, C., Mulloy, B. In *Spectroscopic Methods and Analyses*; Jones, C.; Mulloy, B.; Thomas, A. H., Eds.; Humana Press: Totowa, NJ, 1993; pp 149−167.
21. Lemercinier, X.; Jones, C. *Biologicals* **2000**, *28*, 175–183.
22. Jones, C.; Lemercinier, X. *J. Pharm. Biomed. Anal.* **2002**, *30*, 1233–1247.
23. Lemercinier, X.; Martinez-Cabrera, I.; Jones, C. *Biologicals* **2000**, *28*, 17–24.
24. Pinto, V.; Berti, F. *J. Pharm. Biomed. Anal.* **2014**, *98*, 9–15.
25. Jones, C.; Currie, F. *Carbohydr. Res.* **1988**, *184*, 279–284.
26. Jones, C.; Lemercinier, X. *Carbohydr. Res.* **2005**, *340*, 403–409.
27. Lin, F. L.; Vinogradov, E.; Deng, C.; Zeller, S.; Phelan, L.; Green, B. A.; Jansen, K. U.; Pavliak, V. *Carbohydr. Res.* **2014**, *383*, 97–104.
28. Jones, C. *Carbohydr. Res.* **2005**, *340*, 1097–1106.
29. Wessels, M. R.; Pozsgay, V.; Kasper, D. L.; Jennings, H. J. *J. Biol. Chem.* **1987**, *262*, 8262–8267.
30. Park, I. H.; Pritchard, D. G.; Cartee, R.; Brandao, A.; Brandileone, M. C.; Nahm, M. H. *J. Clin. Microbiol.* **2007**, *45*, 1225–1233.
31. Cai, P.; Moran, J.; Pavliak, V.; Deng, C.; Khoury, N.; Marcq, O.; Ruppen, M. E. *Carbohydr. Res.* **2012**, *351*, 98–107.
32. Vipond, C.; Swann, C. J.; Dougall, T. W.; Rigsby, P.; Gao, F.; Beresford, N. J.; Bolgiano, B. *Biologicals* **2017**, *47*, 33–45.
33. Li, W. *Analyst* **2006**, *131*, 777–781.
34. Zandomeneghi, G.; Ilg, K.; Aebi, M.; Meier, B. H. *J. Am. Chem. Soc.* **2012**, *134*, 17513–17519.

35. Gudlavalleti, S. K.; Szymanski, C. M.; Jarrell, H. C.; Stephens, D. S. *Carbohydr. Res.* **2006**, *341*, 557–562.
36. Ravenscroft, N.; Cescutti, P.; Gavini, M.; Stefanetti, G.; MacLennan, C. A.; Martin, L. B.; Micoli, F. *Carbohydr. Res.* **2015**, *404*, 108–116.
37. Adamo, R.; Romano, M. R.; Berti, F.; Leuzzi, R.; Tontini, M.; Danieli, E.; Cappelletti, E.; Cakici, O. S.; Swennen, E.; Pinto, V.; Brogioni, B.; Proietti, D.; Galeotti, C. L.; Lay, L.; Monteiro, M. A.; Scarselli, M.; Costantino, P. *ACS Chem. Biol.* **2012**, *7*, 1420–1428.
38. Reid, C. W.; Vinogradov, E.; Li, J.; Jarrell, H. C.; Logan, S. M.; Brisson, J. R. *Carbohydr. Res.* **2012**, *354*, 65–73.
39. Lewis, A. L.; Nizet, V.; Varki, A. *Proc. Natl. Acad. Sci. U.S.A.* **2004**, *101*, 11123–11128.
40. Pannaraj, P. S.; Edwards, M. S.; Ewing, K. T.; Lewis, A. L.; Rench, M. A.; Baker, C. J. *Vaccine* **2009**, *27*, 4452–4456.
41. Degeest, B.; De Vuyst, L. *Appl. Environ. Microbiol.* **1999**, *65*, 2863–2870.
42. McNeely, T. B.; Staub, J. M.; Rusk, C. M.; Blum, M. J.; Donnelly, J. J. *Infect. Immun.* **1998**, *66*, 3705–3710.
43. Michon, F.; Huang, C. H.; Farley, E. K.; Hronowski, L.; Di, J.; Fusco, P. C. *Dev. Biol.* **2000**, *103*, 151–160.
44. *Recommendations to Assure the Quality, Safety and Efficacy of Pneumococcal Conjugate Vaccines*; No. 977, Annex 5; World Health Organization: Geneva, Switzerland, 2013.
45. *Requirements for Meningococcal Polysaccharide Vaccine*; No. 594, Annex 2; World Health Organization: Geneva, Switzerland, 1975.
46. Frasch, C. E. *Adv. Biotechnol. Processes* **1990**, *13*, 123–145.
47. *Vaccines for Human Use – Polysaccharide and Glycoconjugate Vaccines*; U.S. Pharmacopeia and the National Formulary: Rockville, MD, 2015; p 38.
48. Gairola, S.; Gautam, M.; Patil, D.; Manoj Kumar, K.; Shinde, P.; Jana, S. K.; Dhere, R.; Jadhav, S. *Biologicals* **2016**, *44*, 517–525.
49. Li, Y.; Lander, R.; Manger, W.; Lee, A. *J. Chromatogr. B: Anal. Technol. Biomed. Life Sci.* **2004**, *804*, 353–358.
50. Maitra, S. K.; Nachum, R.; Pearson, F. C. *Appl. Environ. Microbiol.* **1986**, *52*, 510–514.
51. Yother, J. *Annu. Rev. Microbiol.* **2011**, *65*, 563–581.
52. Beno, D. W.; Mathews, H. L. *Infect. Immun.* **1992**, *60*, 853–863.
53. Xu, Q.; Abeygunawardana, C.; Ng, A. S.; Sturgess, A. W.; Harmon, B. J.; Hennessey, J. P., Jr. *Anal. Biochem.* **2005**, *336*, 262–272.
54. Talaga, P.; Bellamy, L.; Moreau, M. *Vaccine* **2001**, *19*, 2987–2994.
55. Ricci, S.; Fabbri, B.; Giannini, S.; Bernardini, S.; Marcello, M. A.; D'Ascenzi, S. *Appl. Ion Chromatogr. Pharm. Biol. Prod.* **2012**, 423–434.
56. Verez-Bencomo, V.; Fernandez-Santana, V.; Hardy, E.; Toledo, M. E.; Rodriguez, M. C.; Heynngnezz, L.; Rodriguez, A.; Baly, A.; Herrera, L.; Izquierdo, M.; Villar, A.; Valdes, Y.; Cosme, K.; Deler, M. L.; Montane, M.; Garcia, E.; Ramos, A.; Aguilar, A.; Medina, E.; Torano, G.; Sosa, I.; Hernandez, I.; Martinez, R.; Muzachio, A.; Carmenates, A.; Costa, L.; Cardoso, F.; Campa, C.; Diaz, M.; Roy, R. *Science* **2004**, *305*, 522–525.

57. Gao, Q.; Tontini, M.; Brogioni, G.; Nilo, A.; Filippini, S.; Harfouche, C.; Polito, L.; Romano, M. R.; Costantino, P.; Berti, F.; Adamo, R.; Lay, L. *ACS Chem. Biol.* **2013**, *8*, 2561–2567.
58. Zon, G.; Szu, S. C.; Egan, W.; Robbins, J. D.; Robbins, J. B. *Infect. Immun.* **1982**, *37*, 89–103.
59. Sweeney, J. A.; Sumner, J. S.; Hennessey, J. P., Jr. *Dev. Biol. (Basel)* **2000**, *103*, 11–26.
60. Wong, K. H.; Barrera, O.; Sutton, A.; May, J.; Hochstein, D. H.; Robbins, J. D.; Robbins, J. B.; Parkman, P. D.; Seligmann, E. B., Jr. *J. Biol. Stand.* **1977**, *5*, 197–215.
61. Tiesjema, R. H.; Beuvery, E. C.; te Pas, B. J. *Bull. World Health Organ.* **1977**, *55*, 43–48.
62. Egan, W.; Schneerson, R.; Werner, K. E.; Zon, G. *J. Am. Chem. Soc.* **1982**, *104*, 2898–2910.
63. Sturgess, A. W.; Rush, K.; Charbonneau, R. J.; Lee, J. I.; West, D. J.; Sitrin, R. D.; Hennessey, J. P., Jr. *Vaccine* **1999**, *17*, 1169–1178.
64. Bednar, B.; Hennessey, J. P., Jr. *Carbohydr. Res.* **1993**, *243*, 115–130.
65. MacNair, J. E.; Desai, T.; Teyral, J.; Abeygunawardana, C.; Hennessey, J. P., Jr. *Biologicals* **2005**, *33*, 49–58.
66. Thiebaud, J.; Fanget, I.; Jaudinaud, I.; Fourrichon, L.; Sabouraud, A.; Talaga, P.; Uhlrich, S. *Anal. Biochem.* **2014**, *453*, 22–28.
67. Hadidi, M.; Buckley, J. J.; Zydney, A. L. *Carbohydr. Polym.* **2016**, *152*, 12–18.
68. Harding, S. E.; Abdelhameed, A. S.; Morris, G. A.; Adams, G.; Laloux, O.; Cerny, L.; Bonnier, B.; Duvivier, P.; Conrath, K.; Lenfant, C. *Carbohydr. Polym.* **2012**, *90*, 237–242.
69. Kuttel, M. M., Ravenscroft, N. The Role of Molecular Modeling in Predicting Carbohydrate Antigen Conformation and Understanding Vaccine Immunogenicity. In *Carbohydrate Vaccines: From Concept to Clinic*; Prasad, A. K., Ed.; ACS Symposium Series 1290; American Chemical Society: Washington, DC, 2018; Chapter 7.
70. Avery, O. T.; Goebel, W. F. *J. Exp. Med.* **1929**, *50*, 533–550.
71. Goebel, W. F.; Avery, O. T. *J. Exp. Med.* **1931**, *54*, 431–436.
72. Pobre, K.; Tashani, M.; Ridda, I.; Rashid, H.; Wong, M.; Booy, R. *Vaccine* **2014**, *32*, 1423–1430.
73. Pollard, A. J.; Perrett, K. P.; Beverley, P. C. *Nat. Rev. Immunol.* **2009**, *9*, 213–220.
74. Kelly, D. F.; Pollard, A. J.; Moxon, E. R. *JAMA, J. Am. Med. Assoc.* **2005**, *294*, 3019–3023.
75. Szu, S. C.; Li, X. R.; Schneerson, R.; Vickers, J. H.; Bryla, D.; Robbins, J. B. *Infect. Immun.* **1989**, *57*, 3823–3827.
76. Shinefield, H.; Black, S.; Fattom, A.; Horwith, G.; Rasgon, S.; Ordonez, J.; Yeoh, H.; Law, D.; Robbins, J. B.; Schneerson, R.; Muenz, L.; Fuller, S.; Johnson, J.; Fireman, B.; Alcorn, H.; Naso, R. *N. Engl. J. Med.* **2002**, *346*, 491–496.
77. Pinto, V. B.; Burden, R.; Wagner, A.; Moran, E. E.; Lee, C. H. *PLoS One* **2013**, *8*, e79304.

78. Luo, Y.; Friese, O. V.; Runnels, H. A.; Khandke, L.; Zlotnick, G.; Aulabaugh, A.; Gore, T.; Vidunas, E.; Raso, S. W.; Novikova, E.; Byrne, E.; Schlittler, M.; Stano, D.; Dufield, R. L.; Kumar, S.; Anderson, A. S.; Jansen, K. U.; Rouse, J. C. *AAPS J.* **2016**, *18*, 1562–1575.
79. Hickey, J. M.; Torani, V. M.; Kaur, K.; Mishra, R. P. N.; Goel, A.; Oganesyan, N.; Lees, A.; Sitrin, R.; Joshi, B.; Volkin, D. B. *J. Pharm. Sci.* **2018**in press.
80. Mendelman, P. M.; Feeley, L.; Bird, S.; Staub, T.; Matthews, H.; Del Beccaro, M. A.; Overturf, G.; Lee, A.; Ellis, R.; Staub, J.; Szymanski, S.; Donnelly, J.; Hennessey, J. P.; Kniskern, P. *Vaccine* **1997**, *15*, 775–781.
81. Kelly, D. F.; Moxon, E. R.; Pollard, A. J. *Immunology* **2004**, *113*, 163–174.
82. Lee, A. W.; Vesikari, T.; Gilbert, C. L.; Klopfer, S. O.; Schodel, F. P.; Bhuyan, P. K. *Vaccine* **2011**, *29*, 7942–7948.
83. Ginsburg, A. S.; Alderson, M. R. *Drugs Today* **2011**, *47*, 207–214.
84. Pozzi, C.; Wilk, K.; Lee, J. C.; Gening, M.; Nifantiev, N.; Pier, G. B. *PLoS One* **2012**, *7*, e46648.
85. Adamo, R.; Nilo, A.; Harfouche, C.; Brogioni, B.; Pecetta, S.; Brogioni, G.; Balducci, E.; Pinto, V.; Filippini, S.; Mori, E.; Tontini, M.; Romano, M. R.; Costantino, P.; Berti, F. *Glycoconjate J.* **2014**, *31*, 637–647.
86. Wacker, M.; Wang, L.; Kowarik, M.; Dowd, M.; Lipowsky, G.; Faridmoayer, A.; Shields, K.; Park, S.; Alaimo, C.; Kelley, K. A.; Braun, M.; Quebatte, J.; Gambillara, V.; Carranza, P.; Steffen, M.; Lee, J. C. *J. Infect. Dis.* **2014**, *209*, 1551–1561.
87. Nilo, A.; Morelli, L.; Passalacqua, I.; Brogioni, B.; Allan, M.; Carboni, F.; Pezzicoli, A.; Zerbini, F.; Maione, D.; Fabbrini, M.; Romano, M. R.; Hu, Q. Y.; Margarit, I.; Berti, F.; Adamo, R. *ACS Chem. Biol.* **2015**, *10*, 1737–1746.
88. Broker, M.; Berti, F.; Schneider, J.; Vojtek, I. *Vaccine* **2017**, *35*, 3286–3294.
89. Prymula, R.; Schuerman, L. *Expert Rev. Vaccines* **2009**, *8*, 1479–1500.
90. Tontini, M.; Berti, F.; Romano, M. R.; Proietti, D.; Zambonelli, C.; Bottomley, M. J.; De Gregorio, E.; Del Giudice, G.; Rappuoli, R.; Costantino, P.; Brogioni, G.; Balocchi, C.; Biancucci, M.; Malito, E. *Vaccine* **2013**, *31*, 4827–4833.
91. Lockyer, K.; Gao, F.; Derrick, J. P.; Bolgiano, B. *Vaccine* **2015**, *33*, 1345–1352.
92. Borrow, R.; Tang, Y.; Yakubu, A.; Kulkarni, P. S.; LaForce, F. M. *Clin. Infect. Dis.* **2015**, *61 Suppl 5*, S570–577.
93. Schmidt, C. S.; White, C. J.; Ibrahim, A. S.; Filler, S. G.; Fu, Y.; Yeaman, M. R.; Edwards, J. E., Jr; Hennessey, J. P., Jr *Vaccine* **2012**, *30*, 7594–7560.
94. Carboni, F.; Adamo, R.; Fabbrini, M.; De Ricco, R.; Cattaneo, V.; Brogioni, B.; Veggi, D.; Pinto, V.; Passalacqua, I.; Oldrini, D.; Rappuoli, R.; Malito, E.; Margarit, I. Y. R.; Berti, F. *Proc. Natl. Acad. Sci. U.S.A.* **2017**, *114*, 5017–5022.
95. Schutze, M. P.; Deriaud, E.; Przewlocki, G.; LeClerc, C. *J. Immunol.* **1989**, *142*, 2635–2640.
96. Findlow, H.; Borrow, R. *Hum. Vaccines Immunother.* **2016**, *12*, 226–230.
97. Dagan, R.; Poolman, J.; Siegrist, C. A. *Vaccine* **2010**, *28*, 5513–5523.

98. Borrow, R.; Dagan, R.; Zepp, F.; Hallander, H.; Poolman, J. *Expert Rev. Vaccines* **2011**, *10*, 1621–1631.
99. Dagan, R.; Eskola, J.; Leclerc, C.; Leroy, O. *Infect. Immun.* **1998**, *66*, 2093–2098.
100. Dagan, R.; Goldblatt, D.; Maleckar, J. R.; Yaich, M.; Eskola, J. *Infect. Immun.* **2004**, *72*, 5383–5391.
101. Park, D. E.; Johnson, T. S.; Nonyane, B. A.; Chandir, S.; Conklin, L.; Fleming-Dutra, K. E.; Loo, J. D.; Goldblatt, D.; Whitney, C. G.; O'Brien, K. L.; Deloria Knoll, M. *Pediatr. Infect. Dis. J.* **2014**, *33 Suppl 2*, S130–139.
102. Pecetta, S.; Lo Surdo, P.; Tontini, M.; Proietti, D.; Zambonelli, C.; Bottomley, M. J.; Biagini, M.; Berti, F.; Costantino, P.; Romano, M. R. *Vaccine* **2015**, *33*, 314–320.
103. Pecetta, S.; Tontini, M.; Faenzi, E.; Cioncada, R.; Proietti, D.; Seubert, A.; Nuti, S.; Berti, F.; Romano, M. R. *Vaccine* **2016**, *34*, 2334–2341.
104. Peeters, C. C.; Tenbergen-Meekes, A. M.; Poolman, J. T.; Beurret, M.; Zegers, B. J.; Rijkers, G. T. *Infect. Immun.* **1991**, *59*, 3504–3510.
105. Usonis, V.; Bakasenas, V.; Lockhart, S.; Baker, S.; Gruber, W.; Laudat, F. *Vaccine* **2008**, *26*, 4602–4607.
106. LaForce, M. Development Update on a New African Pentavalent ACYWX Conjugate Vaccine. http://www.who.int/immunization/research/meetings_workshops/Kulkarni_LaForce_MVPconf16.pdf?ua=1 (accessed February 1, 2018).
107. Brady, C.; Killeen, K. P.; Taylor, W.; Patkar, A.; Lees, A. *BioProcess Int.* **2012**, *10*, 50–55.
108. Stefan, A.; Conti, M.; Rubboli, D.; Ravagli, L.; Presta, E.; Hochkoeppler, A. *J. Biotechnol.* **2011**, *156*, 245–252.
109. Mahamad, P.; Boonchird, C.; Panbangred, W. *Appl. Microbiol. Biotechnol.* **2016**, *100*, 6319–6330.
110. Goffin, P.; Dewerchin, M.; De Rop, P.; Blais, N.; Dehottay, P. *Biotechnol. J.* **2017**, *12*.
111. Patkar, A. *Genet. Eng. Biotechnol. News* **2012**, *32*, 56–57.
112. Ho, M. M.; Bolgiano, B.; Corbel, M. J. *Vaccine* **2000**, *19*, 716–725.
113. Ho, M. M.; Mawas, F.; Bolgiano, B.; Lemercinier, X.; Crane, D. T.; Huskisson, R.; Corbel, M. J. *Vaccine* **2002**, *20*, 3509–3522.
114. Prasad, A. K., Kim, J.-h., Gu, J. Design and Development of Glycoconjugate Vaccines. In *Carbohydrate Vaccines: From Concept to Clinic*; Prasad, A. K., Ed.; ACS Symposium Series 1290; American Chemical Society: Washington, DC, 2018; Chapter 4.
115. Anderson, P. W.; Pichichero, M. E.; Insel, R. A.; Betts, R.; Eby, R.; Smith, D. H. *J. Immunol.* **1986**, *137*, 1181–1186.
116. Shafer, D. E.; Toll, B.; Schuman, R. F.; Nelson, B. L.; Mond, J. J.; Lees, A. *Vaccine* **2000**, *18*, 1273–1281.
117. Biemans, R. L., Garcon, N. M. J., Hermand, P. V., Poolman, J., Van Mechelen, M. P. U.S. Patent 9,107,872 B2, 2015.
118. Schneerson, R.; Barrera, O.; Sutton, A.; Robbins, J. B. *J. Exp. Med.* **1980**, *152*, 361–376.

119. Chu, C.; Schneerson, R.; Robbins, J. B.; Rastogi, S. C. *Infect. Immun.* **1983**, *40*, 245–256.
120. Vella, P. P.; Staub, J. M.; Armstrong, J.; Dolan, K. T.; Rusk, C. M.; Szymanski, S.; Greer, W. E.; Marburg, S.; Kniskern, P. J.; Schofield, T. L.; Tolman, R. L.; Hartner, R.; Pan, S.; Gerety, R. J.; Ellis, R. W. *Pediatrics* **1990**, *85*, 668–675.
121. Lee, C. H.; Kuo, W. C.; Beri, S.; Kapre, S.; Joshi, J. S.; Bouveret, N.; LaForce, F. M.; Frasch, C. E. *Vaccine* **2009**, *27*, 726–732.
122. Kabanova, A.; Margarit, I.; Berti, F.; Romano, M. R.; Grandi, G.; Bensi, G.; Chiarot, E.; Proietti, D.; Swennen, E.; Cappelletti, E.; Fontani, P.; Casini, D.; Adamo, R.; Pinto, V.; Skibinski, D.; Capo, S.; Buffi, G.; Gallotta, M.; Christ, W. J.; Campbell, A. S.; Pena, J.; Seeberger, P. H.; Rappuoli, R.; Costantino, P. *Vaccine* **2010**, *29*, 104–114.
123. Costantino, P. European Patent 1401489B2, 2001.
124. Costantino, P.; Rappuoli, R.; Berti, F. *Expert Opin. Drug Discovery* **2011**, *6*, 1045–1066.
125. Ihssen, J.; Kowarik, M.; Dilettoso, S.; Tanner, C.; Wacker, M.; Thony-Meyer, L. *Microb. Cell Fact.* **2010**, *9*, 61.
126. Turner, A. E. B.; Gerson, J. E.; So, H. Y.; Krasznai, D. J.; St Hilaire, A. J.; Gerson, D. F. *Synth. Syst. Biotechnol.* **2017**, *2*, 49–58.
127. Zawada, J. F.; Yin, G.; Steiner, A. R.; Yang, J.; Naresh, A.; Roy, S. M.; Gold, D. S.; Heinsohn, H. G.; Murray, C. J. *Biotechnol. Bioeng.* **2011**, *108*, 1570–1578.
128. Kapoor, N.; Vanjak, I.; Rozzelle, J.; Berges, A.; Chan, W.; Yin, G.; Tran, C.; Sato, A. K.; Steiner, A. R.; Pham, T. P.; Birkett, A. J.; Long, C. A.; Fairman, J.; Miura, K. *Biochemistry (Moscow)* **2018**, *57*, 516–519.
129. Pozsgay, V. *Curr. Trends Med. Chem.* **2008**, *8*, 126–140.
130. Zhang, F.; Lu, Y. J.; Malley, R. *Proc. Natl. Acad. Sci. U.S.A.* **2013**, *110*, 13564–13569.
131. Thanawastien, A.; Cartee, R. T.; Griffin, T. J. t.; Killeen, K. P.; Mekalanos, J. J. *Proc. Natl. Acad. Sci. U.S.A.* **2015**, *112*, E1143–1151.
132. Jones, C. H.; Zhang, G.; Nayerhoda, R.; Beitelshees, M.; Hill, A.; Rostami, P.; Li, Y.; Davidson, B. A.; Knight, P., 3rd; Pfeifer, B. A. *Sci. Adv.* **2017**, *3*, e1701797.
133. Broker, M.; Dull, P. M.; Rappuoli, R.; Costantino, P. *Vaccine* **2009**, *27*, 5574–5580.
134. Broker, M.; Costantino, P.; DeTora, L.; McIntosh, E. D.; Rappuoli, R. *Biologicals* **2011**, *39*, 195–204.
135. Beurret, M.; Hamidi, A.; Kreeftenberg, H. *Vaccine* **2012**, *30*, 4897–4906.
136. Costantino, P.; Norelli, F.; Giannozzi, A.; D'Ascenzi, S.; Bartoloni, A.; Kaur, S.; Tang, D.; Seid, R.; Viti, S.; Paffetti, R.; Bigio, M.; Pennatini, C.; Averani, G.; Guarnieri, V.; Gallo, E.; Ravenscroft, N.; Lazzeroni, C.; Rappuoli, R.; Ceccarini, C. *Vaccine* **1999**, *17*, 1251–1263.
137. Steinhoff, M. C.; Edwards, K.; Keyserling, H.; Thoms, M. L.; Johnson, C.; Madore, D.; Hogerman, D. *Pediatr. Infect. Dis. J.* **1994**, *13*, 368–372.

138. Daum, R. S.; Hogerman, D.; Rennels, M. B.; Bewley, K.; Malinoski, F.; Rothstein, E.; Reisinger, K.; Block, S.; Keyserling, H.; Steinhoff, M. *J. Infect. Dis.* **1997**, *176*, 445–455.
139. Anderson, P. W.; Pichichero, M. E.; Stein, E. C.; Porcelli, S.; Betts, R. F.; Connuck, D. M.; Korones, D.; Insel, R. A.; Zahradnik, J. M.; Eby, R. *J. Immunol.* **1989**, *142*, 2464–2468.
140. Pozsgay, V.; Chu, C.; Pannell, L.; Wolfe, J.; Robbins, J. B.; Schneerson, R. *Proc. Natl. Acad. Sci. U.S.A.* **1999**, *96*, 5194–5197.
141. Phalipon, A.; Tanguy, M.; Grandjean, C.; Guerreiro, C.; Belot, F.; Cohen, D.; Sansonetti, P. J.; Mulard, L. A. *J. Immunol.* **2009**, *182*, 2241–2247.
142. Frasch, C. E. *Vaccine* **2009**, *27*, 6468–6470.
143. Egan, W.; Frasch, C. E.; Anthony, B. F. *JAMA, J. Am. Med. Assoc.* **1995**, *273*, 888–889.
144. Xu, Q.; Klees, J.; Teyral, J.; Capen, R.; Huang, M.; Sturgess, A. W.; Hennessey, J. P., Jr.; Washabaugh, M.; Sitrin, R.; Abeygunawardana, C. *Anal. Biochem.* **2005**, *337*, 235–245.
145. Lefeber, D. J.; Gutierrez Gallego, R.; Grun, C. H.; Proietti, D.; D'Ascenzi, S.; Costantino, P.; Kamerling, J. P.; Vliegenthart, J. F. *Carbohydr. Res.* **2002**, *337*, 819–825.
146. Porro, M.; Costantino, P.; Viti, S.; Vannozzi, F.; Naggi, A.; Torri, G. *Mol. Immunol.* **1985**, *22*, 907–919.
147. Cai, X.; Lei, Q. P.; Lamb, D. H.; Shannon, A.; Jacoby, J.; Kruk, J.; Kensinger, R. D.; Ryall, R.; Zablackis, E.; Cash, P. *Anal. Chem.* **2004**, *76*, 7387–7390.
148. Arcuri, M.; Di Benedetto, R.; Cunningham, A. F.; Saul, A.; MacLennan, C. A.; Micoli, F. *PLoS One* **2017**, *12*, e0189100.
149. Kim, J. S.; Laskowich, E. R.; Michon, F.; Kaiser, R. E.; Arumugham, R. G. *Anal. Biochem.* **2006**, *358*, 136–142.
150. Beuvery, E. C.; vd Kaaden, A.; Kanhai, V.; Leussink, A. B. *Vaccine* **1983**, *1*, 31–36.
151. D'Ambra, A.; Baugher, J. E.; Concannon, P. E.; Pon, R. A.; Michon, F. *Anal. Biochem.* **1997**, *250*, 228–236.
152. Bardotti, A.; Averani, G.; Berti, F.; Berti, S.; Galli, C.; Giannini, S.; Fabbri, B.; Proietti, D.; Ravenscroft, N.; Ricci, S. *Vaccine* **2005**, *23*, 1887–1899.
153. Ravenscroft, N.; Averani, G.; Bartoloni, A.; Berti, S.; Bigio, M.; Carinci, V.; Costantino, P.; D'Ascenzi, S.; Giannozzi, A.; Norelli, F.; Pennatini, C.; Proietti, D.; Ceccarini, C.; Cescutti, P. *Vaccine* **1999**, *17*, 2802–2816.
154. Bhushan, R.; Anthony, B. F.; Frasch, C. E. *Infect. Immun.* **1998**, *66*, 5848–5853.
155. Crotti, S.; Zhai, H.; Zhou, J.; Allan, M.; Proietti, D.; Pansegrau, W.; Hu, Q.-Y.; Berti, F.; Adamo, R. *ChemBioChem* **2014**, *15*, 836–843.
156. Fairweather, N. F.; Lyness, V. A. *Nucleic Acids Res.* **1986**, *14*, 7809–7812.
157. Porro, M.; Costantino, P.; Giovannoni, F.; Pellegrini, V.; Tagliaferri, L.; Vannozzi, F.; Viti, S. *Mol. Immunol.* **1986**, *23*, 385–391.

158. Bongat, A. F.; Saksena, R.; Adamo, R.; Fujimoto, Y.; Shiokawa, Z.; Peterson, D. C.; Fukase, K.; Vann, W. F.; Kovac, P. *Glycoconjate J.* **2010**, *27*, 69–77.
159. Mendez, Y.; Chang, J.; Humpierre, A. R.; Zanuy, A.; Garrido, R.; Vasco, A. V.; Pedroso, J. L.; Santana, D.; Rodriguez, L. M.; Garcia-Rivera, D.; Valdes, Y. R.; Verez-Bencomo, V.; Rivera, D. G. *Chem. Sci.* **2018**, *9*, 2581–2588.
160. Peeters, C. C.; Tenbergen-Meekes, A. M.; Poolman, J. T.; Zegers, B. J.; Rijkers, G. T. *Vaccine* **1992**, *10*, 833–840.
161. Kowarik, M.; Young, N. M.; Numao, S.; Schulz, B. L.; Hug, I.; Callewaert, N.; Mills, D. C.; Watson, D. C.; Hernandez, M.; Kelly, J. F.; Wacker, M.; Aebi, M. *EMBO J.* **2006**, *25*, 1957–1966.
162. Feldman, M. F.; Wacker, M.; Hernandez, M.; Hitchen, P. G.; Marolda, C. L.; Kowarik, M.; Morris, H. R.; Dell, A.; Valvano, M. A.; Aebi, M. *Proc. Natl. Acad. Sci. U.S.A.* **2005**, *102*, 3016–3021.
163. Hatz, C. F.; Bally, B.; Rohrer, S.; Steffen, R.; Kramme, S.; Siegrist, C. A.; Wacker, M.; Alaimo, C.; Fonck, V. G. *Vaccine* **2015**, *33*, 4594–4601.
164. Huttner, A.; Hatz, C.; van den Dobbelsteen, G.; Abbanat, D.; Hornacek, A.; Frolich, R.; Dreyer, A. M.; Martin, P.; Davies, T.; Fae, K.; van den Nieuwenhof, I.; Thoelen, S.; de Valliere, S.; Kuhn, A.; Bernasconi, E.; Viereck, V.; Kavvadias, T.; Kling, K.; Ryu, G.; Hulder, T.; Groger, S.; Scheiner, D.; Alaimo, C.; Harbarth, S.; Poolman, J.; Fonck, V. G. *Lancet Infect. Dis.* **2017**, *17*, 528–537.
165. Ravenscroft, N., Richardson, N., Serventi, F., Lizak, C., Lipowsky, G., Faridmoayer, A. NMR Characterization of Pneumococcal Bioconjugate Vaccines. In *19th European Carbohydrate Symposium*; European Carbohydrate Organization: Barcelona, Spain, 2017; p 697.
166. Ravenscroft, N.; Haeuptle, M. A.; Kowarik, M.; Fernandez, F. S.; Carranza, P.; Brunner, A.; Steffen, M.; Wetter, M.; Keller, S.; Ruch, C.; Wacker, M. *Glycobiology* **2016**, *26*, 51–62.
167. Ihssen, J.; Haas, J.; Kowarik, M.; Wiesli, L.; Wacker, M.; Schwede, T.; Thony-Meyer, L. *Open Biol.* **2015**, *5*, 140227.
168. Anderson, P.; Pichichero, M.; Insel, R.; Farsad, P.; Santosham, M. *J. Infect. Dis.* **1985**, *152*, 634–636.
169. Killeen, K. P. Development of Typhax, a Salmonella typhi Vi polysaccharide protein capsular matrix vaccine. In *Vaccine Technology VI*; Palomares, L.; Cox, M.; Garcon, N., Eds.; Engineering Conferences International: Albufeira, Portugal, 2016.
170. Li, Y.; Hill, A.; Beitelshees, M.; Shao, S.; Lovell, J. F.; Davidson, B. A.; Knight, P. R., 3rd; Hakansson, A. P.; Pfeifer, B. A.; Jones, C. H. *Proc. Natl. Acad. Sci. U.S.A.* **2016**, *113*, 6898–6903.
171. *Recommendations for the Production and Control of Pneumococcal Conjugate Vaccines*; No. 927Annex 2; World Health Organization: Geneva, Switzerland, 2005.
172. *Requirements for Menigococcal Polysaccharide Vaccine*; No. 658, Annex 6; World Health Organization: Geneva, Switzerland, 1980.

173. Khandke, L., Welsh, J. A., Alderson, M. R. Formulation Development of Glycoconjugate Vaccines for Low- and Middle-Income Countries. In *Carbohydrate Vaccines: From Concept to Clinic*; Prasad, A. K., Ed.; ACS Symposium Series 1290; American Chemical Society: Washington, DC, 2018; Chapter 9.
174. Immunization Standards. World Health Organization. http://www.who.int/immunization_standards/vaccine_quality/pq_145_mmr_10dose_sii/en/ (accessed February 1, 2018).
175. Lee, S. M.; Petermann, R.; Porte, Q.; Berezuk, G.; Crowe, B.; Shirtz, J. *Hum. Vaccines Immunother.* **2007**, *3*, 27–32.
176. Kayhty, H.; Peltola, H.; Karanko, V.; Makela, P. H. *J. Infect. Dis.* **1983**, *147*, 1100.
177. Anderson, P. *J. Infect. Dis.* **1984**, *149*, 1034–1035.
178. Hasija, M.; Li, L.; Rahman, N.; Ausar, S. F. *Vaccine Dev. Ther.* **2013**, *3*, 11–33.
179. von Hunolstein, C.; Parisi, L.; Recchia, S. *Vaccine* **1999**, *17*, 118–125.
180. Naiim, M.; Boualem, A.; Ferre, C.; Jabloun, M.; Jalocha, A.; Ravier, P. *Soft Matter* **2015**, *11*, 28–32.
181. Etzler, F. M.; Deanne, R. *Part. Part. Syst. Charact.* **1997**, *14*, 278–282.
182. *Requirements for Vi Polysaccharide Typhoid Vaccine*; No. 840, Annex 1; World Health Organization: Geneva, Switzerland, 1994.
183. CMC-Vaccines Working Group. https://www.slideshare.net/shivang47/qbd-model-case-study-of-vaccine-avax (accessed February 1, 2018).
184. Junker, B., Zablackis, E., Verch, T., Schofield, T., Douette, P. In *Vaccine Analysis: Strategies, Principles, And Control*; Nunnally, B.; Turula, V. E.; Sitrin, R. D., Eds.; Springer: New York, 2015; pp 479–520.

Editor's Biography

A. Krishna Prasad

Krishna Prasad is currently Director of Early Phase Development within Pfizer Vaccines Research and Development. He received his Ph.D. in Medicinal Chemistry from Central Drug Research Institute, Lucknow, India. After completing postdoctoral fellowships at University of Rhode Island, Kingston, RI, and Case Western Reserve University, Cleveland, OH, he worked as Asst. Research officer at National Research Council Canada. He joined Wyeth/Pfizer in 1998. He has been responsible for successfully advancing many vaccine candidates from early stages of design and development through all phases of clinical trials through commercial licensure, in a leadership role. Notable among these are: the multi-valent Pneumococcal conjugate vaccine (Prevnar13®), Meningococcal vaccines, many prophylactic candidate vaccines which are currently in various stages of clinical trials (*Staphylococcus aureus*, Group B Streptococcus, next generation Pneumococcal vaccine, Respiratory Syncytial Virus), and therapeutic target candidates (Aβ-Peptide based conjugate immunogens against Alzheimer's disease). These efforts culminated in the development of several novel vaccine technologies resulting in more than 100 patents, worldwide, in which he has been named as an inventor or co-inventor.

Indexes

Author Index

Subject Index

D

G

P

S

Printed in the USA/Agawam, MA
August 8, 2019

708483.001